Einstein's General Theory of Relativity

Øyvind Grøn   Sigbjørn Hervik

Einstein's General Theory of Relativity

With Modern Applications in Cosmology

 Springer

LORETTE WILMOT LIBRARY
NAZARETH COLLEGE

Øyvind Grøn
Oslo University College
Faculty of Engineering
PO Box 4, St Olavs Pl.
0130 Oslo
Norway
and
Institute of Physics
University of Oslo
PO Box 1048 Blindern
0136 Oslo
Norway

Sigbjørn Hervik
Department of Mathematics & Statistics
Dalhousie University
B3H 3J5 Halifax, NS
Canada
herviks@mathstat.dal.ca

Library of Congress Control Number: 2007923152

ISBN-13: 978-0-387-69199-2   e-ISBN-13: 978-0-387-69200-5

Printed on acid-free paper.

springer.com

Contents

I INTRODUCTION:
   NEWTONIAN PHYSICS AND SPECIAL RELATIVITY 1

III EINSTEIN'S FIELD EQUATIONS 177

IV COSMOLOGY 265

List of Problems

List of Examples

"Paradoxically, physicists claim that gravity is the weakest of the fundamental forces."

Prof. Hallstein Høgåsen– after having fallen
from a ladder and breaking both his arms

Preface

Many of us have experienced the same, fallen and broken something. Yet, supposedly, gravity is the weakest of the fundamental forces. It is claimed to be 10^{-15} times weaker than electromagnetism. But still, every one of us have more or less a personal relationship with gravity. Gravity is something which we have to consider every day. Whenever we loose something on the floor and whenever we pour something in a cup, gravity is an active participant. Had it not been for gravity, we could not have done anything of the above. Thus gravity is part of our everyday life.

This is basically what this book is about: gravity. We will try to convey the concepts of gravity to the reader as Albert Einstein saw it. Einstein saw upon gravity as nobody else before him had seen it. He saw upon gravity as curved spaces, four-dimensional manifolds and geodesics. All of these concepts will be presented in this book.

The book offers a rigorous introduction to Einstein's general theory of relativity. We start out from the first principles of relativity and present Einstein's theory in a self-contained way. For the readers convenience, we have included a rough flowchart of chapter dependencies in Appendix D. Such a flowchart is particularly useful if this book is used as a textbook for a course in General Relativity.

After introducing Einstein's field equations, we go onto the most important chapter in this book which contains the three classical tests of the theory and introduces the notion of black holes. Recently, cosmology has also proven to be a very important testing arena for the general theory of relativity. We have thus devoted a large part to this subject. We introduce the simplest models decribing an evolving universe, which in spite of their simpleness, can say quite a lot about the universe we live in. We include the cosmological constant and explain in detail the "standard model" in cosmology. After the main issues have been presented we introduce an anisotropic and an inhomogeneous universe model and explain some of their features. Unless one just accepts the cosmological principles as a fact, one is unavoidably led to the study of such anisotropic and inhomogeneous universe models. As an introductory course in general relativity, it is suitable to stop after finishing the chapters with cosmology.

For the more experienced reader, or for people eager to learn more, we have included a part called "Advanced Topics". These topics have been chosen by the authors because they present topics that are important and that have not been highlighted elsewhere in textbooks. Some of them are on the very edge of research, others are older ideas and topics. In particular, the last two chapters deal with Einstein gravity in five dimensions which has been a hot topic of research in recent years.

All of the ideas and matters presented in this book have one thing in common: they are all based on Einstein's classical idea of gravity. We have not considered any quantum mechanics in our presentation, with one exception:

black hole thermodynamics. Black hole thermodynamics is a quantum feature of black holes, but we chose to include it because the study of black holes would have been incomplete without it.

There are several people whom we wish to thank. First of all, we would like to thank Finn Ravndal who gave a thorough introduction to the theory of relativity in a series of lectures during the late seventies. This laid the foundation for further activity in this field at the University of Oslo. We also want to thank Ingunn K. Wehus and Peter Rippis for providing us with a copy of their theses [Weh01, Rip01], and to Svend E. Hjelmeland for computerizing some of the notes in the initial stages of this book. Furthermore, the kind efforts of Kevin Reid, Jasbir Nagi, James Lucietti, Håvard Alnes, Torquil MacDonald Sørensen, Olav Aursjø who read through the manuscript and pointed out to us numerous errors, typos and grammatical blunders. They are gratefully acknowledged. Last but not the least, we would like to thank Lailani Hervik without whom this book would probably never been published.

Oslo, Norway   ØYVIND GRØN
Halifax, Nova Scotia, Canada   SIGBJØRN HERVIK

Notation

We have tried to be as consistent as possible when it comes to notation in this book. There are some exceptions, but as a general rule we use the following notation.

Due to the large number of equations, the most important equations are boxed, like this:

$$\boxed{E = mc^2.}$$

All tensors, including vectors and forms, are written in bold typeface. A general tensor usually has a upper case letter, late in the alphabet; e.g., **T** is a typical tensor. Vectors are usually written in two possible ways. If it is more natural to associate the vector as a *tangent vector* of some curve, then we usually use lower case bold letters like **u** or **v**. If the vectors are more naturally associated with a vector field, then we use upper case bold letters, like **A** or **X**. However, naturally enough, this rule is the most violated concerning the notation in this book. Forms have Greek bold letters, e.g., $\boldsymbol{\omega}$ is typical form. All the components of tensors, vectors and forms have ordinary math italic fonts.

Matrices are written in sans serif, like M, while determinants are written in the usual math style: $\det(\mathsf{M}) = M$. A typical example is the metric tensor, **g**. In the following notation we have:

g : The metric tensor itself.

$g_{\mu\nu}$: The components of the metric tensor.

g : The matrix made up of $g_{\mu\nu}$.

g : The determinant of the metric tensor, $g = \det(\mathsf{g})$.

The metric tensor comes in many guises, each one is useful for different purposes.

Also, for the signature of the metric tensor, the $(-+++)$-convention is used. Thus the time direction has a $-$ while the spatial directions all have $+$.

The abstract index notation

One of the most heavily used notation, both in this book and in the physics literature in general, is the abstract index notation. So it is best that we get this sorted out as early as possible. As a general rule, *repeated indices means summation!* For example,

$$\alpha^\mu \beta_\mu \equiv \sum_\mu \alpha^\mu \beta_\mu,$$

where the sum is over the range of the index μ. Furthermore, the type of index can make a difference. Greek indices usually run over the spacetime manifold, starting with 0 as the time component. Latin indices are usually associated to a hypersurface or the spatial geometry; they start with 1 and run up to the dimension of the manifold. Hence, if we consider the usual four-dimensional spacetime, then $\mu = 0, ..., 3$, while $i = 1, ..., 3$. However, no rule without exceptions, also this rule is violated occasionally. Also, indices inside square brackets, means the antisymmetrical combination, while round brackets means symmetric part. For example,

$$T_{[\mu\nu]} \equiv \frac{1}{2} \left(T_{\mu\nu} - T_{\nu\mu} \right)$$

$$T_{(\mu\nu)} \equiv \frac{1}{2} \left(T_{\mu\nu} + T_{\nu\mu} \right).$$

The following notation is also convenient to get straight right away. Here, $A_{\mu...\nu}$ is an arbitrary tensor (it may have indices upstairs as well).

$\mathbf{e}_\alpha \left(A_{\mu...\nu} \right) = A_{\mu...\nu,\alpha}$	Partial derivative
$\nabla_\alpha A_{\mu...\nu} = A_{\mu...\nu;\alpha}$	Covariant derivative
$\pounds_{\mathbf{X}}$	Lie derivative with respect to $\mathbf{X}$
$\mathbf{d}$	Exterior derivative operator
$\mathbf{d}^\dagger$	Codifferential operator
$\star$	Hodge's star operator
$\Box$	Covariant Laplacian
$\otimes$	Tensor product
$\wedge$	Wedge product, or exterior product
ω^μ	Basis one-forms
$\mathbf{\Omega}^\mu_{\ \nu}$	Connection one-forms
$\mathbf{R}^\mu_{\ \nu}$	Curvature two-forms
$R^\alpha_{\ \beta\mu\nu}$	The Riemann curvature tensor
$R_{\mu\nu}$	The Ricci tensor
R	The Ricci scalar
$E_{\mu\nu}$	The Einstein tensor

Part I

INTRODUCTION:
NEWTONIAN PHYSICS
AND SPECIAL RELATIVITY

1

Relativity Principles and Gravitation

To obtain a mathematical description of physical phenomena, it is advantageous to introduce a reference frame in order to keep track of the position of events in space and time. The choice of reference frame has historically depended upon the view of human beings and their position in the Universe.

1.1 Newtonian mechanics

When describing physical phenomena on Earth, it is natural to use a coordinate system with origin at the centre of the Earth. This coordinate system is, however, not ideal for the description of the motion of the planets around the Sun. A coordinate system with origin at the centre of the Sun is more natural. Since the Sun moves around the centre of the galaxy, there is nothing special about a coordinate system with origin at the Sun's centre. This argument can be continued ad infinitum.

The fundamental reference frame of Newton is called 'absolute space'. The geometrical properties of this space are characterized by ordinary Euclidean geometry. This space can be covered by a regular Cartesian coordinate system. A non-rotating reference frame at rest, or moving uniformly in absolute space is called a Galilean reference frame. With chosen origin and orientation, the system is fixed. Newton also introduced a universal time which ticks at the same rate at all positions in space.

Relative to a Galilean reference frame, all mechanical systems behave according to Newton's three laws:

Newton's 1st law: Free particles move with constant velocity,

$$\boxed{\mathbf{u} = \frac{d\mathbf{r}}{dt} = \text{constant,}}$$

where $\mathbf{r}$ is a position vector.

Newton's 2nd law: The acceleration $\mathbf{a} = d\mathbf{u}/dt$ of a particle is proportional to the force $\mathbf{F}$ acting on it

$$\mathbf{F} = m_i \frac{d\mathbf{u}}{dt},$$ (1.1)

where m_i is the inertial mass of the particle.

Newton's 3rd law: If particle 1 acts on particle 2 with a force $\mathbf{F}_{12}$, then 2 acts on 1 with a force

$$\mathbf{F}_{21} = -\mathbf{F}_{12}.$$

The first law can be considered as a special case of the second with $\mathbf{F} = 0$. Alternatively, the first law can be thought of as restricting the reference frame to be non-accelerating. This is presupposed for the validity of Newton's second law. Such reference frames are called *inertial frames*.

1.2   Galilei–Newton's principle of Relativity

Let Σ be a Galilean reference frame, and Σ' another Galilean frame moving relative to Σ with a constant velocity $\mathbf{v}$ (see Fig. 1.1).

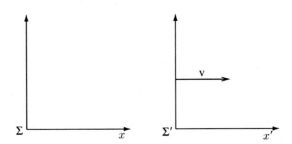

Figure 1.1: Relative translational motion.

We may think of a reference frame as a set of reference particles with given motion. A *comoving coordinate system* in a reference frame is a system in which the reference particles of the frame have constant spatial coordinates.

Let (x, y, z) be the coordinates of a comoving system in Σ, and (x', y', z') those of a comoving system in Σ'. The reference frame Σ moves relative to Σ' with a constant velocity $\mathbf{v}$ along the x-axis. A point with coordinates (x, y, z) in Σ has coordinates

$$x' = x - vt, \quad y' = y, \quad z' = z$$ (1.2)

in Σ', or

$$\mathbf{r}' = \mathbf{r} - \mathbf{v}t.$$ (1.3)

An event at an arbitrary point happens at the same time in Σ and Σ',

$$t' = t.$$ (1.4)

The space coordinate transformations (1.2) or (1.3) with the trivial time transformation (1.4) are called the *Galilei-transformations*.

If the velocity of a particle is u in Σ, then it moves with a velocity

$$\mathbf{u'} = \frac{d\mathbf{r'}}{dt} = \mathbf{u} - \mathbf{v} \tag{1.5}$$

in Σ'.

In Newtonian mechanics one assumes that the inertial mass of a body is independent of the velocity of the body. Thus the mass is the same in Σ as in Σ'. Then the force $\mathbf{F'}$, as measured in Σ', is

$$\mathbf{F'} = m_i \frac{d\mathbf{u'}}{dt'} = m_i \frac{d\mathbf{u}}{dt} = \mathbf{F}. \tag{1.6}$$

hence, the force is the same in Σ' as in Σ. This result may be expressed by saying that Newton's 2nd law is invariant under a Galilei transformation; it is written in the same way in every Galilean reference frame.

All reference frames moving with constant velocity are Galilean, so Newton's laws are valid in these frames. Every mechanical system will therefore behave in the same way in all Galilean frames. This is the *Galilei–Newton principle of relativity*.

It is difficult to find Galilean frames in our world. If, for example, we place a reference frame on the Earth, we must take into account the rotation of the Earth. This reference frame is rotating, and is therefore not Galilean. In such non-Galilean reference frames free particles have accelerated motion. In Newtonian dynamics the acceleration of free particles in rotating reference frames is said to be due to the centrifugal force and the Coriolis force. Such forces, that vanish by transformation to a Galilean reference frame, are called 'fictitious forces'.

A simple example of a non-inertial reference frame is one that has a constant acceleration a. Let Σ' be such a frame. If the position vector of a particle is r in Σ, then its position vector in Σ' is

$$\mathbf{r'} = \mathbf{r} - \frac{1}{2}\mathbf{a}t^2, \tag{1.7}$$

where it is assumed that Σ' was instantaneously at rest relative to Σ at the point of time $t = 0$. Newton's 2nd law is valid in Σ, so that a particle which is acted upon by a force $\mathbf{F}$ in Σ can be described by the equation

$$\mathbf{F} = m_i \frac{d^2\mathbf{r}}{dt^2} = m_i \left(\frac{d^2\mathbf{r'}}{dt^2} + \mathbf{a} \right). \tag{1.8}$$

If this is written as

$$\mathbf{F'} = \mathbf{F} - m_i\mathbf{a} = m_i \frac{d^2\mathbf{r'}}{dt^2}, \tag{1.9}$$

we may formally use Newton's 2nd law in the non-Galilean frame Σ'. This is obtained by a sort of trick, namely by letting the fictitious force act on the particle in addition to the ordinary forces that appear in a Galilean frame.

1.3 The principle of Relativity

At the beginning of the 20th century Einstein realised that Newton's absolute space is a concept without physical content. This concept should therefore be removed from the description of the physical world. This conclusion is in accordance with the negative result of the Michelson–Morley experiment [MM87]. In this experiment one did not succeed in measuring the velocity of the Earth through the so-called 'ether' which was thought of as a 'materialization' of Newton's absolute space.

However, Einstein retained, in his special theory of relativity, the Newtonian idea of the privileged observers at rest in Galilean frames that move with constant velocities relative to each other. Einstein did, however, extend the range of validity of the equivalence of all Galilean frames. While Galilei and Newton had demanded that the laws of *mechanics* are the same in all Galilean frames, Einstein postulated that *all the physical laws governing the behaviour of the material world can be formulated in the same way in all Galilean frames.* This is *Einstein's special principle of relativity.* (Note that in the special theory of relativity it is usual to call the Galilean frames 'inertial frames'. However in the general theory of relativity the concept 'inertial frame' has a somewhat different meaning; it is a freely falling frame. So we will use the term Galilean frames about the frames moving relative to each other with constant velocity.)

Applying the Galilean coordinate transformation to Maxwell's electromagnetic theory, one finds that Maxwell's equations are not invariant under this transformation. The wave-equation has the standard form, with isotropic velocity of electromagnetic waves, only in one 'preferred' Galilean frame. In other frames the velocity relative to the 'preferred' frame appears. Thus Maxwell's electromagnetic theory does not fulfil Galilei–Newton's principle of relativity. The motivation of the Michelson–Morley experiment was to measure the velocity of the Earth relative to the 'preferred' frame.

Einstein demanded that the special principle of relativity should be valid also for Maxwell's electromagnetic theory. This was obtained by replacing the Galilean kinematics by that of the special theory of relativity (see Ch. 2), since Maxwell's equations and Lorentz's force law is invariant under the Lorentz transformations. In particular this implies that the velocity of electromagnetic waves, i.e. of light, is the same in all Galilean frames, $c = 299\ 792.5$ km/s $\approx 3.00 \times 10^8$ m/s.

1.4   Newton's law of Gravitation

Until now we have neglected gravitational forces. Newton found that the force between two point masses M and m at a distance r is given by

$$\boxed{\mathbf{F} = -G\frac{Mm}{r^3}\mathbf{r}.}$$
(1.10)

This is *Newton's law of gravitation.* Here, G is Newton's gravitational constant, $G = 6.67 \times 10^{-11} \text{m}^3/\text{kg s}^2$. The gravitational force on a point mass m at a position $\mathbf{r}$ due to many point masses $M_1, M_2, \ldots, M_n$ at positions $\mathbf{r}'_1, \mathbf{r}'_2, \ldots, \mathbf{r}'_n$ is given by the superposition

$$\mathbf{F} = -mG\sum_{i=1}^{n}\frac{M_i}{|\mathbf{r}-\mathbf{r}'_i|^3}(\mathbf{r}-\mathbf{r}'_i).$$
(1.11)

A continuous distribution of mass with density $\rho(\mathbf{r}')$ so that $dM = \rho(\mathbf{r}')d^3r'$ thus gives rise to a gravitational force at P (see Fig. 1.2)

$$\mathbf{F} = -mG\int \rho(\mathbf{r}')\frac{\mathbf{r}-\mathbf{r}'}{|\mathbf{r}-\mathbf{r}'|^3}d^3r'.$$
(1.12)

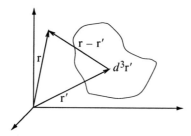

Figure 1.2: Gravitational field from a continuous mass distribution.

Here $\mathbf{r}'$ is associated with positions in the mass distribution, and $\mathbf{r}$ with the position P where the gravitational field is measured.

The gravitational potential $\phi(\mathbf{r})$ at the field point P is defined by

$$\mathbf{F} = -m\boldsymbol{\nabla}\phi(\mathbf{r}). \tag{1.13}$$

Note that the $\boldsymbol{\nabla}$ operator acts on the coordinates of the field point, not of the source point.

Calculating $\phi(\mathbf{r})$ from Eq. (1.12) it will be useful to introduce *Einstein's summation convention*. For arbitrary a and b one has

$$a^j b_j \equiv \sum_{j=1}^n a^j b_j, \tag{1.14}$$

where n is the range of the indices j.

We will also need the Kronecker symbol defined by

$$\delta^i{}_j = \left\{ \begin{array}{ll} 1 & \text{when} \quad i = j \\ 0 & \text{when} \quad i \neq j. \end{array} \right. \tag{1.15}$$

The gradient of $|\mathbf{r} - \mathbf{r}'|^{-1}$ may now be calculated as follows

$$\boldsymbol{\nabla}\frac{1}{|\mathbf{r}-\mathbf{r}'|} = \mathbf{e}_i \frac{\partial}{\partial x_i}\left[(x^j - x^{j'})(x_j - x_j')\right]^{-1/2}$$

$$= -\mathbf{e}_i \frac{(x^j - x^{j'})\frac{\partial x_j}{\partial x_i}}{\left[(x^j - x^{j'})(x_j - x_j')\right]^{3/2}} = -\mathbf{e}_i \frac{(x^j - x^{j'})\delta^i{}_j}{|\mathbf{r}-\mathbf{r}'|^3}$$

$$= -\frac{(x^i - x^{i'})\mathbf{e}_i}{|\mathbf{r}-\mathbf{r}'|^3} = -\frac{(\mathbf{r}-\mathbf{r}')}{|\mathbf{r}-\mathbf{r}'|^3}. \tag{1.16}$$

Comparing with Eqs. (1.12) and (1.13) we see that

$$\phi(\mathbf{r}) = -G \int \rho(\mathbf{r}')\frac{1}{|\mathbf{r}-\mathbf{r}'|}d^3r'. \tag{1.17}$$

When characterizing the mass distribution of a point mass mathematically, it is advantageous to use Dirac's δ-function. This function is defined by the following requirements

$$\delta(\mathbf{r}-\mathbf{r}') = 0, \quad \mathbf{r}' \neq \mathbf{r}, \tag{1.18}$$

and

$$\int_V f(\mathbf{r}')\delta(\mathbf{r} - \mathbf{r}')d^3r' = \begin{cases} f(\mathbf{r}), & \text{when } \mathbf{r}' = \mathbf{r} \text{ is inside } V \\ 0, & \text{when } \mathbf{r}' = \mathbf{r} \text{ is outside } V. \end{cases} \qquad (1.19)$$

A point mass M at a position $\mathbf{r}' = \mathbf{r}_0$ represents a mass density

$$\rho(\mathbf{r}') = M\delta(\mathbf{r}' - \mathbf{r}_0). \qquad (1.20)$$

Substitution into Eq. (1.17) gives the potential of the point mass

$$\boxed{\phi(\mathbf{r}) = -\frac{GM}{|\mathbf{r} - \mathbf{r}_0|}.} \qquad (1.21)$$

1.5   Local form of Newton's Gravitational law

Newton's law of gravitation cannot be a relativistically correct law, because it permits action at a distance. A point mass at one place may then act instantaneously on a point mass at another remote position. According to the special theory of relativity, instantaneous action at a distance is impossible. An action which is instantaneous in one reference frame, is not instantaneous in another frame, moving with respect to the first. This is due to the relativity of simultaneity (see Ch. 2). *Instantaneous action at a distance can only exist in a theory with absolute simultaneity.* As a first step towards a relativistically valid theory of gravitation, we shall give a local form of Newton's law of gravitation.

We shall now show how Newton's law of gravitation leads to a field equation for gravity. Consider a continuous mass-distribution $\rho(\mathbf{r}')$. Equations (1.16) and (1.17) lead to

$$\nabla\phi(\mathbf{r}) = G\int \rho(\mathbf{r}')\frac{(\mathbf{r} - \mathbf{r}')}{|\mathbf{r} - \mathbf{r}'|^3}d^3r', \qquad (1.22)$$

which gives

$$\nabla^2\phi(\mathbf{r}) = G\int \rho(\mathbf{r}')\nabla \cdot \frac{(\mathbf{r} - \mathbf{r}')}{|\mathbf{r} - \mathbf{r}'|^3}d^3r'. \qquad (1.23)$$

Furthermore,

$$\begin{aligned} \nabla \cdot \frac{(\mathbf{r} - \mathbf{r}')}{|\mathbf{r} - \mathbf{r}'|^3} &= \frac{\nabla \cdot \mathbf{r}}{|\mathbf{r} - \mathbf{r}'|^3} + (\mathbf{r} - \mathbf{r}') \cdot \nabla\frac{1}{|\mathbf{r} - \mathbf{r}'|^3} \\ &= \frac{3}{|\mathbf{r} - \mathbf{r}'|^3} + (\mathbf{r} - \mathbf{r}') \cdot \left(-3\frac{\mathbf{r} - \mathbf{r}'}{|\mathbf{r} - \mathbf{r}'|^5}\right) = 0, \quad \mathbf{r} \neq \mathbf{r}'. \end{aligned} \qquad (1.24)$$

In general the volume of integration encompasses the point $\mathbf{r}' = \mathbf{r}$ where the field is measured. Thus, we have to find an expression for $\nabla \cdot (\mathbf{r} - \mathbf{r}')/|\mathbf{r} - \mathbf{r}'|^3$ which is also valid at this point. Equation (1.24) indicates that $\nabla \cdot (\mathbf{r} - \mathbf{r}')/|\mathbf{r} - \mathbf{r}'|^3$ is proportional to Dirac's δ-function. According to Eq. (1.19) the proportionality factor can be found by calculating the integral $\int \nabla \cdot (\mathbf{r} - \mathbf{r}')/|\mathbf{r} - \mathbf{r}'|^3 d^3r'$. We note that $\nabla \cdot (\mathbf{r} - \mathbf{r}')/|\mathbf{r} - \mathbf{r}'|^3 = -\nabla' \cdot (\mathbf{r} - \mathbf{r}')/|\mathbf{r} - \mathbf{r}'|^3$ where ∇' acts on $\mathbf{r}'$. So using Gauss' integral theorem

$$-\int_V \nabla' \cdot \mathbf{A}d^3r' = -\oint_S \mathbf{A} \cdot d\mathbf{S}', \qquad (1.25)$$

where S is the surface enclosing V, we get

$$-\int_V \nabla' \cdot \frac{\mathbf{r}-\mathbf{r}'}{|\mathbf{r}-\mathbf{r}'|^3} d^3 r' = -\oint_S \frac{\mathbf{r}-\mathbf{r}'}{|\mathbf{r}-\mathbf{r}'|^3} \cdot d\mathbf{S}'. \qquad (1.26)$$

Note that the gradient $\nabla\phi$ in Eq. (1.22) is directed away from the source. Thus the divergence of this vector in Eq. (1.23) must be positive. The direction of the surface element $d\mathbf{S}'$ in Fig. 1.3 is chosen to satisfy this criterion.

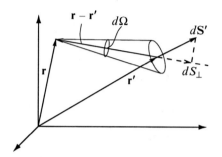

Figure 1.3: Definition of surface elements.

With reference to Fig. 1.3 the solid angle element $d\Omega$ is defined by

$$d\Omega = \frac{dS_\perp}{|\mathbf{r}-\mathbf{r}'|^2}, \qquad (1.27)$$

where

$$dS_\perp = -\frac{\mathbf{r}-\mathbf{r}'}{|\mathbf{r}-\mathbf{r}'|} \cdot d\mathbf{S}'. \qquad (1.28)$$

It follows that

$$d\Omega = -\frac{\mathbf{r}-\mathbf{r}'}{|\mathbf{r}-\mathbf{r}'|^3} \cdot d\mathbf{S}' \qquad (1.29)$$

and

$$-\int_V \nabla' \cdot \frac{\mathbf{r}-\mathbf{r}'}{|\mathbf{r}-\mathbf{r}'|^3} d^3 r' = \int d\Omega = \begin{cases} 4\pi, & P \text{ inside } V \\ 0, & P \text{ outside } V \end{cases} \qquad (1.30)$$

Thus we get

$$\nabla \cdot \frac{\mathbf{r}-\mathbf{r}'}{|\mathbf{r}-\mathbf{r}'|^3} = 4\pi\delta(\mathbf{r}-\mathbf{r}'). \qquad (1.31)$$

Substituting this into Eq. (1.23) and using Eq. (1.19) with $f(\mathbf{r}) = 1$ we have

$$\boxed{\nabla^2 \phi(\mathbf{r}) = 4\pi G\rho(\mathbf{r}).} \qquad (1.32)$$

This *Poisson equation* is the local form of Newton's law of gravitation. Newton's 2nd law applied to a particle falling freely in a gravitational field gives the acceleration of gravity

$$\boxed{\mathbf{g} = -\nabla\phi.} \qquad (1.33)$$

Newton's theory of gravitation can now be summarized in the following way: *Mass generates a gravitational field according to Poisson's equation, and the gravitational field generates acceleration according to Newton's second law.*

1.6   Tidal forces

A tidal force is caused by the difference in the gravitational forces acting on two neighbouring particles in a gravitational field. The tidal force is due to the inhomogeneity of a gravitational field.

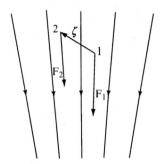

Figure 1.4: Tidal forces.

In Fig. 1.4 two points have a separation vector $\boldsymbol{\zeta}$. The position vectors of the points 1 and 2 are $\mathbf{r}$ and $\mathbf{r} + \boldsymbol{\zeta}$, respectively, where we assume that $|\boldsymbol{\zeta}| \ll |\mathbf{r}|$. The gravitational forces on two equal masses m at 1 and 2 are $\mathbf{F}(\mathbf{r})$ and $\mathbf{F}(\mathbf{r} + \boldsymbol{\zeta})$, respectively. By means of a Taylor expansion to the lowest order in $|\boldsymbol{\zeta}|$, and using Cartesian coordinates, we get for the i-component of the tidal force

$$f_i = F_i(\mathbf{r} + \boldsymbol{\zeta}) - F_i(\mathbf{r}) = \zeta_j \left(\frac{\partial F_i}{\partial x^j} \right)_{\mathbf{r}}. \tag{1.34}$$

The corresponding vector equation is

$$\mathbf{f} = (\boldsymbol{\zeta} \cdot \boldsymbol{\nabla})_{\mathbf{r}} \, \mathbf{F}. \tag{1.35}$$

Given

$$\mathbf{F} = -m \boldsymbol{\nabla} \phi, \tag{1.36}$$

the tidal force may be expressed in terms of the gravitational potential

$$\mathbf{f} = -m \left(\boldsymbol{\zeta} \cdot \boldsymbol{\nabla} \right)_{\mathbf{r}} \boldsymbol{\nabla} \phi. \tag{1.37}$$

It follows that the i-component of the relative acceleration of the particles in Cartesian coordinates is

$$\frac{d^2 \zeta_i}{dt^2} = - \left(\frac{\partial^2 \phi}{\partial x^i \partial x^j} \right)_{\mathbf{r}} \zeta^j. \tag{1.38}$$

Examples

Example 1.1 (Tidal forces on two particles)
Let us first consider the case with vertical separation vector. We introduce a small Cartesian coordinate system at a distance R from a mass M, see Fig. 1.5.
If we place a particle of mass m at a point $(0, 0, z)$, it will, according to Eq. (1.10), be acted upon by a force

$$F_z(z) = -m \frac{GM}{(R + z)^2}, \tag{1.39}$$

while an identical particle at the origin will be acted upon by a force

$$F_z(0) = -m \frac{GM}{R^2}. \tag{1.40}$$

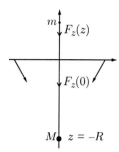

Figure 1.5: Tidal force between vertically separated particles.

If the coordinate system is falling freely together with the particles towards M, an observer at the origin will say that the particle at $(0, 0, z)$ is acted upon by a force (assuming that $z \ll R$)

$$f_z = F_z(z) - F_z(0) = 2mz\frac{GM}{R^3} \tag{1.41}$$

directed away from the origin, along the positive z-axis.

In the same way one finds that particles at the points $(x, 0, 0)$ and at $(0, y, 0)$ are attracted towards the origin by tidal forces

$$f_x = -mx\frac{GM}{R^3} \quad \text{and} \quad f_y = -my\frac{GM}{R^3}. \tag{1.42}$$

Eqs. (1.41) and (1.42) have among others the following consequence. If an elastic circular ring is falling freely in the gravitational field of the Earth, as shown in Fig. 1.6, it will be stretched in the vertical direction and compressed in the horizontal direction.

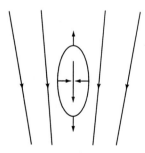

Figure 1.6: Deformation due to tidal forces.

In general, tidal forces cause changes of shape.

Example 1.2 (Flood and ebb on the Earth)
The tidal forces from the Sun and the Moon cause flood and ebb on the Earth. Let M be the mass of the Moon (or the Sun).

The potential in the gravitational field of M at a point P on the surface of the Earth is (see Fig. 1.7)

$$\phi(\mathbf{r}) = -\frac{GM}{(D^2 + R^2 - 2RD\cos\theta)^{1/2}}, \tag{1.43}$$

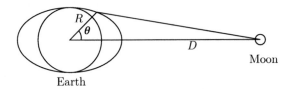

Figure 1.7: Tidal forces from the Moon on a point P on the Earth.

where R is the radius of the Earth, and D the distance from the centre of the Moon (or Sun) to the centre of the Earth. Making a series expansion to 2nd order in R/D we get

$$\phi = -\frac{GM}{D}\left(1 + \frac{R}{D}\cos\theta - \frac{1}{2}\frac{R^2}{D^2} + \frac{3}{2}\frac{R^2}{D^2}\cos^2\theta\right). \tag{1.44}$$

If the gravitational field of the Moon (and the Sun) was homogeneous near the Earth, there would be no tides. At an arbitrary position P on the surface of the Earth the acceleration of gravity in the field of the Moon, would then be the same as at the centre of the Earth

$$g_{\text{Moon}} = -|\nabla\phi_1| = \frac{GM}{D^2}, \tag{1.45}$$

where $\phi_1 = -GM/D$ is the leading order term of Eq. (1.44). The height difference between the point P and the centre of the Earth in the gravitational field of the Moon, is $\Delta H = R\cos\theta$. The 'reference potential' of P representing the potential of P if there were no tides, is

$$\phi_2 = \phi_1 - g_{\text{Moon}}\Delta H = 1 - \frac{GM}{D}\left(1 + \frac{R}{D}\cos\theta\right). \tag{1.46}$$

The tidal potential ϕ_T is the difference between the actual potential at P, given in Eq. (1.43) or to second order in R/D by Eq. (1.44), and the reference potential,

$$\phi_T = \phi - \phi_2 \approx \frac{GMR^2}{2D^3}\left(1 - 3\cos^2\theta\right). \tag{1.47}$$

A water particle at the surface of the Earth is acted upon also by the gravitational field of the Earth. Let **g** be the acceleration of gravity at P. If the water is in static equilibrium, the surface of the water represents an equipotential surface, given by

$$gh + \frac{GMR^2}{2D^3}\left(1 - 3\cos^2\theta\right) = \text{constant}. \tag{1.48}$$

This equation gives the height of the water surface as a function of the angle θ. The difference between flood at $\theta = 0$ and ebb at $\theta = \pi/2$, is

$$\Delta h = \frac{3GMR^2}{2D^3 g}. \tag{1.49}$$

Inserting numerical data for the Moon and the Sun gives $\Delta h_{\text{Moon}} = 0.53$ m and $\Delta h_{\text{Sun}} = 0.23$ m.

Example 1.3 (A tidal force pendulum)
Two particles each with mass m are connected by a rigid rod of length 2ℓ. The system is free to oscillate in any vertical plane about its centre of mass. The mass of the rod is negligible relative to m. The pendulum is at a distance R from the centre of a spherical distribution of matter with mass M (Fig. 1.8).
   The oscillation of the pendulum is determined by the equation of motion

$$|\ell \times (\mathbf{F}_1 - \mathbf{F}_2)| = I\ddot{\theta} \tag{1.50}$$

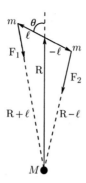

Figure 1.8: Geometry of a tidal force pendulum.

where $I = 2m\ell^2$ is the moment of inertia of the pendulum.

By Newton's law of gravitation

$$\mathbf{F}_1 = -GMm\frac{\mathbf{R}+\boldsymbol{\ell}}{|\mathbf{R}+\boldsymbol{\ell}|^3}, \quad \mathbf{F}_2 = -GMm\frac{\mathbf{R}-\boldsymbol{\ell}}{|\mathbf{R}-\boldsymbol{\ell}|^3}. \tag{1.51}$$

Thus

$$GMm|\boldsymbol{\ell} \times \mathbf{R}|(|\mathbf{R}-\boldsymbol{\ell}|^{-3} - |\mathbf{R}+\boldsymbol{\ell}|^{-3}) = 2m\ell^2\ddot{\theta}. \tag{1.52}$$

From Fig. 1.8 it is seen that

$$|\boldsymbol{\ell} \times \mathbf{R}| = \ell R \sin\theta. \tag{1.53}$$

It is now assumed that $\ell \ll R$. Then we have, to first order in ℓ/R,

$$|\mathbf{R}-\boldsymbol{\ell}|^{-3} - |\mathbf{R}+\boldsymbol{\ell}|^{-3} = \frac{6\ell}{R^4}\cos\theta. \tag{1.54}$$

The equation of motion of the pendulum now takes the form

$$2\ddot{\theta} + \frac{3GM}{R^3}\sin 2\theta = 0. \tag{1.55}$$

This is the equation of motion of a simple pendulum in the variable 2θ, instead of as usual with θ as variable. The equation shows that the pendulum oscillates about a vertical equilibrium position. The reason for 2θ instead of the usual θ, is that the tidal pendulum is invariant under a change $\theta \to \theta + \pi$, while the simple pendulum is invariant under a change $\theta \to \theta + 2\pi$.

Assuming small angular displacements leads to

$$\ddot{\theta} + \frac{3GM}{R^3}\theta = 0.$$

This is the equation of a harmonic oscillator with period

$$T = 2\pi\left(\frac{R^3}{3GM}\right)^{1/2}.$$

Note that the period of the tidal force pendulum is independent of its length. This means that tidal forces can be observed on systems of arbitrarily small size. Also, from the equation of motion it is seen that in a uniform field, where $\mathbf{F}_1 = \mathbf{F}_2$, the pendulum does not oscillate.

The acceleration of gravity at the position of the pendulum is $g = GM/R^2$, so that the period of the tidal pendulum may be written

$$T = 2\pi\left(\frac{R}{3g}\right)^{1/2}.$$

The mass of a spherical body with mean density ρ is $M = (4\pi/3)\rho R^3$, which gives for the period of a tidal pendulum at its surface

$$T = \left(\frac{\pi G}{\rho}\right)^{1/2}.$$

Thus the period depends only upon the density of the body. For a pendulum at the surface of the Earth the period is about 50 minutes. The region in *spacetime* needed in order to measure the tidal force is not arbitrarily small.

1.7 The principle of equivalence

Galilei experimentally investigated the motion of freely falling bodies. He found that they moved in the same way, regardless of mass and of composition.

In Newton's theory of gravitation, mass appears in two different ways:

1. in the law of gravitation as gravitational mass, m_g;

2. in Newton's 2nd law as inertial mass, m_i.

The equation of motion of a freely falling particle in the gravity field of a spherical body with gravitational mass M takes the form

$$\frac{d^2\mathbf{r}}{dt^2} = -\frac{m_g}{m_i}\frac{M}{r^3}\mathbf{r}. \tag{1.56}$$

The result of Galilei's observations, and subsequent measurements that verified his observations, is that that the ratio of gravitational to inertial mass is the same for all bodies. With a suitable choice of units we obtain

$$m_g = m_i. \tag{1.57}$$

Measurements performed by the Hungarian baron Eötvös at the turn of the 20th century, indicated that this equality holds with an accuracy better than 10^{-8}. Recent experiments have given the result $|m_i - m_g|/m_i < 9 \times 10^{-33}$.

Einstein assumed the exact validity of Eq. (1.57) for all kinds of particles. He did not consider this a coincidence, but rather as an expression of a fundamental principle, *the principle of equivalence*.

A consequence of this *universality of free fall* is the possibility of removing the effect of a gravitational force by being in free fall. In order to clarify this, Einstein considered a homogeneous gravitational field in which the acceleration of gravity, **g**, is independent of the position. In a freely falling, non-rotating reference frame in this field, all free particles move according to

$$m_i\frac{d^2\mathbf{r}'}{dt^2} = (m_g - m_i)\mathbf{g} = 0, \tag{1.58}$$

where Eqs. (1.6) and (1.57) have been used.

This means that an observer in such a freely falling reference frame will say that the particles around him are not acted upon by any forces. They move with constant velocities along straight paths. In the general theory of relativity such a reference frame is said to be *inertial*.

Einstein's heuristic reasoning also suggested full equivalence between Galilean frames in regions far from mass distributions, where there are no gravitational fields, and inertial frames falling freely in a gravitational field. Due to this equivalence, the Galilean frames of the special theory of relativity, which presupposes a spacetime free of gravitational fields, shall hereafter be called inertial reference frames. In the relativistic literature the implied strong principle of equivalence has often been interpreted to mean the physical equivalence between freely falling frames and unaccelerated frames in regions free of gravitational fields. This equivalence has a local validity; it is concerned with measurements in the freely falling frames, restricted in duration and spatial extension so that tidal effects cannot be measured.

The principle of equivalence has also been interpreted in 'the opposite way'. An observer at rest in a homogeneous gravitational field, and an observer in an accelerated reference frame in a region far from any mass-distribution, will obtain identical results when they perform similar experiments. The strong equivalence principle states that *locally the behaviour of matter in an accelerated frame of reference cannot be distinguished from its behaviour in a corresponding gravitational field.* Again, there is a local equivalence in an inhomogeneous gravitational field. The equivalence is manifest inside spacetime regions restricted so that the inhomogeneity of the gravitational field cannot be measured. *An inertial field caused by the acceleration or rotation of the reference frame is equivalent to a gravitational field caused by a mass-distribution (as far as tidal effects can be ignored).* The strong equivalence principle is usually elevated to a global equivalence of all spacetime points so that the result of any local test-experiment (non-gravitational or gravitational) is independent of where and when it is performed.

1.8 The covariance principle

The principle of relativity is a physical principle. It is concerned with physical phenomena. It motivates the introduction of a formal principle called the *covariance principle*: the equations of a physical theory shall have the same form in every coordinate system.

This principle may be fulfilled by every theory by writing the equations in an invariant form. This form is obtained by only using spacetime tensors in the mathematical formulation of the theory.

The covariance principle and the equivalence principle may be used to obtain a description of what happens in the presence of gravity. We start with the physical laws as formulated in the special theory of relativity. The laws are then expressed in a covariant way by writing them as tensor equations. They are then valid in an arbitrary accelerated system, but the inertial field ('fictitious force') in the accelerated frame is equivalent to a non-vanishing acceleration of gravity. One has thereby obtained a description valid in the presence of a gravitational field (as far as non-tidal effects are concerned).

In general, the tensor equations have a coordinate independent form. Yet, such covariant equations need not fulfil the principle of relativity. A physical principle, such as the principle of relativity, is concerned with observable relationships. When one is going to deduce the observable consequences of an equation, one has to establish relations between the tensor-components of the equation and observable physical quantities. Such relations have to be defined; they are not determined by the covariance principle.

From the tensor equations, which are covariant, and the defined relations between the tensor components and the observable physical quantities, one can deduce equations between physical quantities. The special principle of relativity demands that these equations must have the same form in every Galilean reference frame.

The relationships between physical quantities and mathematical objects such as tensors (vectors) are theory-dependent. For example, the relative velocity between two bodies is a vector within Newtonian kinematics. In the relativistic kinematics of four-dimensional spacetime, an ordinary velocity which has only three components, is not a vector. Vectors in spacetime, called 4-vectors, have four components.

Equations between physical quantities are not covariant in general. For example, Maxwell's equations in three-vector form are not invariant under a Lorentz transformation. When these equations are written in tensor-form, they are invariant under a Lorentz-transformation, and all other coordinate transformations.

If all equations in a theory are tensor equations, the theory is said to be given a manifestly covariant form. A theory that is written in a *manifestly covariant* form will automatically fulfil the covariance principle, but it need not fulfil the principle of relativity.

1.9   Mach's principle

Einstein wanted to abandon Newton's idea of an absolute space. He was attracted by the idea that all motion is relative. This may sound simple, but it leads to some highly non-trivial and fundamental questions.

Imagine that the Universe consists of only two particles connected by a spring. What will happen if the two particles rotate about each other? Will the string be stretched due to centrifugal forces? Newton would have confirmed that this is indeed what will happen. However, when there is no longer any absolute space that the particles can rotate relatively to, the answer is not as obvious. To observers rotating around stationary particles, the string would not appear to stretch. This situation is, however, kinematically equivalent to the one with rotating particles and observers at rest, which presumably leads to stretching.

Such problems led Mach to the view that all motion is relative. The motion of a particle in an empty universe is not defined. All motion is motion relative to something else, i.e., relative to other masses. According to Mach this implies that inertial forces must be due to a particle's acceleration relative to the great masses of the Universe. If there were no such cosmic masses, there would exist no inertial forces. In our string example, if there were no cosmic masses that the particles could rotate relatively to, there would be no stretching of the string.

Another example makes use of a carousel. If we stay on this while it rotates, we feel that the centrifugal force leads us outwards. At the same time we observe that the heavenly bodies rotate.

Einstein was impressed by Mach's arguments, which likely influenced Einstein's construction of the general theory of relativity. Yet it is clear that general relativity does not fulfil all requirements set by Mach's principle. There exist, for example, general relativistic, rotating cosmological models, where free particles will tend to rotate relative to the cosmic mass of the model.

Some Machian effects have been shown to follow from the equations of the general theory of relativity. For example, inside a rotating, massive shell the inertial frames, i.e., the free particles, are dragged around and tend to rotate in the same direction as the shell. This was discovered by Lense and Thirring in 1918 [LT18] and is called the Lense–Thirring effect. More recent investigations of this effect by D.R. Brill and J.M. Cohen [BC66] and others, led to the following result:

> A massive shell with radius equal to its Schwarzschild radius [see Ch. 10] has often been used as an idealized model of our Universe. Our result shows that in such models local inertial frames near the centre cannot rotate relatively to the mass of the Universe. In this way our result gives an explanation, in accordance with Mach's principle, of the fact that the 'fixed stars' are at rest on heaven as observed from an inertial reference frame.

It is clear to some extent that local inertial frames are determined by the distribution and motion of mass in the Universe, but in Einstein's General Theory of Relativity one cannot expect that matter alone determines the local inertial frames. The gravitational field itself, e.g. in the form of gravitational waves, may play a significant role.

Problems

1.1. *The strength of gravity compared to the Coulomb force*

(a) Determine the difference in strength between the Newtonian gravitational attraction and the Coulomb force of the interaction of the proton and the electron in a hydrogen atom.

(b) What is the gravitational force of attraction of two objects of 1 kg at a separation of 1 m. Compare with the corresponding electrostatic force of two charges of 1 C at the same distance.

(c) Compute the gravitational force between the Earth and the Sun. If the attractive force was not gravitational but caused by opposite electric charges, then what would the charges be?

1.2. *Falling objects in the gravitational field of the Earth*

(a) Two test particles are in free fall towards the centre of the Earth. They both start from rest at a height of 3 Earth radii and with a horizontal separation of 1 m. How far have the particles fallen when the distance between them is reduced to 0.5 m?

(b) Two new test particles are dropped from the same height with a time separation of 1 s. The first particle is dropped from rest. The second particle is given an initial velocity equal to the instantaneous velocity of the first particle, and it follows after the first one in the same trajectory. How far and how long have the particles fallen when the distance between them is doubled?

1.3. *Newtonian potentials for spherically symmetric bodies*

(a) Calculate the Newtonian potential $\phi(r)$ for a spherical shell of matter. Assume that the thickness of the shell is negligible, and the mass per unit area, σ, is constant on the spherical shell. Find the potential both inside and outside the shell.

(b) Let R and M be the radius and the mass of the Earth. Find the potential $\phi(r)$ for $r < R$ and $r > R$. The mass-density is assumed to be constant for $r < R$. Calculate the gravitational acceleration on the surface of the Earth. Compare with the actual value of $g = 9.81\text{m/s}^2$ ($M = 6.0 \cdot 10^{24}\text{kg}$ and $R = 6.4 \cdot 10^6\text{m}$).

(c) Assume that a hollow tube has been drilled right through the centre of the Earth. A small solid ball is then dropped into the tube from the surface of the Earth. Find the position of the ball as a function of time. What is the period of the oscillations of the ball?

(d) We now assume that the tube is not passing through the centre of the Earth, but at a closest distance s from the centre. Find how the period of the oscillations vary as a function of s. Assume for simplicity that the ball is sliding without friction (i.e., no rotation) in the tube.

1.4. *The Earth-Moon system*

(a) Assume that the Earth and the Moon are point objects and isolated from the rest of the Solar system. Put down the equations of motion for the Earth-Moon system. Show that there is a solution where the Earth and Moon are moving in perfect circular orbits around their common centre of mass. What is the radii of the orbits when we know the mass of the Earth and the Moon, and the orbital period of the Moon?

(b) Find the Newtonian potential along the line connecting the two bodies. Draw the result in a plot, and find the point on the line where the gravitational interactions from the bodies exactly cancel each other.

(c) The Moon acts with a different force on a 1 kilogram measure on the surface of Earth, depending on whether it is closest to or farthest from the Moon. Find the difference in these forces.

1.5. *The Roche-limit*
A spherical moon with a mass m and radius R is orbiting a planet with mass M. Show that if the moon is closer to its parent planet's centre than

$$ r = \left(\frac{2M}{m} \right)^{1/3} R, $$

then loose rocks on the surface of the moon will be elevated due to tidal effects.

1.6. *A Newtonian Black Hole*
In 1783 the English physicist John Michell used Newtonian dynamics and laws of gravity to show that for massive bodies which were small enough, the escape velocity of the bodies are larger than the speed of light. (The same was emphasized by the French mathematician and astronomer Pierre Laplace in 1796).

(a) Assume that the body is spherical with mass M. Find the largest radius, R, that the body can have in order for it to be a "Black Hole", i.e., so that light cannot escape. Assume naively that photons have kinetic energy $\frac{1}{2}mc^2$.

(b) Find the tidal force on two bodies m at the surface of a spherical body, when their internal distance is ζ. What would the tidal force be on the

head and the feet of a 2 m tall human, standing upright, in the following
cases (consider the head and feet as point particles, each weighing 5kg):

1. The human is standing on the surface of a Black Hole with 10 times
   the Solar mass.

2. On the Sun's surface.

3. On the Earth's surface.

1.7. Non-relativistic Kepler orbits

(a) Consider first the Newtonian gravitational potential $\varphi(r)$ at a distance r
from the Sun to be $\varphi(r) = -\frac{GM}{r}$, where M is the solar mass. Write down
the classical Lagrangian in spherical coordinates (r, θ, ϕ) for a planet with
mass m. The Sun is assumed to be stationary.

What is the physical interpretation of the canonical momentum $p_\phi = \ell$?
How can we from the Lagrangian see that it is a constant of motion? Find
the Euler-Lagrange equation for θ and show that it can be written

$$\frac{d}{dt}\left(mr^4\dot{\theta}^2 + \frac{\ell^2}{m\sin^2\theta} \right) = 0. \tag{1.59}$$

Show, using this equation, that the planet can be considered to move in a
plane such that at $t = 0$, $\theta = \pi/2$ and $\dot\theta = 0$.

(b) Find the Euler-Lagrange equation for r and use it to find r as a function
of ϕ. Show that the bound orbits are ellipses. Of circular orbits, what is
the orbital period T in terms of the radius R?

(c) If the Sun is not completely spherical, but slightly squashed at the poles,
then the gravitational potential along the equatorial plane has to be modified to

$$\varphi(r) = -\frac{GM}{r} - \frac{Q}{r^3}, \tag{1.60}$$

where Q is a small constant. We will assume that the planet move in the
plane where this expression is valid. Show that a circular orbit is still
possible. What is the relation between T and R in this case?

(d) Assume that the orbit deviates slightly from a circular orbit; i.e. $r = R+\rho$,
where $\rho \ll R$. Show that ρ varies periodically according to

$$\rho = \rho_0 \sin\left(\frac{2\pi}{T_\rho}t \right). \tag{1.61}$$

Find T_ρ, and show that the orbit precesses slightly during each orbit.
What it the angle $\Delta\phi$ of precession for each orbit?

The constant Q can be written $Q = \frac{1}{2}J_2 GM R_S^2$ where J_2 is the Sun's
quadrupole moment, and R_S is the Sun's radius. Observational data
show that $J_2 \lesssim 3 \cdot 10^{-5}$. What is the maximal precession of $\Delta\phi$ for the
Mercurian orbit?

2

The Special Theory of Relativity

In this chapter we shall give a short introduction to the fundamental principles of the special theory of relativity, and deduce some of the consequences of the theory.

The special theory of relativity was presented by Albert Einstein in 1905. It was founded on two postulates:

1. The laws of physics are the same in all Galilean frames.

2. The velocity of light in empty space is the same in all Galilean frames and independent of the motion of the light source.

Einstein pointed out that these postulates are in conflict with Galilean kinematics, in particular with the Galilean law for addition of velocities. According to Galilean kinematics two observers moving relative to each other cannot measure the same velocity for a certain light signal. Einstein solved this problem by a thorough discussion of how two distant clocks should be synchronized.

2.1 Coordinate systems and Minkowski-diagrams

The most simple physical phenomenon that we can describe is called an event. This is an incident that takes place at a certain point in space and at a certain point in time. A typical example is the flash from a flashbulb.

A complete description of an event is obtained by giving the position of the event in space and time. Assume that our observations are made with reference to a reference frame. We introduce a coordinate system into our reference frame. Usually it is advantageous to employ a Cartesian coordinate system. This may be thought of as a cubic lattice constructed by measuring rods. If one lattice point is chosen as origin, with all coordinates equal to zero, then any other lattice point has three spatial coordinates equal to the distances of that point along the coordinate axes that pass through the origin. The spatial coordinates of an event are the three coordinates of the lattice point at which the event happens.

It is somewhat more difficult to determine the point of time of an event. If an observer is sitting at the origin with a clock, then the point of time when he catches sight of an event is not the point of time when the event happened. This is because the light takes time to pass from the position of the event to the observer at the origin. Since observers at different positions have to make different such corrections, it would be simpler to have (imaginary) observers at each point of the reference frame such that the point of time of an arbitrary event can be measured locally.

But then a new problem appears. One has to synchronize the clocks, so that they show the same time and go at the same rate. This may be performed by letting the observer at the origin send out light signals so that all the other clocks can be adjusted (with correction for light-travel time) to show the same time as the clock at the origin. These clocks show the *coordinate time* of the coordinate system, and they are called *coordinate clocks*.

By means of the lattice of measuring rods and coordinate clocks, it is now easy to determine four coordinates ($x^0 = ct, x, y, z$) for every event. (We have multiplied the time coordinate t by the velocity of light c in order that all four coordinates shall have the same dimension.)

This coordinatization makes it possible to describe an event as a point P in a so-called *Minkowski-diagram*. In this diagram we plot ct along the vertical axis and one of the spatial coordinates along the horizontal axis.

In order to observe particles in motion, we may imagine that each particle is equipped with a flash-light, and that they flash at a constant frequency. The flashes from a particle represent a succession of events. If they are plotted into a Minkowski-diagram, we get a series of points that describe a curve in the continuous limit. Such a curve is called a *world-line* of the particle. The world-line of a free particle is a straight line, as shown to left of the time axis in Fig. 2.1.

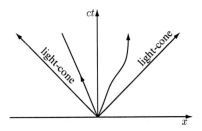

Figure 2.1: World-lines.

A particle acted upon by a net force has a curved world-line as the velocity of the particle changes with time. Since the velocity of every material particle is less than the velocity of light, the tangent of a world line in a Minkowski-diagram will always make an angle less than 45° with the time axis.

A flash of light gives rise to a light-front moving onwards with the velocity of light. If this is plotted in a Minkowski-diagram, the result is a light-cone. In Fig. 2.1 we have drawn a light-cone for a flash at the origin. It is obvious that we could have drawn light-cones at all points in the diagram. An important result is that *the world-line of any particle at a point is inside the light-cone of a flash from that point*. This is an immediate consequence of the special principle of relativity, and is also valid locally in the presence of a gravitational field.

2.2 Synchronization of clocks

There are several equivalent methods that can be used to synchronize clocks. We shall here consider the radar method.

We place a mirror on the x-axis and emit a light signal from the origin at time t_A. This signal is reflected by the mirror at t_B, and received again by the observer at the origin at time t_C. According to the second postulate of the special theory of relativity, the light moves with the same velocity in both directions, giving

$$t_B = \frac{1}{2}(t_A + t_C). \tag{2.1}$$

When this relationship holds we say that the clocks at the origin and at the mirror are *Einstein synchronized*. Such synchronization is presupposed in the special theory of relativity. The situation corresponding to synchronization by the radar method is illustrated in Fig. 2.2.

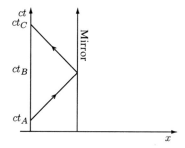

Figure 2.2: Clock synchronization by the radar method.

The radar method can also be used to measure distances. The distance L from the origin to the mirror is given by

$$L = \frac{c}{2}(t_C - t_A). \tag{2.2}$$

Later (Chapter 8) we shall see that when we measure distances in a gravitational field, the results depend upon the measuring technique that is used. For example, measurements made using the radar method differ from those made using measuring rods.

2.3 The Doppler effect

Consider three observers (1, 2, and 3) in an inertial frame. Observers 1 and 3 are at rest, while 2 moves with constant velocity along the x-axis. The situation is illustrated in Fig. 2.3.

Each observer is equipped with a clock. If observer 1 emits light pulses with a constant period τ_1, then observer 2 receives them with a longer period τ_2 according to his or her[1] clock. The fact that these two periods are different is a well-known phenomenon, called the *Doppler effect*. The same effect is

---

[1]For simplicity we shall—without any sexist implications—follow the grammatical convention of using masculine pronouns, instead of the more cumbersome 'his or her'.

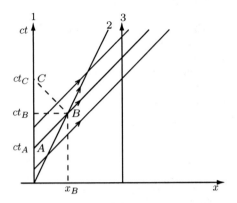

Figure 2.3: The Doppler effect.

observed with sound; the tone of a receding vehicle is lower than that of an approaching one.

We are now going to deduce a relativistic expression for the Doppler effect. Firstly, we see from Fig. 2.3 that the two periods τ_1 and τ_2 are proportional to each other,

$$\tau_2 = K\tau_1. \tag{2.3}$$

The constant $K(v)$ is called Bondi's K-factor. Since observer 3 is at rest, the period τ_3 is equal to τ_1 so that

$$\tau_3 = \frac{1}{K}\tau_2. \tag{2.4}$$

These two equations imply that if 2 moves away from 1, so that $\tau_2 > \tau_1$, then $\tau_3 < \tau_2$. This is because 2 moves towards 3.

The K-factor is most simply determined by placing observer 1 at the origin, while letting the clocks show $t_1 = t_2 = 0$ at the moment when 2 passes the origin. This is done in Fig. 2.3. The light pulse emitted at the point of time t_A, is received by 2 when his clock shows $\tau_2 = Kt_A$. If 2 is equipped with a mirror, the reflected light pulse is received by 1 at a point of time $t_C = K\tau_2 = K^2t_A$. According to Eq. (2.1) the reflection-event then happens at a point of time

$$t_B = \frac{1}{2}(t_C + t_A) = \frac{1}{2}(K^2 + 1)t_A. \tag{2.5}$$

The mirror has then arrived at a distance x_B from the origin, given by Eq. (2.2),

$$x_B = \frac{c}{2}(t_C - t_A) = \frac{c}{2}(K^2 - 1)t_A. \tag{2.6}$$

Thus, the velocity of observer 2 is

$$v = \frac{x_B}{t_B} = c\frac{K^2 - 1}{K^2 + 1}. \tag{2.7}$$

Solving this equation with respect to the K-factor we get

$$K = \left(\frac{c + v}{c - v}\right)^{1/2}. \tag{2.8}$$

This result is relativistically correct. The special theory of relativity was included through the tacit assumption that the velocity of the reflected light is c. This is a consequence of the second postulate of special relativity; the velocity of light is isotropic and independent of the velocity of the light source.

Since the wavelength λ of the light is proportional to the period τ, Eq. (2.3) gives the observed wavelength λ' for the case when the observer moves away from the source,

$$\lambda' = K\lambda = \left(\frac{c+v}{c-v}\right)^{1/2}\lambda. \tag{2.9}$$

This Doppler-effect represents a red-shift of the light. If the light source moves towards the observer, there is a corresponding blue-shift given by K^{-1}.

It is common to express this effect in terms of the relative change of wavelength,

$$z = \frac{\lambda'-\lambda}{\lambda} = K - 1 \tag{2.10}$$

which is positive for red-shift. If $v \ll c$, Eq. (2.9) gives,

$$\frac{\lambda'}{\lambda} = K \approx 1 + \frac{v}{c} \tag{2.11}$$

to lowest order in v/c. The red-shift is then

$$z = \frac{v}{c}. \tag{2.12}$$

This result is well known from non-relativistic physics.

2.4 Relativistic time-dilatation

Every periodic motion can be used as a clock. A particularly simple clock is called the light clock. This is illustrated in Fig. 2.4.

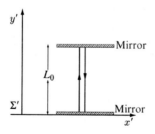

Figure 2.4: Light-clock.

The clock consists of two parallel mirrors that reflect a light pulse back and forth. If the period of the clock is defined as the time interval between each time the light pulse hits the lower mirror, then $\Delta t' = 2L_0/c$.

Assume that the clock is at rest in an inertial reference frame Σ' where it is placed along the y-axis, as shown in Fig. 2.4. If this system moves along the ct-axis with a velocity v relative to another inertial reference frame Σ, the light pulse of the clock will follow a zigzag path as shown in Fig. 2.5.

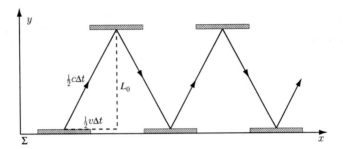

Figure 2.5: Moving light-clock.

The light signal follows a different path in Σ than in Σ'. The period Δt of the clock as observed in Σ is different from the period $\Delta t'$ which is observed in the rest frame. The period Δt is easily found from Fig. 2.5. Since the pulse takes the time $(1/2)\Delta t$ from the lower to the upper mirror and since the light velocity is always the same, we find

$$\left(\frac{1}{2}c\Delta t\right)^2 = L_0^2 + \left(\frac{1}{2}v\Delta t\right)^2, \tag{2.13}$$

i.e.,

$$\Delta t = \gamma\frac{2L_0}{c}, \quad \gamma \equiv \frac{1}{\sqrt{1-\frac{v^2}{c^2}}}. \tag{2.14}$$

The γ factor is a useful short-hand notation for a term which is often used in relativity theory. It is commonly known as the Lorentz factor.

Since the period of the clock in its rest frame is $\Delta t'$, we get

$$\boxed{\Delta t = \gamma\Delta t'.} \tag{2.15}$$

Thus, we have to conclude that the period of the clock when it is observed to move (Δt) is greater that its rest-period ($\Delta t'$). In other words: *a moving clock goes slower than a clock at rest.* This is called *the relativistic time-dilatation.* The period $\Delta t'$ of the clock as observed in its rest frame is called the proper period of the clock. The corresponding time t' is called the proper time of the clock.

One might be tempted to believe that this surprising consequence of the special theory of relativity has something to do with the special type of clock that we have employed. This is not the case. If there had existed a mechanical clock in Σ that did not show the time dilatation, then an observer at rest in Σ might measure his velocity by observing the different rates of his light clock and this mechanical clock. In this way he could measure the absolute velocity of Σ. This would be in conflict with the special principle of relativity.

2.5   The relativity of simultaneity

Events that happen at the same point of time are said to be *simultaneous events.* We shall now show that according to the special theory of relativity, events that are simultaneous in one reference frame are not simultaneous in another reference frame moving with respect to the first. This is what is meant by the expression "the relativity of simultaneity".

Consider again two mirrors connected by a line along the x'-axis, as shown in Fig. 2.6. Halfway between the mirrors there is a flash-lamp emitting a spherical wave front at a point of time t_C.

The points at which the light front reaches the left-hand and the right-hand mirrors are denoted by A and B, respectively. In the reference frame Σ' of Fig. 2.6 the events A and B are simultaneous.

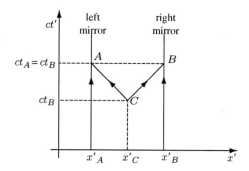

Figure 2.6: Simultaneous events A and B.

If we describe the same course of events from another reference frame (Σ), where the mirror moves with constant velocity v in the positive x-direction we find the Minkowski-diagram shown in Fig. 2.7. Note that the light follows

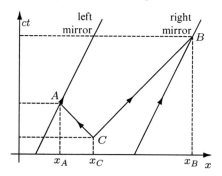

Figure 2.7: The simultaneous events in another frame.

world lines making an angle of $45°$ with the axes. This is the case in every inertial frame.

In Σ the light pulse reaches the left mirror, which moves towards the light, before it reaches the right mirror, which moves in the same direction as the light. In this reference frame the events when the light pulses hit the mirrors are not simultaneous.

As an example illustrating the relativity of simultaneity, Einstein imagined that the events A, B and C happen in a train which moves past the platform with a velocity v. The event C represents the flash of a lamp at the mid-point of a wagon. A and B are the events when the light is received at the back end and at the front end of the wagon respectively. This situation is illustrated in Fig. 2.8.

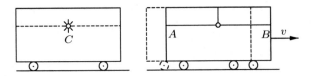

Figure 2.8: Light flash in a moving train.

As observed in the wagon, A and B happen simultaneously. As observed from the platform the rear end of the wagon moves towards the light which moves backwards, while the light moving forwards has to catch up with the front end. Thus, as observed from the platform A will happen before B.

The time difference between A and B as observed from the platform will now be calculated. The length of the wagon, as observed from the platform, will be denoted by L. The time coordinate is chosen such that $t_C = 0$. The light moving backwards hits the rear wall at a point of time t_A. During the time t_A the wall has moved a distance vt_A forwards, and the light has moved a distance ct_A backwards. Since the distance between the wall and the emitter is $L/2$, we get

$$\frac{L}{2} = vt_A + ct_A. \tag{2.16}$$

Thus

$$t_A = \frac{L}{2(c+v)}. \tag{2.17}$$

In the same manner one finds

$$t_B = \frac{L}{2(c-v)}. \tag{2.18}$$

It follows that the time difference between A and B as observed from the platform is

$$\Delta t = t_B - t_A = \frac{\gamma^2 vL}{c^2}. \tag{2.19}$$

As observed from the wagon A and B are simultaneous. As observed from the platform the rear event A happens a time interval Δt before the event B. This is the relativity of simultaneity.

2.6 The Lorentz-contraction

During the first part of the nineteenth century the so-called luminiferous ether was introduced into physics to account for the propagation and properties of light. After J.C. Maxwell showed that light is electromagnetic waves the ether was still needed as a medium in which electromagnetic waves propagated [Ros64].

It was shown that Maxwell's equations do not obey the principle of relativity, when coordinates are changed using the Galilean transformations. If it is assumed that the Galilean transformations are correct, then Maxwell's equations can only be valid in one coordinate system. This coordinate system was the one in which the ether was at rest. Hence, Maxwell's equations in combination with the Galilean transformations implied the concept of 'absolute rest'. This made the measurement of the velocity of the Earth relative to the ether of great importance.

An experiment sufficiently accurate to measure this velocity to order v^2/c^2 was carried out by Michelson and Morley in 1887. A simple illustration of the experiment is shown in Fig. 2.9.

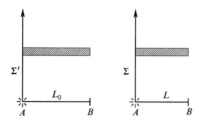

Figure 2.9: Length contraction.

Our earlier photon clock is supplied by a mirror at a distance L along the x-axis from the emitter. The apparatus moves in the x-direction with a velocity v. In the rest-frame (Σ') of the apparatus, the distance between A and B is equal to the distance between A and C. This distance is denoted by L_0 and is called the *rest length* between A and B.

Light is emitted from A. Since the velocity of light is isotropic and the distances to B and C are equal in Σ', the light reflected from B and that reflected from C have the same travelling time. This was the result of the Michelson–Morley experiment, and it seems that we need no special effects such as the Lorentz-contraction to explain the experiment.

However, before 1905 people believed in the physical reality of absolute velocity. The Earth was considered to move though an "ether" with a velocity that changed with the seasons. The experiment should therefore be described under the assumption that the apparatus is moving.

Let us therefore describe a experiment from our reference frame Σ, which may be thought of as at rest in the "ether". Then according to Eq. (2.14) the travel time of the light being reflected at C is

$$\Delta t_C = \gamma \frac{2L_0}{c}. \tag{2.20}$$

For the light moving from A to B we may use Eq. (2.18), and for the light from B to A Eq. (2.17). This gives

$$\Delta t_B = \frac{L}{c-v} + \frac{L}{c+v} = \gamma^2 \frac{2L}{c}. \tag{2.21}$$

If length is independent of velocity, then $L = L_0$. In this case the travelling times of the light signals will be different. The travelling time difference is

$$\Delta t_B - \Delta t_C = \gamma(\gamma - 1)\frac{2L_0}{c}. \tag{2.22}$$

To the lowest order in v/c, $\gamma \approx 1 + \frac{1}{2}(v/c)^2$, so that

$$\Delta t_B - \Delta t_C \approx \frac{1}{2}\left(\frac{v}{c}\right)^2, \tag{2.23}$$

which depends upon the velocity of the apparatus.

According to the ideas involving an absolute velocity of the Earth through the ether, if one lets the light reflected at B interfere with the light reflected at C (at the position A) then the interference pattern should vary with the season. This was not observed. On the contrary, observations showed that $\Delta t_B = \Delta t_C$.

Assuming that length varies with velocity, Eqs. (2.20) and (2.21), together with this observation, gives

$$\boxed{L = \gamma^{-1} L_0.} \tag{2.24}$$

The result that $L < L_0$ (i.e. the length of a rod is less when it moves than when it is at rest) is called the *Lorentz-contraction*.

2.7 The Lorentz transformation

An event P has coordinates $(t', x', 0, 0)$ in a Cartesian coordinate system associated with a reference frame Σ'. Thus the distance from the origin of Σ' to P measured with a measuring rod at rest in Σ' is x'. If the distance between the origin of Σ' and the position at the x-axis where P took place is measured with measuring rods at rest in a reference frame moving with velocity v in the x-direction relative to Σ', one finds the length $\gamma^{-1} x$ due to the Lorentz contraction. Assuming that the origin of Σ and Σ' coincided at the point of time $t = 0$, the origin of Σ' has an x-coordinate vt at a point of time t. The event P thus has an x-coordinate

$$x = vt + \gamma^{-1} x' \tag{2.25}$$

or

$$x' = \gamma(x - vt). \tag{2.26}$$

The x-coordinate may be expressed in terms of t' and x' by letting $v \to -v$,

$$\boxed{x = \gamma(x' + vt').} \tag{2.27}$$

The y and z coordinates are associated with axes directed perpendicular to the direction of motion. Therefore, they are the same in the two coordinate systems

$$y = y' \quad \text{and} \quad z = z'. \tag{2.28}$$

Substituting x' from Eq. (2.26) into Eq. (2.27) reveals the connection between the time coordinates of the two coordinate systems,

$$t' = \gamma \left(t - \frac{vx}{c^2} \right) \tag{2.29}$$

and

$$\boxed{t = \gamma \left(t' + \frac{vx'}{c^2} \right).} \tag{2.30}$$

The latter term in this equation is nothing but the deviation from simultaneity in Σ for two events that are simultaneous in Σ'.

The relations (2.27)–(2.30) between the coordinates of Σ and Σ' represent a special case of the *Lorentz transformations*. The above relations are special since the two coordinate systems have the same spatial orientation, and the x and

x'-axes are aligned along the relative velocity vector of the associated frames. Such transformations are called *boosts*.

For non-relativistic velocities $v \ll c$, the Lorentz transformations (2.27)–(2.30) pass over into the corresponding Galilei-transformations, (1.2) and (1.4).

The Lorentz transformation gives a connection between the relativity of simultaneity and the Lorentz contraction. The *length* of a body is defined as the difference between the coordinates of its end points, *as measured by simultaneity in the rest-frame of the observer*.

Consider the wagon of Section 2.5. Its rest length is $L_0 = x'_B - x'_A$. The difference between the coordinates of the wagon's end-points, $x_A - x_B$ as measured in Σ, is given implicitly by the Lorentz transformation

$$x'_B - x'_A = \gamma \left[x_B - x_A - v(t_B - t_A) \right]. \qquad (2.31)$$

According to the above definition the length (L) of the moving wagon is given by $L = x_B - x_A$ with $t_B = t_A$.

From Eq. (2.31) we then get

$$L_0 = \gamma L, \qquad (2.32)$$

which is equivalent to Eq. (2.24).

The Lorentz transformation will now be used to deduce the relativistic formulae for velocity addition. Consider a particle moving with velocity u along the x'-axis of Σ'. If the particle was at the origin at $t' = 0$, its position at t' is $x' = u't'$. Using this relation together with Eqs. (2.27) and (2.28) we find the velocity of the particle as observed in Σ

$$\boxed{u = \frac{x}{t} = \frac{u' + v}{1 + \frac{u'v}{c^2}}.} \qquad (2.33)$$

A remarkable property of this expression is that by adding velocities less than c one cannot obtain a velocity greater than c. For example, if a particle moves with a velocity c in Σ' then its velocity in Σ is also c regardless of Σ''s velocity relative to Σ'.

Equation (2.33) may be written in a geometric form by introducing the so-called *rapidity* η defined by

$$\boxed{\tanh \eta = \frac{u}{c}} \qquad (2.34)$$

for a particle with velocity u. Similarly the rapidity of Σ' relative to Σ is

$$\tanh \theta = \frac{v}{c}. \qquad (2.35)$$

Since

$$\tanh(\eta' + \theta) = \frac{\tanh \eta' + \tanh \theta}{1 + \tanh \eta' \tanh \theta}, \qquad (2.36)$$

the relativistic velocity addition formula, Eq. (2.33), may be written

$$\eta = \eta' + \theta. \qquad (2.37)$$

Since rapidities are additive, their introduction simplifies some calculations and they have often been used as variables in elementary particle physics.

With these new hyperbolic variables we can write the Lorentz transformation in a particularly simple way. Using Eq. (2.35) in Eqs. (2.27) and (2.30) we find

$$\boxed{x = x' \cosh \theta + ct' \sinh \theta, \quad ct = x' \sinh \theta + ct' \cosh \theta.} \qquad (2.38)$$

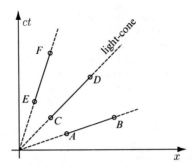

Figure 2.10: The interval between A and B is space-like, between C and D light-like, and between E and F time-like.

2.8   Lorentz-invariant interval

Let two events be given. The coordinates of the events, as referred to two different reference frames Σ and Σ' are connected by a Lorentz transformation. The coordinate differences are therefore connected by

$$\Delta t = \gamma(\Delta t' + \tfrac{v}{c^2}\Delta x'), \quad \Delta x = \gamma(\Delta x' + v\Delta t'),$$
$$\Delta y = \Delta y', \qquad\qquad \Delta z = \Delta z'. \tag{2.39}$$

Just like $(\Delta y)^2 + (\Delta z)^2$ is invariant under a rotation about the x-axis, $-(c\Delta t)^2 + (\Delta x)^2 + (\Delta y)^2 + (\Delta z)^2$ is invariant under a Lorentz transformation, i.e.,

$$(\Delta s)^2 = -(c\Delta t)^2 + (\Delta x)^2 + (\Delta y)^2 + (\Delta z)^2$$
$$= -(c\Delta t')^2 + (\Delta x')^2 + (\Delta y')^2 + (\Delta z')^2. \tag{2.40}$$

This combination of squared coordinate-intervals is called the spacetime interval, or the *interval*. It is invariant under both rotations and Lorentz transformations.

Due to the minus-sign in Eq. (2.40), the interval between two events may be positive, zero or negative. These three types of intervals are called:

$$\begin{array}{ll} (\Delta s)^2 > 0 & \text{space-like} \\ (\Delta s)^2 = 0 & \text{light-like} \\ (\Delta s)^2 < 0 & \text{time-like} \end{array} \tag{2.41}$$

The reasons for these names are the following. Given two events with a space-like interval (A and B in Fig. 2.10), then there exists a Lorentz transformation to a new reference frame where A and B happen simultaneously. In this frame the distance between the events is purely spatial. Two events with a light-like interval (C and D in Fig. 2.10), can be connected by a light signal, i.e. one can send a photon from C to D. The events E and F have a time-like interval between them, and can be observed from a reference frame in which they have the same spatial position, but occur at different points of time.

Since all material particles move with a velocity less than that of light, the points on the world-line of a particle are separated by time-like intervals. The curve is then said to be time-like. All time-like curves through a point pass inside the light-cone from that point.

If the velocity of a particle is $u = \Delta x / \Delta t$ along the x-axis, Eq. (2.40) gives

$$(\Delta s)^2 = -\left(1 - \frac{u^2}{c^2}\right)(c\Delta t)^2. \tag{2.42}$$

In the rest-frame Σ' of the particle, $\Delta x' = 0$, giving

$$(\Delta s)^2 = -(c\Delta t')^2. \tag{2.43}$$

The time t' in the rest-frame of the particle is the same as the time measured on a clock carried by the particle. It is called the *proper time* of the particle, and denoted by τ. From Eqs. (2.42) and (2.43) it follows that

$$\Delta\tau = \sqrt{1 - \frac{u^2}{c^2}}\,\Delta t = \gamma^{-1}\Delta t \tag{2.44}$$

which is an expression of the relativistic time-dilatation.

Equation (2.43) is important. It gives the physical interpretation of a time-like interval between two events. The interval is a measure of the proper time interval between the events. This time is measured on a clock that moves such that it is present at both events. In the limit $u \to c$ (the limit of a light signal), $\Delta\tau = 0$. This shows that $(\Delta s)^2 = 0$ for a light-like interval.

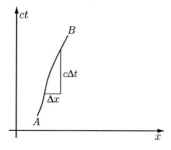

Figure 2.11: World-line of an accelerating particle.

Consider a particle with a variable velocity, $u(t)$, as indicated in Fig. 2.11. In this situation we can specify the velocity at an arbitrary point of the world-line. Eq. (2.44) can be used with this velocity, in an infinitesimal interval around this point,

$$d\tau = \sqrt{1 - \frac{u^2(t)}{c^2}}\,dt. \tag{2.45}$$

This equation means that the acceleration has no local effect upon the proper-time of the clock. Here the word "local" means as measured by an observer at the position of the clock. Such clocks are called *standard clocks*.

If a particle moves from A to B in Fig. 2.11, the proper-time as measured on a standard clock following the particle is found by integrating Eq. (2.45)

$$\tau_B - \tau_A = \int_A^B \sqrt{1 - \frac{u^2(t)}{c^2}}\,dt. \tag{2.46}$$

The relativistic time-dilatation has been verified with great accuracy by observations of unstable elementary particles with short life-times [FS63].

An infinitesimal spacetime interval

$$ds^2 = -c^2 dt^2 + dx^2 + dy^2 + dz^2 \tag{2.47}$$

is called a *line-element*. The physical interpretation of the line-element between two infinitesimally close events on a time-like curve is

$$\boxed{ds^2 = -c^2 d\tau^2,} \tag{2.48}$$

where $d\tau$ is the proper-time interval between the events, measured with a clock following the curve. The spacetime interval between two events is given by the integral (2.46). It follows that *the proper-time interval between two events is path dependent*. This leads to the following surprising result: A time-like interval between two events is *greatest* along the straightest possible curve between them.

2.9   The twin-paradox

Rather than discussing the life-time of elementary particles, we may as well apply Eq. (2.46) to a person. Let her name be Eva. Assume that Eva is rapidly acceleration from rest at the point of time $t = 0$ at origin to a velocity v along the x-axis of a (ct, x) coordinate system in an inertial reference frame Σ. (See Fig. 2.12.)

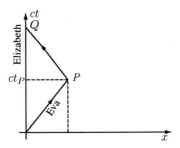

Figure 2.12: World-lines of the twin sisters Eva and Elizabeth.

At a point of time t_P she has come to a position x_P. She then rapidly decelerates until reaching a velocity v in the negative x-direction. At a point of time t_Q, as measured on clocks at rest in Σ, she has returned to her starting location. If we neglect the brief periods of acceleration, Eva's travelling time as measured on a clock which she carries with her is

$$t_{\text{Eva}} = \left(1 - \frac{v^2}{c^2}\right)^{1/2} t_Q. \tag{2.49}$$

Now assume that Eva has a twin-sister named Elizabeth who remains at rest at the origin of Σ.

Elizabeth has become older by $\tau_{\text{Elizabeth}} = t_Q$ during Eva's travel, so that

$$\tau_{\text{Eva}} = \left(1 - \frac{v^2}{c^2}\right)^{1/2} \tau_{\text{Elizabeth}}. \tag{2.50}$$

For example, if Eva travelled to Proxima Centauri (the Sun's nearest neighbour at four light years) with a velocity $v = 0.8c$, she would be gone for ten years as measured by Elizabeth. Therefore Elizabeth has aged 10 years during Eva's travel. According to Eq. (2.50), Eva has only aged 6 years. According to Elizabeth, Eva has aged less than herself during her travels.

The principle of relativity, however, tells that Eva can consider herself as at rest and Elizabeth as the traveller. According to Eva it is Elizabeth who has only aged by 6 years, while Eva has aged by 10 years during the time they are apart.

What happens? How can the twin-sisters arrive at the same prediction as to how much each of them age during the travel? In order to arrive at a clear answer to these questions, we shall have to use a result from the general theory of relativity. The twin-paradox will be taken up again in Chapter 5.

2.10 Hyperbolic motion

With reference to an inertial reference frame it is easy to describe relativistic accelerated motion. The special theory of relativity is in no way limited to describe motion with constant velocity.

Let a particle move with a variable velocity $u(t) = dx/dt$ along the x-axis in Σ. The frame Σ' moves with velocity v in the same direction relative to Σ. In this frame the particle-velocity is $u'(t') = dx'/dt'$. At every moment the velocities u and u' are connected by the relativistic formula for velocity addition, Eq. (2.33). Thus, a velocity change du' in Σ' and the corresponding velocity change du in Σ are related – using Eq. (2.30) – by

$$dt = \frac{dt' + \frac{v}{c^2}dx'}{\sqrt{1 - \frac{v^2}{c^2}}} = \frac{1 + \frac{u'v}{c^2}}{\sqrt{1 - \frac{v^2}{c^2}}}dt'. \qquad (2.51)$$

Combining these expressions we obtain the relationship between the acceleration of the particle as measured in Σ and in Σ'

$$a = \frac{du}{dt} = \frac{\left(1 - \frac{v^2}{c^2}\right)^{3/2}}{\left(1 + \frac{u'v}{c^2}\right)^3}a'. \qquad (2.52)$$

Until now the reference frame Σ' has had an arbitrary velocity. Now we choose $v = u(t)$ so that Σ' is the instantaneous rest frame of the particle at a point of time t. At this moment $u' = 0$. Then Eq. (2.52) reduces to

$$a = \left(1 - \frac{u^2}{c^2}\right)^{3/2} a'. \qquad (2.53)$$

Here a' is the acceleration of the particle as measured in its instantaneous rest frame. It is called *the rest acceleration* of the particle. Eq. (2.53) can be integrated if we know how the rest acceleration of the particle varies with time.

We shall now focus on the case where the particle has uniformly accelerated motion and moves along a straight path in space. The rest acceleration of the particle is constant, say $a' = g$. Integration of Eq. (2.53) with $u(0) = 0$ then gives

$$u = \left[1 + \frac{g^2}{c^2}t^2\right]^{-1/2} gt. \qquad (2.54)$$

Integrating once more gives

$$x = \frac{c^2}{g}\left[1 + \frac{g^2}{c^2}t^2\right]^{1/2} + x_0 - \frac{c^2}{g}, \tag{2.55}$$

where x_0 is a constant of integration corresponding to the position at $t = 0$.
Equation (2.55) can be given the form

$$\boxed{\left(x - x_0 + \frac{c^2}{g}\right)^2 - c^2t^2 = \frac{c^4}{g^2}.} \tag{2.56}$$

As shown in Fig. 2.13, this is the equation of a hyperbola in the Minkowski-diagram.

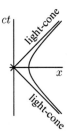

Figure 2.13: World line of particle with constant rest acceleration.

Since the world-line of a particle with uniformly accelerated, rectilinear motion has the shape of a hyperbola, this type of motion is called *hyperbolic motion*.

Using the proper-time τ of the particle as a parameter, we may obtain a simple parametric representation of its world-line. Substituting Eq. (2.54) into Eq. (2.45) we get

$$d\tau = \frac{dt}{\sqrt{1 + \frac{g^2}{c^2}t^2}}. \tag{2.57}$$

Integration with $\tau(0) = 0$ gives

$$\tau = \frac{c}{g}\,\mathrm{arsinh}\left(\frac{gt}{c}\right), \tag{2.58}$$

or

$$\boxed{t = \frac{c}{g}\sinh\left(\frac{g\tau}{c}\right).} \tag{2.59}$$

Inserting this expression into Eq. (2.55), we get

$$\boxed{x = \frac{c^2}{g}\cosh\left(\frac{g\tau}{c}\right) + x_0 - \frac{c^2}{g}.} \tag{2.60}$$

These expressions shall be used later when describing uniformly accelerated reference frames.

Note that *hyperbolic motion* results when the particle moves with *constant rest acceleration*. Such motion is usually called *uniformly accelerated motion*. Motion with constant acceleration as measured in the "laboratory frame" Σ gives rise to the usual parabolic motion.

2.11 Energy and mass

The existence of an electromagnetic radiation pressure was well known before Einstein formulated the special theory of relativity. In black body radiation with energy density ρ there is an isotropic pressure $p = (1/3)\rho c^2$. If the radiation moves in a certain direction (laser), then the pressure in this direction is $p = \rho c^2$.

Einstein gave several deductions of the famous equation connecting the inertial mass of a body with its energy content. A deduction he presented in 1906 is as follows.

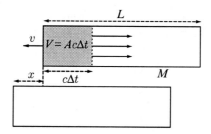

Figure 2.14: Light pulse in a box.

Consider a box with a light source at one end. A light pulse with radiation energy E is emitted to the other end where it is absorbed (see Fig. 2.14).

The box has a mass M and a length L. Due to the radiation pressure of the shooting light pulse the box receives a recoil. The pulse is emitted during a time interval Δt. During this time the radiation pressure is

$$p = \rho c^2 = \frac{E}{V} = \frac{E}{Ac\Delta t} \tag{2.61}$$

where V is the volume of the radiation pulse and A the area of a cross-section of the box. The recoil velocity of the box is

$$\Delta v = -a\Delta t = -\frac{F}{M}\Delta t = -\frac{pA}{M}\Delta t$$
$$= -\left(\frac{E}{Ac\Delta t}\right)\left(\frac{A\Delta t}{M}\right) = -\frac{E}{Mc}. \tag{2.62}$$

The pulse takes the time L/c to move to the other side of the box. During this time the box moves a distance

$$\Delta x = \Delta v\frac{L}{c} = -\frac{EL}{Mc^2}. \tag{2.63}$$

Then the box is stopped by the radiation pressure caused by the light pulse hitting the wall at the other end of the box.

Let m be the mass of the radiation. Before Einstein one would put $m = 0$. Einstein, however, reasoned as follows. Since the box and its contents represents an isolated system, the mass-centre has not moved. The mass centre of the box with mass M has moved a distance Δx to the left, the radiation with mass m has moved a distance L to the right. Thus

$$mL + M\Delta x = 0 \tag{2.64}$$

which gives

$$m = -\frac{M}{L}\Delta x = -\left(\frac{M}{L}\right)\left(-\frac{EL}{Mc^2}\right) = \frac{E}{c^2} \tag{2.65}$$

or

$$\boxed{E = mc^2.} \tag{2.66}$$

Here we have shown that radiation energy has an innate mass given by Eq. (2.65). Einstein derived Eq. (2.66) using several different methods showing that it is valid in general for all types of systems.

The energy content of even small bodies is enormous. For example, by transforming one gram of matter to heat, one may heat 300,000 metric tons of water from room temperature to the boiling point. (The energy corresponding to a mass m is enough to change the temperature by ΔT of an object of mass M and specific heat capacity c_V: $mc^2 = Mc_V\Delta T$.)

2.12   Relativistic increase of mass

In the special theory of relativity, force is defined as rate of change of momentum. We consider a body that gets a change of energy dE due to the work performed on it by a force F. According to Eq. (2.66) and the definition of work (force times distance) the body gets a change of mass dm, given by

$$c^2 dm = dE = Fds = Fvdt = vd(mv) = mvdv + v^2 dm, \tag{2.67}$$

which gives

$$\int_{m_0}^{m} \frac{dm}{m} = \int_0^v \frac{vdv}{c^2 - v^2}, \tag{2.68}$$

where m_0 is the rest mass of the body – i.e. its mass as measured by an observer comoving with the body – and m its mass when its velocity is equal to v. Integration gives

$$m = \frac{m_0}{\sqrt{1 - \frac{v^2}{c^2}}} = \gamma m_0. \tag{2.69}$$

In the case of small velocities compared to the velocity of light we may use the approximation

$$\sqrt{1 - \frac{v^2}{c^2}} \approx 1 + \frac{1}{2}\frac{v^2}{c^2}. \tag{2.70}$$

With this approximation Eqs. (2.66) and (2.69) give

$$E \approx m_0 c^2 + \frac{1}{2}m_0 v^2. \tag{2.71}$$

This equation shows that the total energy of a body encompasses its rest-energy m_0 and its kinetic energy. In the non-relativistic limit the kinetic energy is $m_0 v^2/2$. The relativistic expression for the kinetic energy is

$$E_K = E - m_0 c^2 = (\gamma - 1)m_0 c^2. \tag{2.72}$$

Note that $E_K \to \infty$ when $v \to c$.

According to Eq. (2.33), it is not possible to obtain a velocity greater than that of light by adding velocities. Equation (2.72) gives a dynamical reason that material particles cannot be accelerated up to and above the velocity of light.

2.13 Tachyons

Particles cannot pass the velocity-barrier represented by the velocity of light. However, the special theory of relativity permits the existence of particles that have *always* moved with a velocity $v > c$. Such particles are called *tachyons* [Rec78].

Tachyons have special properties that have been used in the experimental searches for them. There is currently no observational evidence for the physical existence of tachyons [Kre73].

There are also certain theoretical difficulties with the existence of tachyons. The special theory of relativity, applied to tachyons leads to the following paradox. Using a tachyon telephone a person, A, emits a tachyon to B at a point of time t_1. B moves away from A. The tachyon is reflected by B and reach A before it was emitted, see Fig. 2.15. If the tachyon could carry information it might bring an order to destroy the tachyon emitter when it arrives back at A.

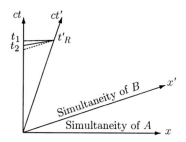

Figure 2.15: A emits a tachyon at the point of time t_1. It is reflected by B and arrives at A at a point of time t_2 before t_1. Note that the arrival event at A is later than the reflection event as measured by B.

To avoid similar problems in regards to the energy-exchange between tachyons and ordinary matter, a reinterpretation principle is introduced for tachyons. For certain observers a tachyon will move backwards in time, i.e. the observer finds that the tachyon is received before it was emitted. Special relativity tells us that such a tachyon is always observed to have negative energy.

According to the reinterpretation principle, the observer will interpret his observations to mean that a tachyon with positive energy moves forward in time. In this way, one finds that the energy-exchange between tachyons and ordinary matter proceeds in accordance with the principle of causality [BDS62].

However, the reinterpretation principle cannot be used to remove the problems associated with exchange of information between tachyons and ordinary matter. The tachyon telephone paradox cannot be resolved by means of the reinterpretation principle. The conclusion is that if tachyons exist, they cannot be carriers of information in our slowly-moving world.

2.14   Magnetism as a relativistic second-order effect

Electricity and magnetism are described completely by Maxwell's equations
of the electromagnetic field,

$$\nabla \cdot \mathbf{E} = \frac{1}{\varepsilon_0}\rho_q \tag{2.73}$$

$$\nabla \cdot \mathbf{B} = 0 \tag{2.74}$$

$$\nabla \times \mathbf{E} = -\frac{\partial \mathbf{B}}{\partial t} \tag{2.75}$$

$$\nabla \times \mathbf{B} = \mu_0 \mathbf{j} + \frac{1}{c^2}\frac{\partial \mathbf{E}}{\partial t} \tag{2.76}$$

together with Lorentz's force-law

$$\mathbf{F} = q(\mathbf{E} + \mathbf{v} \times \mathbf{B}). \tag{2.77}$$

However, the relation between the magnetic and the electric force was not
fully understood until Einstein had constructed the special theory of relativity.
Only then could one clearly see the relationship between the magnetic force
on a charge moving near a current carrying wire and the electric force between
charges.

We shall consider a simple model of a current carrying wire in which we
assume that the positive ions are at rest while the conducting electrons move
with the velocity v. The charge per unit length for each type of charged particle is $\hat{\lambda} = Sne$ where S is the cross-sectional area of the wire, n the number
of particles of one type per unit length and e the charge of one particle. The
current in the wire is

$$J = Snev = \hat{\lambda}v. \tag{2.78}$$

The wire is at rest in an inertial frame $\hat{\Sigma}$. As observed in $\hat{\Sigma}$ it is electrically
neutral. Let a charge q move with a velocity u along the wire in the opposite
direction of the electrons. The rest frame of q is Σ. The wire will now be
described from Σ (see Fig. 2.16 and 2.17).

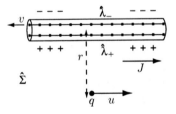

Figure 2.16: Wire seen from its own rest frame.

Note that the charge per unit length of the particles as measured in their
own rest-frames, Σ_0, is

$$\lambda_{0-} = \hat{\lambda}\left(1 - \frac{v^2}{c^2}\right)^{1/2}, \quad \lambda_{0+} = \hat{\lambda} \tag{2.79}$$

since the distance between the electrons is Lorentz contracted in $\hat{\Sigma}$ compared
to their distances in Σ_0.

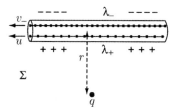

Figure 2.17: Wire seen from rest frame of moving charge.

The velocities of the particles as measured in Σ are

$$v_- = -\frac{v + u}{1 + \frac{uv}{c^2}} \quad \text{and} \quad v_+ = -u. \tag{2.80}$$

The charge per unit length of the negative particles as measured in Σ, is

$$\lambda_- = \left(1 - \frac{v_-^2}{c^2}\right)^{-1/2} \lambda_0. \tag{2.81}$$

Substitution from Eq. (2.79) and (2.80) gives

$$\lambda_- = \gamma \left(1 + \frac{uv}{c^2}\right) \hat{\lambda} \tag{2.82}$$

where $\gamma = \left(1 - u^2/c^2\right)^{-1/2}$. In a similar manner, the charge per unit length of the positive particles as measured in Σ is found to be

$$\lambda_+ = \gamma \hat{\lambda}. \tag{2.83}$$

Thus, as observed in the rest-frame of q the wire has a net charge per unit length

$$\lambda = \lambda_- - \lambda_+ = \frac{\gamma u v}{c^2} \hat{\lambda}. \tag{2.84}$$

As a result of the different Lorentz contractions of the positive and negative ions when we transform from their respective rest frames to Σ, a current carrying wire which is electrically neutral in the laboratory frame, is observed to be electrically charged in the rest frame of the charge q.

As observed in this frame there is a radial electrical field with field strength

$$E = \frac{\lambda}{2\pi\epsilon_0 r}. \tag{2.85}$$

Then a force F acts on q, this is given by

$$F = qE = \frac{q\lambda}{2\pi\epsilon_0 r} = \frac{\hat{\lambda} v}{2\pi\epsilon_0 c^2 r} \gamma q u. \tag{2.86}$$

If a force acts upon q as observed in $\hat{\Sigma}$ then a force also acts on q as observed in Σ. According to the relativistic transformation of a force component in the same direction as the relative velocity between $\hat{\Sigma}$ and Σ, this force is

$$\hat{F} = \gamma^{-1} F = \frac{\hat{\lambda} v}{2\pi\epsilon_0 c^2 r} q u. \tag{2.87}$$

Inserting $J = \hat{\lambda}v$ from Eq. (2.78) and using $c^2 = (\epsilon_0\mu_0)^{-1}$ (where μ_0 is the permeability of a vacuum) we obtain

$$\hat{F} = \frac{\mu_0 J}{2\pi r}qu. \tag{2.88}$$

This is exactly the expression obtained if we calculate the magnetic flux-density $\hat{B}$ around the current carrying wire using Ampere's circuit law

$$\hat{B} = \mu_0 \frac{J}{2\pi r} \tag{2.89}$$

and use the force-law (Eq. (2.77)) for a charge moving in a magnetic field

$$\hat{F} = qu\hat{B}. \tag{2.90}$$

We have seen here how a magnetic force appears as a result of an electrostatic force and the special theory of relativity. The considerations above have also demonstrated that a force which is identified as electrostatic in one frame of reference is observed as a magnetic force in another frame. In other words, the electric and the magnetic force are really the same. What an observer names it depends upon his state of motion.

Problems

2.1. *Two successive boosts in different directions*
Let us consider Lorentz transformations without rotation ("boosts"). A boost in the x-direction is given by

$$x = \gamma(x' + \beta ct'), \quad y = y', \quad z = z', \quad t = \gamma(t' + \beta x'/c)$$
$$\gamma = \frac{1}{\sqrt{1-\beta^2}}, \quad \beta = \frac{v}{c} \tag{2.91}$$

This can be written as

$$x^\mu = \Lambda^\mu_{\mu'}x^{\mu'} \tag{2.92}$$

where $\Lambda^\mu_{\mu'}$ is the matrix

$$\Lambda^\mu_{\mu'} = \begin{bmatrix} \gamma & \gamma\beta & 0 & 0 \\ \gamma\beta & \gamma & 0 & 0 \\ 0 & 0 & 1 & 0 \\ 0 & 0 & 0 & 1 \end{bmatrix} \tag{2.93}$$

(a) Show that (2.92) and (2.93) yield (2.91). Find the transformation matrix, $\bar{\Lambda}^\mu_{\mu'}$, for a boost in the negative y-direction.

(b) Two successive Lorentz transformations are given by the matrix product of each matrix. Find $\Lambda^\mu_\alpha \Lambda^\alpha_{\mu'}$ and $\Lambda^\mu_\alpha \bar{\Lambda}^\alpha_{\mu'}$. Are the product of two boosts a boost? The matrix for a general boost in arbitrary direction is given by

$$\begin{aligned} \Lambda^0_{\ 0} &= \gamma, \\ \Lambda^0_{\ m} &= \Lambda^m_{\ 0} = \gamma\beta_m, \\ \Lambda^m_{\ m'} &= \delta^m_{m'} + \frac{\beta_m\beta_{m'}}{\beta^2}(\gamma - 1), \\ \gamma &= \frac{1}{\sqrt{1-\beta^2}}, \quad \beta^2 = \beta^m\beta_m, \quad m, m' = 1,2,3 \end{aligned} \tag{2.94}$$

Does the set of all possible boosts form a group?

2.2. *Length-contraction and time-dilatation*

(a) A rod with length ℓ is moving with constant velocity $\mathbf{v}$ with respect to the inertial frame Σ. The length of the rod is parallel to $\mathbf{v}$, which we will for simplicity's sake assume is parallel to the x-axis. At time $t = 0$, the rear end of the rod is in the origin of Σ. What do we mean by the length of such a moving rod? Describe how an observer can find this length. Draw the rod in a Minkowski diagram and explain how the length of the rod can be read of the diagram. Using the Lorentz transformations, calculate the position of the endpoints of the rod as a function of time t. Show that the length of the rod, as measured in Σ, is shorter than its rest length ℓ.

(b) The rod has the same velocity as before, but now the rod makes an angle with $\mathbf{v}$. In an inertial frame which follows the movement of the rod (Σ'), this angle is $\alpha' = 45°$ (with the x-axis in Σ). What is the angle between the velocity $\mathbf{v}$ and the rod when measured in Σ? What is the length of the rod as a function of α', as measured from Σ?

(c) We again assume that $\alpha' = 0$. At the centre of the rod there is a flash that sends light signals with a time interval τ_0 between every flash. In the frame Σ', the light signals will reach the two ends simultaneously. Show that these two events are not simultaneous in Σ. Find the time difference between these two events.

Show that the time interval τ measured from Σ between each flash, is larger than the interval τ_0 measured in Σ'.

An observer in Σ is located at the origin. He measures the time-interval Δt between every time he receives a light signal. Find Δt in terms of the speed v, and check whether Δt is greater or less than τ.

(d) The length of the rod is now considered to be $\ell = 1\text{m}$ and its speed, as measured in Σ, is $v = \frac{3}{5}c$. As before, we assume that the rod is moving parallel to the x-axis, but this time at a distance of $y = 10\text{m}$ from the axis. A measuring ribbon is stretched out along the trajectory of the rod. This ribbon is at rest in Σ. An observer at the origin sees the rod move along the background ribbon. The ribbon has tick-marks along it which correspond to the x coordinates. The rods length can be measured by taking a photograph of the rod and the ribbon. Is the length that is directly measured from the photograph identical to the length of the rod in Σ?

In one of the photographs the rod is symmetrically centered with respect to $x = 0$. What is the length of the rod as measured using this photograph? Another photograph shows the rod with its trailing edge at $x = 10\text{m}$. At what point will the leading edge of the rod be on this photograph? Compare with the length of the rod in the Σ frame.

(e) At one point along the trajectory the rod passes through a box which is open at both ends and stationary in Σ. This box is shorter than the rest length of the rod, but longer that the length of the rod as measured in Σ. At a certain time in Σ, the entire rod is therefore inside the box. At this time the box is closed at both ends, trapping the rod inside. The rod is also brought to rest. It is assumed that the box is strong enough to withstand the impact with the rod.

What happens to the rod? Describe what happens as observed from Σ and Σ'. Draw a Minkowski diagram. This is an example of why the theory of relativity has difficulty with the concept of absolute rigid bodies. What is the reason for this difficulty?

2.3. *Faster than the speed of light?*

The quasar 3C273 emits a jet of matter that moves with the speed v_0 towards

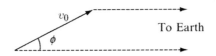

Figure 2.18: A Quasar emitting a jet of matter.

Earth making an angle ϕ to the line of sight (see Fig. 2.18). The observed (the transverse) speed of the light-source is $v = 10c$. Find v_0 when we assume that $\phi = 10°$. What is the largest possible ϕ?

2.4. *Reflection angles of moving mirrors*

(a) The reflection angle of light equals the incidence angle of the light. Show that this is also the case for mirrors that are moving parallel to the reflection surface.

(b) A mirror is moving with a speed v in a direction orthogonal to the reflection surface. Light is sent towards the mirror with an angle ϕ. Find the angle of the reflected light as a function of v and ϕ. What is the frequency to the reflected light expressed in terms of its original frequency f?

2.5. *Minkowski-diagram*

The reference frame Σ' is moving relative to the frame Σ at a speed of $v = 0.6c$. The movement is parallel to the x-axes of the two frames.

Draw the x' and the ct'-axis in the Minkowski-diagram of Σ. Points separated by 1m are marked along both axes. Draw these points in the Minkowski-diagram as for both frames.

Show where the lines of simultaneity for Σ' are in the diagram. Also show where the $x' = $ constant line is.

Assume that the frames are equipped with measuring rods and clocks that are at rest in their respective frames. How can we use the Minkowski-diagram to measure the length-contraction of the rod that is in rest at Σ'? Similarly, how can we measure the length-contraction of the rod in Σ when measured from Σ? Show how the time-dilatation of the clocks can be measured from the diagram.

2.6. *Robb's Lorentz invariant spacetime interval formula* (A.A. Robb, 1936)

Show that the spacetime interval between the origin event and the reflection event in Fig. 2.2 is $s = c\sqrt{t_A t_B}$.

2.7. *The Doppler effect*

A radar antenna emits radio pulses with a wavelength of $\lambda = 1.0$cm, at a time-interval $\tau = 1.0$s. An approaching spacecraft is being registered by the radar.

Draw a Minkowski-diagram for the reference frame Σ. The antenna is at rest in this frame. In this diagram, indicate the position of

1. the antenna,

2. the spacecraft, and

3. the outgoing and reflected radar pulses.

Calculate the time difference Δt_1 between two subsequent pulses as measured in the spacecraft. What is the wavelength of these signals?

Calculate the time difference Δt_2 between two reflected signals, as it is measured from the antenna's receiver? At what wavelength will these signals be?

2.8. *Abberation and Doppler effect*
We shall describe light emitted from a spherical surface that expands with ultra-relativistic velocity. Consider a surface element dA with velocity $v = \beta c$ in the laboratory frame F (i.e. the rest frame of the observer), as shown in Fig. 2.19.

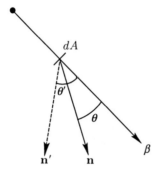

Figure 2.19: Light is emitted in the direction $\mathbf{n}'$ as measured in the rest frame F' of the emitting surface element. The light is measured to propagate in the $\mathbf{n}$-direction in the rest frame F of the observer.

(a) Show by means of the relativistic formula for velocity addition that the relationship between the directions of propagation measured in F and F' is

$$\cos\theta = \frac{\cos\theta' + \beta}{1 + \beta\cos\theta'}. \tag{2.95}$$

This is the abberation formula.

(b) Show that an observer far away from the surface will only observe light from a spherical cap with opening angle (see Fig. 2.20)

$$\theta_0 = \arccos\beta = \arcsin\frac{1}{\gamma} \approx \frac{1}{\gamma} \text{ for } \gamma \gg 1. \tag{2.96}$$

(c) Assume that the expanding shell emits monocromatic light with frequency ν' in F'. Show that the observer in F will measure an angle-dependent frequency

$$\nu = \frac{\nu'}{\gamma(1 - \beta\cos\theta)} = \gamma(1 + \beta\cos\theta')\nu'. \tag{2.97}$$

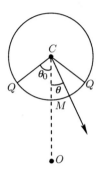

Figure 2.20: The far-away observer, O, can only see light from the spherical cap with opening angle θ_0.

(d) Let the measured frequency of light from M and Q be ν_M and ν_Q, respectively. This is the maximal and minimal frequency. Show that the expansion velocity can be found from these measurements, as

$$v = \frac{\nu_M - \nu_Q}{\nu_Q} c. \tag{2.98}$$

2.9. A traffic problem
A driver is in court for driving through a red light. In his defence, the driver claims that the traffic signal appeared green as he was approaching the junction. The judge says that this does not make his case stronger as he would have been travelling at the speed of ...

At what speed would the driver have to travel for the red traffic signal ($\lambda = 6000$ Å) to Doppler shift to a green signal ($\lambda = 5000$ Å)?

2.10. The twin-paradox
On New Years day 2004, an astronaut (A) leaves Earth on an interstellar journey. He is travelling in a spacecraft at the speed of $v = 4/5c$ heading towards α-Centauri. This star is at a distance of 4 l.y.(l.y. =light years) measured from the reference frame of the Earth. As A reaches the star, he immediately turns around and heads home. He reaches the Earth New Years day 2016 (in Earth's time frame).

The astronaut has a brother (B), who remains on Earth during the entire journey. The brothers have agreed to send each other a greeting every new years day with the aid of radio-telescope.

(a) Show that A only sends 6 greetings (including the last day of travel), while B sends 10.

(b) Draw a Minkowski-diagram where A's journey is depicted with respect to the Earth's reference frame. Include all the greetings that B is sending. Show with the aid of the diagram that while A is outbound, he only receives 1 greeting, while on his way home he receives 9.

(c) Draw a new diagram, still with respect to Earth's reference frame, where A's journey is depicted. Include the greetings that A is sending to B. Show that B is receiving one greeting every 3rd year the first 9 years after A has left, while the last year before his return he receives 3.

(d) Show how the results from (b) and (c) can be deduced from the Doppler-effect.

2.11. *Work and rotation*
A circular ring is initially at rest. It has radius r, rest mass m, and a constant of elasticity k. Find the work that has to be done to give the ring an angular velocity ω. We assume that the ring is accelerated in such a way that its radius is constant. Compare with the non-relativistic case. How can we understand that in the relativistic case we also have to do elastic work?

2.12. *Muon experiment*
How many of the ten million muons created 10km above sea level will reach the Earth? If there are initially n_0 muons, $n = n_0 2^{-t/T}$ will survive for a time t (T is the half-life time).

(a) Compute the non-relativistic result.

(b) What is the result of a relativistic calculation by an Earth observer?

(c) Make a corresponding calculation from the point of view of an observer comoving with the muon. The muon has a rest half-life time $T = 1.56 \cdot 10^{-6}$s and moves with a velocity $v = 0.98c$.

2.13. *Cerenkov radiation*
When a particle moves through a medium with a velocity greater than the velocity of light in the medium, it emits a cone of radiation with a half angle θ given by $\cos\theta = c/nv$ (see Fig. 2.21).

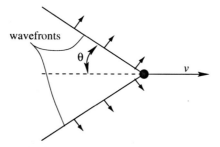

wavefronts

θ

v

Figure 2.21: Cerenkov radiation from a particle.

(a) What is the threshold kinetic energy (in MeV) of an electron moving through water in order that it shall emit Cerenkov radiation? The index of refraction of water is $n = 1.3$. The rest energy of an electron is $m_e = 0.511$ MeV.

(b) What is the limiting half angle of the cone for high speed particles moving through water?

Part II

THE MATHEMATICS OF THE GENERAL THEORY OF RELATIVITY

3

Vectors, Tensors, and Forms

We shall present the theory of differential forms in a way so that the structure of the theory appears as clearly as possibly. In later chapters this formalism will be used to give a mathematical formulation of the fundamental principles of the general theory of relativity. It will also be employed to give an invariant formulation of Maxwell's equations so that the equations can be applied with reference to an arbitrary basis in curved spacetime.

3.1 Vectors

A vector is usually defined as a quantity with magnitude and direction, and is denoted by a letter with an arrow above it, for example $\vec{v}$, or by boldface letters, for example $\mathbf{v}$. We shall use the latter notation.

Vectors can also be defined as quantities fulfilling certain axioms. An example of such an axiom is the following. If a and b are real numbers, and if $\mathbf{u}$ and $\mathbf{v}$ are vectors, then $a\mathbf{u} + b\mathbf{v}$ is a vector.

An expression of the form $a^{\mu}\mathbf{e}_{\mu}$ where a^{μ} (with $\mu \in \{1, \ldots n\}$) are real numbers, is called a linear combination of the vectors $\mathbf{e}_{\mu}$. The vectors $\mathbf{e}_1, \ldots \mathbf{e}_n$ are said to be *linearly independent* if no real numbers $a^{\mu} \neq 0$ exist so that $a^{\mu}\mathbf{e}_{\mu} = 0$. A geometrical interpretation is that the vectors are linearly independent if it is not possible to construct a closed polygon by means of the vectors (see Fig. 3.1).

A set of linearly independent vectors $\{\mathbf{e}_{\mu}\}$ is said to be *maximally* linearly independent if for all vectors $\mathbf{v}$ the set of vectors $\{\mathbf{e}_{\mu}, \mathbf{v}\}$ is linearly dependent. Then there exist non-zero real numbers a^{μ} so that let stand as is

$$a^{\mu}\mathbf{e}_{\mu} + \mathbf{v} = 0. \tag{3.1}$$

A vector-basis for a space V is defined as a set of vectors in V that are maximally linearly independent. The number of vectors in the basis is called the *dimension* of V. For example a vector-basis in spacetime consists of four vectors.

Let the vector set $\{\mathbf{e}_1, \ldots, \mathbf{e}_n\}$ be a basis in an n-dimensional space. Setting $a^{\mu} = -v^{\mu}$ in Eq. (3.1) we get

$$\mathbf{v} = v^{\mu}\mathbf{e}_{\mu}. \tag{3.2}$$

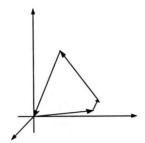

Figure 3.1: Linearly dependent vectors.

The numbers v^μ are called the *components* of **v** relative to the basis $\{e_\mu\}$.

3.2 Four-vectors

Spacetime is four-dimensional. At every point in spacetime we can place four linearly independent basis vectors e_μ. Thus, a vector in spacetime has four components. Such vectors are called four-vectors.

A flat spacetime can be mapped by a global Cartesian coordinate system, with coordinates (t, x, y, z). The basis vectors in this system are denoted by $\{e_t, e_x, e_y, e_z\}$. They are mutually orthogonal unit vectors. Such a basis of orthogonal unit vectors is called an *orthonormal basis*.

The ordinary velocity of a particle is

$$\mathbf{u} = u^x \mathbf{e}_x + u^y \mathbf{e}_y + u^z \mathbf{e}_z = \frac{dx}{dt}\mathbf{e}_x + \frac{dy}{dt}\mathbf{e}_y + \frac{dz}{dt}\mathbf{e}_z. \tag{3.3}$$

According to the Galilean and Newtonian kinematics, all particles move in a three-dimensional space, and the velocity **u** is a vector.

According to the relativistic description, however, particles exist in a four-dimensional spacetime. In this description the ordinary velocity of a particle is not a vector.

Instead one defines a *four-velocity*

$$\mathbf{U} = c\frac{dt}{d\tau}\mathbf{e}_t + \frac{dx}{d\tau}\mathbf{e}_x + \frac{dy}{d\tau}\mathbf{e}_y + \frac{dz}{d\tau}\mathbf{e}_z, \tag{3.4}$$

where τ is the proper time of the particle, i.e. the time measured by a (hypothetical) standard clock carried by the particle.[1] Using Einstein's summation convention we may write

$$\mathbf{U} = U^\mu \mathbf{e}_\mu = \frac{dx^\mu}{d\tau}\mathbf{e}_\mu, \quad x^\mu \in \{x^0, x^1, x^2, x^3\}, \tag{3.5}$$

where $x^0 = ct$, $x^1 = x$, $x^2 = y$, and $x^3 = z$. Since $dt/d\tau = \gamma$ according to Eq. (2.44), the components of the four-velocity are given in terms of the components of the ordinary velocity as

$$\mathbf{U} = \gamma(c, u^x, u^y, u^z), \tag{3.6}$$

---

[1] A 'particle' in this context is an entity so small that its spatial extension can be neglected within the accuracy of the description of spacetime.

which is often written as

$$\boxed{\mathbf{U} = \gamma(c, \mathbf{u}).}\qquad(3.7)$$

Below we shall use this notation when giving the component-form of four-vectors.

In the rest frame of the particle, $u = 0$ and $\gamma = 1$. Hence, the four-velocity reduces to

$$\mathbf{U} = c\mathbf{e}_t.\qquad(3.8)$$

In this frame the particle moves in the time direction with the speed of light.

One often uses units so that $c \equiv 1$. If this is done, both time and space are measured in units of length. In such *geometrical units* of measurement the particle moves with unit velocity in the time-direction in its own rest frame.

The *four-momentum*, $\mathbf{P}$, of a particle with rest-mass m_0 is defined by

$$\mathbf{P} = m_0\mathbf{U}.\qquad(3.9)$$

Referring to the rest-frame of the particle and using units so that $c = 1$, we see that the magnitude of the four-momentum is equal to the rest-mass of the particle.

The ordinary (three-dimensional) relativistic momentum of the particle is

$$\mathbf{p} = m\mathbf{u} = \gamma m_0 \mathbf{u}.\qquad(3.10)$$

From Eqs. (2.65) and (3.8)–(3.10), follows

$$\boxed{\mathbf{P} = (E/c, \mathbf{p})}\qquad(3.11)$$

where E is the total energy of the particle.

The *four-force*, or the *Minkowski force* $\mathbf{F}$, is defined by

$$\mathbf{F} = \frac{d\mathbf{P}}{d\tau},\qquad(3.12)$$

while the ordinary force $\mathbf{f}$ is

$$\mathbf{f} = \frac{d\mathbf{p}}{dt}.\qquad(3.13)$$

It follows that

$$\boxed{\mathbf{F} = \gamma\left(\frac{dE}{cdt}, \frac{d\mathbf{p}}{dt}\right) = \gamma\left(\frac{\mathbf{f}\cdot\mathbf{u}}{c}, \mathbf{f}\right).}\qquad(3.14)$$

In the rest frame of the particle

$$\mathbf{F}_0 = (0, \mathbf{f}_0),\qquad(3.15)$$

where $\mathbf{f}_0$ is the Newtonian force on the particle.

The *four-acceleration* $\mathbf{A}$ of the particle is

$$\boxed{\mathbf{A} = \frac{d\mathbf{U}}{d\tau}.}\qquad(3.16)$$

In the case that the rest-mass is constant we get

$$\mathbf{A} = \frac{1}{m_0}\mathbf{F}.\qquad(3.17)$$

The ordinary acceleration **a** is

$$\mathbf{a} = \frac{d\mathbf{u}}{dt},\tag{3.18}$$

so using that

$$\mathbf{f} = m_0 \frac{d}{dt}(\gamma \mathbf{u}) = \gamma m_0 \left(\mathbf{a} + \gamma^2 \frac{\mathbf{u} \cdot \mathbf{a}}{c^2} \mathbf{u} \right),\tag{3.19}$$

we find from Eqs. (3.14) and (3.16)–(3.19),

$$\mathbf{A} = \gamma^2 \left(\gamma^2 \frac{\mathbf{u} \cdot \mathbf{a}}{c}, \mathbf{a} + \gamma^2 \frac{\mathbf{u} \cdot \mathbf{a}}{c^2} \mathbf{u} \right).\tag{3.20}$$

In the rest frame of the particle this reduces to

$$\mathbf{A}_0 = (0, \mathbf{a}_0),\tag{3.21}$$

where $\mathbf{a}_0$ is the rest-acceleration of the particle. The physical significance of the four-acceleration is that it vanishes for a freely falling particle.

The component expressions (3.14) and (3.20) are valid only with respect to an orthonormal basis field. It will be shown in Chapter 8 that in curved space one can always introduce *local* Cartesian coordinate systems with orthonormal basis coordinate vector fields.

3.3 One-forms

Let the set of real numbers be denoted by $\mathbb{R}$ and let V be a vectorspace.

A function f is said to be linear if

$$f(a\mathbf{u} + b\mathbf{v}) = af(\mathbf{u}) + bf(\mathbf{v}),\tag{3.22}$$

where $a, b \in \mathbb{R}$ and $\mathbf{u}, \mathbf{v} \in V$.

A *one-form*, α is defined as a linear function from V into $\mathbb{R}$; i.e., $\alpha : V \mapsto \mathbb{R}$. In other words, a one-form, α, acts on a vector, $\mathbf{v}$, and gives out a real number, $\alpha(\mathbf{v})$.

The sum of the two one-forms, α and β, and the product of a real number, a, and a one-form is defined in the usual way for real functions

$$(\alpha + \beta)(\mathbf{v}) = \alpha(\mathbf{v}) + \beta(\mathbf{v}),\tag{3.23}$$

and

$$(a\alpha)(\mathbf{v}) = a[\alpha(\mathbf{v})].\tag{3.24}$$

In order to be able to write a form in component-form, we have to define a one-form basis $\{\omega^\mu\}$. The basis is defined by

$$\omega^\mu(\mathbf{e}_\nu) = \delta^\mu{}_\nu,\tag{3.25}$$

where $\delta^\mu{}_\nu$ is the Kronecker-symbol defined in Eq. (1.15). We can now write a one-form as a linear combination of the basis-forms

$$\alpha = \alpha_\mu \omega^\mu.\tag{3.26}$$

The numbers α_μ are called *components* of α relative to the basis $\{\omega^\mu\}$.

By means of Eqs. (3.25) and (3.26) we get

$$\alpha(\mathbf{e}_\mu) = \alpha_\nu \omega^\nu(\mathbf{e}_\mu) = \alpha_\nu \delta^\nu{}_\mu = \alpha_\mu.\tag{3.27}$$

The number $\alpha(\mathbf{v})$ is called the *contraction* or *interior product* of α with $\mathbf{v}$, which we will write as

$$\boxed{\iota_{\mathbf{v}}\alpha \equiv \alpha(\mathbf{v})} \tag{3.28}$$

Eq. (3.27) says that the components of a one-form are given by the contractions of the form with the basis-vectors.

The number $\alpha(\mathbf{v})$ may now be expressed by the components of α and $\mathbf{v}$,

$$\alpha(\mathbf{v}) = \alpha(v^\mu \mathbf{e}_\mu) = v^\mu \alpha(\mathbf{e}_\mu) = v^\mu \alpha_\mu. \tag{3.29}$$

This is just the same number that is obtained by taking the scalar product of two vectors $\mathbf{v}$ and α. One-forms correspond to vectors and the contraction of a one-form by a vector to the scalar-product of two vectors.

Just like forms, vectors can also be perceived as linear functions. If a vector $\mathbf{v}$ acts on a form α, it gives out the number $\mathbf{v}(\alpha) = \alpha_\mu v^\mu$. Since this is equal to $v^\mu \alpha_\mu$ we have $\mathbf{v}(\alpha) = \alpha(\mathbf{v})$, which corresponds to the symmetry of the scalar-product of two vectors. It follows that the vector components v^μ can be expressed as

$$v^\mu = \mathbf{v}(\omega^\mu). \tag{3.30}$$

The components of a vector are the contractions of the vector with the basis forms.

Like the vectors, one-forms satisfy the axioms of a vector space. Therefore, one-forms are sometimes referred to as dual vectors. In Dirac's 'bra-ket' notation in quantum mechanics, the 'kets' $|\psi\rangle$ are the vectors and the 'bras' $\langle\psi|$ are the forms.

3.4 Tensors

We shall now consider functions of several variables. A *multi-linear* function f is a function that is linear in all its arguments. A *tensor* is a multi-linear function that maps vectors and one-forms into $\mathbb{R}$. We distinguish between covariant, contravariant and mixed tensors:

- A *covariant* tensor maps vectors, only.

- A *contravariant* tensor maps one-forms, only.

- A *mixed* tensor maps both vectors and one-forms.

A tensor of *rank* $\{{}^n_{n'}\}$ maps n one-forms and n' vectors into $\mathbb{R}$. (Less precise, of rank $n + n'$).

In order to be able to write a tensor in component-form we need a tensor basis. For this purpose we must introduce the tensor product, which is denoted by $\otimes$. The tensor product between two covariant tensors $\mathbf{T}$ and $\mathbf{S}$ of rank m and n, respectively, is defined by

$$\boxed{\mathbf{T} \otimes \mathbf{S}(\mathbf{u}_1, \ldots, \mathbf{u}_m, \mathbf{v}_1, \ldots, \mathbf{v}_n) = \mathbf{T}(\mathbf{u}_1, \ldots \mathbf{u}_m)\mathbf{S}(\mathbf{v}_1, \ldots \mathbf{v}_n).} \tag{3.31}$$

Corresponding expressions are valid for all types of tensors. The tensor product is distributive and bilinear, but not commutative, $\mathbf{T} \otimes \mathbf{S} \neq \mathbf{S} \otimes \mathbf{T}$.

Example

Example 3.1 (Tensor product between two vectors)
Given two vectors $\mathbf{u}$ and $\mathbf{v}$. The tensor product between $\mathbf{u}$ and $\mathbf{v}$ is a tensor $\mathbf{T}$ given by its action on two arbitrary one-forms α and β

$$\mathbf{T}(\alpha,\beta) = (\mathbf{u} \otimes \mathbf{v})(\alpha,\beta) = \mathbf{u}(\alpha)\mathbf{v}(\beta) = u^{\mu}\alpha_{\mu}v^{\nu}\beta_{\nu}. \tag{3.32}$$

Since $\mathbf{T}$ takes two one-form arguments, it is a tensor of rank $\{^2_0\}$.

A basis for contravariant vectors of rank q is defined as a maximally linearly independent set of basis-elements $\{\mathbf{e}_{\mu_1}, \ldots \mathbf{e}_{\mu_q}\}$. (The reason that the indices μ_i have indices themselves, is that there are $n \geq q$ different indices in an n-dimensional space.) The component form of a contravariant tensor of rank q is

$$\mathbf{R} = R^{\mu_1 \cdots \mu_q} \mathbf{e}_{\mu_1} \otimes \cdots \otimes \mathbf{e}_{\mu_q}. \tag{3.33}$$

The tensor-components are defined as the values of $\mathbf{R}$ when $\mathbf{R}$ is applied to the basis-forms,

$$R^{\mu_1 \cdots \mu_q} \equiv \mathbf{R}(\omega^{\mu_1}, \ldots, \omega^{\mu_q}). \tag{3.34}$$

A covariant tensor $\mathbf{S}$ is expressed in the same way:

$$\mathbf{S} = S_{\mu_1 \ldots \mu_q} \omega^{\mu_1} \otimes \cdots \otimes \omega^{\mu_q}, \tag{3.35}$$

where

$$S_{\mu_1 \ldots \mu_q} \equiv \mathbf{S}(\mathbf{e}_{\mu_1}, \ldots, \mathbf{e}_{\mu_q}). \tag{3.36}$$

A mixed tensor, $\mathbf{T}$, of rank $\{^q_p\}$ is written as the number $T(q$ one-forms, p vectors) and is expressed by the components as

$$\boxed{\mathbf{T} = T^{\mu_1 \cdots \mu_q}{}_{\nu_1 \ldots \nu_p} \mathbf{e}_{\mu_1} \otimes \cdots \otimes \mathbf{e}_{\mu_q} \otimes \omega^{\nu_1} \otimes \cdots \otimes \omega^{\nu_p}.} \tag{3.37}$$

If, for example, $q = p = 1$ we get

$$\mathbf{T}(\mathbf{u}, \alpha) = T_{\mu}{}^{\nu} u^{\mu} \alpha_{\nu}. \tag{3.38}$$

Thus contraction of the mixed tensor $\mathbf{T}$ with the vector $\mathbf{u}$ and the one-form α is a scalar.

Example

Example 3.2 (Tensor-components)
Let $\mathbf{u}$ and $\mathbf{v}$ be two vectors and α and β two one-forms. The tensor components of the tensors $\mathbf{R} = \mathbf{u} \otimes \mathbf{v}, \mathbf{S} = \alpha \otimes \mathbf{v}, \mathbf{T} = \alpha \otimes \beta$ are

$$R^{\mu\nu} = (\mathbf{u} \otimes \mathbf{v})(\omega^{\mu}, \omega^{\nu}) = \mathbf{u}(\omega^{\mu})\mathbf{v}(\omega^{\nu}) = u^{\mu}v^{\nu},$$
$$S_{\mu}{}^{\nu} = (\alpha \otimes \mathbf{v})(\mathbf{e}_{\mu}, \omega^{\nu}) = \alpha(\mathbf{e}_{\mu})\mathbf{v}(\omega^{\nu}) = \alpha_{\mu}v^{\nu},$$
$$T_{\mu\nu} = (\alpha \otimes \beta)(\mathbf{e}_{\mu}, \mathbf{e}_{\nu}) = \alpha(\mathbf{e}_{\mu})\beta(\mathbf{e}_{\nu}) = \alpha_{\mu}\beta_{\nu},$$

when expressed in the bases $\{\mathbf{e}_{\mu}\}$ and $\{\omega^{\mu}\}$.

3.5 Forms

An *antisymmetric tensor*, $\mathbf{A}$, is a tensor that is antisymmetric under exchange of two arbitrary arguments

$$\mathbf{A}(\ldots \mathbf{u}, \ldots \mathbf{v}, \ldots) = -\mathbf{A}(\ldots \mathbf{v}, \ldots \mathbf{u}, \ldots). \tag{3.39}$$

Only purely covariant or contravariant tensors can be antisymmetric, not mixed tensors.

A *p-form* is defined as a covariant antisymmetric tensor of rank p. Since

$$A_{\ldots \mu \ldots \nu \ldots} = \mathbf{A}(\ldots \mathbf{e}_\mu, \ldots \mathbf{e}_\nu, \ldots) = -\mathbf{A}(\ldots \mathbf{e}_\nu, \ldots \mathbf{e}_\mu, \ldots)$$
$$= -A_{\ldots \nu \ldots \mu \ldots}, \tag{3.40}$$

the tensor-components of a form are antisymmetric under exchange of two arbitrary indices.

In order to write a form in component-form we need an antisymmetric tensor basis. The antisymmetric combination of a tensor basis $\omega^{\mu_1} \otimes \cdots \otimes \omega^{\mu_p}$ is denoted by $\omega^{[\mu_1} \otimes \cdots \otimes \omega^{\mu_p]}$ and is defined by

$$\omega^{[\mu_1} \otimes \cdots \otimes \omega^{\mu_p]} = \frac{1}{p!} \sum_{i=1}^{p!} (-1)^{\pi(i)} \omega^{\mu_1} \otimes \cdots \otimes \omega^{\mu_p}, \tag{3.41}$$

where $\pi(i)$ is a function over the $p!$ different permutations of indices μ_1 to μ_p defined by

$$\pi(i) = \begin{cases} 0 & \text{if the permutation is even} \\ 1 & \text{if the permutation is odd.} \end{cases} \tag{3.42}$$

Let us now consider a two-form in a three-dimensional space, and see how it can be written in component form.

$$\alpha = \alpha_{12}\omega^1 \otimes \omega^2 + \alpha_{21}\omega^2 \otimes \omega^1 + \alpha_{13}\omega^1 \otimes \omega^3$$
$$+ \alpha_{31}\omega^3 \otimes \omega^1 + \alpha_{23}\omega^2 \otimes \omega^3 + \alpha_{32}\omega^3 \otimes \omega^2 \tag{3.43}$$

since $\alpha_{11} = \alpha_{22} = \alpha_{33} = 0$ due to the antisymmetry of α. The antisymmetry can also be used to express the form as:

$$\alpha = \alpha_{12}(\omega^1 \otimes \omega^2 - \omega^2 \otimes \omega^1) + \alpha_{13}(\omega^1 \otimes \omega^3 - \omega^3 \otimes \omega^1)$$
$$+ \alpha_{23}(\omega^2 \otimes \omega^3 - \omega^3 \otimes \omega^2) = \alpha_{\mu\nu}\omega^{[\mu} \otimes \omega^{\nu]}. \tag{3.44}$$

An arbitrary p-form α may now be written in component form as

$$\boxed{\alpha = \alpha_{\mu_1 \ldots \mu_p} \omega^{[\mu_1} \otimes \cdots \otimes \omega^{\mu_p]}.} \tag{3.45}$$

Note that a *zero-form* α is only a pure number, $\alpha = \alpha$. The antisymmetry is now trivially satisfied since a zero-form does not have any arguments.

An antisymmetric tensor product, denoted by $\wedge$ and called *the exterior product* or *wedge product*, is defined by

$$\omega^{[\mu_1} \otimes \cdots \otimes \omega^{\mu_p]} \wedge \omega^{[\nu_1} \otimes \cdots \otimes \omega^{\nu_q]}$$
$$= \frac{(p+q)!}{p!q!} \omega^{[\mu_1} \otimes \cdots \otimes \omega^{\mu_p} \otimes \omega^{\nu_1} \otimes \cdots \otimes \omega^{\nu_q]}. \tag{3.46}$$

The exterior product is linear

$$(a\alpha + b\beta) \wedge \gamma = a(\alpha \wedge \gamma) + b(\beta \wedge \gamma), \tag{3.47}$$

$$\alpha \wedge (a\beta + b\gamma) = a(\alpha \wedge \beta) + b(\alpha \wedge \gamma), \tag{3.48}$$

and associative

$$\alpha \wedge (\beta \wedge \gamma) = (\alpha \wedge \beta) \wedge \gamma. \tag{3.49}$$

Therefore we need not include the brackets in products like that in Eq. (3.49).

The antisymmetric basis in Eqs. (3.41) and (3.42) will now be expressed by the exterior product. Putting $q = p = 1$ in Eq. (3.46) we get

$$\omega^{\mu_1} \wedge \omega^{\nu_1} = 2! \, \omega^{[\mu_1} \otimes \omega^{\nu_1]}. \tag{3.50}$$

Using Eq. (3.46) once more we find

$$\omega^{\mu_1} \wedge \omega^{\nu_1} \wedge \omega^{\nu_2} = 2\omega^{[\mu_1} \otimes \omega^{\nu_1]} \wedge \omega^{\nu_2}$$

$$= 3! \, \omega^{[\mu_1} \otimes \omega^{\nu_1} \otimes \omega^{\nu_2]}. \tag{3.51}$$

Continuing in the same manner we get

$$p! \, \omega^{[\mu_1} \otimes \ldots \otimes \omega^{\mu_p]} = \omega^{\mu_1} \wedge \omega^{\mu_2} \wedge \ldots \wedge \omega^{\mu_p}. \tag{3.52}$$

According to Eqs. (3.45) and (3.52) an arbitrary p-form α may be written as

$$\boxed{\alpha = \frac{1}{p!} \alpha_{\mu_1 \ldots \mu_p} \omega^{\mu_1} \wedge \ldots \wedge \omega^{\mu_p}.} \tag{3.53}$$

The reason for $p!$ in the denominator of this expression is that every term is included $p!$ times due to the summation with both increasing and decreasing indices.

From the definition (3.46) follows that

$$\omega^{\mu} \wedge \omega^{\nu} = -\omega^{\nu} \wedge \omega^{\mu}. \tag{3.54}$$

An exchange of two basis forms in Eq. (3.46) involves an odd number of permutations of two neighbouring forms. Thus the exterior product (3.46) is antisymmetric under exchange of two arbitrary one-forms. An important consequence is that an exterior product is zero if it contains two equal basis one-forms. It follows that *in a space of n dimensions there do not exist non-trivial forms with rank higher than n*, since there are only n linearly independent one-forms in such a space. So, *in spacetime there are only zero, one, two, three and four-forms*.

Another consequence of Eq. (3.54) is the equation

$$\alpha \wedge \beta = (-1)^{pq} \beta \wedge \alpha, \tag{3.55}$$

where α is a q-form and β a p-form.

Before we proceed further in the theory of forms, we shall deduce a useful calculational result. Consider the quantities $A_{\mu_1\mu_2}$ and $B^{\mu_1\mu_2}$. Assume that $A_{\mu_1\mu_2}$ is antisymmetric and $B^{\mu_1\mu_2}$ is symmetric. Then,

$$A_{\mu_1\mu_2} B^{\mu_1\mu_2} = \frac{1}{2} A_{\mu_1\mu_2} B^{\mu_1\mu_2} - \frac{1}{2} A_{\mu_2\mu_1} B^{\mu_1\mu_2}$$

$$= \frac{1}{2} A_{\mu_1\mu_2} B^{\mu_1\mu_2} - \frac{1}{2} A_{\mu_2\mu_1} B^{\mu_2\mu_1} = 0, \tag{3.56}$$

since we may exchange the names of the dummy-indices μ_1 and μ_2 in the last term. In general we find that *summation over the indices in a product of an antisymmetric and a symmetric quantity gives zero,*

$$A_{[\mu_1 \cdots \mu_p]} B^{(\mu_1 \cdots \mu_p)} = 0, \tag{3.57}$$

where ( ) denotes a symmetric combination.

For a covariant tensor of rank two, for example, we have

$$T_{\mu\nu} = \frac{1}{2}(T_{\mu\nu} - T_{\nu\mu}) + \frac{1}{2}(T_{\mu\nu} + T_{\nu\mu}) = T_{[\mu\nu]} + T_{(\mu\nu)}; \tag{3.58}$$

hence, every covariant or contravariant tensor of rank two can be separated in an antisymmetric and a symmetric part.

From a covariant tensor, $\mathbf{T}$, without any symmetry, one can construct a form, $\boldsymbol{\tau}$. This consists of the antisymmetric part of $\mathbf{T}$. Thus

$$\boldsymbol{\tau} = T_{[\mu_1 \cdots \mu_p]} \boldsymbol{\omega}^{\mu_1} \otimes \cdots \otimes \boldsymbol{\omega}^{\mu_p}$$
$$= T_{[\mu_1 \cdots \mu_p]} \boldsymbol{\omega}^{\mu_1} \wedge \cdots \wedge \boldsymbol{\omega}^{\mu_p}. \tag{3.59}$$

Note that *the tensor-equation* $\mathbf{T} = 0$ gives the component equations $T_{\mu_1 \cdots \mu_p} = 0$ while *the form-equation* $\boldsymbol{\tau} = 0$ gives the component equations $T_{[\mu_1 \cdots \mu_p]} = 0$.

From Eq. (3.46) follows that the exterior product between a p-form, $\boldsymbol{\alpha}$, and a q-form, $\boldsymbol{\beta}$, is given by

$$(\boldsymbol{\alpha} \wedge \boldsymbol{\beta})_{\mu_1 \cdots \mu_p \mu_{p+1} \cdots \mu_{p+q}} = \frac{(p+q)!}{p! q!} \alpha_{[\mu_1 \cdots \mu_p} \beta_{\mu_{p+1} \cdots \mu_{p+q}]}. \tag{3.60}$$

Until now we have only considered antisymmetric covariant tensors. We may go through the same procedure step for step, with antisymmetric contravariant tensors. Such a tensor of rank p is called a *p-vector*, and has the component form

$$\mathbf{A} = \frac{1}{p!} A^{\mu_1 \cdots \mu_p} \mathbf{e}_{\mu_1} \wedge \cdots \wedge \mathbf{e}_{\mu_p}. \tag{3.61}$$

One-vectors are usual vectors. Two-vectors are called bi-vectors.

The exterior product of p vectors $\mathbf{A}_i$ is a p-vector with components

$$(\mathbf{A}_1 \wedge \ldots \wedge \mathbf{A}_p)^{\mu_1 \cdots \mu_p} = p! A_1^{[\mu_1} A_2^{\mu_2} \ldots A_p^{\mu_p]}. \tag{3.62}$$

The corresponding expression for forms is

$$(\boldsymbol{\alpha}^1 \wedge \ldots \wedge \boldsymbol{\alpha}^p)_{\mu_1 \cdots \mu_p} = p! \alpha^1_{[\mu_1} \alpha^2_{\mu_2} \ldots \alpha^p_{\mu_p]}. \tag{3.63}$$

Example 3.3 (Exterior product and vector product)   Example
Let $\mathbf{A}$ and $\mathbf{B}$ be two vectors in a 3-dimensional space. Then

$$\mathbf{A} \wedge \mathbf{B} = 2! A^{[\mu_1} B^{\mu_2]} \mathbf{e}_{\mu_1} \wedge \mathbf{e}_{\mu_2}$$
$$= (A^1 B^2 - A^2 B^1) \mathbf{e}_1 \wedge \mathbf{e}_2 + (A^1 B^3 - A^3 B^1) \mathbf{e}_1 \wedge \mathbf{e}_3 + (A^2 B^3 - A^3 B^2) \mathbf{e}_2 \wedge \mathbf{e}_3. \tag{3.64}$$

Thus,
$$(\mathbf{A} \wedge \mathbf{B})^k = (\mathbf{A} \times \mathbf{B})^k. \tag{3.65}$$

The exterior product of two vectors has the same components as the vector product. So $\mathbf{A} \wedge \mathbf{B}$ gives the area and orientation of the surface defined by $\mathbf{A}$ and $\mathbf{B}$. Also, if

$\mathbf{A} \wedge \mathbf{B} = 0$ then $\mathbf{A}$ and $\mathbf{B}$ are parallel to each other.

Given a p-form α and a q-vector $\mathbf{A}$ in a space with n dimensions, where $p \geq q$. Then the *contraction* of α with $\mathbf{A}$ is a $(p-q)$-form defined by

$$\iota_{\mathbf{A}} \alpha = \frac{1}{q!(p-q)!} \alpha_{\nu_1 \ldots \nu_q \mu_1 \ldots \mu_{p-q}} A^{\nu_1 \ldots \nu_q} \omega^{\mu_1} \wedge \ldots \wedge \omega^{\mu_{p-q}}. \qquad (3.66)$$

If $q = p$, then $\iota_{\mathbf{A}} \alpha$ is the scalar function

$$\iota_{\mathbf{A}} \alpha = \alpha(\mathbf{A}) = \frac{1}{p!} \alpha_{\mu_1 \ldots \mu_p} A^{\mu_1 \ldots \mu_p}. \qquad (3.67)$$

All covariant tensors can similarly be contracted with a vector. In this case the wedge product in Eq. (3.66) is just replaced by a tensor product. For instance, if $\mathbf{g}$ is a covariant vector of rank 2, then for a vector $\mathbf{v}$ we get

$$\iota_{\mathbf{v}} \mathbf{g} = \mathbf{g}(\mathbf{v}, -) = v^{\mu} g_{\mu\nu} \omega^{\nu}. \qquad (3.68)$$

So a contraction with a vector is just applying the vector using the first slot in the tensor; hence, the rank is reduced with one.

Problems

3.1. *The tensor product*

(a) Given one-forms α and β. Assume that the components of α and β are $(1,1,0,0)$ and $(-1,0,1,0)$, respectively. Show – by using two vectors as arguments – that $\alpha \otimes \beta \neq \beta \otimes \alpha$. Find also the components of $\alpha \otimes \beta$.

(b) Find also the components of the symmetric and anti-symmetric part of $\alpha \otimes \beta$, defined above.

3.2. *Wedge products of forms*
Given the one-forms

$$\alpha = x^2 \omega^1 - y\omega^2, \quad \beta = y\omega^1 - xz\omega^2 + y^2\omega^3, \quad \sigma = y^2 z \omega^2,$$

the two-form

$$\eta = xy\omega^1 \wedge \omega^3 + x\omega^2 \wedge \omega^3,$$

and the three-form

$$\theta = xyz\omega^1 \wedge \omega^2 \wedge \omega^3.$$

Calculate the wedge products

$$\alpha \wedge \beta, \quad \alpha \wedge \beta \wedge \sigma, \quad \alpha \wedge \eta, \quad \alpha \wedge \theta.$$

3.3. *Contractions of tensors*
Assume that $\mathbf{A}$ is an anti-symmetric tensor of rank $\{^2_0\}$, $\mathbf{B}$ a symmetric tensor of rank $\{^0_2\}$, $\mathbf{C}$ an arbitrary tensor of rank $\{^0_2\}$, and $\mathbf{D}$ an arbitrary tensor of rank $\{^2_0\}$. Show that

$$A^{\alpha\beta} B_{\alpha\beta} = 0,$$
$$A^{\alpha\beta} C_{\alpha\beta} = A^{\alpha\beta} C_{[\alpha\beta]},$$

and

$$B_{\alpha\beta} D^{\alpha\beta} = B_{\alpha\beta} D^{(\alpha\beta)}.$$

3.4. *Four-vectors*

(a) Given three four-vectors:

$$\begin{aligned}
\mathbf{A} &= 4\mathbf{e}_t + 3\mathbf{e}_x + 2\mathbf{e}_y + \mathbf{e}_z \\
\mathbf{B} &= 5\mathbf{e}_t + 4\mathbf{e}_x + 3\mathbf{e}_y \\
\mathbf{C} &= \mathbf{e}_t + 2\mathbf{e}_x + 3\mathbf{e}_y + 4\mathbf{e}_z,
\end{aligned}$$

where

$$\mathbf{e}_x \cdot \mathbf{e}_x = \mathbf{e}_y \cdot \mathbf{e}_y = \mathbf{e}_z \cdot \mathbf{e}_z = 1,$$

while

$$\mathbf{e}_t \cdot \mathbf{e}_t = -1.$$

Show that $\mathbf{A}$ is time-like, $\mathbf{B}$ is light-like and $\mathbf{C}$ is space-like.

(b) Assume that $\mathbf{A}$ and $\mathbf{B}$ are two non-zero orthogonal four-vectors, $\mathbf{A} \cdot \mathbf{B} = 0$. Show the following:

- If $\mathbf{A}$ is time-like, then $\mathbf{B}$ is space-like.
- If $\mathbf{A}$ is light-like, then $\mathbf{B}$ is space-like or light-like.
- If $\mathbf{A}$ and $\mathbf{B}$ is light-like, then they are proportional.
- If $\mathbf{A}$ is space-like, then $\mathbf{B}$ is time-like, light-like or space-like.

Illustrate this in a three-dimensional Minkowski-diagram.

(c) A change of basis is given by

$$\begin{aligned}
\mathbf{e}_{t'} &= \cosh \alpha \, \mathbf{e}_t + \sinh \alpha \, \mathbf{e}_x, \\
\mathbf{e}_{x'} &= \sinh \alpha \, \mathbf{e}_t + \cosh \alpha \, \mathbf{e}_x, \\
\mathbf{e}_{y'} &= \mathbf{e}_y, \quad \mathbf{e}_{z'} = \mathbf{e}_z.
\end{aligned}$$

Show that this describes a Lorentz-transformation along the x-axis, where the relative velocity v between the reference frames, are given by $v = \tanh \alpha$. Draw the vectors in a two-dimensional Minkowski-diagram and find what type of curves the $\mathbf{e}_{t'}$ and $\mathbf{e}_{x'}$ describe as α varies.

(d) The three-vector $\mathbf{v}$ describing the velocity of a particle is defined *with respect to an observer*. Explain why the four-velocity $\mathbf{u}$ is defined *independent* of any observer.

The four-momentum of a particle, with rest mass m, is defined by $\mathbf{p} = m\mathbf{u} = m d\mathbf{r}/d\tau$, where τ is the co-moving time of the particle. Show that $\mathbf{p}$ is time-like, and that $\mathbf{p} \cdot \mathbf{p} = -m^2$. Draw, in a Minkowski-diagram, the curve to which $\mathbf{p}$ must be tangent to, and explain how this is altered as $m \longrightarrow 0$.

Assume that the energy of the particle is being observed by an observer with four-velocity $\mathbf{u}$. Show that the energy he measures is given by

$$E = -\mathbf{p} \cdot \mathbf{u}. \tag{3.69}$$

This is an expression which is very useful when one wants to calculate the energy of a particle in an arbitrary reference frame.

3.5. *The Lorentz-Abraham-Dirac equation*

(a) Show that the Lorentz's force-law, eq. (2.77), can be written as the four-vector equation

$$m\frac{du^\mu}{d\tau} = qF^\mu_{\ \nu}u^\nu, \tag{3.70}$$

where m is the rest mass of a particle, q its charge, and τ its proper time. Here, $F^\mu_{\ \nu}$ are the components of the electromagnetic field tensor,

$$F^\mu_{\ \nu} = \begin{bmatrix} 0 & E_1 & E_2 & E_3 \\ E_1 & 0 & B_3 & -B_2 \\ E_2 & -B_3 & 0 & B_1 \\ E_3 & B_2 & -B_1 & 0 \end{bmatrix}. \tag{3.71}$$

Since an accelerated charge radiates one expects that the electromagnetic field produced by the charge acts upon the charge. This is not taken into account by the Lorentz force-law. Hence one is lead to modify the equation of motion of the charge as

$$m\frac{du^\mu}{d\tau} = qF^\mu_{\ \nu}u^\nu + \Gamma^\mu, \tag{3.72}$$

where Γ^μ is the field reaction four-force. According to Larmor's formula the energy radiated by the charge per unit proper time is $(2/3)\alpha A^\beta A_\beta$ where $\alpha = q^2/(4\pi c)$ and A^β is the four-acceleration of the charge. The radiated four-momentum per unit proper time is $(2/3)\alpha A^\beta A_\beta u^\mu$. This acts back on the charge. Assuming that the particle radiates for a finite time one may thus require that

$$\Gamma^\mu = -\frac{2}{3}\alpha A^\beta A_\beta u^\mu + \frac{dc^\mu}{d\tau}, \tag{3.73}$$

for some vector c^μ, because the second term does not contribute to the total change in four-momentum $\int_{-\infty}^{\infty} \Gamma^\mu d\tau$.

(b) Use the four-velocity identity, $u_\beta A^\beta = 0$, and the antisymmetry of $F_{\mu\nu}$ to show that $u_\beta \Gamma^\beta = 0$, and deduce that

$$\Gamma^\mu = \frac{2}{3}\frac{\alpha}{c^2}\left(\frac{d^2u^\mu}{d\tau^2} - \frac{1}{c^2}A^\beta A_\beta u^\mu\right). \tag{3.74}$$

The equation of motion of a charged particle with this expression for the field reaction four-force is called *the Lorentz-Abraham-Dirac equation*.

(c) Deduce the non-relativistic limit of the Lorentz-Abraham-Dirac equation. Is this equation invariant against reversal of the time direction?

4

Basis Vector Fields
and the Metric Tensor

In this chapter we are going to introduce the basic concepts necessary to grasp the geometrical significance of the metric tensor.

4.1 Manifolds and their coordinate-systems

Let $\mathbb{R}^n$ denote a succession of n real numbers $(x^1, \ldots, x^n)$. A map f, from a space M to a space N is a rule that to each point, x, in M associates one point, $f(x)$, in N (this is usually denoted by $f : M \mapsto N$). For the funtion to be one-to-one we require that different points M are mapped to different points in N, i.e., $x \neq y \Rightarrow f(x) \neq f(y)$. This requirement implies that the map f has a well-defined inverse, $f^{-1} : N \mapsto M$. These concepts are illustrated in Fig. 4.1.

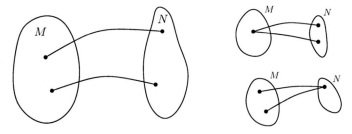

Figure 4.1: The mapping to the left is a one-to-one mapping. Those to the right are not one-to-one mappings.

A *manifold* M is a space satisfying the following properties.

1. There exists a family of open neighbourhoods U_i together with continuous one-to-one mappings $f_i : U_i \longmapsto \mathbb{R}^n$ with a continuous inverse for a number n.

2. The family of open neighbourhoods cover the whole of M; i.e.

$$\bigcup_i U_i = M.$$

The definition of M does only involve open sub-spaces in M because we do not want to restrict the topological properties of M. The *whole* surface of a sphere, for example, cannot be mapped onto $\mathbb{R}^2$. In particular, spherical co-ordinates do not represent a *one-to-one* mapping on $\mathbb{R}^2$. On the other hand, using a family of open neighbourhoods we can cover the sphere where each of the neighbourhoods can be mapped onto the plane $\mathbb{R}^2$. Hence, the sphere is a manifold.

According to the definition of a manifold M there exist mappings $\phi : U \to \mathbb{R}^n$, where U is an open region in M. If P is a point in M, then $\phi(P) = (x^1, ..., x^n)$ will be a vector in $\mathbb{R}^n$. Such a mapping is called a *coordinate system*, and U is called a coordinate region of M. A coordinate system consists therefore of a set of maps $\{x^\mu\}_{\mu=1,...,n}$, and the coordinate system is a representation of points, P, in U by n-tuples $(x^1, \ldots, x^n)$. In two dimensions it may be represented by a net of squares, and in three dimensions by a cubic network and so forth (see Fig. 4.2).

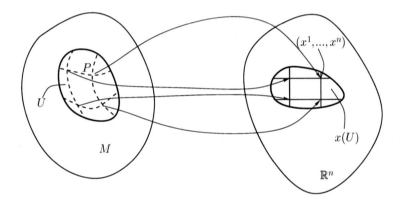

Figure 4.2: The coordinate system is a mapping from the manifold into a Euclidean space.

If two regions U and V have non-empty intersection $U \cap V \neq \emptyset$ with coordinates $\{x^\mu\}$ and $\{x^{\mu'}\}$, then we can define an invertible *coordinate transformation*

$$x^\mu = x^{\mu'}(x^\mu) \tag{4.1}$$

in $U \cap V$ (see Fig. 4.3). Unless otherwise explicitly stated, we will assume that such coordinate transformations can be differentiated an arbitrary number of times. Functions with this property that they can be differentiated an arbitrary number of times, and having continuous derivatives at all levels, are called *smooth functions*. Moreover, if a manifold has smooth coordinate mappings, then the manifold is called a *smooth manifold*.

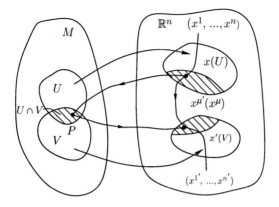

Figure 4.3: A coordinate transformation between two sets of coordinates.

Example 4.1 (Transformation between plane polar-coordinates and Cartesian coordinates)   Example

The connection between the plane polar coordinates (r, θ) and the Cartesian coordinates (x, y) is shown in Fig. 4.4

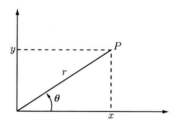

Figure 4.4: Polar coordinates in the plane.

From the figure follows that in this case the transformation equation (4.1) is

$$x = r \cos \theta, \quad y = r \sin \theta.$$

The inverse transformation is

$$r = (x^2 + y^2)^{\frac{1}{2}}, \quad \theta = \arctan(y/x).$$

4.2   Tangent vector fields and the coordinate basis vector fields

Let us consider the special case that the manifold M is a curved surface in a three-dimensional Euclidean space $\mathbb{R}^3$. Then it is possible to introduce a position vector $\mathbf{v}$ in $\mathbb{R}^3$ to an arbitrary point P in M. Let $\mathbf{r}(\lambda)$ be a curve in M

with parameter λ. The *tangent vector* $\mathbf{t}(\lambda_0)$ of this curve at a point λ_0 is defined by

$$\mathbf{t}(\lambda_0) = \left(\frac{d\mathbf{r}}{d\lambda}\right)_{\lambda=\lambda_0}. \tag{4.2}$$

This is illustrated in Fig. 4.5

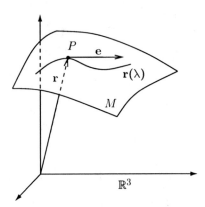

Figure 4.5: Tangent vectors.

The definition given above presupposes that the manifold M is embedded in a higher-dimensional Euclidean space. If this is not possible, one cannot define a finite position vector $\mathbf{r}$. This is because vectors do not exist in a curved space M, but in a *tangent space* T_P which is defined as follows (see Fig. 4.6). The tangent space T_P of a space M at point P is generated by the tangent

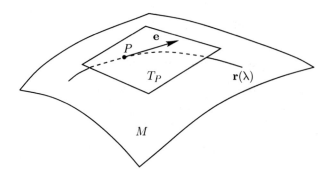

Figure 4.6: The tangent space of a point P.

vectors of all possible curves in M through P.

Different points in M have different tangent spaces. Vector addition is therefore possible only for vectors at one and the same point.

The *coordinate basis vectors*, $\mathbf{e}_\mu$, of a coordinate system, $\{x^\mu\}$, in M are defined by

$$\mathbf{e}_\mu = \frac{\partial \mathbf{r}}{\partial x^\mu}. \tag{4.3}$$

(This definition will be generalized below so that the position vector can be disposed of).

Hence, at every point P of the manifold M we can define a vector space T_P. The union of all these spaces over each point P is called the *tangent space of M*, or the *tangent bundle TM*:

$$TM \equiv \bigcup_{P \in M} T_P. \tag{4.4}$$

A *vector field*[1] is a continuum of vectors in TM, with components that are continuous and differentiable functions of the coordinates, x^μ. In general the basis vectors $\mathbf{e}_\mu$ define a *basis vector field* in a neighbourhood of a point $P \in M$.

A tangent vector $\mathbf{v}$ to a curve $\mathbf{r}(\lambda)$ can be expressed in component form, relatively to an arbitrary basis vector field, as

$$\mathbf{v} = \frac{d\mathbf{r}}{d\lambda} = \frac{dx^\mu}{d\lambda}\frac{\partial \mathbf{r}}{\partial x^\mu} = \frac{dx^\mu}{d\lambda}\mathbf{e}_\mu. \tag{4.5}$$

The coordinate basis vectors are tangent vectors to the coordinate curves, with the coordinates as curve parameters.

The basis vectors $\mathbf{e}_\mu$ are linearly independent. The number of vectors in a basis is equal to the number of coordinates, which is equal to the dimension of the manifold, M.

The relation between the coordinate basis vectors of two different coordinate systems $\{x^\mu\}$ and $\{x^{\mu'}\}$ is

$$\mathbf{e}_{\mu'} = \frac{\partial \mathbf{r}}{\partial x^{\mu'}} = \frac{\partial \mathbf{r}}{\partial x^\mu}\frac{\partial x^\mu}{\partial x^{\mu'}} = \mathbf{e}_\mu \frac{\partial x^\mu}{\partial x^{\mu'}}, \tag{4.6}$$

and

$$\mathbf{e}_\mu = \frac{\partial \mathbf{r}}{\partial x^\mu} = \frac{\partial \mathbf{r}}{\partial x^{\mu'}}\frac{\partial x^{\mu'}}{\partial x^\mu} = \mathbf{e}_{\mu'}\frac{\partial x^{\mu'}}{\partial x^\mu}. \tag{4.7}$$

For an arbitrary vector $\mathbf{v}$ we find

$$\mathbf{v} = \mathbf{e}_{\mu'}v^{\mu'} = \mathbf{e}_\mu v^\mu = \mathbf{e}_{\mu'}\frac{\partial x^{\mu'}}{\partial x^\mu}v^\mu. \tag{4.8}$$

Hence, vector components transform as

$$v^{\mu'} = \frac{\partial x^{\mu'}}{\partial x^\mu}v^\mu. \tag{4.9}$$

Consider now the directional differential operator along a curve with parameter λ:

$$\frac{d}{d\lambda} = \frac{dx^\mu}{d\lambda}\frac{\partial}{\partial x^\mu}. \tag{4.10}$$

The directional derivative along the coordinate curves are the partial derivatives

$$\partial_\mu = \frac{\partial}{\partial x^\mu}. \tag{4.11}$$

In an n-dimensional space, M, there are n linearly independent directional derivatives. They transform in the same way as the basis vectors

$$\frac{\partial}{\partial x^{\mu'}} = \frac{\partial x^\mu}{\partial x^{\mu'}}\frac{\partial}{\partial x^\mu}. \tag{4.12}$$

---

[1]In the mathematical literature, vector fields are often called *sections*.

Thus, the directional derivative can be used as a basis of M

$$e_\mu = \frac{\partial}{\partial x^\mu}. \tag{4.13}$$

This is *the general definition of a coordinate basis vector* and does not rely on the existence of a finite position vector. The definition is equally valid in curved space as in flat space.

Since arbitrary vectors can be written in component form as linear combinations of basis vectors, eq.(4.13) implies that an arbitrary vector can be thought of as a differential operator, too.

Examples

Example 4.2 (The coordinate basis vector field of plane polar coordinates)
(See Example 4.1)
From the transformation equation

$$x = r\cos\theta, \quad y = r\sin\theta,$$

we find the coordinate basis vectors of the polar coordinate system

$$e_r = \frac{\partial}{\partial r} = \frac{\partial x}{\partial r}\frac{\partial}{\partial x} + \frac{\partial y}{\partial r}\frac{\partial}{\partial y} = \cos\theta e_x + \sin\theta e_y,$$

$$e_\theta = \frac{\partial}{\partial \theta} = \frac{\partial x}{\partial \theta}\frac{\partial}{\partial x} + \frac{\partial y}{\partial \theta}\frac{\partial}{\partial y} = -r\sin\theta e_x + r\cos\theta e_y.$$

The basis vectors e_r and e_θ are shown in Fig. 4.7.

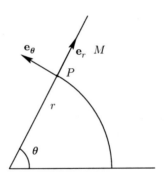

Figure 4.7: Basis vectors in polar coordinates.

A point concerned with practical calculations should be mentioned. If the transformation of basis vectors and vector components is calculated by means of matrix multiplication, the basis vectors, $\{e_\mu\}$, must be written as row matrix, and the vector components $\{v^\mu\}$ as column matrix, respectively.

Example 4.3 (The velocity vector of a particle moving along a circular path)
We consider a particle moving with constant velocity along a circular path on a plane surface. Then the position of the particle can be described by a position vector r on the surface. A system of plane polar-coordinates (r,θ) on the surface with origin at the centre of the circle is introduced. According to the results of the preceding two examples, the position vector may be expressed as

$$r = xe_x + ye_y = r\cos\theta e_x + r\sin\theta e_y = re_r$$

(Note that the formula $\mathbf{r} = x^i \mathbf{e}_i$, with summation over all coordinates, is *not* generally valid in a curved coordinate system). The velocity vector,

$$\mathbf{v} = \frac{d\mathbf{r}}{dt} = \frac{dr}{dt}\mathbf{e}_r + r\frac{d\mathbf{e}_r}{dt},$$

is tangent vector to the circular path. In the present case $r = r_0$, $dr/dt = 0$ and

$$\mathbf{v} = r_0\frac{d\mathbf{e}_r}{dt}.$$

Using the expression for $\mathbf{e}_r$ and $\mathbf{e}_\theta$ from example 4.2, and that $\theta = \omega t$, $\omega = $ constant, we get

$$\mathbf{v} = -r_0\omega \sin \omega t \mathbf{e}_x + r_0\omega \cos \omega t \mathbf{e}_y = \omega \mathbf{e}_\theta.$$

This result follows immediately from eq. (4.5)

$$\mathbf{v} = \frac{dr}{dt}\mathbf{e}_r + \frac{d\theta}{dt}\mathbf{e}_\theta,$$

with $r = r_0$ and $\theta = \omega t$.

Example 4.4 (Transformation of coordinate basis vectors and vector components)
Consider the transformation:

$$(\mathbf{e}_{1'}, \mathbf{e}_{2'}) = (\mathbf{e}_1, \mathbf{e}_2)\begin{bmatrix} \frac{\partial x^1}{\partial x^{1'}} & \frac{\partial x^1}{\partial x^{2'}} \\ \frac{\partial x^2}{\partial x^{1'}} & \frac{\partial x^2}{\partial x^{2'}} \end{bmatrix}.$$

This gives

$$\mathbf{e}_{1'} = \mathbf{e}_1\frac{\partial x^1}{\partial x^{1'}} + \mathbf{e}_2\frac{\partial x^2}{\partial x^{1'}}; \quad \mathbf{e}_{2'} = \mathbf{e}_1\frac{\partial x^1}{\partial x^{2'}} + \mathbf{e}_2\frac{\partial x^2}{\partial x^{2'}}.$$

Furthermore,

$$\begin{bmatrix} v^{1'} \\ v^{2'} \end{bmatrix} = \begin{bmatrix} \frac{\partial x^{1'}}{\partial x^1} & \frac{\partial x^{1'}}{\partial x^2} \\ \frac{\partial x^{2'}}{\partial x^1} & \frac{\partial x^{2'}}{\partial x^2} \end{bmatrix}\begin{bmatrix} v^1 \\ v^2 \end{bmatrix},$$

which gives

$$v^{1'} = \frac{\partial x^{1'}}{\partial x^1}v^1 + \frac{\partial x^{1'}}{\partial x^2}v^2; \quad v^{2'} = \frac{\partial x^{2'}}{\partial x^1}v^1 + \frac{\partial x^{2'}}{\partial x^2}v^2.$$

Example 4.5 (Some transformation matrices)

A: Transformation from plane polar coordinates to Cartesian coordinates

$$x = r\cos\theta; \quad y = r\sin\theta.$$

$$(M^\mu{}_{\mu'}) = \left(\frac{\partial x^\mu}{\partial x^{\mu'}}\right) = \begin{bmatrix} \frac{\partial x}{\partial r} & \frac{\partial x}{\partial \theta} \\ \frac{\partial y}{\partial r} & \frac{\partial y}{\partial \theta} \end{bmatrix} = \begin{bmatrix} \cos\theta & -r\sin\theta \\ \sin\theta & r\cos\theta \end{bmatrix}.$$

B: Rotation of a Cartesian coordinate system From Fig. 4.8 is seen that

$$x = x'\cos\alpha - y'\sin\alpha; \quad y = x'\sin\alpha + y'\cos\alpha$$

this gives

$$(M^\mu{}_{\mu'}) = \begin{bmatrix} \cos\alpha & -\sin\alpha \\ \sin\alpha & \cos\alpha \end{bmatrix}.$$

Note that the transformation matrix of a rotation has the property $M^T = M^{-1}$.

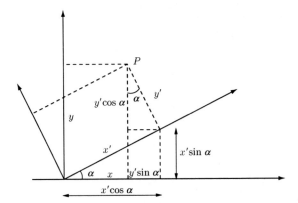

Figure 4.8: Rotation of a Cartesian coordinate system.

C: A Lorentz transformation Let R be a laboratory frame with a local Cartesian coordinate system (t, x, y, z). The frame R' is moving in the negative x-direction with velocity v relative to R, and has comoving coordinate system (t', x', y', z'). The Lorentz transformation between (t', x', y', z') and (t, x, y, z) is

$$t = \gamma(t' + \frac{v}{c^2}x'), \quad x = \gamma(x' + vt'), \quad y = y', \quad z = z'.$$

Differentiation gives the transformation matrix

$$\left(\frac{\partial x^\mu}{\partial x^{\mu'}}\right) = \begin{bmatrix} \gamma & \gamma\frac{v}{c} & 0 & 0 \\ \gamma\frac{v}{c} & \gamma & 0 & 0 \\ 0 & 0 & 1 & 0 \\ 0 & 0 & 0 & 1 \end{bmatrix}.$$

Applying this boost to the components of a four vector $\mathbf{A} = A^\mu \mathbf{e}_{\mu'}$, we get

$$A^t = \gamma\left(A^{t'} + \frac{v}{c}A^{x'}\right), \quad A^x = \gamma\left(A^{x'} + \frac{v}{c}A^{t'}\right),$$

$$A^y = A^{y'}, \quad A^z = A^{z'}.$$

Identifying $\mathbf{A}$ with the four-momentum, eq. (3.9), we find the transformation formula for energy and ordinary momentum,

$$E = \gamma\left(E' + vp^{x'}\right), \quad p^x = \gamma\left(p^{x'} + \frac{v}{c^2}E'\right),$$

$$p^y = p^{y'}, \quad p^z = p^{z'}.$$

Let R' be the rest-frame of the particle. Then the particle moves with velocity in the x-direction in R, with energy and momentum given by

$$E = \gamma E_0, \quad p^x = \gamma\frac{v}{c^2}E_0, \quad p^y = p^z = 0.$$

Now, letting $\mathbf{A} = \mathbf{F}$ where $\mathbf{F}$ is the four-force (3.12), we find the force components

$$f^x = f_0^x, \quad f^y = \gamma^{-1}f_0^y, \quad f^z = \gamma^{-1}f_0^z.$$

Finally, letting $\mathbf{A}$ be the four-acceleration, eq. (3.16), we find the components of the ordinary acceleration of the moving particle in terms of the components of its rest acceleration

$$a^x = \gamma^{-3}a_0^x, \quad a^y = \gamma^{-2}a_0^y, \quad a^z = \gamma^{-2}a_0^z.$$

The basis vectors have been defined as differential operators in eq. (4.12). In order to see from an expression, when a basis vector is to be applied as a differential operator, we shall indicate this by bracketing the argument. Thus, if f is a scalar function, then

$$\mathbf{e}_\mu(f) = \frac{\partial f}{\partial x^\mu}. \tag{4.14}$$

We shall also use a simple notation for partial derivatives introduced by Einstein,

$$\frac{\partial f}{\partial x^\mu} \equiv f_{,\mu} \quad \text{and} \quad \frac{\partial^2 f}{\partial x^\mu \partial x^\nu} \equiv f_{,\mu\nu}. \tag{4.15}$$

4.3   Structure coefficients

In an arbitrary (non-coordinate) basis the vectors $\mathbf{e}_\mu$ are not simply partial derivatives, but they are still linear combinations of partial derivatives.

Consider a vector field $\mathbf{u} = u^\mu \mathbf{e}_\mu$. If f is a scalar function we have

$$\boxed{\mathbf{u}(f) = u^\mu \mathbf{e}_\mu(f),} \tag{4.16}$$

where $\mathbf{e}_\mu$ is a first order differential operator.

We define an operator product $\mathbf{uv}$ by

$$\mathbf{uv}(f) = u^\mu \mathbf{e}_\mu(v^\nu \mathbf{e}_\nu(f)). \tag{4.17}$$

This may be written

$$\mathbf{uv}(f) = u^\mu \mathbf{e}_\mu(v^\nu)\mathbf{e}_\nu(f) + u^\mu v^\nu \mathbf{e}_\mu \mathbf{e}_\nu(f). \tag{4.18}$$

The operator $\mathbf{uv}$ is not a vector since it contains second order derivatives.

The *commutator* (or Lie-product) of two vectors $\mathbf{u}$ and $\mathbf{v}$, denoted by $[\mathbf{u}, \mathbf{v}]$, is defined by

$$\boxed{[\mathbf{u}, \mathbf{v}] = \mathbf{uv} - \mathbf{vu}.} \tag{4.19}$$

Using eq. (4.18) we get

$$[\mathbf{u}, \mathbf{v}] = \{u^\mu \mathbf{e}_\mu(v^\nu) - v^\mu \mathbf{e}_\mu(u^\nu)\}\mathbf{e}_\nu + u^\mu v^\nu [\mathbf{e}_\mu, \mathbf{e}_\nu]. \tag{4.20}$$

Since $f_{,\mu\nu} = f_{,\nu\mu}$ and $\mathbf{e}_\mu$ are linear combinations of partial derivatives, the terms with second order derivatives will cancel. Thus, $[\mathbf{u}, \mathbf{v}]$ is a vector.

The *structure coefficients*, $c^\rho{}_{\mu\nu}$, of an arbitrary basis field $\{\mathbf{e}_\mu\}$ are defined by

$$\boxed{[\mathbf{e}_\mu, \mathbf{e}_\nu] = c^\rho{}_{\mu\nu}\mathbf{e}_\rho.} \tag{4.21}$$

Then eq. (4.20) takes the form

$$\boxed{[\mathbf{u}, \mathbf{v}] = [u^\mu \mathbf{e}_\mu(v^\nu) - v^\mu \mathbf{e}_\mu(u^\nu)]\mathbf{e}_\nu + u^\mu v^\nu c^\rho{}_{\mu\nu}\mathbf{e}_\rho.} \tag{4.22}$$

For two coordinate basis vectors we get

$$[\mathbf{e}_\mu, \mathbf{e}_\nu] = \left[\frac{\partial}{\partial x^\mu}, \frac{\partial}{\partial x^\nu}\right] = \frac{\partial^2}{\partial x^\mu \partial x^\nu} - \frac{\partial^2}{\partial x^\nu \partial x^\mu} = 0. \tag{4.23}$$

Thus the structure coefficients vanish in a coordinate basis. In this case eq. (4.22) reduces to

$$[\mathbf{u}, \mathbf{v}] = (u^\mu v^\nu{}_{,\mu} - v^\mu u^\nu{}_{,\mu})\mathbf{e}_\nu. \tag{4.24}$$

4.4   General basis transformations

The transformation between two arbitrary basis-fields $\{e_\mu\}$ and $\{e_{\mu'}\}$ is written

$$\mathbf{e}_{\mu'} = \mathbf{e}_\mu M^\mu{}_{\mu'}, \tag{4.25}$$

$$\mathbf{e}_\mu = \mathbf{e}_{\mu'} M^{\mu'}{}_\mu, \tag{4.26}$$

where the transformation matrix $(M^{\mu'}{}_\mu)$ is inverse to the matrix $(M^\mu{}_{\mu'})$, i.e.

$$M^\mu{}_{\mu'} M^{\mu'}{}_\nu = \delta^\mu{}_\nu. \tag{4.27}$$

The elements of the transformation matrix, $M^\mu{}_{\mu'}$, are the components of the basis-vectors $e_{\mu'}$ as decomposed in the basis $\{e_\mu\}$.

In the special case of a *transformation between two coordinate basis fields,*

$$M^\mu{}_{\mu'} = \frac{\partial x^\mu}{\partial x^{\mu'}}, \tag{4.28}$$

so that

$$M^{\mu'}{}_{\mu,\nu} = M^{\mu'}{}_{\nu,\mu}. \tag{4.29}$$

These equations are not valid in general.

The transformation equation for vector components follows immediately from eq. (4.26)

$$\mathbf{v} = \mathbf{e}_\mu v^\mu = \mathbf{e}_{\mu'} M^{\mu'}{}_\mu v^\mu = \mathbf{e}_{\mu'} v^{\mu'}, \tag{4.30}$$

which gives

$$v^{\mu'} = M^{\mu'}{}_\mu v^\mu. \tag{4.31}$$

Tensor components with upper indices transform according to eq. (4.31) and will be called *contravariant components*. Components with lower indices transform in the same way as the basis vectors. They are called *covariant components*.

Basis one-forms have upper indices and transform contravariantly

$$\boldsymbol{\omega}^{\mu'} = M^{\mu'}{}_\mu \boldsymbol{\omega}^\mu. \tag{4.32}$$

The components of one-forms have lower indices and transform covariantly

$$\alpha_{\mu'} = \alpha_\mu M^\mu{}_{\mu'}. \tag{4.33}$$

Corresponding formulae are valid for components of tensors of arbitrary rank. For a mixed tensor of rank $\{{}^1_2\}$, for example,

$$T^{\alpha'}{}_{\mu'\nu'} = M^{\alpha'}{}_\alpha M^\mu{}_{\mu'} M^\nu{}_{\nu'} T^\alpha{}_{\mu\nu}. \tag{4.34}$$

The components of a tensor transform homogeneously. The transformed components are proportional to the original ones. A non-vanishing tensor has at least one component different from zero. It follows that there is at least one transformed component different from zero, too. It is not possible to transform away a tensor, and it is not possible to obtain a non-vanishing tensor from a vanishing tensor. *Tensors have, in general, a coordinate independent existence.*

The fact that one can transform away the ordinary velocity of a particle by going into its rest frame, shows that the three-velocity is not a vector. The four-velocity, on the contrary, is a vector. It cannot be transformed away.

If we contract two of the indices of a component of a mixed tensor, we obtain a quantity with two indices less. Let us consider the transformation properties of these new quantities. Summing over α' and μ' in eq. (4.34), and using eq. (4.27), we get

$$T^{\alpha'}_{\alpha'\nu'} = M^{\alpha'}_{\alpha} M^{\mu}_{\alpha'} M^{\nu}_{\nu'} T^{\alpha}_{\mu\nu} = \delta^{\mu}_{\alpha} M^{\nu}_{\nu'} T^{\alpha}_{\mu\nu} = M^{\nu}_{\nu'} T^{\alpha}_{\alpha\nu}. \qquad (4.35)$$

This equation shows that the quantities $T^{\alpha}_{\alpha\nu}$ transform as a tensor of rank $\{^0_1\}$. The reduction of the rank of a tensor by two by summing over a contravariant and covariant index is called *contraction of the tensor*. The contracted tensor is a new tensor compared to the original one. The contraction of a tensor of a tensor of rank $\{^1_1\}$ gives a scalar function, equal to the trace of the matrix made up of its components.

The push-forward

We will now define a common notation which is used in the literature and will also be useful later on. Consider a coordinate transformation $f = (x'^1, ..., x'^n)$, then we can for any vector $\mathbf{v} = v^{\mu}\mathbf{e}_{\mu}$, define the derivative

$$\boxed{f_*\mathbf{v} \equiv v^{\mu}\frac{\partial x^{\mu'}}{\partial x^{\mu}}\mathbf{e}_{\mu'}.} \qquad (4.36)$$

In general the map f does not need to be a coordinate transformation, it only needs to be a map from the manifold M where $\mathbf{v}$ lives. The *push-forward* will then be the mapping f_* as defined above. Thus it can be considered as the linear map, with the Jacobian matrix

$$(f_*)^{\mu'}_{\mu} = \frac{\partial x^{\mu'}}{\partial x^{\mu}}. \qquad (4.37)$$

If $\mathbf{v} = \frac{\partial}{\partial x^{\alpha}}$, then

$$f_*\mathbf{v} = \frac{\partial x^{\mu'}}{\partial x^{\alpha}}\frac{\partial}{\partial x^{\mu'}}, \qquad (4.38)$$

so this is nothing but the "chain rule for partial derivatives". If $f : M \longmapsto N$ and g is a function $g : N \longmapsto Q$, then we can form the composition $(g \circ f) : M \longmapsto Q$. The chain rule now says that

$$(g \circ f)_* \frac{\partial}{\partial x^{\alpha}} = \frac{\partial x^{\mu'}}{\partial x^{\alpha}}\frac{\partial x^{\mu''}}{\partial x^{\mu'}}\frac{\partial}{\partial x^{\mu''}}. \qquad (4.39)$$

We also find that

$$g_* f_* \frac{\partial}{\partial x^{\alpha}} = \frac{\partial x^{\mu'}}{\partial x^{\alpha}}\frac{\partial x^{\mu''}}{\partial x^{\mu'}}\frac{\partial}{\partial x^{\mu''}}. \qquad (4.40)$$

Since the push-forward is linear we have

$$(g \circ f)_* = g_* f_*. \qquad (4.41)$$

This is just the chain rule in a more modern setting.

4.5   The metric tensor

In our development of the theory of tensors we have not yet been able to define formally the absolute value of a vector. Therefore the scalar product $\mathbf{u} \cdot \mathbf{v}$, between two vectors could not be calculated from the elementary formula $\mathbf{u} \cdot \mathbf{v} = |\mathbf{u}||\mathbf{v}| \cos \alpha$, where α is the angle between $\mathbf{u}$ and $\mathbf{v}$.

However, since every vector is a linear combination of basis vectors, the scalar product between two arbitrary vectors can be defined by specifying the values of all scalar products between the basis vectors in a basis $\{\mathbf{e}_\mu\}$.

The scalar product between two vectors $\mathbf{u}$ and $\mathbf{v}$ is denoted by $\mathbf{g}(\mathbf{u}, \mathbf{v})$ and is defined as a *symmetric bilinear mapping*, which for every pair of vectors gives a scalar. It follows from the definition of tensors that this mapping is a covariant, symmetrical tensor of rank $\{^0_2\}$. It is called *the metric tensor*. Thus

$$\mathbf{v} \cdot \mathbf{u} = \mathbf{u} \cdot \mathbf{v} = \mathbf{g}(\mathbf{u}, \mathbf{v}) = g_{\mu\nu} u^\mu v^\nu, \tag{4.42}$$

where

$$\boxed{g_{\nu\mu} = g_{\mu\nu} = \mathbf{g}(\mathbf{e}_\mu, \mathbf{e}_\nu) = \mathbf{e}_\mu \cdot \mathbf{e}_\nu.} \tag{4.43}$$

The absolute value or *norm* of a vector is defined by

$$|\mathbf{v}| = [g(\mathbf{v}, \mathbf{v})]^{1/2} = (g_{\mu\nu} v^\mu v^\nu)^{1/2}. \tag{4.44}$$

The scalar product between two vectors can now be written

$$\mathbf{u} \cdot \mathbf{v} = |\mathbf{u}||\mathbf{v}| \cos \alpha. \tag{4.45}$$

The contravariant components, $g^{\mu\nu}$, of the metric tensor are defined as the elements of the inverse matrix to the one made up of the covariant components, i.e.,

$$\boxed{g^{\mu\alpha} g_{\alpha\nu} = \delta^\mu_{\ \nu}.} \tag{4.46}$$

By a basis transformation the metric tensor gets new components

$$g_{\mu'\nu'} = g_{\mu\nu} M^\mu_{\ \mu'} M^\nu_{\ \nu'}. \tag{4.47}$$

The *transpose* of a matrix, $\mathbf{M}^T$ is defined as the matrix obtained by interchanging the rows and columns of the matrix $\mathbf{M}$. The transformation equation (4.47) can be written in matrix form as

$$\mathbf{g}' = \mathbf{M}^T \mathbf{g} \mathbf{M}, \tag{4.48}$$

where $\mathbf{g}'$ and $\mathbf{g}$ are the matrices made up of the components $g_{\mu'\nu'}$ and $g_{\mu\nu}$ of the metric tensor respectively. By means of the metric tensor we can define linear one-to-one mappings between tensors of different type (covariant or contravariant), but with the same rank. We can for example map a vector on a one-form, $\boldsymbol{\omega} = \iota_\mathbf{v} \mathbf{g}$, with components

$$\begin{aligned}
v_\mu &= \mathbf{g}(\mathbf{v}, \mathbf{e}_\mu) = \mathbf{g}(v^\nu \mathbf{e}_\nu, \mathbf{e}_\mu) = v^\nu \mathbf{g}(\mathbf{e}_\nu, \mathbf{e}_\mu) \\
&= v^\nu g_{\nu\mu} = g_{\mu\nu} v^\nu.
\end{aligned} \tag{4.49}$$

This mapping is called *lowering of an index*. The *raising of an index* is given by

$$v^\mu = \delta^\mu_{\ \nu} v^\nu = g^{\mu\alpha} g_{\alpha\nu} v^\nu = g^{\mu\alpha} v_\alpha. \tag{4.50}$$

Corresponding expressions are valid for tensors of arbitrary rank, for example,

$$T^{\nu}{}_{\mu\gamma} = g_{\mu\alpha}g^{\nu\beta}T^{\alpha}{}_{\beta\gamma}. \tag{4.51}$$

Equation (4.46) can now be written

$$g^{\mu}{}_{\nu} = \delta^{\mu}{}_{\nu}. \tag{4.52}$$

Thus, the mixed components of the metric tensor are equal to the Kronecker symbols. In this sense the metric tensor can be thought of as the unit tensor of rank two.

The metric tensor will now be used to define the *distance* along a curve. Consider an infinitesimal distance, ds, between two points on a curve $x^{\mu}(\lambda)$, at λ_0 and $\lambda_0 + d\lambda$. Let $\mathbf{v}$ be a tangent vector field of the curve. Then

$$\boxed{ds^2 = \mathbf{g}(\mathbf{v},\mathbf{v})d\lambda^2 = g_{\mu\nu}v^{\mu}v^{\nu}d\lambda^2 = g_{\mu\nu}dx^{\mu}dx^{\nu}} \tag{4.53}$$

since $v^{\mu} = dx^{\mu}/d\lambda$. The quantity ds is called the *line-element* associated with the metric tensor $g_{\mu\nu}$.

The finite distance along a curve, between two points λ_0 and λ is calculated from the line integral

$$s = \int_{\lambda_0}^{\lambda}\sqrt{|g_{\mu\nu}v^{\mu}v^{\nu}|}d\lambda, \quad v^{\mu} = \frac{dx^{\mu}}{d\lambda}. \tag{4.54}$$

The physical interpretation of the line-element along time-like curves in space-time was discussed in section (2.8).

Example 4.6 (The line-element of flat 3-space in spherical coordinates)   Example
The line-element of Euclidean 3-space in Cartesian coordinates is

$$ds^2 = dx^2 + dy^2 + dz^2. \tag{4.55}$$

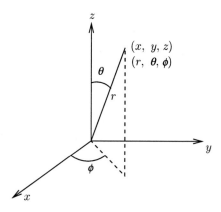

Figure 4.9: Relationship between Cartesian and spherical coordinates.

From Fig. 4.9 is seen that the transformation between spherical and the Cartesian coordinates is

$$x = r\cos\phi\sin\theta, \quad y = r\sin\phi\sin\theta, \quad z = r\cos\theta. \tag{4.56}$$

Differentiating and substituting into eq. (4.55) we obtain the line-element of Euclidean 3-space in spherical coordinates,

$$ds^2 = dr^2 + r^2(d\theta^2 + \sin^2\theta d\phi^2).$$ (4.57)

4.6 Orthonormal basis

For an arbitrary metric one can introduce a field of basis-vectors consisting of *orthogonal unit vectors*. Such a basis is called an *orthonormal basis*, and fulfills

$$\mathbf{e}_{\hat{\mu}} \cdot \mathbf{e}_{\hat{\nu}} = \eta_{\hat{\mu}\hat{\nu}} = \begin{cases} \pm 1 & \text{, for } \hat{\mu} = \hat{\nu} \\ 0 & \text{, for } \hat{\mu} \neq \hat{\nu}. \end{cases}$$ (4.58)

The components of the metric tensor relative to an orthonormal basis, are

$$g_{\hat{\mu}\hat{\nu}} = \text{diag}(-1,\ldots,-1,1,\ldots,1).$$ (4.59)

The sum of the diagonal components, $g_{\mu\mu}$, is called the *signature* of the metric tensor, and is denoted by $\text{sgn}(g)$. We usually only write the sign of the diagonal elements for the signature of the metric. If the signature for a space is $(++\ldots+)$, then we call the space *Riemannian*. If the signature is $(-++\ldots+)$, then we call the space *Lorentzian*. Hence, spacetime is a Lorentzian space, while the spatial surfaces are Riemannian.

In an Euclidean space one can introduce a Cartesian coordinate-system, with an orthonormal coordinate basis vector field. The components of the metric tensor are then

$$g_{\hat{\mu}\hat{\nu}} = \delta_{\hat{\mu}\hat{\nu}},$$ (4.60)

or, in matrix form

$$\mathbf{g} = 1,$$ (4.61)

where 1 is the unit matrix.

A transformation matrix, M_C, between two Cartesian coordinate systems, must satisfy

$$1 = M_C^T \cdot 1 \cdot M_C,$$ (4.62)

which requires

$$M_C^T = M_C^{-1}.$$ (4.63)

Thus the transformation matrices between Cartesian coordinate systems are orthogonal. These transformation matrices form a group called the orthonormal group.

In the following three examples we shall consider a two-dimensional Euclidean plane, with a system of plane polar coordinates. Some differences between the coordinate basis vectors of this system and the corresponding orthonormal basis vector field, are demonstrated.

Examples

Example 4.7 (Basis vector field in a system of plane polar coordinates)
In Example 4.2 it was shown that the coordinate basis vectors of the polar coordinate system, as decomposed in a Cartesian coordinate system, are

$$\begin{aligned} \mathbf{e}_r &= \cos\theta\,\mathbf{e}_x + \sin\theta\,\mathbf{e}_y, \\ \mathbf{e}_\theta &= -r\sin\theta\,\mathbf{e}_x + r\cos\theta\,\mathbf{e}_y. \end{aligned}$$

The components of the metric tensor are

$$g_{rr} = \mathbf{e}_r \cdot \mathbf{e}_r = 1, \quad g_{\theta\theta} = \mathbf{e}_\theta \cdot \mathbf{e}_\theta = r^2, \quad g_{r\theta} = \mathbf{e}_r \cdot \mathbf{e}_\theta = 0.$$

This gives the line-element

$$ds^2 = dr^2 + r^2 d\theta^2,$$

which represents the Pythagorean theorem as expressed in plane polar coordinates.
The absolute values of the coordinate basis vectors are

$$|\mathbf{e}_r| = (\mathbf{e}_r \cdot \mathbf{e}_r)^{1/2} = 1, \quad |\mathbf{e}_\theta| = (\mathbf{e}_\theta \cdot \mathbf{e}_\theta)^{1/2} = r.$$

Thus $\mathbf{e}_\theta$ is not a unit vector.
The corresponding orthonormal basis field is

$$\mathbf{e}_{\hat{r}} = \mathbf{e}_r, \quad \mathbf{e}_{\hat{\theta}} = \frac{1}{r}\mathbf{e}_\theta.$$

Example 4.8 (Velocity field in plane polar coordinates)
As decomposed in a coordinate system with plane polar coordinates (r, θ) the velocity vector, $\mathbf{v}$, of a particle is

$$\mathbf{v} = v^\mu \mathbf{e}_\mu = v^r \mathbf{e}_r + v^\theta \mathbf{e}_\theta,$$

where

$$v^r = \frac{dr}{dt}, \quad v^\theta = \frac{d\theta}{dt},$$

are the *coordinate components* of the velocity.

These components do not have the same dimension. While v^r is a velocity, the component v^θ is an angular velocity.

A common dimension of the velocity components is obtained if the velocity vector is decomposed in an orthonormal basis field

$$\mathbf{v} = v^{\hat{\mu}} \mathbf{e}_{\hat{\mu}} = v^{\hat{r}} \mathbf{e}_{\hat{r}} + v^{\hat{\theta}} \mathbf{e}_{\hat{\theta}},$$

where

$$v^{\hat{r}} = \frac{dr}{dt}, \quad v^{\hat{\theta}} = r\frac{d\theta}{dt}.$$

The component $v^{\hat{\theta}}$ is the velocity in the $\mathbf{e}_\theta$-direction. The "physical components" $v^{\hat{r}}$ and $v^{\hat{\theta}}$ both have the dimension length/time.

The physical meaning of a calculation is often easier to see in an orthonormal basis than in a coordinate basis.

Example 4.9 (Structure coefficients of an orthonormal basis field associated with plane polar coordinates)
(See example 4.7)

$$
\begin{aligned}
[\mathbf{e}_{\hat{r}}, \mathbf{e}_{\hat{\theta}}] = [\mathbf{e}_r, \frac{1}{r}\mathbf{e}_\theta] &= [\frac{\partial}{\partial r}, \frac{1}{r}\frac{\partial}{\partial\theta}] \\
&= \frac{\partial}{\partial r}\left(\frac{1}{r}\frac{\partial}{\partial\theta}\right) - \frac{1}{r}\frac{\partial}{\partial\theta}\left(\frac{\partial}{\partial r}\right) \\
&= -\frac{1}{r^2}\frac{\partial}{\partial\theta} + \frac{1}{r}\frac{\partial^2}{\partial r\partial\theta} - \frac{1}{r}\frac{\partial^2}{\partial\theta\partial r} \\
&= -\frac{1}{r^2}\mathbf{e}_\theta = -\frac{1}{r}\mathbf{e}_{\hat{\theta}} = c^{\hat{\theta}}{}_{\hat{r}\hat{\theta}}\mathbf{e}_{\hat{\theta}},
\end{aligned}
$$

where eq. (4.21) has been used. This gives

$$c^{\hat{\theta}}{}_{\hat{r}\hat{\theta}} = -c^{\hat{\theta}}{}_{\hat{\theta}\hat{r}} = -\frac{1}{r}.$$

Spacetime is four-dimensional. An orthonormal basis $\{e_{\hat{\mu}}\}$ for spacetime has one time-like vector, $e_{\hat{t}}$, and three space-like vectors, $e_{\hat{i}}$, $\hat{i} = 1, 2, 3 \ldots$. Such a basis, and the corresponding one-form basis, will be called a *tetrad*.

The metric tensor of an arbitrary tetrad is denoted by η and its components are given by

$$\eta_{\hat{\mu}\hat{\nu}} = \text{diag}(-1, 1, 1, 1). \tag{4.64}$$

A transformation Λ between two tetrads must fulfill the equation

$$\boxed{\eta = \Lambda^T \eta \Lambda.} \tag{4.65}$$

These are just the *Lorentz transformations*.

An arbitrary coordinate transformation, $x^\mu = x^\mu(x^{\mu'})$, is in general different from the corresponding transformation of vector components, eq. (4.31). Coordinates *do not* in general transform like vector components. However, from the chain-rule for differentiation follows that the coordinate *differentials* transform as vector components.

In the special case of linear transformations with constant elements of the transformation matrix, the coordinates themselves transform like the coordinate differentials. This is the case for the Lorentz transformations.

4.7   Spatial geometry

Three fundamental kinematical concepts are *position, direction,* and *motion*. To each of these concepts there correspond an independent type of reference. The position of a particle is referred to a *coordinate system*. The direction of a rod is referred to a *basis vector field*, and the motion of a body is referred to a *reference frame*.

These types of reference can be introduced in a physical description independently of each other, and there are several sorts of each type. What type, and which sort of reference that one introduces, is a matter of convenience. In general relativity there is a much used alternative to introduce basis vectors, namely to use basis one-forms. This corresponds to characterizing the direction of a rod by the plane normal to the rod.

A coordinate system K covering a region of spacetime, is a continuum of four variables $\{x^\mu\}$ that uniquely label every event in the region. We define a reference frame R as a continuum of non-crossing time-like or light-like curves in spacetime. According to this definition a reference frame may be thought of as a continuum of world-lines of particles, called reference particles or observers.

A *comoving coordinate system* in a reference frame is defined by the requirement that the reference particles of the frame have constant spatial coordinates.

In general the observers of a frame need not move freely. For example the observers of a hyperbolically accelerated reference frame in flat spacetime are not inertial. However, the kinematical properties of cosmological models, their expansion, shear and rotation, usually are defined with reference to observers that move freely. This class of observers will be called *inertial observers*.

An orthonormal tetrad field can be associated with a reference frame R, where e_0 is the unit tangent vector field of the world lines of the fundamental observers in R. In other words $u = e_0$, where u is the four-velocity field of these observers.

A set of simultaneous events, as measured by Einstein-synchronized clocks at rest relative to an observer, defines a local 3-dimensional space, which we call *the rest space of the observer*. This space is orthogonal to the four-velocity vector u of the observer.

We shall now describe the spatial geometry in an arbitrary reference frame R. Let $\mathbf{e}_0$ be a tangent vector field to the world lines of the fundamental observers of R. The space-like basis vectors $\mathbf{e}_i$ will not in general be orthogonal to $\mathbf{e}_0$. Let the vectors $\mathbf{e}_{i\perp}$ be the projections of $\mathbf{e}_i$ orthogonal to $\mathbf{e}_0$, i.e.,

$$\mathbf{e}_{i\perp} \cdot \mathbf{e}_0 = 0. \tag{4.66}$$

The *spatial metric tensor* is defined by

$$\gamma_{ij} = \mathbf{e}_{i\perp} \cdot \mathbf{e}_{j\perp}, \quad \gamma_{i0} = 0, \quad \gamma_{00} = 0. \tag{4.67}$$

Since

$$\mathbf{e}_{i\perp} = \mathbf{e}_i - \mathbf{e}_{i\parallel}, \tag{4.68}$$

where

$$\mathbf{e}_{i\parallel} = \frac{\mathbf{e}_i \cdot \mathbf{e}_0}{\mathbf{e}_0 \cdot \mathbf{e}_0} \mathbf{e}_0 = \frac{g_{i0}}{g_{00}} \mathbf{e}_0, \tag{4.69}$$

we get

$$\boxed{\gamma_{ij} = g_{ij} - \frac{g_{i0}g_{j0}}{g_{00}}.} \tag{4.70}$$

The *spatial line-element* is defined by

$$\boxed{dl^2 = \gamma_{ij} dx^i dx^j.} \tag{4.71}$$

Consider a transformation of the form

$$x^0 = x^0(x^{\mu'}), \quad x^i = x^i(x^{i'}), \tag{4.72}$$

where $\mu = 0, 1, 2, 3$ and $i = 1, 2, 3$. Using eq. (4.28), transforming $g_{\mu\nu}$ according to eq. (4.47) and noting that $M^i_{0'} = 0$ for the transformation (4.72), we get

$$\gamma_{i'j'} = M^i_{i'} M^j_{j'} \gamma_{ij}. \tag{4.73}$$

This shows that the quantities γ_{ij} transform as tensor components under a transformation of the form (4.72).

From the transformation (4.72) we find

$$\frac{\partial}{\partial x^{0'}} = \frac{\partial x^\mu}{\partial x^{0'}} \frac{\partial}{\partial x^\mu} = \frac{\partial x^0}{\partial x^{0'}} \frac{\partial}{\partial x^0} \tag{4.74}$$

since $\partial x^i / \partial x^{0'} = 0$. Thus

$$\mathbf{e}_{0'} = \frac{\partial x^0}{\partial x^{0'}} \mathbf{e}_0, \tag{4.75}$$

showing that $\mathbf{e}_{0'}$, is parallel to $\mathbf{e}_0$. It follows that the four-velocity fields of particles with fixed coordinates in two coordinate systems connected by a transformation of the form (4.72), are identical. Eq. (4.72) thus represents coordinate transformations between different comoving coordinate systems in a single reference frame R. Such coordinate transformations are called *internal coordinate transformations*.

Eq. (4.73) shows that the spatial metric tensor transforms like a tensor, and the spatial line-element is invariant under internal coordinate transformations.

The line-element of spacetime may be written

$$ds^2 = -d\hat{t}^2 + dl^2, \tag{4.76}$$

where

$$d\hat{t} = \sqrt{-g_{00}}\left[dt + \frac{g_{i0}}{g_{00}}dx^i\right]. \tag{4.77}$$

Here, $d\hat{t} = 0$ represents the local 3-dimensional space of simultaneity orthogonal to e_0. Eq. (4.76) shows that the spatial metric tensor describes the geometry of this space.

The difference in coordinate time dt of two simultaneous events, $d\hat{t} = 0$, with spatial coordinates x^i and $x^i + dx^i$ is

$$dt = -\frac{g_{i0}}{g_{00}}dx^i, \tag{4.78}$$

which generally is not an exact differential[2]. In general the line integral of dt around a closed path will not vanish. This means that one cannot always synchronize clocks along closed path, or globally in space. However, if $g_{i0} = 0$, $i = 1, 2, 3$, then this *is* possible.

We have seen that if $g_{i0} \neq 0$, there does not exist a single space of simultaneity encompassing the "rest spaces" of all observers in an arbitrary reference frame. In this sense the 3-dimensional space described by the spatial metrical tensor is local.

4.8 The tetrad field of a comoving coordinate system

Let K be a comoving coordinate system of a reference frame R, with a coordinate basis vector field $\{e_\mu\}$. It is assumed that the space-like vectors $\{e_i\}$ are orthogonal to each other, but not necessarily to the vector e_0. We shall find a tetrad-field $\{e_{\hat{\mu}}\}$ so that $e_{\hat{0}}$ is parallel to e_0. The vector $e_{\hat{0}}$ is the four-velocity of the reference particles of R.

Since the absolute value of e_0 is $(-g_{00})^{1/2}$ the vector e_0 is given by

$$e_{\hat{0}} = (-g_{00})^{-1/2}e_0. \tag{4.79}$$

From eqs. (4.68) and (4.69) follow

$$e_{i\perp} = e_i - \frac{g_{i0}}{g_{00}}e_0. \tag{4.80}$$

According to eq. (4.67) the absolute value of $e_{i\perp}$ is $\gamma_{ii}^{1/2}$, thus the ith vector of the tetrad is given by

$$e_{\hat{i}} = \gamma_{ii}^{-1/2}\left[e_i - \frac{g_{i0}}{g_{00}}e_0\right]. \tag{4.81}$$

Another space-like vector $e_{\hat{j}}$ of the tetrad is chosen so that $e_{\hat{j}} \cdot e_{\hat{i}} = 0$, $e_{\hat{j}} \cdot e_{\hat{0}} = 0$. The last one is given by $e_{\hat{k}} = e_{\hat{i}} \times e_{\hat{j}}$, where $\times$ denotes the vector product.

---

[2]What we mean by exact differential will be more rigorously defined in chapter 6.

The corresponding orthonormal form-basis is given by

$$\omega^{\hat{\mu}}(\mathbf{e}_{\hat{\nu}}) = \delta^{\hat{\mu}}_{\hat{\nu}}, \tag{4.82}$$

giving

$$\omega^{\hat{0}} = \sqrt{-g_{00}} \left[\omega^0 + \frac{g_{i0}}{g_{00}}\omega^i\right], \tag{4.83}$$

$$\omega^{\hat{i}} = \sqrt{\gamma_{ii}}\omega^i. \tag{4.84}$$

Applying a basis-form ω^μ to an infinitesimal displacement-vector

$$d\mathbf{r} = dr^\nu \mathbf{e}_\nu, \tag{4.85}$$

leads to

$$dr^\mu = \omega^\mu(d\mathbf{r}). \tag{4.86}$$

If $\mathbf{e}_\nu$ is a coordinate basis vector, i.e., $\mathbf{e}_\nu$ is a tangent vector to a coordinate curve, then $dr^\mu = dx^\mu$.

The components of a tensor relative to an orthonormal basis are called the tetrad components of the tensor. They are invariant under an internal coordinate transformation that does not change the orientation of the space-like basis vectors, but Lorentz transform when the reference frame is changed.

The tetrad components of a basis vector $\mathbf{e}_\nu$ are denoted by $e^{\hat{\mu}}_\nu$, and are given by

$$\boxed{\mathbf{e}_\nu = e^{\hat{\mu}}_\nu \mathbf{e}_{\hat{\mu}}.} \tag{4.87}$$

It follows that the metric tensor of an arbitrary basis $\{\mathbf{e}_\mu\}$ are given in terms of the tetrad components as

$$g_{\mu\nu} = e^{\hat{\alpha}}_\mu e^{\hat{\beta}}_\nu \eta_{\hat{\alpha}\hat{\beta}}. \tag{4.88}$$

4.9   The volume form

The antisymmetric *Levi-Civitá symbol* is defined by

$$\varepsilon_{\mu_1\ldots\mu_n} = \text{sgn}(g)\varepsilon^{\mu_1\cdots\mu_n} = \begin{cases} 1 & \text{if } \mu_1\ldots\mu_n \text{ is an even permutation of } 1\ldots n \\ -1 & \text{if } \mu_1\ldots\mu_n \text{ is an odd permutation of } 1\ldots n \\ 0 & \text{otherwise.} \end{cases} \tag{4.89}$$

It follows that $\varepsilon_{\mu_1\ldots\mu_n} = 0$ if two indices are equal.

The determinant of an $n \times n$-matrix A with elements $A^{\mu\nu}$ may be written

$$A = \det(\mathbf{A}) = \varepsilon_{\mu_1\ldots\mu_n} A^{1\mu_1} A^{2\mu_2} \ldots A^{n\mu_n}. \tag{4.90}$$

For example, for $n = 2$ this equation gives

$$A = \varepsilon_{\mu_1\mu_2} A^{1\mu_1} A^{2\mu_2} = \varepsilon_{12} A^{11} A^{22} + \varepsilon_{21} A^{12} A^{21} = A^{11} A^{22} - A^{12} A^{21}. \tag{4.91}$$

We shall now consider an n-dimensional space with a metric tensor. Let $\{\omega^{\hat{\mu}}\}$ be a tetrad basis of one-forms. The *volume form* ϵ is defined by

$$\epsilon = \omega^{\hat{1}} \wedge \ldots \wedge \omega^{\hat{n}}. \tag{4.92}$$

Let $(M^{\mu}_{\hat{\mu}})$ be the transformation matrix to arbitrary basis $\omega^{\mu} = M^{\mu}_{\hat{\mu}}\omega^{\hat{\mu}}$. Then

$$
\begin{aligned}
\epsilon &= M^{\hat{1}}_{\mu_1}\ldots M^{\hat{n}}_{\mu_n}\omega^{\mu_1}\wedge\ldots\wedge\omega^{\mu_n} \\
&= M^{\hat{1}}_{\mu_1}\ldots M^{\hat{n}}_{\mu_n}\varepsilon^{\mu_1\ldots\mu_n}\omega^1\wedge\ldots\wedge\omega^n \\
&= M\omega^1\wedge\ldots\wedge\omega^n,
\end{aligned}
\tag{4.93}
$$

where M is the determinant of the transformation matrix.

Since the determinant of a matrix is equal to the determinant of the transposed matrix (rows and columns interchanged), the transformation equation (4.48) for the components of the metric tensor leads to the determinant equation

$$
g = M^2\hat{g},
\tag{4.94}
$$

where $\hat{g}$ is the determinant of the metric tensor relative to a tetrad basis. From eq. (4.59) follows that $\hat{g} = \pm 1$, where the sign depends upon the signature of g.

Inserting the positive square root of M from eq. (4.94) into eq. (4.93), the volume form can be written

$$
\boxed{\epsilon = \sqrt{|g|}\omega^1\wedge\ldots\wedge\omega^n = \frac{1}{n!}\sqrt{|g|}\varepsilon_{\mu_1\ldots\mu_n}\omega^{\mu_1}\wedge\ldots\wedge\omega^{\mu_n},}
\tag{4.95}
$$

where $|g|$ is the absolute value of the determinant of the metric tensor. The volume form describes an *oriented* n-dimensional parallelepiped. If the orientation of the vector basis is changed, so that for example e_1 and e_2 are exchanged, then the sign of ϵ is changed. A transformation that does not change the sign of ϵ preserves the orientation of the basis, or in the case of coordinate basis, of the coordinate system. The tensor components of the volume form are

$$
\epsilon_{\mu_1\ldots\mu_n} = |g|^{1/2}\varepsilon_{\mu_1\ldots\mu_n}.
\tag{4.96}
$$

The volume form represents an invariant volume element. The corresponding invariant distance in the $\mu\mu$-direction is

$$
\epsilon_{\mu} = \sqrt{|g_{\mu\mu}|}\omega^{\mu}.
\tag{4.97}
$$

4.10   Dual forms

Let the p-vector $\mathbf{A}$ have contravariant components found by raising the indices of a p-form α. The *dual* of the form α in an n-dimensional space is designated by $\star\alpha$ and is defined as the contraction of ϵ with $\mathbf{A}$,

$$
\boxed{\star\alpha = \iota_{\mathbf{A}}\epsilon.}
\tag{4.98}
$$

The star $\star$ is called *Hodge's star operator* . From the definitions (4.98) and (3.66) follows that $\star\alpha$ is a $(n-p)$-form given by

$$
\boxed{\star\alpha = \frac{1}{p!(n-p)!}\epsilon_{\nu_1\ldots\nu_p\mu_1\ldots\mu_{n-p}}\alpha^{\nu_1\ldots\nu_p}\omega^{\mu_1}\wedge\ldots\wedge\omega^{\mu_{n-p}}.}
\tag{4.99}
$$

The dual of an orthogonal basis p-form is

$$
\star(\omega^{\nu_1}\wedge\ldots\wedge\omega^{\nu_p}) = \frac{1}{(n-p)!}\sqrt{|g|}g_p^{-1}\varepsilon_{\nu_1\ldots\nu_p\mu_1\ldots\mu_{n-p}}\omega^{\mu_1}\wedge\ldots\wedge\omega^{\mu_{n-p}},
\tag{4.100}
$$

where g_p is the determinant of the metric tensor associated with the space of the p-form α, and g is the determinant of the metric in the n-dimensional space.

Example 4.10 (Spherical coordinates in Euclidean 3-space)   Example
The transformation from spherical coordinates (r, θ, ϕ) to Cartesian coordinates (x, y, z)
is

$$x = r \cos \phi \sin \theta, \quad y = r \sin \phi \sin \theta, \quad z = r \cos \theta.$$

By differentiation one finds the basis vectors

$$
\begin{aligned}
\mathbf{e}_r &= \mathbf{e}_x \sin \theta \cos \phi + \mathbf{e}_y \sin \theta \sin \phi + \mathbf{e}_z \cos \theta, \\
\mathbf{e}_\theta &= \mathbf{e}_x r \cos \theta \cos \phi + \mathbf{e}_y r \cos \theta \sin \phi - \mathbf{e}_z r \sin \theta, \\
\mathbf{e}_\phi &= -\mathbf{e}_x r \sin \phi \sin \theta + \mathbf{e}_y r \cos \phi \sin \theta.
\end{aligned}
$$

From eq. (4.43) we now find the non-vanishing components of the metric tensor

$$g_{rr} = 1, \quad g_{\theta\theta} = r^2, \quad g_{\phi\phi} = r^2 \sin^2 \theta.$$

The line-element takes the form

$$dl^2 = dr^2 + r^2 d\theta^2 + r^2 \sin^2 \theta d\phi^2,$$

so the volume form is

$$\epsilon = r^2 \sin \theta \omega^r \wedge \omega^\theta \wedge \omega^\phi.$$

The dual of a one-form ω^ν is

$$\star\omega^\nu = \frac{1}{2} |g|^{1/2} |g_\nu|^{-1} \epsilon_{\nu\mu_1\mu_2} \omega^{\mu_1} \wedge \omega^{\mu_2},$$

with

$$g_r = 1, \quad g_\theta = r^2, \quad g_\phi = r^2 \sin \theta.$$

Letting $(x^1, x^2, x^3) = (r, \theta, \phi)$ gives

$$\star\omega^r = r^2 \sin \theta \varepsilon_{123} \omega^2 \wedge \omega^3 = r^2 \sin \theta \omega^\theta \wedge \omega^\phi,$$

and, in the same way,

$$
\begin{aligned}
\star\omega^\theta &= \sin \theta \omega^\phi \wedge \omega^r, \\
\star\omega^\phi &= \omega^r \wedge \omega^\theta.
\end{aligned}
$$

The double dual is given by

$$\star\star\alpha = \hat{g}(-1)^{p(n-p)} \alpha. \tag{4.101}$$

Hence, the double dual operator is the identity up to a sign:

$$\star^2 = \star\star = \pm 1. \tag{4.102}$$

The dual of the volume form is

$$\star\epsilon = \frac{1}{n!} \epsilon_{\mu_1 \ldots \mu_n} \epsilon^{\mu_1 \ldots \mu_n} = \hat{g} = \pm 1. \tag{4.103}$$

Note that the equations (4.101) and (4.103) gives a useful expression for the
volume form:

$$\boxed{\epsilon = \star 1.} \tag{4.104}$$

Let α and β be p-forms with corresponding vectors $\mathbf{A}$ and $\mathbf{B}$ respectively.
Then

$$(\star\alpha) \wedge \beta = \frac{1}{p!} \alpha^{\mu_1 \ldots \mu_p} \beta_{\mu_1 \ldots \mu_p} \epsilon_{1 \ldots n} \omega^1 \wedge \ldots \wedge \omega^n = (\mathbf{A} \cdot \mathbf{B})\epsilon. \tag{4.105}$$

0-form:	3-form:
$\phi = \phi$	$\star\phi : (\star\phi)_{123} = \sqrt{g}\phi$
1-form:	2-form:
$\mathbf{E} : [E_1, E_2, E_3]$	$\star\mathbf{E} : \sqrt{g}\begin{bmatrix} 0 & E^3 & -E^2 \\ -E^3 & 0 & E^1 \\ E^2 & -E^1 & 0 \end{bmatrix}$
2-form:	1-form:
$\mathbf{B} : \begin{bmatrix} 0 & B_{12} & -B_{31} \\ -B_{12} & 0 & B_{23} \\ B_{31} & -B_{23} & 0 \end{bmatrix}$	$\star\mathbf{B} : \sqrt{g}[B^{23}, B^{31}, B^{12}]$
3-form:	0-form:
$\mathbf{G} : (\mathbf{G})_{123} = G$	$\star\mathbf{G} = g^{-\frac{1}{2}}G$

Table 4.1: Dual forms in 3-dimensional space with $g > 0$.

Furthermore,

$$(\star\alpha) \wedge \beta = \alpha \wedge (\star\beta). \qquad (4.106)$$

The following connection for $n = 3$ between the wedge product of one forms and the vector product of vectors should be noted

$$\star(\alpha \wedge \beta) = \iota_{\mathbf{A}\wedge\mathbf{B}}\epsilon = \frac{1}{2}\epsilon_{\nu\lambda\mu}(\mathbf{A} \wedge \mathbf{B})^{\nu\lambda}\omega^\mu = (\mathbf{A} \times \mathbf{B})_\mu\omega^\mu. \qquad (4.107)$$

See Tables 4.1 and 4.2 for examples of dual forms in 3 and 4 dimensions.

Problems

4.1. *Coordinate-transformations in a two-dimensional Euclidean plane*
In this problem we will investigate vectors $\mathbf{x}$ in the two-dimensional Euclidean plane $\mathbb{E}^2$. The set $\{\mathbf{e}_m | m = x, y\}$ is an orthonormal basis in $\mathbb{E}^2$, i.e.

$$\mathbf{e}_m \cdot \mathbf{e}_n = \delta_{mn}.$$

The components of a vector $\mathbf{x}$ in this basis is given by x and y, or x^m:

$$\mathbf{x} = x^m\mathbf{e}_m = x\mathbf{e}_x + y\mathbf{e}_y.$$

A skew basis set, $\{\mathbf{e}_\mu | \mu = 1, 2\}$, is also given. In this basis

$$\mathbf{x} = x^\mu\mathbf{e}_\mu = x^1\mathbf{e}_1 + x^2\mathbf{e}_2.$$

The transformation between these to coordinates are

$$\begin{aligned} x^1 &= 2x - y, \\ x^2 &= x + y. \end{aligned}$$

0-form:	4-form:
$\phi = \phi$	$\star\phi : (\star\phi)_{0123} = \sqrt{-g}\phi$

1-form:	3-form:
$\mathbf{A} : [A_0, A_1, A_2, A_3]$	$\star\mathbf{A} : (\star\mathbf{A})_{012} = -\sqrt{-g}A^3$ etc.

2-form:	2-form:
$\mathbf{F} : \begin{bmatrix} 0 & F_{01} & F_{02} & F_{03} \\ -F_{01} & 0 & F_{12} & F_{13} \\ -F_{02} & -F_{12} & 0 & F_{23} \\ -F_{03} & -F_{13} & -F_{23} & 0 \end{bmatrix}$	$\star\mathbf{F} : \sqrt{-g}\begin{bmatrix} 0 & F^{23} & -F^{13} & F^{12} \\ -F^{23} & 0 & F^{03} & -F^{02} \\ F^{13} & -F^{03} & 0 & F^{01} \\ -F^{12} & F^{02} & -F^{01} & 0 \end{bmatrix}$

3-form:	1-form:
$\mathbf{G} : (\mathbf{G})_{\alpha\beta\gamma}$	$\star\mathbf{G} : \sqrt{-g}[-G^{123}, G^{230}, -G^{301}, G^{012}]$

4-form:	0-form:
$\mathbf{H} : (\mathbf{H})_{0123} = H$	$\star\mathbf{H} = -(-g)^{-1/2}H$

Table 4.2: Dual forms in 4-dimensional space with $g < 0$.

(a) Find $\mathbf{e}_1$ and $\mathbf{e_2}$ expressed in terms of $\mathbf{e}_x$ and $\mathbf{e}_y$. Determine the transformation matrix M, defined by

$$x^m = M^m_\mu x^\mu.$$

What is M^{-1}?

(b) The metric tensor **g** is given by

$$ds^2 = g_{\mu\nu}dx^\mu dx^\nu = g_{mn}dx^m dx^n,$$

where ds is the distance between $\mathbf{x}$ and $\mathbf{x} + \mathbf{dx}$. Show that we have

$$\mathbf{e}_\mu \cdot \mathbf{e}_\nu = g_{\mu\nu}.$$

What is the relation between the matrices $(g_{\mu\nu})$ and (g_{mn}) and the transformation matrix M?

The scalar product between two vectors can therefore be expressed as

$$\mathbf{v} \cdot \mathbf{u} = g_{\mu\nu}v^\mu u^\nu = g_{mn}v^m u^n.$$

Verify this equation for the case $\mathbf{u} = 2\mathbf{e}_1$ and $\mathbf{v} = 3\mathbf{e}_2$.

(c) Using the basis vectors e_μ, we can define a new set ω^μ by

$$\omega^\mu \cdot e_\nu = \delta^\mu_{\ \nu}.$$

Find ω^1 and ω^2 expressed in terms of e_x and e_y. Why is $\omega^m = e_m$, while $\omega^\mu \neq e_\mu$?

A vector $\mathbf{x}$ can now be expressed as

$$\mathbf{x} = x^\mu e_\mu = x_\mu \omega^\mu.$$

What is the relation between the contravariant components x^μ and the covariant components x_μ? Determine both set of components for the vector $\mathbf{A} = 3e_x + e_y$.

In a (x, y)-diagram, draw the the three set of basis vectors $\{e_\mu\}$, $\{e_m\}$ and $\{\omega_\mu\}$. What is the geometrical interpretation of the relation between the two sets $\{e_\mu\}$ and $\{\omega_\mu\}$? Depict also the vector $\mathbf{A}$ and explain how the components of $\mathbf{A}$ in the three basis sets can be seen from the diagram.

(d) Find the matrix $(g^{\mu\nu})$ defined by

$$\omega^\mu \cdot \omega^\nu = g^{\mu\nu}.$$

Verify that this matrix is the inverse to $(g_{\mu\nu})$.

The metric tensor is a symmetric tensor of rank 2, and can therefore be expressed with the basis vectors $e_m \otimes e_n$ in the tensor product space $\mathbb{E}^2 \otimes \mathbb{E}^2$,

$$\mathbf{g} = g_{mn} e_m \otimes e_n.$$

Show that we also can express it as

$$\mathbf{g} = g_{\mu\nu} \omega^\mu \otimes \omega^\nu,$$

and

$$\mathbf{g} = g^{\mu\nu} e_\mu \otimes e_\nu.$$

What is the dimension of the space spanned by the vectors $e_m \otimes e_n$?

The antisymmetric tensors span a one-dimensional subspace. Show this by showing that an antisymmetric tensor A_{mn} is a linear combination of the basis vector

$$e_x \wedge e_y = e_x \otimes e_y - e_y \otimes e_x$$

Find $\mathbf{u} \wedge \mathbf{v}$ where $\mathbf{u}$ and $\mathbf{v}$ are the vectors from (b), expressed in terms of the basis vector $e_x \wedge e_y$. What is the relation between this and the area that is spanned by $\mathbf{u}$ and $\mathbf{v}$? Calculate also $\omega^1 \wedge \omega^2$.

4.2. *Covariant and contravariant components*

(a) In a two-dimensional space the metric is given in covariant components as

$$(g_{\mu\nu}) = \begin{bmatrix} 1 & 2 \\ 2 & 3 \end{bmatrix}.$$

Find the covariant components to the vector $\mathbf{v} = 3e_1 - 4e_2$.

(b) The tensor $\mathbf{T} = T^{\mu\nu} e_\mu \otimes e_\nu$ has the contravariant components given by

$$(T^{\mu\nu}) = \begin{bmatrix} -1 & 2 \\ 0 & 3 \end{bmatrix}.$$

Calculate the mixed components $T^\mu_{\ \nu}$ and $T_\mu^{\ \nu}$ and the covariant components $T_{\mu\nu}$.

4.3. *The Levi-Civitá symbol*

The three-dimensional Levi-Civitá symbol ε_{ijk}, can be defined by

i) $\varepsilon_{xyz} = +1$,

ii) ε_{ijk} is antisymmetric in any exchange of indices.

(a) Use this calculate all of the 27 components of the Levi-Civitá symbol.

(b) Show that the Levi-Civitá symbol satisfies the following relations

$$\varepsilon_{ijk}\varepsilon^i{}_{mn} = \delta_{jm}\delta_{kn} - \delta_{jn}\delta_{km},$$
$$\varepsilon_{ijk}\varepsilon^{ij}{}_m = 2\delta_{km}.$$

(c) Show how the components to a cross-product $\mathbf{A} \times \mathbf{B}$ can be expressed with the use of ε_{ijk}, and use this, together with the properties of the Levi-Civitá symbol to calculate the following expressions:

$$\mathbf{A} \times (\mathbf{B} \times \mathbf{C}), \quad (\mathbf{A} \times \mathbf{B}) \cdot (\mathbf{C} \times \mathbf{D}), \quad (\mathbf{A} \times \mathbf{B}) \times (\mathbf{C} \times \mathbf{D}),$$
$$\nabla \times (\phi\mathbf{A}), \quad \nabla \cdot (\mathbf{A} \times \mathbf{B}), \quad \nabla \times (\mathbf{A} \times \mathbf{B}), \quad \nabla \times (\nabla \times \mathbf{A}).$$

(d) The cofactor determinant, $\mathrm{Cof}(M_{ij})$, of the matrix element M_{ij} in a 3×3-matrix M, is defined by

$$\mathrm{Cof}(M_{ij}) = \frac{1}{2}\varepsilon_{ikl}\varepsilon_{jmn}M_{km}M_{ln}.$$

Show that the inverse matrix M^{-1} is given by

$$(\mathsf{M}^{-1})_{ij} = \frac{\mathrm{Cof}(M_{ij})}{|\mathsf{M}|}.$$

4.4. *Properties of transformations of a basis*

Consider a transformation in 3-dimensional space, given in terms of the basis transformation $\mathbf{e}_m \to \mathbf{e}_\mu$:

$$\mathbf{e}_\mu = M^m{}_\mu \mathbf{e}_m.$$

(a) What is the corresponding transformation for the one-forms ω^μ defined by $\omega^\mu(\mathbf{e}_\nu) = \delta^\mu_\nu$? Find also the transformation of the components of the metric tensor $g_{\mu\nu}$ and $g^{\mu\nu}$, and the transformation of the components a^μ and a_ν for an arbitrary vector a.

(b) A quantity that is independent of any choice of basis is called an invariant. Show, using the transformations above, that

$$ds^2 = g_{\mu\nu}\omega^\mu \otimes \omega^\nu,$$

is an invariant.

The volume element dV spanned by three vectors da, db and dc, can be expressed as

$$dV \mathbf{e}_x \wedge \mathbf{e}_y \wedge \mathbf{e}_z = da^\mu db^\nu dc^\rho \mathbf{e}_\mu \wedge \mathbf{e}_\nu \wedge \mathbf{e}_\rho,$$

where $\{\mathbf{e}_x, \mathbf{e}_y, \mathbf{e}_z\}$ is an orthonormal basis. Show, using this equation, that dV is an invariant. In particular, show that when $\mathbf{da} = dx^1\mathbf{e}_1$, $\mathbf{db} = dx^2\mathbf{e}_2$, $\mathbf{dc} = dx^3\mathbf{e}_3$, we have

$$dV = \det(M^m{}_\mu)dx^1 dx^2 dx^3 = \sqrt{|g|}dx^1 dx^2 dx^3,$$

where g is the determinant of the matrix $(g_{\mu\nu})$.

(c) The fundamental contraction in a Cartesian system between the basis one-forms ω^m and the basis vectors $\mathbf{e}_n$ is $\omega^m(\mathbf{e}_n) = \delta^m_n$. Assume also that in the transformed system we have $\omega^\mu(\mathbf{e}_\nu) = \delta^\mu_\nu$. How must the basis one-forms transform?

Given the one-form σ which in the Cartesian system has components σ_m: $\sigma = \sigma_m \omega^m$. In the transformed system it has components σ_μ: $\sigma = \sigma_\mu \omega^\mu$. Show that the components transform covariantly; i.e., $\sigma_\mu = \sigma_m M^m_\mu$.

4.5. Dual forms
Let $\{\mathbf{e}_i\}$ be a Cartesian basis in the three-dimensional Euclidean space. Using a vector $\mathbf{a} = a^i \mathbf{e}_i$ there are two ways of constructing a form:

i) By constructing a one-form from its covariant components $a_j = g_{ji} a^i$:

$$\mathbf{A} = a_i \mathbf{dx}^i.$$

ii) By constructing a two-form from its dual components, defined by $\alpha_{ij} = \varepsilon_{ijk} a^k$:

$$\alpha = \frac{1}{2} \alpha_{ij} \mathbf{dx}^i \wedge \mathbf{dx}^j.$$

We write this form as $\alpha = \star\mathbf{A}$ where $\star$ means to take the dual form.

(a) Given the vectors $\mathbf{a} = \mathbf{e}_x + 2\mathbf{e}_y - \mathbf{e}_z$ and $\mathbf{b} = 2\mathbf{e}_x - 3\mathbf{e}_y + \mathbf{e}_z$. Find the corresponding one-forms $\mathbf{A}$ and $\mathbf{B}$, and the dual two-forms $\alpha = \star\mathbf{A}$ and $\beta = \star\mathbf{B}$. Find also the dual form θ to the one-form $\sigma = \mathbf{dx} - 2\mathbf{dy}$.

(b) Take the exterior product $\mathbf{A} \wedge \mathbf{B}$, and show that

$$\theta_{ij} = \varepsilon_{ijk} C^k.$$

where $\theta = \mathbf{A} \wedge \mathbf{B}$, and $\mathbf{C} = \mathbf{a} \times \mathbf{b}$. Show also that the exterior product $\mathbf{A} \wedge \star\mathbf{B}$ is given by the three-form

$$\mathbf{A} \wedge \star\mathbf{B} = (\mathbf{a} \cdot \mathbf{b}) \, \mathbf{dx} \wedge \mathbf{dy} \wedge \mathbf{dz}.$$

4.6. Wedge product
Consider the two functions $u = u(x, y)$ and $v = v(x, y)$ and assume that this is a coordinate transformation between two coordinate systems (x, y) and (u, v). Show that

$$\mathbf{du} \wedge \mathbf{dv} = J \mathbf{dx} \wedge \mathbf{dy},$$

where J is the Jacobian

$$J = \frac{\partial(u, v)}{\partial(x, y)} = \det\left(\begin{bmatrix} \frac{\partial u}{\partial x} & \frac{\partial u}{\partial y} \\ \frac{\partial v}{\partial x} & \frac{\partial v}{\partial y} \end{bmatrix}\right).$$

5

Non-inertial Reference Frames

In this chapter we shall consider some consequences of the formalism developed so far, by studying the relativistic kinematics in two types of non-inertial reference frames: the rotating reference frame and the uniformly accelerating reference frame.

5.1 Spatial geometry in rotating reference frames

Let IF be an inertial reference frame with a cylindrical coordinate system (T, R, ϑ, Z). In this system the line-element of spacetime is

$$ds^2 = -c^2 dT^2 + dR^2 + R^2 d\vartheta^2 + dZ^2. \tag{5.1}$$

A reference frame RF with cylindrical coordinates (t, r, θ, z) rotates with constant angular velocity ω relative to IF. The coordinate clocks of RF are synchronized and adjusted so that they show the same time as those in IF.

The transformation between the comoving coordinates of IF and RF is

$$t = T, \quad r = R, \quad \theta = \vartheta - \omega T, \quad z = Z. \tag{5.2}$$

Differentiating and substituting into eq. (5.1) we get the line-element

$$\boxed{ds^2 = -\left(1 - \frac{r^2 \omega^2}{c^2}\right) c^2 dt^2 + dr^2 + 2r^2 \omega\, dt d\theta + r^2 d\theta^2 + dz^2.} \tag{5.3}$$

Thus the non-vanishing components of the metric tensor are

$$g_{tt} = -c^2 \gamma^{-2}, \quad g_{rr} = 1, \quad g_{t\theta} = r^2 \omega, \quad g_{\theta\theta} = r^2, \quad g_{zz} = 1, \tag{5.4}$$

where

$$\gamma = \left(1 - \frac{r^2 \omega^2}{c^2}\right)^{-1/2}. \tag{5.5}$$

The transformation between the coordinate basis vectors of IF and RF follows from eqs. (4.26) and (4.28)

$$\mathbf{e}_t = \mathbf{e}_T + \omega\mathbf{e}_\vartheta, \quad \mathbf{e}_r = \mathbf{e}_R, \quad \mathbf{e}_\theta = \mathbf{e}_\vartheta, \quad \mathbf{e}_z = \mathbf{e}_Z \qquad (5.6)$$

Even if $t = T$ the basis vectors $\mathbf{e}_t$ and $\mathbf{e}_T$ have different directions. The vector field $\mathbf{e}_T$ is directed along the world lines of the reference particles of IF, while the vector field $\mathbf{e}_t$ is directed along the world lines of the particles of RF. The *rest-space* of IF is orthogonal to $\mathbf{e}_T$, while the *rest-space* of RF is a succession of 3-dimensional simultaneity planes locally orthogonal to $\mathbf{e}_t$. From eqs. (4.79), (4.70), (4.81) and (5.6) we find a comoving orthonormal basis field in RF,

$$\mathbf{e}_{\hat{t}} = \gamma c^{-1}\mathbf{e}_t, \quad \mathbf{e}_{\hat{r}} = \mathbf{e}_r, \quad \mathbf{e}_{\hat{\theta}} = \gamma^{-1}r^{-1}\mathbf{e}_\theta + \gamma r\omega c^{-1}\mathbf{e}_t, \quad \mathbf{e}_{\hat{z}} = \mathbf{e}_z. \qquad (5.7)$$

The simultaneity planes of RF are shown on Fig. 5.1. Inserting the expressions (5.7) into eqs. (4.67), (4.71) one finds the spatial line-element in the comoving rotating coordinate system

$$dl^2 = dr^2 + \gamma^2 r^2 d\theta^2 + dz^2. \qquad (5.8)$$

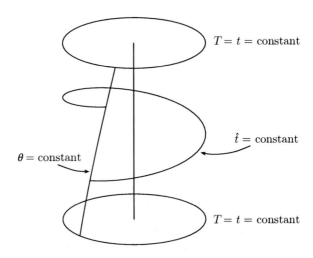

Figure 5.1: Planes of simultaneity in the rotating frame.

The distance between two points (t, r, θ, z) and $(t, r, \theta + d\theta, z)$, as measured with standard rods at rest in RF is $dl_\theta = \gamma r d\theta$. Thus the length of a circle with coordinate radius r about the axis in RF is $\gamma 2\pi r$. The distance from (t, r, θ, z) to $(t, r + dr, \theta, z)$ is $dl_r = dr$, so that the measured radius from the axis to (t, r, θ, z) is r. Thus the quotient between the measured periphery and radius is $2\gamma\pi$, which is greater than 2π. This means that the surface $d\hat{t} = 0$ (see eq. (4.77)), $z = $ constant has negative curvature (see chapter 7).

5.2   Ehrenfest's paradox

Ehrenfest formulated his paradox as follows: "Let r' be the radius of the rotating disk, as observed in the inertial frame IF, and r the radius of the disk when it is at rest. Then r' must fulfill the following two requirements:

1. The periphery of the disk must be Lorentz contracted: $2\pi r' < 2\pi r$.

2. Since the radial line is moving normally to its direction, it is not Lorentz contracted: $r' = r$.

The kinematical resolution of this paradox depends upon the relativity of simultaneity. Consider n equally spaced points around the periphery, see Fig. 5.2. In order to fulfill the requirement (1) one is to realize an acceleration

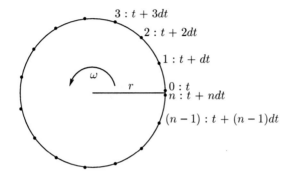

Figure 5.2: Simultaneous events in the rotating frame.

program so that the *rest* distance between the particles remain constant. Then the distance between them will be Lorentz contracted. Hence, for all pairs of neighbouring particles, at θ and $\theta + d\theta$, the two particles have to be accelerated simultaneously as observed in their instantaneous inertial rest frames, IF'. A Lorentz transformation from IF' to IF shows that, as observed in IF, the rear point is accelerated $dt = \gamma(\omega r^2/c^2)d\theta'$ earlier than the front point. Going around the periphery one finds that the point $n - 1$ is to be accelerated $\Delta t = 2\pi\gamma(\omega r^2/c^2)$ later than the point n. However, it should also be accelerated $dt = \gamma(\omega r^2/c^2)d\theta'$ earlier than the point n. So it appears that *due to the relativity of simultaneity the acceleration program that would realize an angular acceleration of the disk, while keeping the rest length between neighbouring points on the periphery constant, represents kinematically self-contradicting boundary conditions.*

Thus, the motion corresponding to the condition given in point (1) of Ehrenfest cannot be realized according to the special relativistic kinematics. This is the kinematic resolution of Ehrenfest's paradox.

We have seen that it is impossible to define locally simultaneous events for all comoving observers along the periphery of a rotating disk. This leads to an inconsistency as referred to the laboratory frame IF. If the comoving observers in RF should synchronize their clocks, they must be able to define a set of events which is locally simultaneous for each observer. Since such a set of events does not exist, *it is impossible to synchronize clocks in a rotating reference frame.*

Let us now see how the non-Euclidean spatial geometry develops as the disk is given an angular velocity. The geometry is measured by *standard measuring rods* in the "rest space" of RF. If arbitrary measuring rods are kept in a fixed position relative to an accelerated system of reference, they will generally be submitted to forces that will cause deformations of the rods. These deformations will, however, depend upon the elastic properties of the rods, and all such deformations can therefore be corrected for. In general the *standard* measuring rods are subjected to Lorentz contractions only, which means that they must move so that their rest lengths remain constant. Such motion is called Born rigid. In other words: *all standard measuring rods are assumed to perform Born rigid motions.*

In order to obtain this, n rods are assumed to rest on the disk without friction, being kept in place by a frictionless rim on the circumference of the disk, each rod being fastened to the disk at one end only, at points $P_{k'}$ so that they just cover the circumference when the disk is not rotating, as shown in Fig. 5.3.

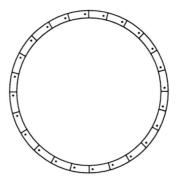

Figure 5.3: A disk at rest with the periphery covered by measuring rods.

Now we regard the process of accelerating the disk with the rods, so that it acquires an angular velocity. At a certain moment, the disk has an angular velocity ω, which is to be increased. The acceleration of the rods and the disk must be prescribed so that (*a*) the proper length L_0 of the rods remains unchanged, and (*b*) no kinematic inconsistencies result.

Condition (*a*) demands that in the instantaneous rest frame $IF_{k'}$ of each rod, every point of the rod with which this inertial frame is associated is accelerated simultaneously. According to the Lorentz transformations from $IF_{k'}$ to IF one observes in IF that the front end of each rod is accelerated at a time $(\omega r/c^2)L_0$ later than the rear end of it. Thus each rod gets an increased Lorentz contraction due to the acceleration. When the disk has an angular velocity ω, every rod is observed in IF with a length $L = L_0(1 - \omega^2 r^2/c^2)^{1/2}$.

The only isotropic way of giving the disk an angular velocity is to accelerate all points $P_{k'}$ simultaneously as measured in IF. In $IF_{k'}$ one then measures that the point $P_{k'}$ is accelerated at a point of time

$$\Delta t_{k'} = \frac{\gamma \omega r^2}{c^2} L_0 \tag{5.9}$$

earlier than the point $P_{k'+1}$. Thus the distance between these points, that is the point at the front of one measuring rod and at the front of the next, increases, as observed in $IF_{k'}$. However, as the measuring rods are moving rigidly, their proper lengths remain unchanged. Accordingly the rods separate from each other as the disk accelerates. The velocity change, as observed from $IF_{k'}$, is $(1 - \omega^2/r^2)^{-1} r d\omega$. Then the distance between two neighbouring rods increases by

$$ds_{k'} = \frac{\gamma^3 \omega r^2}{c^2} L_0 d\omega. \tag{5.10}$$

Integrating, one finds the distance between the rods, as measured in IF_k, when the disk rotates with an angular velocity ω:

$$s_{k'} = (\gamma - 1)L_0. \tag{5.11}$$

Thus the distance as measured in IF is

$$s = L_0 - L_0\sqrt{1 - \frac{\omega^2 r^2}{c^2}}, \tag{5.12}$$

in accordance with the fact that the measuring rods are Lorentz contracted, while the circumference of the disk is not. The view of the rotating disk and the measuring rods, as seen from IF, is shown in Fig. 5.4. The result of this

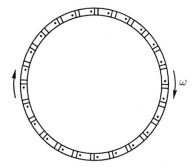

Figure 5.4: A rotating disk with measuring rods that have been Lorentz contracted.

analysis is that the ratio of the measured length of the periphery and the radius of the rotating is $2\pi\gamma$, which is consistent with the calculation based upon the spatial metric tensor.

5.3   The Sagnac effect

An emitter is placed at a position (r, θ, z) in RF. Light is emitted in the positive and negative θ-directions, and absorbed at the position of the emitter, in such a way that the light having traversed the circumference in opposite directions, interferes. By observing how the interference pattern depends upon the radius of the circular path and the angular velocity of RF, one finds that the difference in travel times of the paths is given by

$$\Delta t = \frac{4\pi\gamma^2 r^2 \omega}{c^2}. \tag{5.13}$$

This result will now be deduced in two ways: first with reference to IF, and then as described in RF.

As referred to IF the velocity of light is the same in two directions, but the absorber moves a distance $r\omega t$ where T is the travelling time. Thus the travelling times for light moving in the two directions are given by

$$2\pi r + r\omega t_1 = ct_1, \quad 2\pi r - r\omega t_2 = ct_2, \tag{5.14}$$

which gives the travelling time difference

$$\delta t = t_1 - t_2 = \frac{4\gamma^2 A\omega}{c^2},\tag{5.15}$$

where $A = \pi r^2$ is the area enclosed by the light path.

In RF the absorber is at rest, but the light moves with different velocities in the two directions. From eq. (5.3) with $ds = dr = dz = 0$, we get

$$r^2 d\theta^2 + 2r^2\omega dt d\theta - (c^2 - r^2\omega^2)dt^2 = 0.\tag{5.16}$$

The light velocities in the two directions are

$$v_\pm = r\frac{d\theta}{dt} = -r\omega \pm c,\tag{5.17}$$

which, again, leads to eq. (5.15). This is called the *Sagnac effect* [Sag13] and is a first order effect in the angular velocity.

Note also that the Sagnac effect provides an optical means by which one can measure the angular velocity of the apparatus, i.e. of the laboratory, by observations inside the laboratory. This means that in the special theory of relativity, at least, where spacetime is flat and unchangeable, angular velocity has an absolute character. In the special theory of relativity every non-accelerated observer can consider the laboratory to be at rest with respect to translational velocity, but not with respect to angular velocity. The angular velocity of the laboratory can be locally measured optically, by means of the Sagnac effect, as well as mechanically, by means of a Fouceault pendulum.

The status of angular velocity, with respect to the principle of relativity, is not so obvious in general relativity, due to the dynamical character of spacetime in this theory. The moving matter in the universe may act upon the spacetime in the laboratory in such a way that the Sagnac effect results.

5.4   Gravitational time dilatation

The coordinate clocks in RF are everywhere showing the same time as the clocks in IF. Thus the coordinate time in RF represents a position independent rate of time.

Consider now a standard clock in RF at a distance r from the axis. As observed in IF the clock moves with a velocity $r\omega$. From eq. (2.44) follows that the time shown by the clock is

$$\tau = \sqrt{1 - \frac{r^2\omega^2}{c^2}}\tau_0,\tag{5.18}$$

where τ_0 is the time shown by a standard clock at rest in IF, say at the axis of RF. The standard clocks of RF go at a slower rate, the farther they are from the axis. An observer in IF would ascribe this to the velocity dependent special-relativistic time dilatation.

However, as observed in RF, these clocks are at rest. Yet, the fact that a standard clock at $r > 0$ goes slower that a standard clock at $r = 0$, must be equally true from this point of view.

This is immediately verified from eqs. (5.3) and (2.48) with $dr = d\theta = dz = 0$, which gives

$$\boxed{d\tau = \sqrt{1 - \frac{r^2\omega^2}{c^2}}dt.}\tag{5.19}$$

Since the rate of coordinate time t is position independent, this equation is equivalent to eq. (5.18). The interpretation of eq. (5.19) however, must be quite different from that of eq. (5.18) since no velocities are involved as observed from RF. In general, *the explanation of an effect depends upon the frame of reference.*

According to Newtonian dynamics there is a centrifugal field in RF. The centrifugal field is an inertial field causing free particles in RF to accelerate away from the axis of rotation. As stated in section 1.7 the principle of equivalence says that an inertial field caused by the acceleration or rotation of the reference frame is locally equivalent to a gravitational field caused by a mass distribution. This is one of the fundamental principles of the general theory of relativity. Hence, in this theory the centrifugal field of Newtonian physics is reckoned as a genuine gravitational field.

The gravitational potential at r, with zero at the axis, is

$$\phi = - \int_0^r r\omega^2 \, dr = -\frac{1}{2} r^2 \omega^2, \tag{5.20}$$

so eq. (5.18) can be written

$$d\tau = \sqrt{1 + \frac{2\phi}{c^2}} \, d\tau_0. \tag{5.21}$$

The interpretation of this equation is that the rate of time is position dependent in a gravitational field. Since ϕ is less (more negative) farther down the field, we conclude: *time goes slower farther down in a gravitational field.* Conversely, time goes faster higher up in a gravitational field.

5.5 Uniformly accelerated reference frame

Let (T, X, Y, Z) be the Cartesian coordinates of an inertial frame IF_0. A particle moves along the X-axis with constant rest-acceleration g. It performs hyperbolic motion, as discussed in section 2.10. The position X_0 of the particle is given in terms of its proper time τ_0 by eq. (2.60):

$$1 + \frac{gX_0}{c^2} = \cosh\left(\frac{g\tau_0}{c}\right), \tag{5.22}$$

with $X_0(0) = 0$. The coordinate time T_0 at a point of time τ_0 is given by eq. (2.59)

$$\frac{gT_0}{c} = \sinh\left(\frac{g\tau_0}{c}\right), \tag{5.23}$$

with $T_0(0) = 0$.

We now introduce a field of particles rigidly comoving with the one considered above. They are reference particles of a uniformly accelerated reference frame UA, with coordinate (t, x, y, z). The position of the above particle, P_0, is $(t, 0, 0, 0)$ in this system.

The coordinate time t is defined by

$$t = \tau_0; \tag{5.24}$$

i.e., the coordinate clocks in UA show the same time as a standard clock at the spatial origin of UA. The coordinate time t represents a position independent rate of time.

Let $\mathbf{X}_0$ be the position vector of P_0. Its components in IF_0 are given by

$$\mathbf{X}_0 = \left(\frac{c^2}{g} \sinh\left(\frac{gt}{c}\right), \frac{c^2}{g}\left[\cosh\left(\frac{gt}{c}\right) - 1\right], 0, 0 \right). \qquad (5.25)$$

Consider now an event P in an instantaneous simultaneity space of P_0, as shown in Fig. 5.5. Let $\hat{\Sigma}$ be a comoving orthonormal tetrad basis for P_0. The position vector of P relative to $\hat{\Sigma}$ is orthogonal to the time-like basis vector of $\hat{\Sigma}$.

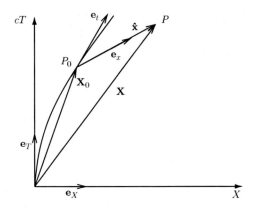

Figure 5.5: The simultaneity space of a uniformly accelerated reference frame.

Thus $\hat{\mathbf{x}} = (0, \hat{x}, \hat{y}, \hat{z})$. The spatial coordinates (x, y, z) are defined by

$$x = \hat{x}, \quad y = \hat{y}, \quad z = \hat{z}. \qquad (5.26)$$

The position vector of P is

$$\mathbf{X} = \mathbf{X}_0 + \hat{\mathbf{x}}. \qquad (5.27)$$

The connection between the basis vectors in IF and in $\hat{\Sigma}$ is given by a Lorentz transformation along the x-axis. From eq. (2.38) we have

$$\mathbf{e}_{\hat{\mu}} = \mathbf{e}_\mu \frac{\partial x^\mu}{\partial x^{\hat{\mu}}} = (\mathbf{e}_T, \mathbf{e}_X, \mathbf{e}_Y, \mathbf{e}_Z) \begin{bmatrix} \cosh\theta & \sinh\theta & 0 & 0 \\ \sinh\theta & \cosh\theta & 0 & 0 \\ 0 & 0 & 1 & 0 \\ 0 & 0 & 0 & 1 \end{bmatrix}, \qquad (5.28)$$

where θ is the rapidity of P_0 as defined in eq.(2.35). Thus

$$\tanh\theta = \frac{v_0}{c} = \frac{1}{c}\frac{dX_0}{dT_0}. \qquad (5.29)$$

Differentiation of the expressions (5.22) and (5.23) gives

$$\frac{dX_0}{dT_0} = c\tanh\left(\frac{gt}{c}\right), \qquad (5.30)$$

which shows that the rapidity of P_0 is

$$\theta = \frac{gt}{c}. \qquad (5.31)$$

The Lorentz transformation of the basis vectors may now be written

$$\mathbf{e}_{\hat{t}} = \mathbf{e}_T \cosh\left(\frac{gt}{c}\right) + \mathbf{e}_X \sinh\left(\frac{gt}{c}\right),$$

$$\mathbf{e}_{\hat{x}} = \mathbf{e}_T \sinh\left(\frac{gt}{c}\right) + \mathbf{e}_X \cosh\left(\frac{gt}{c}\right), \qquad (5.32)$$

$$\mathbf{e}_{\hat{y}} = \mathbf{e}_Y, \qquad \mathbf{e}_{\hat{z}} = \mathbf{e}_Z.$$

Substituting this into eq. (5.27) and using eq. (5.25) we find the coordinate transformation

$$\frac{gT}{c} = \left(1 + \frac{gx}{c^2}\right)\sinh\left(\frac{gt}{c}\right),$$

$$1 + \frac{gX}{c^2} = \left(1 + \frac{gx}{c^2}\right)\cosh\left(\frac{gt}{c}\right),$$

$$Y = y, \qquad Z = z. \qquad (5.33)$$

The first two equations give

$$\frac{gT}{c} = \left(1 + \frac{gX}{c^2}\right)\tanh\left(\frac{gt}{c}\right). \qquad (5.34)$$

This equation shows that in a (cT, X)-Minkowski diagram the coordinate curves $t = $ constant are straight lines through the point $(0, -c^2/g)$. Applying the identity $\cosh^2\theta - \sinh^2\theta = 1$ to the first two equations (5.33) gives

$$\boxed{\left(1 + \frac{gX}{c^2}\right)^2 - \left(\frac{gT}{c}\right)^2 = \left(1 + \frac{gx}{c^2}\right)^2} \qquad (5.35)$$

Thus, the curves $x = $ constant are hyperbolae with asymptotes $\pm cT = X + c^2/g$. The coordinate curves $t = $ constant and $x = $ constant as drawn in the (cT, X)-Minkowski diagram are shown in Fig. 5.6. The hyperbolic curves $x = $ constant are world lines of the reference particles in UA. The lines $t = $ constant are planes of simultaneity of these particles.

From the Minkowski diagram Fig. 5.6 is seen that an observer in UA cannot receive information from an emitter to the left of the asymptote $cT = X + c^2/g$. This asymptote is therefore called an *event horizon* of UA. It is located at the position $x = -c^2/g$ in UA.

Since infinitely many coordinate lines $t = $ constant pass through the point $(0, -c^2/g)$, there is a *coordinate singularity* at this point. The coordinate system cannot be continued through this point. In general, event horizons and coordinate singularities appear in comoving coordinate systems of accelerated reference frames.

Differentiating eq. (5.33) we obtain the line element

$$ds^2 = -c^2 dT^2 + dX^2 + dY^2 + dZ^2$$

$$= -\left(1 + \frac{gx}{c^2}\right)^2 c^2 dt^2 + dx^2 + dy^2 + dz^2. \qquad (5.36)$$

The "rest space" of UA, $dt = 0$, has Euclidean geometry.

The rate of time as measured on standard clocks at rest in UA is given by

$$d\tau = \left(1 + \frac{gx}{c^2}\right)dt = \left(1 + \frac{gx}{c^2}\right)d\tau_0, \qquad (5.37)$$

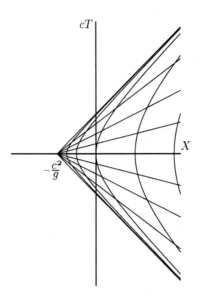

Figure 5.6: Minkowski diagram for the comoving coordinates of a uniformly
accelerated reference frame.

where τ_0 is the proper time at $x = 0$. This equation shows that $d\tau > d\tau_0$ for
$x > 0$.

Since UA accelerates in the positive x−direction an observer in UA experiences a gravitational field in the negative x−direction. Thus the x−axis points
upwards in this gravitational field. Eq. (5.37) shows that a standard clock at
$x > 0$ measures a larger time interval between two events than a clock at
$x = 0$. Thus time goes faster farther upwards in the gravitational field of UA.
This is again the position dependent rate of time a gravitational field, which
we also found in the rotating reference frame.

5.6 Covariant Lagrangian dynamics

Consider a particle which moves along a world-line between two points P_1
and P_2. Let the curve be described by an invariant parameter λ. The Lagrangian L of the particle is a function of the coordinates and their deriva-
tives with respect to λ; $L = L(x^\mu, \dot{x}^\mu)$, $\dot{x}^\mu = dx^\mu/d\lambda$. The action integral is
$S = \int L(x^\mu, \dot{x}^\mu)d\lambda$. The world line of the particle is determined by the condition that S has a stationary value for all infinitesimal variations of the curve
connecting the fixed points $P_1 = x^\mu(\lambda_1)$, $P_2 = x^\mu(\lambda_2)$. Thus the curve is determined by

$$\delta \int_{\lambda_1}^{\lambda_2} L(x^\mu, \dot{x}^\mu)d\lambda = 0, \qquad (5.38)$$

for all variations $\delta x^\mu(\lambda)$ satisfying the boundary conditions

$$\delta x^\mu(\lambda_1) = \delta x^\mu(\lambda_2) = 0. \qquad (5.39)$$

Furthermore,

$$
\delta \int\limits_{\lambda_1}^{\lambda_2} L d\lambda = \int\limits_{\lambda_1}^{\lambda_2} \left[\frac{\partial L}{\partial x^\mu} \delta x^\mu(\lambda) + \frac{\partial L}{\partial \dot{x}^\mu} \delta \dot{x}^\mu \right] d\lambda. \tag{5.40}
$$

Performing a partial integration of the last term, using that $\delta \dot{x}^\mu = d(\delta x^\mu)/d\lambda$, and taking into account the boundary condition (5.39), we obtain

$$
\delta \int\limits_{\lambda_1}^{\lambda_2} L d\lambda = \int\limits_{\lambda_1}^{\lambda_2} \left[\frac{\partial L}{\partial x^\mu} - \frac{d}{d\lambda} \left(\frac{\partial L}{\partial \dot{x}^\mu} \right) \right] \delta x^\mu(\lambda). \tag{5.41}
$$

In order that this integral shall vanish for all variations $\delta x^\mu(\lambda)$, the factor in the square bracket must be zero along the curve. The variational (5.38) thus leads to the *Euler-Lagrange's equations of motion*

$$
\boxed{\frac{d}{d\lambda} \left(\frac{\partial L}{\partial \dot{x}^\mu} \right) - \frac{\partial L}{\partial x^\mu} = 0.} \tag{5.42}
$$

The covariant momentum conjugate to a coordinate x^μ is defined by

$$
p_\mu = \frac{\partial L}{\partial \dot{x}^\mu}. \tag{5.43}
$$

The Euler-Lagrange equations can then be written as

$$
\frac{dp_\mu}{d\lambda} = \frac{\partial L}{\partial x^\mu}. \tag{5.44}
$$

If the Lagrange-function L is independent of a coordinate x^μ, then this coordinate is said to be cyclic. It follows that *the covariant momentum conjugate to a cyclic coordinate is a constant of motion.*

For material particles the parameter λ is usually chosen to be the proper time of the particle. In the case of a photon, λ is kept arbitrary.

In non-relativistic dynamics the Lagrange-function of a free particle is equal to its kinetic energy. The corresponding relativistic Lagrangian is a scalar-function depending upon the square of the particle's four-velocity, **V**.

For a material particle we choose

$$
L = \frac{1}{2} \mathbf{V} \cdot \mathbf{V} = \frac{1}{2} \dot{x}_\mu \dot{x}^\mu = \frac{1}{2} g_{\mu\nu} \dot{x}^\mu \dot{x}^\nu, \tag{5.45}
$$

where $\dot{x}^\mu = dx^\mu/d\tau$, and τ is the proper time of the particle.

Thus

$$
\frac{\partial L}{\partial \dot{x}^\beta} = g_{\beta\nu} \dot{x}^\nu, \tag{5.46}
$$

$$
\frac{\partial L}{\partial x^\beta} = \frac{1}{2} g_{\mu\nu,\beta} \dot{x}^\mu \dot{x}^\nu, \tag{5.47}
$$

and

$$
\frac{d}{d\tau} (g_{\beta\nu} \dot{x}^\nu) = g_{\beta\nu,\mu} \dot{x}^\mu \dot{x}^\nu + g_{\beta\nu} \ddot{x}^\nu. \tag{5.48}
$$

Since $\dot{x}^\mu \dot{x}^\nu$ is symmetric in μ and ν, only the symmetric part of $g_{\beta\nu,\mu}$ contributes to the first term at the right hand side of eq. (5.48). Hence, the Euler-Lagrange equations (5.42) for a free particle takes the form

$$
g_{\alpha\nu} \ddot{x}^\nu + \frac{1}{2} (g_{\alpha\mu,\nu} + g_{\alpha\nu,\mu} - g_{\mu\nu,\alpha}) \dot{x}^\mu \dot{x}^\nu = 0. \tag{5.49}
$$

Using that $\ddot{x}^\alpha = g^{\alpha\beta} g_{\beta\nu} \ddot{x}^\nu$ we get

$$\ddot{x}^\alpha + \Gamma^\alpha{}_{\mu\nu} \dot{x}^\mu \dot{x}^\nu = 0, \tag{5.50}$$

where

$$\Gamma^\alpha{}_{\mu\nu} = \frac{1}{2} g^{\alpha\beta} (g_{\beta\mu,\nu} + g_{\beta\nu,\mu} - g_{\mu\nu,\beta}). \tag{5.51}$$

are called *Christoffel symbols*. These symbols and their geometrical importance will be discussed in the next chapter.

The four-velocity identity

$$\dot{x}_\mu \dot{x}^\mu = -c^2, \tag{5.52}$$

is a first integral of the relativistic Euler-Lagrange equations for free material particles.

For photons

$$L = \frac{1}{2} \mathbf{P} \cdot \mathbf{P} = \frac{1}{2} g_{\mu\nu} P^\mu P^\nu, \tag{5.53}$$

where the four-momentum $\mathbf{P}$ is given in eq. (3.11) with

$$E = \hbar\omega, \quad \mathbf{p} = (\hbar\omega/c)\mathbf{n}. \tag{5.54}$$

Here, $\hbar$ is Planck's constant divided by 2π, ω is the frequency of the light and $\mathbf{n}$ is a unit vector in the direction of motion of the photon. The Lagrange-function for a photon may also be written as in eq. (5.45), where the dot designates differentiation with respect to a (non-vanishing) invariant parameter. In this case

$$\dot{x}_\mu \dot{x}^\mu = 0, \tag{5.55}$$

which follows from the fact that $\mathbf{P}$ of eq. (5.53) is a light-like vector.

Example

Example 5.1 (Vertical free motion in a uniformly accelerated reference frame)
In the comoving coordinate system of a uniformly accelerated reference frame, with the line-element (5.36), there is Minkowski metric at $x = 0$. Let a particle with unit rest mass be shot upwards from the origin, so that it moves in the x-direction with an initial velocity v. Then its four-velocity at $x = 0$ is

$$\mathbf{u} = (u^0, u^x, 0, 0) = \gamma(c, v, 0, 0), \quad \gamma = \frac{1}{\sqrt{1 - \frac{v^2}{c^2}}}. \tag{5.56}$$

We shall calculate the maximal height x_M reached by the particle.

The Lagrange function of the particle is

$$L = -\frac{1}{2} \left(1 + \frac{gx}{c^2}\right)^2 c^2 \dot{t}^2 + \frac{1}{2} \dot{x}^2, \tag{5.57}$$

where the dot designates differentiation with respect to the particle's proper time. From the four-velocity identity (5.52) follows

$$\dot{x}^2 = \left(1 + \frac{gx}{c^2}\right)^2 c^2 \dot{t}^2 - c^2. \tag{5.58}$$

Since t is a cyclic coordinate the covariant momentum p_t conjugate to t is a constant of motion:

$$p_t = \frac{\partial L}{c \partial \dot{t}} = -\left(1 + \frac{gx}{c^2}\right)^2 c\dot{t} = u^0. \tag{5.59}$$

Inserting this into eq. (5.58) gives

$$\dot{x} = \sqrt{\left(\frac{u^0}{1 + \frac{gx}{c^2}}\right)^2 - c^2}. \tag{5.60}$$

The maximal height is reached when $\dot{x} = 0$, giving

$$x_M = \frac{c}{g}(u^0 - c) = \frac{c^2}{g}(\gamma - 1). \tag{5.61}$$

For velocities $v \ll c$ we have

$$\gamma \approx 1 + \frac{v^2}{2c^2}, \tag{5.62}$$

giving

$$x_M \sim \frac{v^2}{2g}, \tag{5.63}$$

which is the usual, non-relativistic result.

Consider now a particle falling from rest at $x = 0$. Then $u^t = c$, so that

$$\dot{x} = c\sqrt{\left(1 + \frac{gx}{c^2}\right)^{-2} - 1}. \tag{5.64}$$

Integration gives

$$\tau = \frac{c}{g}\sqrt{1 - \left(1 + \frac{gx}{c^2}\right)^2}. \tag{5.65}$$

The proper time taken by the particle to reach the horizon at $x = -c^2/g$ is $\tau_H = c/g$, which is finite.

The coordinate time is found from

$$dt = \dot{t}d\tau = \frac{d\tau}{\left(1 + \frac{gx}{c^2}\right)^2}. \tag{5.66}$$

Differentiating the above expression for τ and integrating the resulting expression for dt leads to

$$t = \frac{c}{g}\ln\left[\frac{1 + \sqrt{1 - \left(1 + \frac{gx}{c^2}\right)^2}}{1 + \frac{gx}{c^2}}\right], \tag{5.67}$$

which gives $t(-c^2/g) = \infty$. As measured by an observer at $x = 0$, the particle takes an infinitely long time to reach the horizon.

Resolution of the twin-paradox

The twin-paradox was considered in section 2.9. Elizabeth was at home and Eva travelled to Proxima Centauri and back with a velocity $v = 0.8c$, using ten years as measured by Elizabeth, and six years as measured by her own clock. According to Elizabeth this is due to the velocity dependent time dilatation.

The principle of relativity tells, however, that Eva can consider herself as at rest and Elizabeth as the traveller. Let us see how Eva calculates her own and Elizabeth's aging during the travel.

Eva observes that the Earth and Proxima Centauri moves with a velocity $v = 0.8c$. Since the rest-distance between these bodies is $s_0 = 4$ l.y., she observes a Lorentz contracted distance $s = s_0\sqrt{1 - v^2/c^2} = 2.4$ l.y., and she ages by $t = s/v = 3$ years during Elizabeth's travel, just as Elizabeth found for her travel.

But what about Elizabeth? Eva observes that Elizabeth moves away with a velocity $v = 0.8c$ for a time $\Delta t = 3$ years as measured on her own clock. The corresponding time measured on Elizabeth's clock is $\Delta t_{\text{Elizabeth}} = \Delta t \sqrt{1 - v^2/c^2} = 1.8$ years.

Then Eva feels a gravitational field with an acceleration of gravity g. If the rest-acceleration of Eva is constant, we can associate a uniformly accelerated reference frame UA with Eva, in which Eva is an observer P_0 at the spatial origin.

Eva observes that Elizabeth (and the Earth) moves with constant velocity until she is at a distance $x_1 = 2.4$ l.y. from Eva. Then Eva experiences a gravitational field, directed away from Elizabeth. Eva is at rest in the field, but Elizabeth falls freely in it. Then the velocity of Elizabeth is retarded, she comes to rest at $x_2 = 4$ l.y., and then accelerates towards Eva.

The aging of Elizabeth as calculated by Eva, during the time that Eva experiences the gravitational field, is found by applying the equations of Example (5.1) to Elizabeth.

From eqs. (5.58) and (5.59) follow that the constant momentum conjugate to the time-coordinate is given by

$$p_t = \sqrt{c^2 + \dot{x}^2}\left(1 + \frac{gx}{c^2}\right). \tag{5.68}$$

Since $\dot{x} = 0$ for $x = x_2$ we get

$$p_t = c\left(1 + \frac{gx_2}{c^2}\right). \tag{5.69}$$

Eq. (5.60) may be written

$$d\tau = \frac{1 + \frac{gx}{c^2}}{\sqrt{p_t^2 - c^2\left(1 + \frac{gx}{c^2}\right)^2}}dx. \tag{5.70}$$

Integration from x_1 to x_2 gives

$$\tau_{1-2} = \frac{1}{g}\left[\sqrt{p_t^2 - c^2\left(1 + \frac{gx_1}{c^2}\right)^2} - \sqrt{p_t^2 - c^2\left(1 + \frac{gx_2}{c^2}\right)^2}\right]. \tag{5.71}$$

Because of eq. (5.69) the last term vanishes, so that

$$\tau_{1-2} = \frac{c}{g}\left[\left(1 + \frac{gx_2}{c^2}\right)^2 - \left(1 + \frac{gx_1}{c^2}\right)^2\right]^{1/2}, \tag{5.72}$$

which gives

$$\lim_{g \to \infty} \tau_{1-2} = \frac{1}{c}\left(x_2^2 - x_1^2\right)^{1/2}. \tag{5.73}$$

Inserting $x_1 = 2.4$ l.y. and $x_2 = 4$ l.y. gives Elizabeth's aging during the turning, as calculated by Eva

$$\delta\tau_{\text{Elizabeth}} = 2\lim_{g \to \infty} \tau_{1-2} = 6.4 \text{ l.y.} \tag{5.74}$$

Eva thus finds that Elizabeth ages by $2 \cdot 1.8$ years $+ 6.4$ years $= 10$ years during the travel, in accordance with the expectation from Elizabeth's point of view. The explanation given by Eva, that Elizabeth is older than herself when they meet again, in spite of the velocity dependent time-dilation during the outward and inward parts of the journey, is that Elizabeth ages incredibly fast during the short time Eva herself experiences the gravitational field which makes Elizabeth move back again.

Example 5.2 (The path of a photon in uniformly accelerated reference frame)   Example
Let a photon be emitted in the y-direction from the origin of the coordinate system of
section 5.5. The Lagrange function is then

$$L = -\frac{1}{2}\left(1 + \frac{gx}{c^2}\right)^2 c^2 \dot{t}^2 + \frac{1}{2}\dot{x}^2 + \frac{1}{2}\dot{y}^2, \tag{5.75}$$

where the dot designates differentiation with respect to an invariant parameter. From
eq. (5.55) then follows

$$\dot{x}^2 = \left(1 + \frac{gx}{c^2}\right)^2 c^2 \dot{t}^2 - \dot{y}^2. \tag{5.76}$$

The momentum p_t and p_y conjugate to t and y are constants of motion

$$p_t = \frac{\partial L}{c\partial \dot{t}} \quad = \quad -\left(1 + \frac{gx}{c^2}\right)^2 c\dot{t} = -c\dot{t}(0), \tag{5.77}$$

$$p_y \quad = \quad \frac{\partial L}{\partial \dot{y}} = \dot{y}. \tag{5.78}$$

Choosing the coordinate time at the origin as parameter gives $-p_t = p_y = c$, so that

$$\dot{t} = -\left(1 + \frac{gx}{c^2}\right)^{-2}, \quad \dot{y} = c. \tag{5.79}$$

Inserting these expressions into the equation for $\dot{x}$ gives

$$\dot{x} = c\frac{\sqrt{1 - \left(1 + \frac{gx}{c^2}\right)^2}}{1 + \frac{gx}{c^2}}. \tag{5.80}$$

The equation for the path of the photon is

$$\frac{dy}{dx} = \frac{\dot{y}}{\dot{x}} = \frac{1 + \frac{gx}{c^2}}{\sqrt{1 - \left(1 + \frac{gx}{c^2}\right)^2}}. \tag{5.81}$$

Integration with $y(0) = 0$ gives

$$\left(x + \frac{c^2}{g}\right)^2 + y^2 = \frac{c^4}{g^2}. \tag{5.82}$$

This shows that the photon follows a circular path as shown in Fig. 5.7. This path illustrates an interesting and non-trivial property concerning the kinematics of flat spacetime as referred to a uniformly accelerated reference frame. Even though 3-space is
Euclidean a photon starting out with velocity c in the y-direction ends up moving into
the horizon at $x = -c^2/g$ without any motion at all in the y-direction. This is possible because of the gravitational time-dilatation in a gravitational field. The velocity of
light is constant and equal to c *as measured locally*, but an observer for example at $x = 0$
will measure a decreasing light velocity, $(\dot{x}^2 + \dot{y}^2)^{1/2}/\dot{t} = (1 + gx/c^2)c$, as the photon
approaches the horizon, in accordance with redshift-measurements that show to him
that time goes slower far down in the gravitational field near the horizon.

5.7   A general equation for the Doppler effect

The four-momentum of a particle with relativistic energy E and spatial velocity **w** is given in eq. (3.11), which may be written

$$\mathbf{P} = E(c^{-1}, \mathbf{w}). \tag{5.83}$$

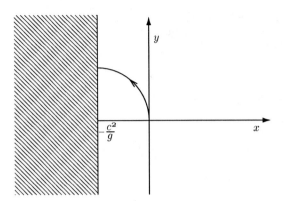

Figure 5.7: A photon path in a uniformly accelerated reference frame. The line at
$x = -\frac{c^2}{g}$, is a horizon for the observer.

Let $\mathbf{U}$ be the four-velocity of an observer. In the comoving reference frame CF
of the observer his four velocity is given by eq. (3.8), and

$$\mathbf{U} \cdot \mathbf{P} = -\hat{E}, \tag{5.84}$$

where $\hat{E}$ is the energy of the particle as measured with instruments at the
position of the observer, and at rest in CF. In short, $\hat{E} = -\mathbf{U} \cdot \mathbf{P}$ is the energy
of the particle as measured locally by the observer.

Since $\mathbf{U} \cdot \mathbf{P}$ is a scalar, the value of $\hat{E}$ may be evaluated in an arbitrary
reference frame and in arbitrary coordinates; the result is always the same as
in $\hat{\Sigma}$. However, $\mathbf{U}$ and $\mathbf{P}$ are still fixed four-vectors associated with a certain
observer and a certain particle. In this case, changing the reference frame does
not mean that the observer is changed.

Let $E_s = -(\mathbf{U} \cdot \mathbf{P})_s$ and $E_a = -(\mathbf{U} \cdot \mathbf{P})_a$ be the energy of a photon with
four-momentum $\mathbf{P}$ as seen by the source and observer (an absorber) with four-
velocities $\mathbf{U}_s$ and $\mathbf{U}_a$, respectively. One immediately has

$$\frac{E_s}{(\mathbf{U} \cdot \mathbf{P})_s} = \frac{E_a}{(\mathbf{U} \cdot \mathbf{P})_a}. \tag{5.85}$$

The frequencies ω_s and ω_a of the light, as measured at the source and observer
respectively, are given by $\omega_s = E_s/\hbar$ and $\omega_a = E_a/\hbar$, which leads to

$$\omega_a = \frac{(\mathbf{U} \cdot \mathbf{P})_a}{(\mathbf{U} \cdot \mathbf{P})_s} \omega_s. \tag{5.86}$$

This is the equation for the gravitational and kinematical Doppler effect, and
it is generally valid.

The frequency shift will now be expressed by the components of the metric
tensor, the direction of the velocity of light and the velocities of source and
observer.

The proper time interval $d\tau$ measured on a clock which moves with three-
velocity $\mathbf{v}$ in an arbitrary coordinate system Σ, where the elements of the

metric tensor are $g_{\mu\nu}$, is

$$
\begin{aligned}
d\tau &= (-g_{\mu\nu}dx^\mu dx^\nu)^{1/2} \\
&= (-g_{00} - 2g_{i0}v^i - g_{ij}v^iv^j)^{1/2}dx^0,
\end{aligned}
\tag{5.87}
$$

where $v^i = dx^i/dx^0$, and dx^0 is the coordinate time interval. The four-velocity of an observer (who may be accelerated or not) carrying the clock is

$$
\mathbf{U} = \frac{d\mathbf{x}}{d\tau} = (-g_{00} - 2g_{i0}v^i - g_{ij}v^iv^j)^{-1/2}(1, \mathbf{v}).
\tag{5.88}
$$

Evaluating $\mathbf{U} \cdot \mathbf{P}$ in the arbitrary coordinate system Σ gives

$$
\begin{aligned}
\hat{E} &= -\mathbf{U} \cdot \mathbf{P} \\
&= -g_{00}U^0P^0 - g_{i0}U^iP^0 - g_{i0}U^0P^i - g_{ij}U^iP^j,
\end{aligned}
\tag{5.89}
$$

where we have used the fact that the metric tensor is symmetric. Use of $U^i = U^0v^i$ and $P^i = P^0w^i$, together with the equation (5.88) gives

$$
\hat{E} = \frac{g_{00} + g_{i0}v^i + g_{i0}w^i + g_{ij}v^iw^j}{(-g_{00} - 2g_{i0}v^i - g_{ij}v^iv^j)^{1/2}}P^0.
\tag{5.90}
$$

We now assume that the metric is stationary, which means that there exists a coordinate system such that the metric tensor is independent of the time coordinate x^0. Let us consider a freely moving particle in a time-independent metric. Its relativistic Lagrangian is then independent of the coordinate x^0. In other words x^0 is a cyclic coordinate. From the equations of motion then follows that the covariant momentum conjugate to the time coordinate, P_0, is a constant of motion for the particle.

The connection between P_0 and P^0 is

$$
P_0 = g_{0\nu}P^\nu = g_{00}P^0 + g_{i0}P^i = (g_{00} + g_{i0}w^i)P^0.
\tag{5.91}
$$

Substitution into equation (5.90) gives

$$
\hat{E} = -\frac{(g_{00} + g_{i0}v^i + g_{i0}w^i + g_{ij}v^iw^j)}{(-g_{00} - 2g_{i0}v^i - g_{ij}v^iv^j)^{1/2}(g_{00} + g_{i0}w^i)}P_0.
\tag{5.92}
$$

Using equation (5.84), (5.86) and that P_0 is a constant of motion gives the desired equation

$$
\boxed{D_a\omega_a = D_s\omega_s,}
\tag{5.93}
$$

where D is a general Doppler shift factor,

$$
D = \frac{(-g_{00} - 2g_{i0}v^i - g_{ij}v^iv^j)^{1/2}(g_{00} + g_{i0}w^i)}{g_{00} + g_{i0}v^i + g_{i0}w^i + g_{ij}v^iw^j},
\tag{5.94}
$$

and s designates source and a absorber.

We shall now consider some special cases.

Minkowski metric

In the Minkowski metric $g_{00} = -c^2$, $g_{i0} = 0$, $g_{ij} = 0$ for $i \neq j$, $g_{ii} = 1$. There is no deflection of the light, and the magnitude of the velocity of light is constant, so $\mathbf{w}_a = \mathbf{w}_s = \mathbf{n}$, where $\mathbf{n}$ is a unit vector in the direction of propagation of the light. In this case equations (5.93) and (5.94) give

$$\omega_a = \frac{\gamma_a \left(1 - \frac{\mathbf{v}_a \cdot \mathbf{n}}{c}\right)}{\gamma_s \left(1 - \frac{\mathbf{v} \cdot \mathbf{n}}{c}\right)} \omega_s, \qquad \gamma = \left(1 - \frac{v^2}{c^2}\right)^{-1/2}. \tag{5.95}$$

With the absorber at rest this equation gives

$$\omega_a = \sqrt{1 - \frac{v_s^2}{c^2}} \left(1 - \frac{\mathbf{v} \cdot \mathbf{n}}{c}\right) \omega_s. \tag{5.96}$$

In order for the light to reach the absorber, $\mathbf{n}$ must be pointing from the source towards the absorber.

If the source moves in a direction orthogonal to this line, $\mathbf{v} \cdot \mathbf{n} = 0$, and

$$\boxed{\omega_a = \sqrt{1 - \frac{v_s^2}{c^2}} \omega_s.} \tag{5.97}$$

This frequency shift is called the transverse Doppler effect. It is an expression of the special-relativistic time dilatation.

On the other hand, if the source moves towards the absorber, $\mathbf{v}_s \cdot \mathbf{n} = v_s$, which gives

$$\boxed{\omega_a = \sqrt{\frac{c + v_s}{c - v_s}} \omega_s.} \tag{5.98}$$

This is the longitudinal shift.

Source and absorber at rest in an arbitrary stationary metric

In this case $\mathbf{v}_a = \mathbf{v}_s = 0$. Equations (5.93) and (5.94) give

$$\omega_a = \left(\frac{g_{00}^s}{g_{00}^a}\right)^{1/2} \omega_s. \tag{5.99}$$

This frequency shift is termed the gravitational Doppler effect.

As applied to a source and absorber at rest in a uniformly accelerated reference frame this equation gives

$$\omega_a = \frac{1 + \frac{g x_s}{c^2}}{1 + \frac{g x_a}{c^2}} \omega_s. \tag{5.100}$$

For $x_a < x_s$ eq. (5.100) gives $\omega_a > \omega_s$. In the case of a height difference $h = x_s - x_a = 20$ m at the surface of the Earth, $g = 10$ m/s^2, the relative frequency shift is $(\omega_a - \omega_s)/\omega_s \sim gh/c^2 = 2 \cdot 10^{-15}$. This frequency shift has been measured by Pound and Rebka [PRj60] using the Mössbauer effect, and the prediction of eq. (5.100) was verified.

If the source and absorber are both at rest in a rotating reference frame RF, at distance r_s and r_a from the axis, respectively, eqs. (5.100) and (5.3) give

$$\boxed{\omega_a = \sqrt{\frac{1 - \frac{r_s^2 \omega^2}{c^2}}{1 - \frac{r_a^2 \omega^2}{c^2}}} \omega_s.} \tag{5.101}$$

The prediction of this equation has been confirmed experimentally by Champeney *et al.* [CIK65].

Problems

5.1. *Geodetic curves in space*

(a) In the two-dimensional Euclidean plane the line-element is given by $ds^2 = dx^2 + dy^2$. A curve $y = y(x)$ connects two points A and B in the plane. The distance between A and B along the curve is therefore

$$S = \int_A^B ds = \int_A^B \left[1 + \left(\frac{dy}{dx} \right)^2 \right]^{\frac{1}{2}} dx. \tag{5.102}$$

If we vary the shape of this curve slightly, but keeping the end points A and B fixed, it would lead to a change δS of the length of the curve. Whenever $\delta S = 0$ for all small arbitrary variations with respect to a given curve, then the curve is a geodetic curve. Find the Euler-Lagrange equation which corresponds to $\delta S = 0$ and show that geodetic curves in the plane are straight lines.

(b) A particle with mass m is moving without friction on a two-dimensional surface embedded in the three-dimensional space. Write down the expressions for the Lagrangian, L, and the corresponding Euler-Lagrange equations for the particle. Show that L is a constant of motion and explain this by referring to the forces acting on the particle.

The geodetic curves are found by variation of $S = \int_A^B ds$. Show, using the Euler-Lagrange equations, that the particle is moving along a geodetic curve with constant speed.

(c) A particle is moving without friction on a sphere. Express the Lagrange function in terms of the polar angles θ and ϕ and find the corresponding Euler-Lagrange equations.

The coordinate axes can be chosen so that at the time $t = 0$, $\theta = \pi/2$ and $\dot{\theta} = 0$. Show, using the Euler-Lagrange equations, that this implies that θ is constant and equal to $\pi/2$ for all t. Hence, the particle is moving on a great circle, i.e. on a geodetic curve on the sphere.

Assume further that at $t = 0$, $\theta = \pi/2$ and $\phi = 0$, and at $t = t_1 > 0$, $\phi = \theta = \pi/2$. Along what type of different curves can the particle have travelled for $0 < t < t_1$ so that $\delta \int_0^{t_1} L dt = 0$ for the different curves? Find the action integral $S = \int_0^{t_1} L dt$ for the different curves. Do the all the curves correspond to local minima for the total length S?

5.2. *Free particle in a hyperbolic reference frame*
The metric for a two-dimensional space is given by

$$ds^2 = -V^2 dU^2 + dV^2. \tag{5.103}$$

(a) Find the Euler-Lagrange equations for the motion of a free particle using this metric. Show that they admit the following solutions:

$$\frac{1}{V} = \frac{1}{V_0} \cosh(U - U_0).$$

What is the physical interpretation of the constants V_0 and U_0?

(b) Show that these are straight lines in the coordinate system (t, x) given by

$$\begin{aligned} x &= V \cosh U \\ t &= V \sinh U. \end{aligned} \tag{5.104}$$

Express the speed of the particle in terms of U_0, and its x-component at $t = 0$ in terms of V_0 and U_0. Find the interval ds^2 expressed in terms of x and t and show that the space in which the particle is moving, is a Minkowski space with one time and one spatial dimension.

(c) Express the covariant component p_U of the momentum using $p_t = -E$ and $p_x = p$, and show that it is a constant of motion. How can this fact be directly extracted from the metric? Show further that the contravariant component p^U is not a constant of motion. Are p_V or p^V constants of motion?

5.3. *Spatial geodesics in a rotating RF*

We studied a rotating reference frame in the beginning of this chapter and will now consider spatial geodesics in this reference frame. Consider the spatial metric

$$d\ell^2 = dr^2 + \frac{r^2}{1 - \frac{\omega^2 r^2}{c^2}} d\theta^2 + dz^2. \tag{5.105}$$

Using the Lagrangian $L = \frac{1}{2} \left(\frac{d\ell}{d\lambda} \right)^2$ the shortest distance curves between points will be calculated. We will for the sake of simplicity assume that $\frac{dz}{d\lambda} = 0$, i.e., the curve is *planar*.

(a) Assume that the parameter λ is the arc-length of the curve. What is the "three-velocity" identity in this case?

(b) The system possesses a cyclic coordinate. Which coordinate is that? Set down the expression for the corresponding constant of motion.

(c) Find the expressions for $\frac{dr}{d\lambda}$ and $\frac{d\theta}{d\lambda}$ as a function of r. Deduce the differential equation for the curve.

(d) Use the initial condition $\frac{dr}{d\lambda} = 0$ for $r = r_0$ and show that

$$\frac{p_\theta}{r_0} = \sqrt{1 + \frac{\omega^2 p_\theta^2}{c^2}}.$$

(e) Show that the differential equation can be written

$$\frac{dr}{r\sqrt{r^2 - r_0^2}} - \frac{\omega^2}{c^2} \frac{r\, dr}{\sqrt{r^2 - r_0^2}} = \frac{d\theta}{r_0}. \tag{5.106}$$

Integrate this equation and find the equation for the curve. Finally, draw the curve.

6

Differentiation, Connections, and Integration

Forms prove to be a powerful tool in differential geometry and in physics. They have many wonderful properties that we shall explore further in this chapter. We know that in physics and mathematics, integration and differentiation are important, if not essential, operations that appear in almost all physical theories. In this chapter we will explore differentiation on curved manifolds and reveal several interesting properties.

6.1 Exterior Differentiation of forms

First we must find a way to differentiate forms. Since forms are antisymmetric by construction, it is advantageous to define a differentiation that preserves this antisymmetrical property.

Let us first consider 0-forms. The *exterior derivative*, denoted by **d**, is a local operator, and for a 0-form f (a function) it is defined by

$$\boxed{\mathbf{d}f = \frac{\partial f}{\partial x^\mu}\mathbf{dx}^\mu.}$$

(6.1)

That this expression is invariant under a coordinate transformation can be quite easily checked. Since

$$\frac{\partial f}{\partial x^\mu} = \frac{\partial f}{\partial x^{\mu'}}\frac{\partial x^{\mu'}}{\partial x^\mu},$$

(6.2)

and

$$\mathbf{dx}^\mu = \mathbf{dx}^{\nu'}\frac{\partial x^\mu}{\partial x^{\nu'}},$$

(6.3)

we have that

$$\mathbf{d}f = \frac{\partial f}{\partial x^\mu}\mathbf{dx}^\mu = \frac{\partial f}{\partial x^{\mu'}}\mathbf{dx}^{\mu'}.$$

(6.4)

Hence, df has the same form in any coordinate system, and is thus independent of the coordinate system chosen. When we take the exterior derivative of a 0-form we obtain a one-form. Thus if $\mathbf{X} = X^\mu \mathbf{e}_\mu$ is a vector defined at a point, where $\mathbf{e}_\mu$ is a coordinate basis, then we can form the *directional derivative*

$$\boxed{df(\mathbf{X}) = X^\mu \frac{\partial f}{\partial x^\mu} = \mathbf{X}(f).}$$
(6.5)

The directional derivative can be interpreted as the rate of change of the function in the direction of the vector.

Similarly, we define the exterior derivative of a p-form, α, by

$$\boxed{d\alpha = \frac{1}{p!} \frac{\partial \alpha_{\mu_1 \ldots \mu_p}}{\partial x^\nu} \mathbf{dx}^\nu \wedge \mathbf{dx}^{\mu_1} \wedge \ldots \wedge \mathbf{dx}^{\mu_p}.}$$
(6.6)

This object is an antisymmetric tensor. Note that due to the antisymmetry of the basis only the antisymmetric combination $\alpha_{[\mu_1 \ldots \mu_p, \nu]}$ contributes in the expression for $d\alpha$. Hence, the component equations corresponding to the form-equation $d\alpha = 0$ are $\alpha_{[\mu_1 \ldots \mu_p, \nu]} = 0$ (and not $\alpha_{\mu_1 \ldots \mu_p, \nu} = 0$).

The exterior derivative of a p-form is a $(p + 1)$-form. An immediate consequence of this is that in an n-dimensional space, any n-form α yields $d\alpha = 0$. This is due to the fact that there are no non-trivial $(n + 1)$-forms in an n-dimensional space.

Example

Example 6.1 (Exterior differentiation in 3-space.)
Consider a one-form $\mathbf{A} = A_1 \mathbf{dx}^1 + A_2 \mathbf{dx}^2 + A_3 \mathbf{dx}^3$ and a two-form $\mathbf{F} = F_1 \mathbf{dx}^2 \wedge \mathbf{dx}^3 + F_2 \mathbf{dx}^3 \wedge \mathbf{dx}^1 + F_3 \mathbf{dx}^1 \wedge \mathbf{dx}^2$.

Then the exterior derivatives are

$$\begin{aligned}
\mathbf{dA} = {} & \left(\frac{\partial A_2}{\partial x^1} - \frac{\partial A_1}{\partial x^2} \right) \mathbf{dx}^1 \wedge \mathbf{dx}^2 + \left(\frac{\partial A_3}{\partial x^2} - \frac{\partial A_2}{\partial x^3} \right) \mathbf{dx}^2 \wedge \mathbf{dx}^3 \\
& + \left(\frac{\partial A_1}{\partial x^3} - \frac{\partial A_3}{\partial x^1} \right) \mathbf{dx}^3 \wedge \mathbf{dx}^1
\end{aligned}$$
(6.7)

$$\mathbf{dF} = \left(\frac{\partial F_1}{\partial x^1} + \frac{\partial F_2}{\partial x^2} + \frac{\partial F_3}{\partial x^3} \right) \mathbf{dx}^1 \wedge \mathbf{dx}^2 \wedge \mathbf{dx}^3.$$
(6.8)

Comparing these expressions to the corresponding expressions for *curl*, $\nabla \times \vec{A}$, and *divergence*, $\nabla \cdot \vec{F}$, we note that each component coincide:

$$\varepsilon_{ijk}(\nabla \times \vec{A})^i \cong (\mathbf{dA})_{jk}$$
(6.9)
$$\nabla \cdot \vec{F} \cong (\mathbf{dF})_{123}.$$
(6.10)

If we set $\vec{F} = \nabla \times \vec{A}$ we will, because of the identity

$$\nabla \cdot (\nabla \times \vec{A}) = 0,$$

get

$$\mathbf{ddA} = \mathbf{d}^2 \mathbf{A} = 0,$$
(6.11)

which can be easily checked. As we will show in what follows, this is by no means a coincidence, in fact it is a very useful and powerful result of the exterior derivative.

Inspired by the results of the previous example we will take the exterior derivative of a p-form twice.

From the definition of the exterior derivative (6.6) we get for a p-form ω

$$
\begin{aligned}
\mathbf{d}^2\omega &= \frac{1}{p!}\frac{\partial^2\omega_{\mu_1\ldots\mu_p}}{\partial x^\alpha\partial x^\beta}\mathbf{dx}^\alpha\wedge\mathbf{dx}^\beta\wedge\mathbf{dx}^{\mu_1}\wedge\ldots\wedge\mathbf{dx}^{\mu_p}\\
&=\frac{1}{2\,p!}\left(\frac{\partial^2\omega_{\mu_1\ldots\mu_p}}{\partial x^\alpha\partial x^\beta}-\frac{\partial^2\omega_{\mu_1\ldots\mu_p}}{\partial x^\beta\partial x^\alpha}\right)\mathbf{dx}^\alpha\wedge\mathbf{dx}^\beta\wedge\mathbf{dx}^{\mu_1}\wedge\ldots\wedge\mathbf{dx}^{\mu_p}\quad (6.12)\\
&=0,
\end{aligned}
$$

since partial derivatives commute. Since this result holds for all coordinate systems, we may state

$$\boxed{\mathbf{d}^2 = 0.}\qquad(6.13)$$

This is related to Poincaré's Lemma (see below) and is valid for pure forms. We will see later that this is not true when we define d on "vector valued forms".

One might contemplate whether the opposite is true, i.e., if α is a p-form and $\mathbf{d}\alpha = 0$, does there always exist a $(p-1)$-form β such that $\alpha = \mathbf{d}\beta$? The answer in general will be no, but this is by no means trivial. The general answer to this question is beyond the scope of this book, but we will mention a special case. We introduce a couple of important concepts related to this. If α is a p-form, then we call α *closed* if

$$\mathbf{d}\alpha = 0,\qquad(6.14)$$

and *exact* if

$$\alpha = \mathbf{d}\beta.\qquad(6.15)$$

Thus all exact forms are closed (but not all closed forms are exact).

There is one important case when the opposite is true:

Theorem: *For any "star shaped[1]" open set U there will exist, for any closed p-form α, a $(p-1)$-form β such that*

$$\alpha = \mathbf{d}\beta\qquad(6.16)$$

when restricted to U.

This is called *Poincaré's Lemma* and is true *locally*. This is often sufficient in various problems and simplifies our calculation considerably in many cases.

We have also a rule for the differentiation of a wedge product. Let α and β be a p-form and a q-form respectively. Then

$$\mathbf{d}(\alpha\wedge\beta) = \mathbf{d}\alpha\wedge\beta + (-1)^p\alpha\wedge\mathbf{d}\beta.\qquad(6.17)$$

Notice the sign in the last term.

The equation (6.17) has some important consequences, for example:

$$\mathbf{d}(\alpha\wedge\mathbf{d}\beta) = \mathbf{d}\alpha\wedge\mathbf{d}\beta,\qquad(6.18)$$

and also note that

$$\mathbf{d}(\mathbf{d}\alpha\wedge\mathbf{d}\beta) = 0.\qquad(6.19)$$

---

[1]By star shaped we mean a region that is homomorphic to a region in an Euclidean space that has a point that can be connected to any other point in the region by a straight line.

The Codifferential operator

We shall now see how we can define a similar operator to the differential operator $\mathbf{d}$, but which work in the "opposite" direction. For an n-dimensional metric space we define the *coderivative* of a p-form α, denoted $\mathbf{d}^\dagger\alpha$, by

$$\boxed{\mathbf{d}^\dagger\alpha = \text{sgn}(g)(-1)^{n(p+1)+1} \star \mathbf{d} \star \alpha,} \tag{6.20}$$

where $\star$ is Hodge's star operator defined in eq. (4.98). The coderivative of a p-form is a $(p-1)$-form. We note that

$$\left(\mathbf{d}^\dagger\right)^2\alpha = \mathbf{d}^\dagger\mathbf{d}^\dagger\alpha = \pm \star \mathbf{d} \star \star \mathbf{d} \star \alpha = \pm \star \mathbf{dd} \star \alpha = 0. \tag{6.21}$$

Hence, we have that

$$\boxed{\left(\mathbf{d}^\dagger\right)^2 = 0.} \tag{6.22}$$

Let us now see how we can define a covariant divergence using the coderivative. Using eq. (4.99) we have that for a one-form

$$\star\alpha = \frac{\sqrt{|g|}}{(n-1)!}\varepsilon_{\mu\mu_1\dots\mu_{n-1}}\alpha^\mu \mathbf{dx}^{\mu_1} \wedge \dots \wedge \mathbf{dx}^{\mu_{n-1}}, \tag{6.23}$$

where $\alpha^\mu = g^{\mu\nu}\alpha_\nu$ are the components of the vector $\mathbf{A} \equiv \alpha^\mu\mathbf{e}_\mu$. Exterior differentiation gives

$$\mathbf{d}\star\alpha = \pm\frac{1}{\sqrt{|g|}}\left(\sqrt{|g|}\alpha^\mu\right)_{,\mu}\epsilon, \tag{6.24}$$

where $\epsilon = \sqrt{|g|}\varepsilon_{|\mu_1\dots\mu_n|}\mathbf{dx}^{\mu_1} \wedge\dots\wedge \mathbf{dx}^{\mu_n}$ is the volume form. Taking the Hodge dual, we find

$$\boxed{\nabla \cdot \mathbf{A} = -\mathbf{d}^\dagger\alpha = \frac{1}{\sqrt{|g|}}\left(\sqrt{|g|}\alpha^\mu\right)_{,\mu}.} \tag{6.25}$$

This expression is called the *covariant divergence* of the vector $\mathbf{A}$.

The Laplacian and d'Alembertian differential operators are both generalized by the second-order differential operator Δ called *de Rahm's operator*

$$\boxed{\Delta \equiv \mathbf{dd}^\dagger + \mathbf{d}^\dagger\mathbf{d}.} \tag{6.26}$$

Because of a sign-convention, we usually use minus this operator, i.e., if we introduce $\square \equiv -\Delta$, then $\square$ is the usual Laplacian used in physics. If we let de Rahm's operator act on a scalar f, we obtain

$$\square f = -\Delta f = \frac{1}{\sqrt{|g|}}\left(\sqrt{|g|}g^{\mu\nu}f_{,\nu}\right)_{,\mu}. \tag{6.27}$$

This expression is valid in a curved space-time, when de Rahm's operator is acting on 0-forms.

Specializing to three-dimensional Euclidean space with Cartesian coordinates we have

$$\square f = \left(\frac{\partial^2}{\partial x^2} + \frac{\partial^2}{\partial y^2} + \frac{\partial^2}{\partial z^2}\right)f. \tag{6.28}$$

In Minkowski space-time we get

$$\square f = \left(\frac{\partial^2}{\partial x^2} + \frac{\partial^2}{\partial y^2} + \frac{\partial^2}{\partial z^2} - \frac{1}{c^2}\frac{\partial^2}{\partial t^2}\right)f. \tag{6.29}$$

6.2 Electromagnetism

The electromagnetic field can be expressed in a very elegant way in terms of forms as this section will show. We have introduced a powerful tool just waiting to be applied to physics. The electromagnetic field has been studied since the time of Faraday, but little did he know of forms. However, forms and exterior derivatives are now known and we will see that the electromagnetic field can be considered as a two-form.

For simplicity's sake we will assume that we are in Minkowski space. We define the electromagnetic field tensor as the 2-form

$$\begin{aligned}\mathbf{F} &= (E_1 \mathbf{dx}^1 + E_2 \mathbf{dx}^2 + E_3 \mathbf{dx}^3) \wedge \mathbf{dt} \\ &\quad + B_1 \mathbf{dx}^2 \wedge \mathbf{dx}^3 + B_2 \mathbf{dx}^3 \wedge \mathbf{dx}^1 + B_3 \mathbf{dx}^1 \wedge \mathbf{dx}^2 \\ &\equiv \mathcal{E} \wedge \mathbf{dt} + \mathcal{B}, \end{aligned} \tag{6.30}$$

where we have defined $\mathcal{E} = E_i \mathbf{dx}^i$ and $\mathcal{B} = \frac{1}{2}\varepsilon_{ijk}B^i \mathbf{dx}^j \wedge \mathbf{dx}^k$. In component form $\mathbf{F} = \frac{1}{2}F_{\mu\nu}\mathbf{dx}^\mu \wedge \mathbf{dx}^\nu$, where

$$F_{\mu\nu} = \begin{bmatrix} 0 & -E_1 & -E_2 & -E_3 \\ E_1 & 0 & B_3 & -B_2 \\ E_2 & -B_3 & 0 & B_1 \\ E_3 & B_2 & -B_1 & 0 \end{bmatrix}. \tag{6.31}$$

We split the exterior derivative in a spatial part $\mathcal{D}$ and a time part D_t:

$$\mathbf{d} = \mathcal{D} + D_t. \tag{6.32}$$

The exterior derivative of $\mathbf{F}$ can now be written

$$\mathbf{dF} = \left(\mathcal{D}\mathcal{E} + \frac{\partial \mathcal{B}}{\partial t}\right) \wedge \mathbf{dt} + \mathcal{D}\mathcal{B}. \tag{6.33}$$

We see that

$$\mathcal{D}\mathcal{E} + \frac{\partial \mathcal{B}}{\partial t} = \frac{1}{2}\left(\nabla \times \mathbf{E} + \frac{\partial \mathbf{B}}{\partial t}\right)^i \varepsilon_{ijk}\mathbf{dx}^j \wedge \mathbf{dx}^k, \tag{6.34}$$

and

$$\mathcal{D}\mathcal{B} = (\nabla \cdot \mathbf{B})\mathbf{dx}^1 \wedge \mathbf{dx}^2 \wedge \mathbf{dx}^3. \tag{6.35}$$

Thus from the homogeneous Maxwell's equations, equations (2.74) and (2.75), we have

$$\mathbf{dF} = 0. \tag{6.36}$$

The electromagnetic field tensor is therefore closed and Poincaré's Lemma ensures us that there exists locally a one-form $\mathbf{A}$ such that

$$\boxed{\mathbf{F} = \mathbf{dA},} \tag{6.37}$$

or in component form

$$\boxed{F_{\mu\nu} = \frac{\partial A_\nu}{\partial x^\mu} - \frac{\partial A_\mu}{\partial x^\nu}.} \tag{6.38}$$

We note that $\mathbf{A}$ is not uniquely determined; for any function f, $\mathbf{A}'$ given by

$$\boxed{\mathbf{A}' = \mathbf{A} + \mathbf{d}f}$$
(6.39)

defines – since $\mathbf{d}^2 = 0$ – the same electromagnetic field tensor. The shift of potential according to eq. (6.39) is called a *gauge transformation* because it leaves the field tensor invariant. The remaining Maxwell's equations can also be written using forms. To simplify notation we will will use the *Hodge star operator* $\star$.

We introduce the current one-form $\mathbf{J}$ with components J_μ which is the corresponding one-form to the four-current vector $(\rho, \vec{J})$. The remaining Maxwell's equations can then be written

$$\mathbf{d} \star \mathbf{F} = \star \mathbf{J}.$$
(6.40)

This can be seen as follows. We write $\star\mathbf{F}$ as

$$\begin{aligned}\star\mathbf{F} &= -(B_1\mathbf{dx}^1 + B_2\mathbf{dx}^2 + B_3\mathbf{dx}^3) \wedge \mathbf{dt} \\ &+ E_1\mathbf{dx}^2 \wedge \mathbf{dx}^3 + E_2\mathbf{dx}^3 \wedge \mathbf{dx}^1 + E_3\mathbf{dx}^1 \wedge \mathbf{dx}^2 \\ &\equiv -\hat{B} \wedge \mathbf{dt} + \hat{\mathcal{E}};\end{aligned}$$
(6.41)

i.e., we can get the form $\star\mathbf{F}$ from $\mathbf{F}$ by the mapping $E_i \mapsto -B_i$ and $B_i \mapsto E_i$. The star operator acting on the current one-form is

$$\begin{aligned}\star\mathbf{J} &= \rho\mathbf{dx}^1 \wedge \mathbf{dx}^2 \wedge \mathbf{dx}^3 \\ &- (J^1\mathbf{dx}^2 \wedge \mathbf{dx}^3 + J^2\mathbf{dx}^3 \wedge \mathbf{dx}^1 + J^3\mathbf{dx}^1 \wedge \mathbf{dx}^2) \wedge \mathbf{dt}.\end{aligned}$$
(6.42)

We can now see that equation (6.40) follows from Maxwell's equations (2.73) and (2.76).

Conservation of charge is expressed by the identity

$$\boxed{\mathbf{d}^2 \star \mathbf{F} = \mathbf{d} \star \mathbf{J} = 0,}$$
(6.43)

because of $\mathbf{d}^2 = 0$. Writing this in component form, we get the familiar form

$$\boxed{J^\mu{}_{,\mu} = 0.}$$
(6.44)

Thus Maxwell's equations can be written

$$\boxed{\begin{aligned}\mathbf{dF} &= 0 \\ \mathbf{d} \star \mathbf{F} &= \star\mathbf{J},\end{aligned}}$$
(6.45)

or in component form

$$\boxed{\begin{aligned}F_{[\mu\nu,\lambda]} &= 0 \\ F^{\mu\nu}{}_{,\mu} &= -J^\nu.\end{aligned}}$$
(6.46)

Let us assume that we have a one-form potential $\mathbf{A}$, such that

$$\mathbf{dA} = \mathbf{F}.$$
(6.47)

As we already have mentioned, this is always the case locally. The first half of Maxwell's equations are now automatically satisfied, while the other half states

$$\mathbf{d} \star \mathbf{dA} = \star\mathbf{J}.$$
(6.48)

Using the Hodge star operator on each side and using eq. (6.20), we get

$$\mathbf{d}^\dagger \mathbf{d}\mathbf{A} = \mathbf{J}. \tag{6.49}$$

We can write this using the Laplacian, eq. (6.26),

$$\Box \mathbf{A} = -\mathbf{J} - \mathbf{d}\mathbf{d}^\dagger \mathbf{A}. \tag{6.50}$$

We can achieve a further simplification by choosing a specific gauge. We have the freedom of changing the potential by a gauge transformation. A particular useful choice of gauge is the *Lorenz gauge* (introduced by the Danish physicist Ludwig Lorenz [Lor67]):

$$\mathbf{d}^\dagger \mathbf{A} = 0. \tag{6.51}$$

This choice of gauge can be achieved as follows. If we for any gauge potential $\mathbf{A}'$, let f be a function that satisfies

$$\Box f = \mathbf{d}^\dagger \mathbf{A}', \tag{6.52}$$

then by the gauge transformation,

$$\mathbf{A} = \mathbf{A}' + \mathbf{d}f, \tag{6.53}$$

the one-form $\mathbf{A}$ will satisfy the Lorenz gauge condition $\mathbf{d}^\dagger \mathbf{A} = 0$. In Lorenz gauge, equation (6.50) simplifies to

$$\boxed{\Box \mathbf{A} = -\mathbf{J}.} \tag{6.54}$$

Hence, Maxwell's equations imply a pure wave-equation. From this equation follows that electromagnetic waves move with the speed of light.

6.3   Integration of forms

The previous sections have shown how to differentiate forms. We will now see how to integrate forms as well.

Let us start with the simplest example, one-forms. Assume that we have a curve c in some n-dimensional space, and let this curve be parameterised by $x^\mu(\lambda)$, $0 \le \lambda \le t$. Given a one-form ω, we can define the line integral

$$\int_c \omega \equiv \int_0^t \omega(c'(\lambda))d\lambda = \int_0^t \omega_\mu \frac{dx^\mu}{d\lambda} d\lambda. \tag{6.55}$$

The tangent vector of the curve is evaluated by the one-form along the curve. The result is then integrated in an invariant manner. Some forms are particularly easy to integrate. Let $\omega = \mathbf{d}f$, where f is a 0-form. Then

$$\boxed{\int_c \omega = \int_0^t \mathbf{d}f = f(c(t)) - f(c(0));} \tag{6.56}$$

i.e., the integration is only dependent on the start and end points. Thus for exact forms the integral does not depend on the path taken between the start

and end points at all. In particular, if the path is a closed loop then we get
zero:

$$\oint \mathbf{d}f = 0, \tag{6.57}$$

no matter how the loop actually look like. If we have two forms $\mathbf{A}'$ and $\mathbf{A}$
which differ only by an exact form, i.e.

$$\mathbf{A}' = \mathbf{A} + \mathbf{d}f, \tag{6.58}$$

then any loop integral will yield

$$\oint \mathbf{A}' = \oint (\mathbf{A} + \mathbf{d}f) = \oint \mathbf{A} + \oint \mathbf{d}f = \oint \mathbf{A}. \tag{6.59}$$

Comparing this to the electromagnetic gauge transformation, these loop integrals give rise to *gauge invariant loop integrals*. In physics these are often given
the name *Wilson loops*.

We have so far only studied integration of one-forms, but we will now
study the integration of another type of forms. If we consider an n-dimensional
space (which could be a submanifold of a higher-dimensional space), there exists only one type of n-forms. All these n-forms are of the form

$$\mathbf{V} = V\mathbf{dx}^1 \wedge ... \wedge \mathbf{dx}^n, \tag{6.60}$$

where V is some function. If M is some bounded region of the space we define
the integral of $\mathbf{V}$ over M by

$$\int_M \mathbf{V} = \int \cdots \int_M V(x^1, ..., x^n)dx^1...dx^n; \tag{6.61}$$

i.e., just a multiple integral over the region with some suitable parameterization. For a metric space $\mathbf{V}$ can typically be the *volume form*. The volume form
is a unique positive form that "measures" the volume of the metric space. For
example, the volume form in Euclidean 3-space in ordinary Cartesian coordinates is

$$\epsilon = \mathbf{dx} \wedge \mathbf{dy} \wedge \mathbf{dz}, \tag{6.62}$$

or in spherical coordinates

$$\epsilon = r^2 \sin \phi \mathbf{dr} \wedge \mathbf{d\theta} \wedge \mathbf{d\phi}. \tag{6.63}$$

Note that we usually write dV for the volume form, but in spite of the notation,
the volume form does by no means need to be exact! In most cases it is not,
but it is always closed.

Using the Hodge star operator the volume form can be expressed

$$\epsilon = \star 1, \tag{6.64}$$

which is very useful for certain cases.

Example 6.2 (Not all closed forms are exact)
We will here give an example of a form that is closed but cannot be exact. Let us choose
the one-form in $\mathbb{R}^2$ given by

$$\omega = \frac{x\mathrm{d}y}{x^2 + y^2} - \frac{y\mathrm{d}x}{x^2 + y^2}, \tag{6.65}$$

and choose the path parameterized by

$$c(\theta) = (\cos\theta, \sin\theta),$$

where $0 \le \theta \le 2\pi$. Let us first check that this form is closed:

$$\begin{aligned}
\mathrm{d}\omega &= \frac{\mathrm{d}x \wedge \mathrm{d}y}{x^2 + y^2} - \frac{2x^2 \mathrm{d}x \wedge \mathrm{d}y}{(x^2 + y^2)^2} - \frac{\mathrm{d}y \wedge \mathrm{d}x}{x^2 + y^2} + \frac{2y^2 \mathrm{d}y \wedge \mathrm{d}x}{(x^2 + y^2)^2} \\
&= \frac{2\mathrm{d}x \wedge \mathrm{d}y}{x^2 + y^2} - \frac{2\mathrm{d}x \wedge \mathrm{d}y}{x^2 + y^2} = 0.
\end{aligned} \tag{6.66}$$

Thus ω is closed. Let us now calculate the integral of ω along $c(\theta)$:

$$\oint \omega = \int_0^{2\pi} \left(\frac{\cos\theta \mathrm{d}(\sin\theta)}{\cos^2\theta + \sin^2\theta} - \frac{\sin\theta \mathrm{d}(\cos\theta)}{\cos^2\theta + \sin^2\theta} \right)$$

$$= \int_0^{2\pi} 1 \cdot \mathrm{d}\theta = 2\pi. \tag{6.67}$$

Hence, according to equation (6.57), ω cannot be exact.

Example 6.3 (The surface area of the sphere)
We shall consider a rather simple example. We will calculate the surface area of the
sphere. Most readers will already know the answer but this simple case can serve as a
good illustration.
    We parameterize the surface of a sphere in $\mathbb{R}^3$ with radius R by

$$\begin{aligned}
x &= R\cos\theta\sin\phi \\
y &= R\sin\theta\sin\phi \\
z &= R\cos\phi,
\end{aligned} \tag{6.68}$$

where $0 \le \theta \le 2\pi$, $0 \le \phi \le \pi$. The volume form on the spherical surface is

$$\epsilon = R^2 \sin\phi \mathrm{d}\theta \wedge \mathrm{d}\phi. \tag{6.69}$$

Integrating we find

$$\int \epsilon = \int_0^\pi \mathrm{d}\phi \int_0^{2\pi} \mathrm{d}\theta R^2 \sin\phi = 4\pi R^2, \tag{6.70}$$

which is of course the correct area for the sphere.

Stoke's Theorem

Similarly can we define the 2-dimensional surface integral over a 2-form, and
in general the p-dimensional integral over a p-form.
    Without proof we will state

Stoke's Theorem: *Let M be a smooth n-dimensional (oriented) compact manifold with intrinsic boundary ∂M, and let α be an $(n-1)$-form. Then*

$$\boxed{\int_M d\alpha = \int_{\partial M} \alpha.}$$ (6.71)

This is the generalisation of Gauss' law for vector calculus.

In particular we note that if $\partial M = 0$ then the integral vanishes. This is why any closed loop integral over an exact one-form yields zero, a loop has no boundary.

We will not get into details in this book, but we will just see how this law can be applied to the electromagnetic case.

Consider a bounded 3-dimensional region M which is purely spatial. From Stoke's theorem and Maxwell's equations we have

$$\int_M d \star \mathbf{F} = \int_M \star \mathbf{J} = \int_{\partial M} \star \mathbf{F}.$$ (6.72)

Since the region is purely spatial, so will its intrinsic boundary be. Thus

$$\int_M \star \mathbf{J} = \int_M \rho \, \mathbf{dx}^1 \wedge \mathbf{dx}^2 \wedge \mathbf{dx}^3 \equiv Q;$$ (6.73)

i.e., Q is the total charge inside the spatial region. On the other hand

$$
\begin{aligned}
\int_{\partial M} \star \mathbf{F} &= \int_{\partial M} \left(E_1 \mathbf{dx}^2 \wedge \mathbf{dx}^3 + E_2 \mathbf{dx}^3 \wedge \mathbf{dx}^1 + E_3 \mathbf{dx}^1 \wedge \mathbf{dx}^2 \right) \\
&= \int_{\partial M} \mathbf{E} \cdot \mathbf{dS},
\end{aligned}
$$ (6.74)

where $\mathbf{E} \cdot \mathbf{dS}$ can be interpreted as the electric flux out of the surface element $\mathbf{dS}$. Thus,

$$\int_{\partial M} \mathbf{E} \cdot \mathbf{dS} = Q.$$ (6.75)

This is the famous Gauss' law in electromagnetism.

The corresponding law for the magnetic field is

$$0 = \int_M d\mathbf{F} = \int_{\partial M} \mathbf{F} = \int_{\partial M} \mathbf{B} \cdot \mathbf{dS},$$ (6.76)

and can be interpreted as the lack of magnetic monopoles in electromagnetism.

Examples   **Example 6.4 (The Electromagnetic Field outside a static point charge)**
Let us investigate the electromagnetic field outside a point charge.

Assume that the field is a function of the radial coordinate, r, only. A pure static electric field can be generated by a field potential $A_0 = \varphi(r)$. Using

$$F_{\mu\nu} = \frac{\partial A_\nu}{\partial x^\mu} - \frac{\partial A_\mu}{\partial x^\nu},$$ (6.77)

the only non-zero components of the field tensor are

$$F_{i0} = \frac{\partial \varphi}{\partial x^i} = \frac{\partial \varphi}{\partial r} \frac{x^i}{r}, \tag{6.78}$$

using the chain rule and $r = \sqrt{(x^1)^2 + (x^2)^2 + (x^3)^2}$. The electric field is now just $E_i = F_{i0}$. If M is the interior of a spherical region so that ∂M is a sphere of radius r, the area surface element $\mathbf{dS}$ has components

$$(\mathbf{dS})_i = \frac{x^i}{r} \cdot r^2 d\Omega. \tag{6.79}$$

Thus Gauss's law, eq. (6.75), gives

$$Q = \int_{\partial M} \mathbf{E} \cdot \mathbf{dS} = \frac{\partial \varphi}{\partial r} r^2 \int d\Omega = \frac{\partial \varphi}{\partial r} 4\pi r^2, \tag{6.80}$$

which can be integrated to yield

$$\varphi(r) = -\frac{Q}{4\pi r}. \tag{6.81}$$

We should note that even though this result was derived in Minkowski coordinates, the result is highly general. Around any point we can introduce geodesic normal coordinates (this will be shown later in this chapter). Hence, this result is valid for all spherically symmetric spacetimes. The only requirement is that the coordinate system is such that the area of a spherical surface with radius r is $4\pi r^2$.

Example 6.5 (Gauss' integral theorem)
Let us apply Stoke's theorem to the one-form α, and let $\mathbf{A}$ be the vector associated with this form. Define $\beta = \star\alpha$ so that according to equations (6.24) and (6.25) we have

$$d\beta = (\nabla \cdot \mathbf{A})\epsilon. \tag{6.82}$$

According to Stoke's theorem

$$\int_M d\beta = \int_M (\nabla \cdot \mathbf{A})\epsilon = \int_{\partial M} \beta. \tag{6.83}$$

Also

$$\beta = \star\alpha = \frac{\sqrt{|g|}}{(n-1)!} \varepsilon_{\mu\mu_1 \cdots \mu_{n-1}} \alpha^\mu \mathbf{dx}^{\mu_1} \wedge \ldots \wedge \mathbf{dx}^{\mu_{n-1}}. \tag{6.84}$$

If we now assume that $x^1, ..., x^{n-1}$ are coordinates on the surface ∂M, and that x^n is the orthogonal direction, then this can be written

$$\beta = \sqrt{\frac{|g|}{g_{nn}}} \varepsilon_{1\ldots(n-1)\hat{n}} \alpha^{\hat{n}} \mathbf{dx}^1 \wedge \ldots \wedge \mathbf{dx}^{n-1}. \tag{6.85}$$

Thus

$$\int_{\partial M} \beta = \int_{\partial M} \mathbf{A} \cdot \mathbf{dS}, \tag{6.86}$$

where $\mathbf{dS} = \hat{\mathbf{n}} \sqrt{\frac{|g|}{g_{nn}}} dx^1 \ldots dx^{n-1}$ is the invariant surface element. In this case Stoke's theorem takes the form

$$\boxed{\int_M (\nabla \cdot \mathbf{A})\epsilon = \int_{\partial M} \mathbf{A} \cdot \mathbf{dS},} \tag{6.87}$$

which is *Gauss' integral theorem*.

6.4   Covariant differentiation of vectors

We will now consider another kind of differentiation. In a curved space we need to know how to differentiate different types of tensors, not only forms. In order to do this we need to introduce *connections*. As the word says these are geometrical operators that give us a rule of how to differentiate, or in more physical terms, how nearby basis-vectors are connected in a basis-vector field. This is a necessary ingredient in the generalized derivative, called the *covariant derivative*. This covariant derivative will in component form be denoted by a semicolon; i.e., for a vector A^μ, $A^\mu{}_{;\nu}$ is its covariant derivative. We will first define it for a vector, then in the subsequent sections generalize it for any tensor field.

Heuristic motivation of the concept 'covariant differentiation'

As we noted in section 4.4 tensors have a coordinate independent existence since the tensor components transform homogeneously. This is the essential property of tensors making it possible to formulate the laws of nature by equations that have the same form in arbitrary coordinate systems. These equations contain generally derivatives of tensor components. Hence, in order to be able to formulate the laws of nature in terms of tensors the derivative of a tensor component must itself be the component of a tensor.

Vectors are tensors of rank one. Let us see if the partial derivative of a vector component transform as a tensor component. Consider a vector **A** with components A^μ in a coordinate basis $\{e_\mu\}$. The partial derivatives of A^μ transform as

$$A^{\mu'}{}_{,\nu'} = \frac{\partial x^\nu}{\partial x^{\nu'}} \frac{\partial}{\partial x^\nu} \left(\frac{\partial x^{\mu'}}{\partial x^\mu} A^\mu \right) = \frac{\partial x^\nu}{\partial x^{\nu'}} \frac{\partial x^{\mu'}}{\partial x^\mu} A^\mu{}_{,\nu} + \frac{\partial x^\nu}{\partial x^{\nu'}} \frac{\partial^2 x^{\mu'}}{\partial x^\nu \partial x^\mu} A^\mu. \quad (6.88)$$

Due to the last term the partial derivatives of vector components do not transform as tensor components. Hence, one needs a generalization of this derivative in a tensor formulation of the laws of nature.

Besides this formal defect of the partial derivative of vector components the meaning of this derivative is not quite appropriate. Although it represents the change of a vector as decomposed in a Cartesian coordinate system, this is not so in an arbitrary coordinate system. Differentiating a vector with respect to an invariant parameter λ yields

$$\frac{d\mathbf{A}}{d\lambda} = \frac{d}{d\lambda} \left(A^\mu e_\mu \right) = \frac{dA^\mu}{d\lambda} e_\mu + A^\mu \frac{de_\mu}{d\lambda} = A^\mu{}_{,\nu} u^\nu e_\mu + A^\mu u^\nu e_{\mu,\nu}, \quad (6.89)$$

where $u^\mu \equiv \frac{dx^\mu}{d\lambda}$. Hence the partial derivative $A^\mu{}_{,\nu}$ does only represent the change of the vector component A^μ, and not the whole vector.

We would like to have a generalized derivative of tensor components that fulfill two requirements. The derivative of tensor components should transform as tensor components, and it should represent the change of the whole vector, not only one of its components. This new derivative will be called the covariant derivative.

The derivative of a scalar function involves the difference between the value of a function at a point and its value at a nearby point. Similarly, the derivative of a vector field involves the difference between its value and direction at two nearby points. However, as we saw in section 4.2, in a curved space the vectors at different points exist in different tangent planes. Hence, in order to compare two different vectors of a vector field one must first parallel transport one vector to the position of the other. This process has not yet been defined in curved spaces, but we know how to parallel transport a vector in flat space. In flat space it is transported so that the components in a Cartesian coordinate system remain unchanged. Hence, in order to obtain an intuitive approach to the concepts of parallel transport and covariant derivative, we shall first consider parallel transport in flat space using arbitrary coordinates.

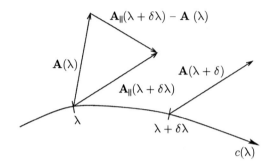

Figure 6.1: The directional derivative of a vector field $\mathbf{A}$ along a curve $c(\lambda)$ with tangent vector field $\mathbf{u} = (dx^\mu/d\lambda)\mathbf{e}_\mu$.

Consider a curve c in flat space parameterized by $x^\mu(\lambda)$ where λ is an invariant parameter. The tangent vector is $\mathbf{u} = u^\mu\mathbf{e}_\mu$ where $u^\mu = dx^\mu/d\lambda$. In figure 6.1 we have drawn the curve $c(\lambda)$ and two vectors of the field $\mathbf{A}$ at two nearby points marked by the parameters λ and $\lambda + \delta\lambda$. Here $\mathbf{A}_{||}(\lambda + \delta\lambda)$ is the vector $\mathbf{A}(\lambda+\delta\lambda)$ parallel transported from $\lambda+\delta\lambda$ to λ. Then $\mathbf{A}(\lambda)$ is subtracted from $\mathbf{A}_{||}(\lambda + \delta\lambda)$ by the usual parallelogram rule.

Parallel transport can be defined in the same way, transporting a vector so that its components remain unchanged, in a local Cartesian coordinate system in curved space. This motivates the following definition of the *directional covariant derivative* of a vector field $\mathbf{A} = A^\mu\mathbf{e}_\mu$ in the direction of a vector $\mathbf{u} = u^\mu\mathbf{e}_\mu$ in curved space,

$$A^\mu{}_{;\nu}u^\nu\mathbf{e}_\mu = \lim_{\delta\lambda\to 0}\frac{\mathbf{A}_{||}(\lambda+\delta\lambda) - \mathbf{A}(\lambda)}{\delta\lambda}. \tag{6.90}$$

In a local Cartesian coordinate system the component A^μ do not change as the vector $\mathbf{A}$ is parallel transported. Hence, evaluating eq. (6.90) in such coordinates is just like evaluating the derivative of a a scalar function. Thus

$$A^\mu{}_{;\nu}u^\nu\mathbf{e}_\mu = A^\mu{}_{,\nu}u^\nu\mathbf{e}_\mu \quad \text{(LCCS)} \tag{6.91}$$

where LCCS means that the equation is only valid in a local Cartesian coordinate system.

This expression must, however, be generalized to arbitrary coordinate systems. Comparing with eq. (6.89) we see that the right hand side of eq. (6.91) is equal to the derivative of the vector components A^μ. However, the derivative of the vector field $\mathbf{A}$ also involves a term representing the change of the basis vectors with position. These changes are proportional to the difference in position of two basis vectors. This implies

$$\frac{de_\mu}{d\lambda} = \Gamma^\alpha_{\mu\nu}\frac{dx^\nu}{d\lambda}e_\alpha = \Gamma^\alpha_{\mu\nu}u^\nu e_\alpha, \tag{6.92}$$

where the functions $\Gamma^\alpha_{\mu\nu}$ are *connection coefficients*. Hence, eq. (6.89) takes the form

$$\frac{d\mathbf{A}}{d\lambda} = \left(A^\mu_{,\nu} + \Gamma^\mu_{\nu\alpha}A^\alpha\right)u^\nu e_\mu. \tag{6.93}$$

The covariant derivative, $A^\mu_{;\nu}$, of the vector components A^μ are defined by

$$\frac{d\mathbf{A}}{d\lambda} \equiv A^\mu_{;\nu}u^\nu e_\mu. \tag{6.94}$$

Comparing with the previous equation we obtain

$$\boxed{A^\mu_{;\nu} \equiv A^\mu_{,\nu} + A^\alpha\Gamma^\mu_{\alpha\nu}.} \tag{6.95}$$

This derivative represents the change of the whole vector $\mathbf{A}$, not only the components A^μ. If the covariant derivative $A^\mu_{;\nu}$ shall transform as components of a tensor of rank $\{^1_1\}$, the connection coefficients have to transform according to

$$\Gamma^\alpha_{\mu\nu} = M^{\nu'}_\nu M^{\mu'}_\mu M^\alpha_{\alpha'}\Gamma^{\alpha'}_{\mu'\nu'} + M^\alpha_{\alpha'}M^{\alpha'}_{\mu,\nu}. \tag{6.96}$$

Hence, the connection coefficients do not transform as tensor components; they transform inhomogeneously. This will turn out to be of physical significance, as will be discussed in the next section.

The covariant derivative was introduced by Christoffel in order to differentiate tensor fields. Today we associate his name to the so-called *Christoffel symbols*. The Christoffel symbols are the (metric) connection coefficients, $\Gamma^\alpha_{\mu\nu}$, *when expressed in a coordinate basis*. The Christoffel symbols possess a special symmetry which can be seen by letting the primed coordinates represent a local Cartesian system with respect to a point. The Christoffel symbols, $\Gamma^{\alpha'}_{\mu'\nu'}$, will then vanish at that point. Then equation (6.96) takes the form

$$\Gamma^\alpha_{\mu\nu} = \frac{\partial x^\alpha}{\partial x^{\alpha'}}\frac{\partial^2 x^{\alpha'}}{\partial x^\mu \partial x^\nu}, \tag{6.97}$$

showing that the Christoffel symbols have to be symmetric in the lower indices:

$$\Gamma^\alpha_{\mu\nu} = \Gamma^\alpha_{\nu\mu}. \tag{6.98}$$

We must stress, however, that this symmetric property is only valid for the Christoffel symbols, not for the generalized connection coefficients which we will come to later.

The Levi-Civitá Connection

The geometrical interpretation of the covariant derivative was first given by Levi-Civitá [LC17], and goes as follows. Consider again Fig. 6.1. If the curve passes through a vector field $\mathbf{A} = A^\mu e_\mu$, we define the *directional covariant derivative* along the curve of the vector field as $A^\mu_{;\nu} u^\nu$. The vectors of the field are said to be *parallel transported* if

$$\boxed{A^\mu_{;\nu} u^\nu = A^\mu_{,\nu} u^\nu + A^\alpha \Gamma^\mu_{\alpha\nu} u^\nu = 0.}$$
(6.99)

According to the chain rule for differentiation, we can write

$$A^\mu_{,\nu} u^\nu = \frac{\partial A^\mu}{\partial x^\nu} \frac{dx^\nu}{d\lambda} = \frac{dA^\mu}{d\lambda}.$$
(6.100)

Thus equation (6.99) may be written as

$$\frac{dA^\mu}{d\lambda} + \Gamma^\mu_{\alpha\nu} A^\alpha \frac{dx^\nu}{d\lambda} = 0.$$
(6.101)

The geometrical interpretation of the covariant derivative is that

$$A^\mu_{;\nu} u^\nu \mathbf{e}_\mu = \lim_{\delta\lambda \to 0} \frac{\mathbf{A}_{\|}(\lambda + \delta\lambda) - \mathbf{A}(\lambda)}{\delta\lambda}.$$
(6.102)

Since the vectors at λ and $\lambda + \delta\lambda$ belong to two different tangent planes, the vector at $\lambda + \delta\lambda$ must be parallel transported to λ before they are subtracted.

Geodesic curves

In Euclidean space we know that the shortest curve connecting two different points is a straight line. In a curved space this is not longer true. Curves that connect points in the shortest or longest possible way are called *geodesic curves*. There are two seemingly equivalent definitions of geodesic curves in a general curved space. Either we can define it as the path with extremal length between any two points (see Fig. 6.2), or we can define it as a straightest possible curve, i.e., a curve whose tangent vectors are connected by parallel transport.

Let us consider the latter definition first. The tangent vector of the curve is given by

$$\mathbf{u} = u^\mu \mathbf{e}_\mu.$$
(6.103)

By the definition the tangent vectors of a geodesic curve are connected by parallel transport, hence

$$\boxed{u^\mu_{;\nu} u^\nu = 0,}$$
(6.104)

or

$$\boxed{\frac{d^2 x^\mu}{d\lambda^2} + \Gamma^\mu_{\alpha\beta} \frac{dx^\alpha}{d\lambda} \frac{dx^\beta}{d\lambda} = 0.}$$
(6.105)

Denoting differentiation with respect to λ by a dot, the geodesic equation can be written as

$$\ddot{x}^\mu + \Gamma^\mu_{\alpha\beta} \dot{x}^\alpha \dot{x}^\beta = 0.$$
(6.106)

In a Cartesian coordinate system the Christoffel symbols vanish, and the solution of the geodesic equation are straight lines. In a curved space, they are the "straightest possible", but they are still curved.

Figure 6.2: A geodesic is the shortest line connecting any two points.

Let us see how the same result can come out of the other definition of a geodesic curve. The variation principle expressing that geodesic curves have extremal "spacetime distance" between two given points, has the form

$$\delta \int_{\lambda_1}^{\lambda_2} ds = 0, \qquad (6.107)$$

where $ds^2 = g_{\mu\nu} dx^\mu dx^\nu$ is the line-element of spacetime. Equation (6.107) may be written as

$$\delta \int_{\lambda_1}^{\lambda_2} L(x^\mu, \dot{x}^\mu) d\lambda = 0, \qquad (6.108)$$

where $L = \sqrt{|g_{\mu\nu} \dot{x}^\mu \dot{x}^\nu|}$. We can now calculate the equations of the geodesic curve by the Lagrange equations:

$$\frac{d}{d\lambda}\left(\frac{\partial L}{\partial \dot{x}^\mu}\right) - \frac{\partial L}{\partial x^\mu} = 0. \qquad (6.109)$$

This gives the equations

$$\ddot{x}^\mu + \frac{1}{2} g^{\mu\lambda}(g_{\lambda\alpha,\nu} + g_{\lambda\nu,\alpha} - g_{\alpha\nu,\lambda})\dot{x}^\alpha \dot{x}^\nu = 0. \qquad (6.110)$$

Comparing with equation (6.106) we see that the Christoffel symbols are given in terms of the components of the metric tensor:

$$\boxed{\Gamma^\mu_{\alpha\nu} = \frac{1}{2} g^{\mu\lambda}(g_{\lambda\alpha,\nu} + g_{\lambda\nu,\alpha} - g_{\alpha\nu,\lambda}).} \qquad (6.111)$$

Comparing with the calculation in section 5.6 we may conclude that *free particles particles follow geodesic curves in spacetime.*

In order to find a physical interpretation of some of the Christoffel symbols we will consider a free particle instantaneously at rest. Since the spatial components of the four-velocity vanish, the geodesic equation then reduces to

$$\ddot{x}^\mu = -\Gamma^\mu_{00}. \qquad (6.112)$$

Hence, the Christoffel symbols Γ^μ_{00} represent the acceleration of gravity in the chosen reference frame.

The generalized connection (the Koszul connection)

The covariant derivative or the connection can be defined in a coordinate free manner. Its independence of the coordinates chosen is then settled once and for all.

We define a (Koszul) *connection* ∇ as a function that associates a vector field $\nabla_{\mathbf{X}}\mathbf{Y}$ to any two vector fields $\mathbf{X}$ and $\mathbf{Y}$, and which satisfies

$$
\begin{align}
(1) \quad & \nabla_{\mathbf{X}_1+\mathbf{X}_2}\mathbf{Y} = \nabla_{\mathbf{X}_1}\mathbf{Y} + \nabla_{\mathbf{X}_2}\mathbf{Y} & (6.113)\\
(2) \quad & \nabla_{\mathbf{X}}(\mathbf{Y}_1 + \mathbf{Y}_2) = \nabla_{\mathbf{X}}\mathbf{Y}_1 + \nabla_{\mathbf{X}}\mathbf{Y}_2 & (6.114)\\
(3) \quad & \nabla_{f\mathbf{X}}\mathbf{Y} = f \cdot \nabla_{\mathbf{X}}\mathbf{Y} & (6.115)\\
(4) \quad & \nabla_{\mathbf{X}}(f\mathbf{Y}) = f \cdot \nabla_{\mathbf{X}}\mathbf{Y} + \mathbf{X}(f) \cdot \mathbf{Y} & (6.116)
\end{align}
$$

where f is a function.

Assume now that we have a set of basis vectors, $\mathbf{e}_\mu$, and for the sake of simplicity we will denote $\nabla_{\mathbf{e}_\mu}$ by ∇_μ. The connection coefficients are defined as the components of the directional derivative of the basis vectors,

$$
\boxed{\nabla_\nu \mathbf{e}_\mu = \Gamma^\alpha_{\ \mu\nu}\mathbf{e}_\alpha.}
\tag{6.117}
$$

Hence, the connection coefficient $\Gamma^\alpha_{\ \mu\nu}$ represents the α-component of the rate of change of $\mathbf{e}_\mu$ by a displacement in the $\mathbf{e}_\nu$ direction. If we have two vector fields $\mathbf{A} = A^\mu \mathbf{e}_\mu$ and $\mathbf{u} = u^\mu \mathbf{e}_\mu$, then according to (6.114), (6.115) and (6.116)

$$
\nabla_{\mathbf{u}}\mathbf{A} = \left(\mathbf{e}_\nu\left(A^\mu\right)u^\nu + A^\alpha \Gamma^\mu_{\ \alpha\nu}u^\nu\right)\mathbf{e}_\mu.
\tag{6.118}
$$

In component form on an arbitrary basis, this turns into

$$
A^\mu_{\ ;\nu} = \mathbf{e}_\nu\left(A^\mu\right) + A^\alpha \Gamma^\mu_{\ \alpha\nu}.
\tag{6.119}
$$

So $\mathbf{A}$ is parallel transported along $\mathbf{u}$ if

$$
\nabla_{\mathbf{u}}\mathbf{A} = 0,
\tag{6.120}
$$

and the curve c is a geodesic if

$$
\nabla_{\mathbf{u}}\mathbf{u} = 0.
\tag{6.121}
$$

Everything is now expressed in a coordinate-free manner, thus the connection has to be independent of the choice of coordinates. The connection coefficients are on the contrary, dependent on the choice of frame.

Example 6.6 (The Christoffel symbols for plane polar coordinates)   Examples
If we decompose the basis vectors in a Cartesian coordinate system, we get the following:

$$
\begin{align}
\mathbf{e}_r &= \cos\theta\,\mathbf{e}_x + \sin\theta\,\mathbf{e}_y,\\
\mathbf{e}_\theta &= -r\sin\theta\,\mathbf{e}_x + r\cos\theta\,\mathbf{e}_y.
\end{align}
\tag{6.122}
$$

We get

$$
\begin{align}
\nabla_\theta \mathbf{e}_r &= -\sin\theta\,\mathbf{e}_x + \cos\theta\,\mathbf{e}_y = \frac{1}{r}\mathbf{e}_\theta = \Gamma^\theta_{\ r\theta}\mathbf{e}_\theta,\\
\nabla_\theta \mathbf{e}_\theta &= -r\cos\theta\,\mathbf{e}_x - r\sin\theta\,\mathbf{e}_y = -r\mathbf{e}_r = \Gamma^r_{\ \theta\theta}\mathbf{e}_r;
\end{align}
\tag{6.123}
$$

hence,

$$
\Gamma^\theta_{\ r\theta} = \Gamma^\theta_{\ \theta r} = \frac{1}{r}, \quad \Gamma^r_{\ \theta\theta} = -r.
\tag{6.124}
$$

Example 6.7 (The acceleration of a particle as expressed in plane polar coordinates)
We consider a particle with velocity much less that the speed of light, so that we can replace the proper time of the particle with the coordinate time. Then the acceleration of the particle is

$$\ddot{\mathbf{r}} = \dot{\mathbf{v}} = \left(\dot{v}^i + \Gamma^i{}_{jk} v^j v^k \right) \mathbf{e}_i, \tag{6.125}$$

where dot denotes derivative with respect to t. Inserting the velocity components from Example 4.8 and the Christoffel symbols from Example 6.6, we get

$$\ddot{\mathbf{r}}_{\text{inert}} = \left(\ddot{r} - r\dot{\theta}^2 \right) \mathbf{e}_r + \left(\ddot{\theta} + \frac{2}{r}\dot{r}\dot{\theta} \right) \mathbf{e}_\theta, \tag{6.126}$$

where the suffix "inert" indicates that the coordinate system is associated with an inertial frame. Introducing an orthonormal basis $\mathbf{e}_{\hat{r}} = \mathbf{e}_r$, $\mathbf{e}_{\hat{\theta}} = (1/r)\mathbf{e}_\theta$, the acceleration is expressed as

$$\ddot{\mathbf{r}}_{\text{inert}} = \left(\ddot{r} - r\dot{\theta}^2 \right) \mathbf{e}_{\hat{r}} + \left(r\ddot{\theta} + 2\dot{r}\dot{\theta} \right) \mathbf{e}_{\hat{\theta}}. \tag{6.127}$$

Example 6.8 (The acceleration of a particle relative to a rotating reference frame)
In the case, finding the Christoffel symbols is left as a problem (see problem 6.4). Using this result and the results from the previous Example, leads to

$$\begin{aligned}
\ddot{\mathbf{r}}_{\text{rot}} &= \left(\ddot{r} - r\dot{\theta}^2 - r\omega^2 - 2r\omega\dot{\theta} \right) \mathbf{e}_{\hat{r}} + \left(r\ddot{\theta} + 2\dot{r}\dot{\theta} + 2\dot{r}\omega \right) \mathbf{e}_{\hat{\theta}} \\
&= \ddot{\mathbf{r}}_{\text{inert}} - \left(r\omega^2 + 2r\omega\dot{\theta} \right) \mathbf{e}_{\hat{r}} + 2\dot{r}\omega\mathbf{e}_{\hat{\theta}}.
\end{aligned} \tag{6.128}$$

With

$$\boldsymbol{\omega} = \omega\mathbf{e}_z, \quad \mathbf{r} = r\mathbf{e}_{\hat{r}}, \quad \dot{\mathbf{r}} = \dot{r}\mathbf{e}_{\hat{r}} + r\dot{\theta}\mathbf{e}_{\hat{\theta}}, \tag{6.129}$$

this equation can be written

$$\ddot{\mathbf{r}}_{\text{rot}} = \ddot{\mathbf{r}}_{\text{inert}} + \boldsymbol{\omega} \times (\boldsymbol{\omega} \times \mathbf{r}) + 2\boldsymbol{\omega} \times \dot{\mathbf{r}}. \tag{6.130}$$

The middle term on the right hand side is the centrifugal acceleration, and the last term the Coriolis acceleration in a rotating reference frame.

We shall now explore further the relation between the connection coefficients and the structure constants. From eqs. (4.20) and (6.118) we obtain an expression for the commutator of two vectors valid in an arbitrary basis,

$$[\mathbf{u}, \mathbf{v}] = \nabla_{\mathbf{u}}\mathbf{v} - \nabla_{\mathbf{v}}\mathbf{u} + \left(\Gamma^\rho{}_{\mu\nu} - \Gamma^\rho{}_{\nu\mu} + c^\rho{}_{\mu\nu} \right) u^\mu v^\nu \mathbf{e}_\rho. \tag{6.131}$$

The *torsion* operator $\mathbf{T}$ is defined by

$$\boxed{\mathbf{T} \left(\mathbf{u} \wedge \mathbf{v} \right) \equiv \nabla_{\mathbf{u}}\mathbf{v} - \nabla_{\mathbf{v}}\mathbf{u} - [\mathbf{u}, \mathbf{v}].} \tag{6.132}$$

The operator $\mathbf{T}$ is a two-form with vector components

$$\mathbf{T} \left(\mathbf{u} \wedge \mathbf{v} \right) = - \left(\Gamma^\rho{}_{\mu\nu} - \Gamma^\rho{}_{\nu\mu} + c^\rho{}_{\mu\nu} \right) u^\mu v^\nu \mathbf{e}_\rho. \tag{6.133}$$

Introducing the scalar torsion components $T^\rho{}_{\mu\nu}$ by (using the sign convention of [MTW73])

$$\mathbf{T} \left(\mathbf{u} \wedge \mathbf{v} \right) = T^\rho{}_{\mu\nu} u^\mu v^\nu \mathbf{e}_\rho, \tag{6.134}$$

we have

$$T^{\rho}_{\ \mu\nu} = \Gamma^{\rho}_{\ \nu\mu} - \Gamma^{\rho}_{\ \mu\nu} - c^{\rho}_{\ \mu\nu}. \tag{6.135}$$

In a coordinate basis $c^{\rho}_{\ \mu\nu} = 0$, so that

$$T^{\rho}_{\ \mu\nu} = \Gamma^{\rho}_{\ \nu\mu} - \Gamma^{\rho}_{\ \mu\nu}. \tag{6.136}$$

The spacetime of the general theory of relativity is assumed to be torsion free (we will later see that in this case the connection is compatible with the metric). Then the connection coefficients are related to the structure constants by

$$\boxed{c^{\alpha}_{\ \mu\nu} = \Gamma^{\alpha}_{\ \nu\mu} - \Gamma^{\alpha}_{\ \mu\nu},} \tag{6.137}$$

and in the special case where we are in a coordinate basis, the structure coefficients vanish, and the Christoffel symbols are symmetric in their lower indices.

The geometrical meaning of the commutator $[\mathbf{u}, \mathbf{v}]$ in a torsion-free space is shown in Fig. 6.3.

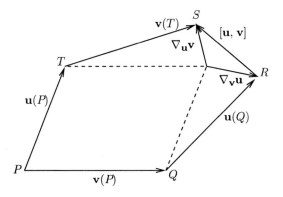

Figure 6.3: Geometrical meaning of the commutator $[\mathbf{u}, \mathbf{v}]$ in a torsion-free space.

Letting $\mathbf{u} = \mathbf{e}_{\mu}$ and $\mathbf{v} = \mathbf{e}_{\nu}$ in eqs. (6.131) and (6.132) we get

$$\nabla_{\mu}\mathbf{e}_{\nu} - \nabla_{\nu}\mathbf{e}_{\mu} = \left(c^{\rho}_{\ \mu\nu} + T^{\rho}_{\ \mu\nu} \right) \mathbf{e}_{\rho}. \tag{6.138}$$

The geometrical meaning of this equation is shown in Fig. 6.4.

6.5   Covariant differentiation of forms and tensors

Let us now look at how the covariant derivative can be generalized so that it can act on any tensor, not only on vectors. First we define the covariant derivative on a scalar function as

$$\nabla_{\mathbf{X}} f = \mathbf{X}(f). \tag{6.139}$$

The covariant directional derivative of a one-form is defined as

$$\left(\nabla_{\mathbf{X}}\alpha\right)(\mathbf{A}) = \nabla_{\mathbf{X}}\left[\alpha(\mathbf{A})\right] - \alpha\left(\nabla_{\mathbf{X}}\mathbf{A}\right) \tag{6.140}$$

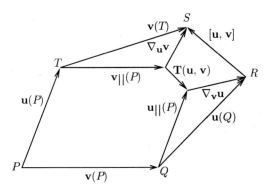

Figure 6.4: Geometrical meaning of torsion. Here $\mathbf{u}_{||}(P)$ means the parallel transported vector.

for any vector field $\mathbf{A}$. The contraction between basis one-forms and the basis vectors are equal to the Kronecker symbols, $\boldsymbol{\omega}^\mu(\mathbf{e}_\nu) = \delta^\mu_\nu$, so their partial derivatives vanish. Thus,

$$(\nabla_\alpha \boldsymbol{\omega}^\mu)(\mathbf{e}_\beta) = -\boldsymbol{\omega}^\mu(\nabla_\alpha \mathbf{e}_\beta) = -\boldsymbol{\omega}^\mu(\Gamma^\lambda_{\beta\alpha}\mathbf{e}_\lambda) = -\Gamma^\mu_{\beta\alpha}. \qquad (6.141)$$

This equation says that the β-component of $\nabla_\alpha \boldsymbol{\omega}^\mu$ equals $-\Gamma^\mu_{\beta\alpha}$, so that

$$\nabla_\alpha \boldsymbol{\omega}^\mu = -\Gamma^\mu_{\beta\alpha}\boldsymbol{\omega}^\beta. \qquad (6.142)$$

The covariant derivative of a one-form $\boldsymbol{\alpha} = \alpha_\mu \boldsymbol{\omega}^\mu$ is then

$$\nabla_\lambda \boldsymbol{\alpha} = (\nabla_\lambda \alpha_\mu)\boldsymbol{\omega}^\mu + \alpha_\mu \nabla_\lambda \boldsymbol{\omega}^\mu = [\mathbf{e}_\lambda(\alpha_\nu) - \alpha_\mu \Gamma^\mu_{\nu\lambda}]\boldsymbol{\omega}^\nu. \qquad (6.143)$$

The covariant derivative of the one-form components α_ν is denoted by $\alpha_{\nu;\lambda}$ and is defined by

$$\nabla_\lambda \boldsymbol{\alpha} = \alpha_{\nu;\lambda}\boldsymbol{\omega}^\nu. \qquad (6.144)$$

It follows that

$$\boxed{\alpha_{\nu;\lambda} = \mathbf{e}_\lambda(\alpha_\nu) - \alpha_\mu \Gamma^\mu_{\nu\lambda}.} \qquad (6.145)$$

We have now found the expression for the covariant derivative of vectors and one-forms. The covariant derivative can now be generalized to arbitrary tensors. Let $\mathbf{A}$ and $\mathbf{B}$ be two tensors of arbitrary rank. Then we define inductively the covariant derivative on the tensor $\mathbf{T} = \mathbf{A} \otimes \mathbf{B}$ by

$$\boxed{\nabla_\mathbf{X}(\mathbf{A} \otimes \mathbf{B}) = (\nabla_\mathbf{X}\mathbf{A}) \otimes \mathbf{B} + \mathbf{A} \otimes (\nabla_\mathbf{X}\mathbf{B}).} \qquad (6.146)$$

As an illustration we shall deduce the expression for the covariant derivative of a covariant tensor $\mathbf{S} = S_{\mu\nu}\boldsymbol{\omega}^\mu \otimes \boldsymbol{\omega}^\nu$. Using (6.146),

$$\begin{aligned}
\nabla_\alpha \mathbf{S} &= \nabla_\alpha(S_{\mu\nu}\boldsymbol{\omega}^\mu \otimes \boldsymbol{\omega}^\nu) \\
&= (\nabla_\alpha S_{\mu\nu})\boldsymbol{\omega}^\mu \otimes \boldsymbol{\omega}^\nu + S_{\mu\nu}(\nabla_\alpha\boldsymbol{\omega}^\mu) \otimes \boldsymbol{\omega}^\nu + S_{\mu\nu}\boldsymbol{\omega}^\mu \otimes (\nabla_\alpha\boldsymbol{\omega}^\nu) \qquad (6.147) \\
&= [\mathbf{e}_\alpha(S_{\mu\nu}) - S_{\beta\nu}\Gamma^\beta_{\mu\alpha} - S_{\mu\beta}\Gamma^\beta_{\nu\alpha}]\boldsymbol{\omega}^\mu \otimes \boldsymbol{\omega}^\nu.
\end{aligned}$$

Thus the components of $\nabla_\alpha \mathbf{S}$ are

$$S_{\mu\nu;\alpha} = \mathbf{e}_\alpha(S_{\mu\nu}) - S_{\beta\nu}\Gamma^\beta{}_{\mu\alpha} - S_{\mu\beta}\Gamma^\beta{}_{\nu\alpha}. \tag{6.148}$$

Since the metric tensor is a covariant tensor of rank 2 we get

$$g_{\mu\nu;\alpha} = \mathbf{e}_\alpha(g_{\mu\nu}) - g_{\beta\nu}\Gamma^\beta{}_{\mu\alpha} - g_{\mu\beta}\Gamma^\beta{}_{\nu\alpha}. \tag{6.149}$$

We now claim that there is a unique connection which is compatible with a given metric in the sense that the metric tensor is covariantly constant:

$$\boxed{\nabla_\mathbf{u}\mathbf{g} = 0,} \tag{6.150}$$

for all $\mathbf{u}$. This is what we call a *metric connection*. If $\mathbf{A}$ and $\mathbf{B}$ are two vectors which is parallel-transported along a vector $\mathbf{u}$, then their inner product is constant as well:

$$\nabla_\mathbf{u}(\mathbf{A} \cdot \mathbf{B}) = (g_{\mu\nu}A^\mu B^\nu)_{;\alpha}u^\alpha = 0. \tag{6.151}$$

Thus both the length of the vectors and the angle between them is preserved under the parallel-transport.

If we are in a coordinate basis we can use the expression from equation (6.111) to check whether this is the correct expression of the Christoffel symbols that makes the metric tensor covariantly constant. Inserting (6.111) into equation (6.149) we find

$$\boxed{g_{\mu\nu;\alpha} = 0.} \tag{6.152}$$

Thus the Christoffel symbols are given by equation (6.96). Furthermore, since the Christoffel symbols in a coordinate basis is symmetric in the lower indices, eq. (6.136) implies that *for the metric connection the torsion tensor vanishes.*

6.6 Exterior differentiation of vectors

Until now we have only defined the exterior derivative of pure p-forms. It is also convenient to define the exterior derivative of vector-valued forms. Consider the mixed tensor

$$\mathbf{T} = T^\mu{}_\nu \mathbf{e}_\mu \otimes \boldsymbol{\omega}^\nu. \tag{6.153}$$

This tensor can be viewed upon as a vectorial one-form as follows; it is linear tensor that to any vector $\mathbf{u}$, assigns a vector given by

$$\mathbf{T}(-, \mathbf{u}) = T^\mu{}_\nu u^\nu \mathbf{e}_\mu. \tag{6.154}$$

Thus this can be considered a *vector-valued one-form* or a *vectorial one-form*. In principle the tensor $\mathbf{T}$ can be interpreted in three ways.

1. A mixed tensor of rank 2.

2. A vectorial one-form.

3. A form valued vector.

A vectorial p-form is defined as the tensor $\mathbf{A} \otimes \boldsymbol{\omega}$ where $\mathbf{A}$ is a vector and $\boldsymbol{\omega}$ is a p-form. It has the basis elements given by

$$\mathbf{e}_\mu \otimes \boldsymbol{\omega}^{\nu_1} \wedge ... \wedge \boldsymbol{\omega}^{\nu_p}. \tag{6.155}$$

A vectorial p-form is antisymmetrical in the covariant part and assigns a vector to every set of p vectors. In particular a pure vector is a vectorial 0-form.

We define the exterior derivative of a basis vector field $\mathbf{e}_\mu$ as

$$\boxed{\mathbf{de}_\mu \equiv \Gamma^\nu{}_{\mu\alpha} \mathbf{e}_\nu \otimes \boldsymbol{\omega}^\alpha.} \tag{6.156}$$

The exterior derivative of a vector field $\mathbf{A}$ is

$$\mathbf{dA} = \mathbf{d}(\mathbf{e}_\mu A^\mu) = \mathbf{e}_\mu \otimes \mathbf{d}A^\nu + A^\mu \mathbf{de}_\mu. \tag{6.157}$$

Using that in an arbitrary basis

$$\mathbf{d}A^\mu = \mathbf{e}_\lambda(A^\mu)\boldsymbol{\omega}^\lambda, \tag{6.158}$$

we get

$$
\begin{aligned}
\mathbf{dA} &= \mathbf{e}_\mu \otimes [\mathbf{e}_\lambda(A^\mu)\boldsymbol{\omega}^\lambda] + A^\mu \Gamma^\nu{}_{\mu\lambda}\mathbf{e}_\nu \otimes \boldsymbol{\omega}^\lambda \\
&= \left[\mathbf{e}_\lambda(A^\nu) + A^\mu \Gamma^\nu{}_{\mu\lambda}\right] \mathbf{e}_\nu \otimes \boldsymbol{\omega}^\lambda.
\end{aligned}
\tag{6.159}
$$

One has to be a bit careful when one calculates the exterior derivative of a vector field. The vector field must be written in component form as $\mathbf{e}_\mu A^\mu$ so that the factors of the tensor product $\mathbf{e}_\nu \otimes \boldsymbol{\omega}^\lambda$ shall not appear in different order in the two terms. Using equation (6.119) we can write

$$\mathbf{dA} = A^\nu{}_{;\lambda}\mathbf{e}_\nu \otimes \boldsymbol{\omega}^\lambda. \tag{6.160}$$

For $\mathbf{S}$ a vectorial p-form and $\mathbf{T}$ a q-form, then we define exterior derivative inductively by

$$\mathbf{d}(\mathbf{S} \wedge \mathbf{T}) = \mathbf{dS} \wedge \mathbf{T} + (-1)^p \mathbf{S} \wedge \mathbf{dT}. \tag{6.161}$$

If $\mathbf{S}$ is a vectorial 0-form, then the wedge product $\mathbf{S}\wedge$ just turns into a tensor product $\mathbf{S}\otimes$.

In a coordinate basis this gives for the exterior derivative of a vectorial one-form $\mathbf{A} = A^\nu{}_\lambda \mathbf{e}_\nu \otimes \boldsymbol{\omega}^\lambda$,

$$
\begin{aligned}
\mathbf{dA} &= \mathbf{d}(\mathbf{e}_\nu A^\mu{}_\lambda \otimes \boldsymbol{\omega}^\lambda) = \mathbf{d}(\mathbf{e}_\nu A^\mu{}_\lambda) \otimes \boldsymbol{\omega}^\lambda \\
&= A^\nu{}_{\lambda,\mu}\mathbf{e}_\nu \otimes \boldsymbol{\omega}^\mu \wedge \boldsymbol{\omega}^\lambda + A^\nu{}_\lambda \mathbf{de}_\nu \otimes \boldsymbol{\omega}^\lambda \\
&= \left(A^\nu{}_{[\lambda,\mu]} + \Gamma^\nu{}_{\tau[\nu}A^\tau{}_{\mu]}\right)\mathbf{e}_\nu \otimes \boldsymbol{\omega}^\mu \wedge \boldsymbol{\omega}^\lambda.
\end{aligned}
\tag{6.162}
$$

Since the Christoffel symbols are symmetrical in the lower indices, we may add a term, $-A^\nu{}_\tau \Gamma^\tau{}_{[\lambda\nu]} = 0$, inside the parenthesis. The equation can then be written

$$\mathbf{dA} = A^\nu{}_{[\lambda;\mu]}\mathbf{e}_\nu \otimes \boldsymbol{\omega}^\mu \wedge \boldsymbol{\omega}^\lambda. \tag{6.163}$$

With a metric connection the double exterior derivative of a vector field is

$$\mathbf{d}^2\mathbf{A} = A^\nu{}_{;[\lambda\mu]}\mathbf{e}_\nu \otimes \boldsymbol{\omega}^\mu \wedge \boldsymbol{\omega}^\lambda. \tag{6.164}$$

Equation (6.164) shows that equation (6.13) in general fails when $\mathbf{d}^2$ is acting on vectorial forms.

The *connection forms* $\mathbf{\Omega}^\nu{}_\mu$ are one-forms defined by

$$\boxed{\mathbf{d}\mathbf{e}_\mu = \mathbf{e}_\nu \otimes \mathbf{\Omega}^\nu{}_\mu.} \tag{6.165}$$

Then according to (6.156) we have

$$\boxed{\mathbf{\Omega}^\nu{}_\mu = \Gamma^\nu{}_{\mu\alpha}\omega^\alpha.} \tag{6.166}$$

Defining the scalar product between a vector $\mathbf{u} = u^\mu \mathbf{e}_\mu$ and a vectorial one-form $\mathbf{A} = A^\nu{}_\lambda \mathbf{e}_\nu \otimes \omega^\lambda$ by

$$\mathbf{u} \cdot \mathbf{A} = u^\mu A^\nu{}_\lambda (\mathbf{e}_\mu \cdot \mathbf{e}_\nu) \omega^\lambda = u^\mu A^\nu{}_\lambda g_{\mu\nu} \omega^\lambda, \tag{6.167}$$

we can calculate the exterior derivative of the components of the metric tensor

$$\begin{aligned}
\mathbf{d}g_{\mu\nu} = \mathbf{d}(\mathbf{e}_\mu \cdot \mathbf{e}_\nu) &= \mathbf{e}_\mu \cdot \mathbf{d}\mathbf{e}_\nu + \mathbf{e}_\nu \cdot \mathbf{d}\mathbf{e}_\mu \\
&= (\mathbf{e}_\mu \cdot \mathbf{e}_\lambda)\mathbf{\Omega}^\lambda{}_\nu + (\mathbf{e}_\nu \cdot \mathbf{e}_\lambda)\mathbf{\Omega}^\lambda{}_\mu \\
&= g_{\mu\lambda}\mathbf{\Omega}^\lambda{}_\nu + g_{\nu\lambda}\mathbf{\Omega}^\lambda{}_\mu = \mathbf{\Omega}_{\mu\nu} + \mathbf{\Omega}_{\nu\mu}.
\end{aligned} \tag{6.168}$$

We now consider a field of *orthonormal* basis vectors. This means that the basis vectors only change their *direction*. Their magnitude and relative angles are constants. The connection forms $\mathbf{\Omega}^{\hat\lambda}{}_{\hat\nu}$ in such an orthonormal basis are called *rotation forms* for that reason. They have, together with the corresponding connection coefficients $\Gamma^{\hat\alpha}{}_{\hat\nu\hat\mu}$ some beautiful properties. Since the components of the metric tensor in an orthonormal basis is everywhere 0 or ± 1, we have

$$\mathbf{d}g_{\mu\nu} = 0, \tag{6.169}$$

and hence,

$$\boxed{\mathbf{\Omega}_{\hat\mu\hat\nu} = -\mathbf{\Omega}_{\hat\nu\hat\mu}, \quad \Gamma_{\hat\nu\hat\mu\hat\alpha} = -\Gamma_{\hat\mu\hat\nu\hat\alpha}.} \tag{6.170}$$

The power of the *orthonormal frame formalism* is due to this antisymmetry. This formalism also have several other nice properties. One is revealed in the so-called *Cartan's first structural equation* which we will derive in what follows.

Let $\alpha = \alpha_\mu \omega^\mu$ be a one-form. Then

$$\alpha([\mathbf{u}, \mathbf{v}]) = u^\mu \alpha_\nu v^\nu{}_{,\mu} - v^\mu \alpha_\nu u^\nu{}_{,\mu}. \tag{6.171}$$

Furthermore,

$$\mathbf{u}(\alpha(\mathbf{v})) = u^\mu v^\nu \alpha_{\nu,\mu} + u^\mu \alpha_\nu v^\nu{}_{,\mu}, \tag{6.172}$$

and

$$\mathbf{v}(\alpha(\mathbf{u})) = v^\mu u^\nu \alpha_{\nu,\mu} + v^\mu \alpha_\nu u^\nu{}_{,\mu}. \tag{6.173}$$

Also

$$\mathbf{d}\alpha(\mathbf{u} \wedge \mathbf{v}) = (\alpha_{\mu,\nu} - \alpha_{\nu,\mu}) u^\nu v^\mu. \tag{6.174}$$

From these equations it follows that

$$\boxed{d\alpha(\mathbf{u} \wedge \mathbf{v}) = \mathbf{u}(\alpha(\mathbf{v})) - \mathbf{v}(\alpha(\mathbf{u})) - \alpha([\mathbf{u}, \mathbf{v}]).} \tag{6.175}$$

This equation in valid in an arbitrary basis. Applying the equation to the basis form $\alpha = \omega^\rho$ and the basis vectors $\mathbf{u} = \mathbf{e}_\mu$ and $\mathbf{v} = \mathbf{e}_\nu$, we get

$$d\omega^\rho(\mathbf{e}_\mu \wedge \mathbf{e}_\nu) = -\omega^\rho([\mathbf{e}_\mu, \mathbf{e}_\nu]) = -c^\rho{}_{\mu\nu}. \tag{6.176}$$

Thus,

$$\boxed{d\omega^\rho = -\frac{1}{2}c^\rho{}_{\mu\nu}\omega^\mu \wedge \omega^\nu.} \tag{6.177}$$

From eqs. (6.134) and (6.135) follow that the torsion operator has the component form

$$\mathbf{T} = \frac{1}{2}\left(\Gamma^\rho{}_{\nu\mu} - \Gamma^\rho{}_{\mu\nu} - c^\rho{}_{\mu\nu}\right)\mathbf{e}_\rho \otimes \omega^\mu \wedge \omega^\nu. \tag{6.178}$$

Inserting eqs. (6.166) and (6.177) we get

$$\mathbf{T} = \mathbf{e}_\rho \otimes \left(d\omega^\rho + \mathbf{\Omega}^\rho{}_\nu \wedge \omega^\nu\right). \tag{6.179}$$

The torsion two-forms $\mathbf{T}^\rho$ are defined by

$$\mathbf{T} \equiv \mathbf{e}_\rho \otimes \mathbf{T}^\rho. \tag{6.180}$$

Hence,

$$\mathbf{T}^\rho = d\omega^\rho + \mathbf{\Omega}^\rho{}_\nu \wedge \omega^\nu. \tag{6.181}$$

This is *Cartan's first structural equation.* With a metric connection (which we will assume is the underlying geometry in most of this book) it reduces to

$$\boxed{d\omega^\rho = -\mathbf{\Omega}^\rho{}_\nu \wedge \omega^\nu.} \tag{6.182}$$

Using eq. (6.166) we can write this equation in terms of the connection coefficients

$$d\omega^\rho = -\frac{1}{2}(\Gamma^\rho{}_{\nu\mu} - \Gamma^\rho{}_{\mu\nu})\omega^\mu \wedge \omega^\nu = -\Gamma^\rho{}_{\nu\mu}\omega^\mu \wedge \omega^\nu. \tag{6.183}$$

This equation can by itself leave only information about the antisymmetric part of the connection. For example, it is not very profitable to use this for computing Christoffel symbols because they are purely symmetric in the last two indices. In order to calculate the connection coefficients in an orthonormal frame, however, this equation turns out to be very useful indeed, as we will see in the next example.

Example   **Example 6.9 (The rotation coefficients of an orthonormal basis field attached to plane polar coordinates)**
Let us look at a plane with polar coordinates

$$ds^2 = dr^2 + r^2 d\theta^2. \tag{6.184}$$

We introduce an orthonormal basis associated with the coordinate system

$$\omega^{\hat{r}} = \mathrm{d}\mathbf{r}, \quad \omega^{\hat{\theta}} = r\mathrm{d}\theta. \tag{6.185}$$

Exterior differentiation of $\omega^{\hat{r}}$ gives

$$\mathrm{d}\omega^{\hat{r}} = 0. \tag{6.186}$$

Comparing with equation (6.182) we get

$$\Omega^{\hat{r}}{}_{\hat{\theta}} \wedge \omega^{\hat{\theta}} = 0. \tag{6.187}$$

Thus,

$$\Omega^{\hat{r}}{}_{\hat{\theta}} = \Gamma^{\hat{r}}{}_{\hat{\theta}\hat{r}}\omega^{\hat{r}} + \Gamma^{\hat{r}}{}_{\hat{\theta}\hat{\theta}}\omega^{\hat{\theta}} = \Gamma^{\hat{r}}{}_{\hat{\theta}\hat{\theta}}\omega^{\hat{\theta}}, \tag{6.188}$$

giving $\Gamma^{\hat{r}}{}_{\hat{\theta}\hat{r}} = 0$, while $\Gamma^{\hat{r}}{}_{\hat{\theta}\hat{\theta}}$ is still undetermined.

Exterior differentiation of $\omega^{\hat{\theta}}$ gives

$$\mathrm{d}\omega^{\hat{\theta}} = -\Omega^{\hat{\theta}}{}_{\hat{r}} \wedge \omega^{\hat{r}} = \frac{1}{r}\omega^{\hat{r}} \wedge \omega^{\hat{\theta}} = -\frac{1}{r}\omega^{\hat{\theta}} \wedge \omega^{\hat{r}}. \tag{6.189}$$

Thus,

$$\Omega^{\hat{\theta}}{}_{\hat{r}} = \Gamma^{\hat{\theta}}{}_{\hat{r}\hat{\theta}}\omega^{\hat{\theta}} + \Gamma^{\hat{\theta}}{}_{\hat{r}\hat{r}}\omega^{\hat{r}} = \frac{1}{r}\omega^{\hat{\theta}} + \Gamma^{\hat{\theta}}{}_{\hat{r}\hat{r}}\omega^{\hat{r}}, \tag{6.190}$$

which gives

$$\Gamma^{\hat{\theta}}{}_{\hat{r}\hat{\theta}} = \frac{1}{r}, \tag{6.191}$$

while $\Gamma^{\hat{\theta}}{}_{\hat{r}\hat{r}}$ is still left undetermined.

The undetermined connection coefficients are determined by means of the antisymmetry equation (6.170). In the orthonormal frame the metric coefficients are that of Minkowski space. Thus antisymmetry implies

$$\Omega^{\hat{r}}{}_{\hat{\theta}} = -\Omega^{\hat{\theta}}{}_{\hat{r}}, \tag{6.192}$$

which shows that

$$\Gamma^{\hat{r}}{}_{\hat{\theta}\hat{\theta}} = -\frac{1}{r}, \quad \Gamma^{\hat{\theta}}{}_{\hat{r}\hat{r}} = 0. \tag{6.193}$$

The non-vanishing connection forms are

$$\Omega^{\hat{r}}{}_{\hat{\theta}} = -\Omega^{\hat{\theta}}{}_{\hat{r}} = -\frac{1}{r}\omega^{\hat{\theta}}. \tag{6.194}$$

6.7   Covariant exterior derivative

In an arbitrary basis $\{\mathbf{e}_{\mu}\}$ the exterior derivative of a function ϕ is given by

$$\mathrm{d}\phi = \mathbf{e}_{\mu}(\phi)\omega^{\mu}. \tag{6.195}$$

The exterior curvature of a one-form $\boldsymbol{\alpha} = \alpha_{\mu}\omega^{\mu}$ is

$$\mathrm{d}\boldsymbol{\alpha} = \mathrm{d}\alpha_{\mu} \wedge \omega^{\mu} + \alpha_{\lambda}\mathrm{d}\omega^{\lambda}. \tag{6.196}$$

Using eq. (6.177) we get

$$\mathbf{d}\alpha = \left(\alpha_{[\mu,\nu]} - \frac{1}{2}a_\lambda c^\lambda_{\nu\mu}\right)\boldsymbol{\omega}^\nu \wedge \boldsymbol{\omega}^\mu. \tag{6.197}$$

From eqs. (6.135) and (6.145) follows

$$\alpha_{[\mu;\nu]} = \alpha_{[\mu,\nu]} + \frac{1}{2}\alpha_\lambda\left(c^\lambda_{\mu\nu} + T^\lambda_{\mu\nu}\right). \tag{6.198}$$

Inserting this into eq. (6.197) we get

$$(\mathbf{d}\alpha)_{\nu\mu} = 2\alpha_{[\mu,\nu]} + a_\lambda c^\lambda_{\mu\nu} = 2\alpha_{[\mu;\nu]} - a_\lambda T^\lambda_{\mu\nu}. \tag{6.199}$$

In Riemannian geometry (with $T^\lambda_{\mu\nu} = 0$) and in an arbitrary basis

$$(\mathbf{d}\alpha)_{\nu\mu} = \alpha_{\mu;\nu} - \alpha_{\nu;\mu}. \tag{6.200}$$

In an arbitrary space, with or without torsion, but with a reference to a coordinate basis, eq. (6.199) reduces to

$$(\mathbf{d}\alpha)_{\nu\mu} = \alpha_{\mu,\nu} - \alpha_{\nu,\mu}. \tag{6.201}$$

The exterior derivative of a 2-form $\mathbf{F} = (1/2)F_{\mu\nu}\boldsymbol{\omega}^\mu \wedge \boldsymbol{\omega}^\nu$ has components

$$(\mathbf{dF})_{\lambda\mu\nu} = 3F_{[\mu\nu,\lambda]} - 3F_{\alpha[\nu}c^\alpha_{\lambda\mu]} = 3F_{[\mu\nu,\lambda]} + 3F_{\alpha[\nu}T^\alpha_{\lambda\mu]}. \tag{6.202}$$

Due to the antisymmetry of $F_{\mu\nu}$ the expression reduces to

$$(\mathbf{dF})_{\lambda\mu\nu} = F_{\mu\nu,\lambda} + F_{\nu\lambda,\mu} + F_{\lambda\mu,\nu}, \tag{6.203}$$

in a coordinate basis, and to

$$(\mathbf{dF})_{\lambda\mu\nu} = F_{\mu\nu;\lambda} + F_{\nu\lambda;\mu} + F_{\lambda\mu;\nu}, \tag{6.204}$$

for a metric connection in an arbitrary basis.

We shall now introduce the covariant exterior derivative. For a p-form α with scalar components, the covariant exterior derivative, $\mathbf{D}\alpha$, is defined as

$$\mathbf{D}\alpha \equiv \mathbf{d}\alpha. \tag{6.205}$$

It follows from the above formulae that the covariant exterior derivative of the elements, $\mathbf{R}^\mu_\nu$, of a matrix of 2-forms is

$$\mathbf{DR}^\mu_\nu = \frac{1}{2}\left(R^\mu_{\nu[\alpha\beta;\lambda]} + R^\mu_{\nu\tau[\alpha}T^\tau_{\beta\lambda]}\right)\boldsymbol{\omega}^\lambda \wedge \boldsymbol{\omega}^\alpha \wedge \boldsymbol{\omega}^\beta. \tag{6.206}$$

Given a vector $\mathbf{A}$ with p-forms as components,

$$\mathbf{A} = \mathbf{e}_\mu \otimes \mathbf{A}^\mu = \frac{1}{p!}A^\mu_{\nu_1\ldots\nu_p}\mathbf{e}_\mu \otimes \boldsymbol{\omega}^{\nu_1} \wedge \cdots \wedge \boldsymbol{\omega}^{\nu_p}. \tag{6.207}$$

The covariant exterior derivative of the form-valued vector-components are defined by

$$\mathbf{e}_\mu \otimes \mathbf{DA}^\mu = \mathbf{dA}. \tag{6.208}$$

For a vector $\mathbf{v} = v^\mu \mathbf{e}_\mu$ with scalar components we have

$$\mathbf{D}v^\mu = v^\mu_{;\nu}\omega^\nu, \tag{6.209}$$

which may be written

$$\mathbf{D}v^\mu = dv^\mu + \Omega^\mu_{\ \nu}v^\nu. \tag{6.210}$$

Let $\mathbf{A}$ be the vector

$$\mathbf{A} = \mathbf{e}_\mu \otimes \mathbf{A}^\mu = A^\mu_{\ \nu}\mathbf{e}_\mu \otimes \omega^\nu, \tag{6.211}$$

with components $\mathbf{A}^\mu = A^\mu_{\ \nu}\omega^\nu$ that are one-forms. Then

$$\mathbf{D}\mathbf{A}^\mu = \left(A^\mu_{\ [\nu;\lambda]} + \frac{1}{2}A^\mu_{\ \tau}T^\tau_{\ \lambda\nu} \right)\omega^\lambda \wedge \omega^\nu. \tag{6.212}$$

Let $\mathbf{A}^\mu_{\ \nu}$ be a matrix of p-forms, and consider a tensor

$$\mathbf{A} = \mathbf{e}_\mu \otimes \mathbf{A}^\mu_{\ \nu} \wedge \omega^\nu. \tag{6.213}$$

This may be interpreted as a tensor of $\{^1_1\}$ with components $\mathbf{A}^\mu_{\ \nu}$ that are p-forms. The covariant exterior derivative of these components is defined by

$$\mathbf{e}_\mu \otimes \mathbf{D}\mathbf{A}^\mu_{\ \nu} \wedge \omega^\nu = d\mathbf{A}. \tag{6.214}$$

Differentiation eq. (6.213) then yields

$$\mathbf{D}\mathbf{A}^\mu_{\ \nu} = d\mathbf{A}^\mu_{\ \nu} + \Omega^\mu_{\ \alpha} \wedge \mathbf{A}^\alpha_{\ \nu} - (-1)^p \mathbf{A}^\mu_{\ \alpha} \wedge \Omega^\alpha_{\ \nu}. \tag{6.215}$$

Let $\mathbf{S} = \mathbf{e}_\mu \otimes \mathbf{S}^\mu$ where $\mathbf{S}^\mu$ are p-forms. Then we define $\mathbf{D}\mathbf{S}^\mu$ by $\mathbf{e}_\mu \otimes \mathbf{D}\mathbf{S}^\mu = d\mathbf{S}$ and obtain

$$\mathbf{D}\mathbf{S}^\mu = d\mathbf{S}^\mu + \Omega^\mu_{\ \alpha} \wedge \mathbf{S}^\alpha. \tag{6.216}$$

This equation is valid for arbitrary p.

The torsion operator has the same form as $\mathbf{S}$ with $p = 2$. Hence, the covariant exterior derivative of the torsion operator is

$$\boxed{\mathbf{D}\mathbf{T}^\rho = d\mathbf{T}^\rho + \Omega^\rho_{\ \nu} \wedge \mathbf{T}^\nu.} \tag{6.217}$$

Cartan's first structural equation may now be written

$$\mathbf{T}^\rho = \mathbf{D}\omega^\rho, \tag{6.218}$$

which with a metric connection reduces to $\mathbf{D}\omega^\rho = 0$.

The quantities $\mathbf{e}_\mu \otimes \Omega^\mu_{\ \nu}$ are vectorial one-forms. It follows that

$$\mathbf{e}_\mu \otimes \mathbf{D}\Omega^\mu_{\ \nu} = d\left(\mathbf{e}_\mu \otimes \Omega^\mu_{\ \nu}\right). \tag{6.219}$$

Thus the covariant exterior derivatives of the connection forms are

$$\boxed{\mathbf{D}\Omega^\mu_{\ \nu} = d\Omega^\mu_{\ \nu} + \Omega^\mu_{\ \alpha} \wedge \Omega^\alpha_{\ \nu}.} \tag{6.220}$$

Example

Example 6.10 (Curl in spherical coordinates)
Let $\mathbf{A} = A^i \mathbf{e}_i$ be a vector in flat 3-space. Noting that the contravariant components of $\nabla \times \mathbf{A}$ correspond to the covariant components of $\star d\mathbf{A}$, and using eq. (6.197) in an orthonormal basis $\mathbf{e}_{\hat{i}}$, we get

$$\nabla \times \mathbf{A} = \varepsilon^{\hat{i}\hat{j}\hat{k}} \left(A_{\hat{j},\hat{i}} - \frac{1}{2} A_{\hat{m}} c^{\hat{m}}_{\hat{i}\hat{j}} \right) \mathbf{e}_{\hat{k}}. \tag{6.221}$$

The orthonormal basis vectors of a coordinate system are

$$\{ \mathbf{e}_{\hat{r}}, \mathbf{e}_{\hat{\theta}}, \mathbf{e}_{\hat{\phi}} \} = \left\{ \frac{\partial}{\partial r}, \frac{1}{r}\frac{\partial}{\partial \theta}, \frac{1}{r\sin\theta}\frac{\partial}{\partial \phi} \right\}. \tag{6.222}$$

Calculating the structure coefficients in the same way as in Example 4.9 we find

$$c^{\hat{\theta}}_{\hat{\theta}\hat{r}} = -c^{\hat{\theta}}_{\hat{r}\hat{\theta}} = c^{\hat{\phi}}_{\hat{\phi}\hat{r}} = -c^{\hat{\phi}}_{\hat{r}\hat{\phi}} = \frac{1}{r},$$
$$c^{\hat{\phi}}_{\hat{\phi}\hat{\theta}} = -c^{\hat{\phi}}_{\hat{\theta}\hat{\phi}} = -\frac{\cot\theta}{r}. \tag{6.223}$$

Inserting these into eq. (6.221) gives the components of the curl in spherical coordinates:

$$(\nabla \times \mathbf{A})^{\hat{r}} = \frac{1}{r\sin\theta} \left[\frac{\partial}{\partial \theta}\left(\sin\theta A^{\hat{\phi}}\right) - \frac{\partial A^{\hat{\theta}}}{\partial \phi} \right],$$
$$(\nabla \times \mathbf{A})^{\hat{\theta}} = \frac{1}{r\sin\theta} \left[\frac{\partial A^{\hat{r}}}{\partial \phi} - \sin\theta\frac{\partial}{\partial r}\left(r A^{\hat{\phi}}\right) \right],$$
$$(\nabla \times \mathbf{A})^{\hat{\phi}} = \frac{1}{r} \left[\frac{\partial}{\partial r}\left(r A^{\hat{\theta}}\right) - \frac{\partial A^{\hat{r}}}{\partial \theta} \right]. \tag{6.224}$$

6.8 Geodesic normal coordinates

We shall now show that there exists a local "Cartesian" coordinate system with "canonical metric" $g_{\mu\nu} = \text{diag}(-1, 1, 1, 1)$ and vanishing Christoffel symbols covering an infinitesimal region about an arbitrary point P in a Lorentzian space-time. Such coordinates are termed *geodesic coordinates*.

The canonical form of the metric is obtained by introducing orthonormal coordinate basis vectors at P. We shall now show that is it always possible to introduce a coordinate system $\{\bar{x}^\mu\}$ with vanishing Christoffel symbols at P.

The transformation of the Christoffel symbols between two coordinate systems $\{\bar{x}^\mu\}$ and $\{x^\mu\}$ are given by eq. (6.96)

$$\bar{\Gamma}^\lambda_{\mu\nu} = \frac{\partial x^\alpha}{\partial \bar{x}^\nu}\frac{\partial x^\beta}{\partial \bar{x}^\mu}\frac{\partial \bar{x}^\lambda}{\partial x^\tau}\Gamma^\tau_{\alpha\beta} + \frac{\partial \bar{x}^\lambda}{\partial x^\tau}\frac{\partial^2 x^\tau}{\partial \bar{x}^\mu \partial \bar{x}^\nu}. \tag{6.225}$$

From an arbitrary coordinate system $\{x^\mu\}$ with origin at point P a new coordinate system $\{\bar{x}^\mu\}$ is introduced via the transformation

$$\bar{x}^\mu = x^\mu + \frac{1}{2}\left(\Gamma^\mu_{\alpha\beta}\right)_P x^\alpha x^\beta, \tag{6.226}$$

where $(\;)_P$ denotes the value at the point P. To second order in the distance from P (in space-time) the inverse transformation is

$$x^\mu = \bar{x}^\mu - \frac{1}{2}\left(\Gamma^\mu{}_{\alpha\beta}\right)_P \bar{x}^\alpha \bar{x}^\beta. \tag{6.227}$$

This leads to

$$\left(\frac{\partial \bar{x}^\alpha}{\partial x^\tau}\right)_P = \delta^\alpha{}_\tau. \tag{6.228}$$

Furthermore,

$$\frac{\partial x^\tau}{\partial \bar{x}^\mu} = \delta^\tau{}_\mu - \frac{1}{2}\left(\Gamma^\tau{}_{\alpha\beta}\right)_P \left(\bar{x}^\alpha \delta^\beta{}_\mu + \bar{x}^\beta \delta^\alpha{}_\mu\right) = \delta^\tau{}_\mu - \frac{1}{2}\left(\Gamma^\tau{}_{\alpha\mu} + \Gamma^\tau{}_{\mu\alpha}\right)_P \bar{x}^\alpha, \tag{6.229}$$

which leads to

$$\left(\frac{\partial x^\tau}{\partial \bar{x}^\mu}\right)_P = \delta^\tau{}_\mu, \tag{6.230}$$

and

$$\left(\frac{\partial^2 x^\tau}{\partial \bar{x}^\mu \partial \bar{x}^\nu}\right)_P = -\frac{1}{2}\left(\Gamma^\tau{}_{\alpha\mu} + \Gamma^\tau{}_{\mu\alpha}\right)_P. \tag{6.231}$$

Inserting these expressions into eq. (6.225) gives

$$\begin{aligned}
\left(\bar{\Gamma}^\lambda{}_{\mu\nu}\right)_P &= \delta^\alpha{}_\mu \delta^\beta{}_\nu \delta^\lambda{}_\tau \left(\Gamma^\lambda{}_{\alpha\beta}\right)_P - \frac{1}{2}\delta^\lambda{}_\tau \left(\Gamma^\tau{}_{\alpha\mu} + \Gamma^\tau{}_{\mu\alpha}\right)_P \\
&= \frac{1}{2}\left(\Gamma^\tau{}_{\alpha\mu} - \Gamma^\tau{}_{\mu\alpha}\right)_P. \tag{6.232}
\end{aligned}$$

Hence, in a Lorentzian space we have

$$\left(\bar{\Gamma}^\lambda{}_{\mu\nu}\right)_P = 0. \tag{6.233}$$

Thus we have a local coordinate system with Minkowski metric and vanishing Christoffel symbols at P.

In section 6.4 we saw that the Christoffel symbols $\Gamma^\mu{}_{00}$ represent the acceleration of gravity in the chosen frame of reference. Hence, the possibility of transforming into local geodesic normal coordinates with vanishing Christoffel symbols means physically that one may transform away the acceleration of gravity locally. This is exactly what one does when staying inside a satellite in orbit about the Earth. The possibility of transforming away the Christoffel symbols locally is thus a mathematical expression of the principle of equivalence.

6.9   One-parameter groups of diffeomorphisms

We will now introduce a special type of diffeomorphisms, or a change of coordinates. These diffeomorphisms are associated with a vector field $\mathbf{X}$ as follows. Consider a vector field $\mathbf{X} = X^\mu \mathbf{e}_\mu$. Then for a point P we define a path $\phi(P, t)$ by

$$\begin{aligned}
\frac{\partial \phi}{\partial t} &= \mathbf{X}_{\phi(P,t)} \\
\phi(P, 0) &= P \quad \text{(Initial condition).} \tag{6.234}
\end{aligned}$$

Hence, $\phi(P, t)$ is the curve, starting at P at $t = 0$, with $\mathbf{X}$ as a tangent vector (see figure 6.5). Let us for a fixed t denote this path by $\phi(P, t) \equiv \phi_t(P)$. Hence, we can consider the map ϕ_t as the diffeomorphism that for every P moves the point along the vector field $\mathbf{X}$. If $t = 0$ then we do not move the point at all, so

$$\phi_0(P) = P. \tag{6.235}$$

This is the trivial diffeomorphism and reflects only the initial condition in eq. (6.234). We also note that if we move P to Q, where $\phi_t(P) = Q$, and then continue to $R = \phi_s(Q)$ then we have

$$\phi_s(\phi_t(P)) = (\phi_s \circ \phi_t)(P) = \phi_{s+t}(P) = (\phi_t \circ \phi_s)(P) = \phi_t(\phi_s(P)). \tag{6.236}$$

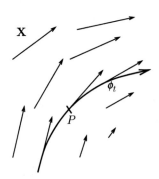

Figure 6.5: A vector field $\mathbf{X}$ determines a unique flow ϕ_t for any given point P.

These diffeomorphisms are very useful in mathematics and physics, because these are special types of coordinate transformations written in a coordinate independent manner. They depend only on the vector field $\mathbf{X}$. We say that ϕ_t is a *local flow generated by the vector field* $\mathbf{X}$. For infinitesimal t, the differential equation (6.234) can be approximated by

$$x'^\mu \approx x^\mu + t X^\mu, \tag{6.237}$$

where x'^μ is to be understood as the μ-component of $\phi_t(P)$ and x^μ the μ component of P in some coordinate system. We can do a "Taylor expansion" of ϕ_t to find a formal solution of (6.234). The result is manifested as follows. We say that the flow ϕ_t is the *exponentiation* of $\mathbf{X}$ and is denoted by

$$x'^\mu = \exp_P(t\mathbf{X}) x^\mu. \tag{6.238}$$

The justification of the name exponentiation, can be given when we look at the Taylor expansion of $\phi_t(P)$ along the curve. To evaluate the coordinate x'^μ of a point which is separated from the initial point $P = \phi_0(P)$ by the parameter distance t along the flow ϕ_t, the coordinate x'^μ corresponding to $\phi_t(P)$ is

$$
\begin{aligned}
x'^\mu &= x^\mu + t \frac{d}{ds} \phi_s^\mu(P) \Big|_{s=0} + \frac{t^2}{2!} \left(\frac{d}{ds} \right)^2 \phi_s^\mu(P) \Big|_{s=0} + \ldots \\
&= \left[1 + t \frac{d}{ds} + \frac{t^2}{2!} \left(\frac{d}{ds} \right)^2 + \ldots \right] \phi_s^\mu(P) \Big|_{s=0} \\
&\equiv \exp\left(t \frac{d}{ds} \right) \phi_s^\mu(P) \Big|_{s=0}.
\end{aligned}
\tag{6.239}
$$

This equation can also be written as in eq. (6.238). More specifically, the exponential function is a formal solution to eq. (6.234).

Pull-backs

Related to these diffeomorphisms are so-called pull-backs. These pull-backs
can be defined for any differentiable function F. If the function F is a function
from the space M to N, $F : M \longmapsto N$, the we can introduce a local set of
coordinates y so that $y^\mu = F^\mu(x)$, or for short, $y = y(x)$. The pull-back F^* is
now defined on *covariant* tensors as follows.

For f a function, the pull-back is simply

$$(F^*f)(x) = (f \circ F)(x) = f(y(x)),\qquad(6.240)$$

i.e., the composition of f with F. If α is a one-form on N, the pull-back of
$\alpha = \alpha_\mu \mathbf{dy}^\mu$ is defined by

$$(F^*\alpha)(\mathbf{v}) = \alpha(F_*\mathbf{v}),\qquad(6.241)$$

for all vectors $\mathbf{v}$ and where F_* is the differential of F. Using the local coordinates, we have that

$$F_*\mathbf{v} = v^\beta \frac{\partial y^\mu}{\partial x^\beta}\frac{\partial}{\partial y^\mu},\qquad(6.242)$$

so

$$(F^*\alpha)(\mathbf{v}) = (\alpha_\nu \mathbf{dy}^\nu)\left(v^\beta \frac{\partial y^\mu}{\partial x^\beta}\frac{\partial}{\partial y^\mu}\right) = v^\beta \frac{\partial y^\mu}{\partial x^\beta}\alpha_\mu.\qquad(6.243)$$

Since this is valid for any $\mathbf{v}$, $F^*\alpha$ is the one-form on M given by

$$F^*\alpha = \frac{\partial y^\mu}{\partial x^\beta}\alpha_\mu \mathbf{dx}^\beta.\qquad(6.244)$$

Here we see the reason why we call it pull-back: The one-form, α, on N is
"pulled back" to a one-form on M.

If now $\alpha = \mathbf{dy}^\nu$ the pull-b ck is only the chain rule

$$F^*\mathbf{dy}^\nu = \frac{\partial y^\nu}{\partial x^\beta}\mathbf{dx}^\beta.\qquad(6.245)$$

Now let α be a covariant tensor of rank $\{^0_p\}$ at y. Then the pull-back is
defined by

$$(F^*\alpha)(\mathbf{v},...,\mathbf{v}) = \alpha(F(\mathbf{v}),...,F(\mathbf{v})).\qquad(6.246)$$

It is easy to prove the following facts

$$F^*(\alpha \wedge \beta) = (F^*\alpha) \wedge (F^*\beta),\qquad(6.247)$$
$$F^*(\alpha \otimes \beta) = (F^*\alpha) \otimes (F^*\beta).\qquad(6.248)$$

We have also that the pull-back commutes with the exterior differentiation

$$F^*(\mathbf{d}\alpha) = \mathbf{d}(F^*\alpha).\qquad(6.249)$$

6.10   The Lie derivative

Another useful derivation on a curved space is the *Lie derivative* which we will
denote $\mathcal{L}_X \mathbf{T}$ where $\mathbf{X}$ is a vector field and $\mathbf{T}$ is a general tensor field. The Lie
derivative transforms tensors of rank $\{^p_q\}$ into tensors of rank $\{^p_q\}$, and it can
be defined in the following way. Consider a vector field $\mathbf{X} = X^\mu \mathbf{e}_\mu$ which
induces the following infinitesimal transformation:

$$x^\mu \longrightarrow x'^\mu = x^\mu + tX^\mu, \qquad (6.250)$$

where t is a small parameter. Under this displacement the tensor field changes
from $\mathbf{T}_x$ to $\mathbf{T}'_{x'}$ where $\mathbf{T}_x$ means the tensor field $\mathbf{T}$ at the point x. Note that
$\mathbf{T}_x \neq \mathbf{T}'_{x'}$ for $x' = x$ because they represent different points in space. The Lie
derivative of $\mathbf{T}$ with respect to $\mathbf{u}$ can now be written as

$$\mathcal{L}_X \mathbf{T} \equiv \lim_{t \longrightarrow 0} \frac{1}{t} \left(\mathbf{T}'_{x'} - \mathbf{T}_x \right). \qquad (6.251)$$

We first have to define how we determine the new tensor $\mathbf{T}'_{x'}$. Consider
a vector field $\mathbf{X}$ and let $\phi(t) = \phi_t$ be the local flow generated by $\mathbf{X}$. The
parameter t can be considered as a "time" parameter so that $\phi_t(x)$ is the point
t seconds along the integral curve of $\mathbf{X}$. Thus if the parameter t is infinitesimal,
we can approximate

$$(\phi_t(x))^\mu \equiv x'^\mu = x^\mu + tX^\mu. \qquad (6.252)$$

If $\mathbf{T}$ is a covariant tensor, then we define the new tensor $\mathbf{T}'_{x'}$ as

$$\mathbf{T}'_{x'} = \phi_t^* \mathbf{T}_{\phi_t(x)}, \qquad (6.253)$$

i.e., the Lie derivative of a *covariant* tensor can be written as

$$\boxed{\mathcal{L}_X \mathbf{T} \equiv \lim_{t \longrightarrow 0} \frac{1}{t} \left(\phi_t^* \mathbf{T}_{\phi_t(x)} - \mathbf{T}_x \right).} \qquad (6.254)$$

Let us first consider a function f, the pull-back is merely the composition
with ϕ_t, $\phi_t^* f = f \circ \phi_t$, i.e. just the value of the function at the point $\phi_t(x)$ rather
than x. Hence, we can Taylor expand the function around x along ϕ_t:

$$f(\phi_t(x)) \approx f(x) + tX^\mu f_{;\mu} = f(x) + t\mathbf{X}(f). \qquad (6.255)$$

The Lie derivative of a function can now be seen to be

$$\mathcal{L}_X f \equiv \lim_{t \longrightarrow 0} \frac{1}{t} \left[f(\phi_t(x)) - f(x) \right] = \mathbf{X}(f) = \nabla_X f. \qquad (6.256)$$

Thus the Lie derivative of a function f with respect to a vector $\mathbf{X}$ is equal to
the directional derivative of f in the $\mathbf{X}$-direction.

We now define the Lie derivative of a vector $\mathbf{Y}$ with respect to the vector
$\mathbf{X}$. If $\mathbf{T}'_{x'}$ is a vector then we define (see Fig. 6.6)

$$\mathbf{T}'_{x'} = \phi_{-t*} \mathbf{T}_{\phi_t(x)}. \qquad (6.257)$$

Thus, the Lie derivative of a vector $\mathbf{Y}$ is

$$\boxed{\mathcal{L}_X \mathbf{Y} \equiv \lim_{t \longrightarrow 0} \frac{1}{t} \left(\phi_{-t*} \mathbf{Y}_{\phi_t(x)} - \mathbf{Y}_x \right).} \qquad (6.258)$$

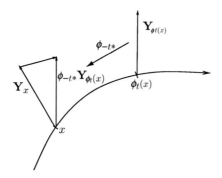

Figure 6.6: The Lie derivative: To compare the vectors $\mathbf{Y}_x$ and $\mathbf{Y}_{\phi_t(x)}$ the latter must be pushed back to x with ϕ_{-t*}.

At $\phi_t(x) = x'$ the vector field has the value (introducing a locally Cartesian coordinate system at x'):

$$\mathbf{Y}_{x'} \;=\; (Y^\mu \mathbf{e}_\mu)_{x'} \approx \left(Y^\mu + tX^\nu Y^\mu_{;\nu}\right)_x (\mathbf{e}_\mu)_{x'} . \tag{6.259}$$

In a locally Cartesian coordinate system $\mathbf{e}_\mu = \frac{\partial}{\partial x^\mu}$, so that

$$(\mathbf{e}_\mu)_{x'} \approx \left(\frac{\partial x^\nu}{\partial x'^\mu} \mathbf{e}_\nu\right)_x . \tag{6.260}$$

From the infinitesimal transformation we can write

$$\frac{\partial x^\nu}{\partial x'^\mu} = \delta^\nu_{\ \mu} - tX^\nu_{\ ;\mu}, \tag{6.261}$$

which gives

$$\boxed{\pounds_{\mathbf{X}}\mathbf{Y} = \left(X^\nu Y^\mu_{;\nu} - Y^\nu X^\mu_{\ ;\nu}\right)\mathbf{e}_\mu = [\mathbf{X}, \mathbf{Y}] .} \tag{6.262}$$

The Lie derivative of $\mathbf{Y}$ with respect to $\mathbf{X}$ is the commutator of $\mathbf{X}$ and $\mathbf{Y}$. It follows that

$$\pounds_{\mathbf{X}}\mathbf{Y} = -\pounds_{\mathbf{Y}}\mathbf{X}. \tag{6.263}$$

Using the commutator relation we find that

$$
\begin{aligned}
\pounds_{\mathbf{X}}\mathbf{e}_\mu &= [\mathbf{X}, \mathbf{e}_\mu] = [X^\nu \mathbf{e}_\nu, \mathbf{e}_\mu] \\
&= X^\nu [\mathbf{e}_\nu, \mathbf{e}_\mu] - \mathbf{e}_\mu(X^\nu)\mathbf{e}_\nu = \left[X^\nu c^\alpha_{\ \nu\mu} - \mathbf{e}_\mu(X^\alpha)\right]\mathbf{e}_\alpha.
\end{aligned}
\tag{6.264}
$$

Thus in a coordinate basis

$$\pounds_{\mathbf{e}_\mu}\mathbf{X} = X^\alpha_{\ ,\mu}\mathbf{e}_\alpha. \tag{6.265}$$

For the Lie derivative, the following rules will apply. Let $\mathbf{S}$ and $\mathbf{T}$ be tensors of arbitrary rank, f, a, b scalar functions, $\mathbf{X}$ and $\mathbf{Y}$ vector fields, and α a one-form. Then

$$
\begin{aligned}
\pounds_{\mathbf{X}}(\mathbf{S} + \mathbf{T}) &= \pounds_{\mathbf{X}}\mathbf{S} + \pounds_{\mathbf{X}}\mathbf{T} & (6.266) \\
\pounds_{\mathbf{X}}(f\mathbf{T}) &= \mathbf{T}\cdot\mathbf{X}(f) + f\pounds_{\mathbf{X}}\mathbf{T} & (6.267) \\
\pounds_{\mathbf{X}}(\alpha(\mathbf{Y})) &= (\pounds_{\mathbf{X}}\alpha)(\mathbf{Y}) + \alpha(\pounds_{\mathbf{X}}\mathbf{Y}) & (6.268) \\
\pounds_{\mathbf{X}}(\mathbf{S}\otimes\mathbf{T}) &= (\pounds_{\mathbf{X}}\mathbf{S})\otimes\mathbf{T} + \mathbf{S}\otimes(\pounds_{\mathbf{X}}\mathbf{T}) & (6.269) \\
\pounds_{a\mathbf{X}+b\mathbf{Y}}\mathbf{T} &= a\pounds_{\mathbf{X}}\mathbf{T} + b\pounds_{\mathbf{Y}}\mathbf{T}. & (6.270)
\end{aligned}
$$

Using equations (6.264) and (6.268) we calculate the Lie derivative of a one-form:

$$\begin{aligned}
(\pounds_{\mathbf{X}}\alpha)_{\nu} &= (\pounds_{\mathbf{X}}\alpha)(\mathbf{e}_{\nu}) = \pounds_{\mathbf{X}}(\alpha(\mathbf{e}_{\nu})) - \alpha(\pounds_{\mathbf{X}}\mathbf{e}_{\nu}) \\
&= \pounds_{\mathbf{X}}\alpha_{\nu} - \alpha\left[\left(X^{\mu}c^{\lambda}_{\mu\nu} - \mathbf{e}_{\nu}(X^{\lambda})\right)\mathbf{e}_{\lambda}\right] \\
&= \mathbf{X}(\alpha_{\nu}) + \alpha_{\lambda}\mathbf{e}_{\nu}(X^{\lambda}) - \alpha_{\lambda}X^{\mu}c^{\lambda}_{\mu\nu}.
\end{aligned} \tag{6.271}$$

In a coordinate basis this expression simplifies to

$$(\pounds_{\mathbf{X}}\alpha)_{\nu} = X^{\mu}\alpha_{\nu,\mu} + \alpha_{\mu}X^{\mu}_{,\nu}. \tag{6.272}$$

An alternative form of equation (6.271) valid in an arbitrary basis is

$$\boxed{(\pounds_{\mathbf{X}}\alpha)_{\nu} = X^{\mu}\alpha_{\nu;\mu} + \alpha_{\mu}X^{\mu}_{;\nu}.} \tag{6.273}$$

For a basis one-form equation (6.271) gives

$$\pounds_{\mathbf{X}}\omega^{\nu} = \left[X^{\mu}c^{\nu}_{\alpha\mu} + \mathbf{e}_{\alpha}(X^{\nu})\right]\omega^{\alpha}. \tag{6.274}$$

In a coordinate basis this equation becomes particularly simple:

$$\pounds_{\mathbf{X}}\mathbf{dx}^{\nu} = X^{\nu}_{,\alpha}\mathbf{dx}^{\alpha} = \mathbf{d}X^{\nu}. \tag{6.275}$$

Let us now derive a very useful relation for one-forms, which turns out to be valid for any p-form. Using

$$\begin{aligned}
\mathbf{d}(\alpha(\mathbf{X})) &= \mathbf{d}(\alpha_{\mu}X^{\mu}) = \left(X^{\mu}\alpha_{\mu,\nu} + \alpha_{\mu}X^{\mu}_{,\nu}\right)\mathbf{dx}^{\nu}, & (6.276) \\
\mathbf{d}\alpha(\mathbf{X}) &= X^{\mu}\left(\alpha_{\nu,\mu} - \alpha_{\mu,\nu}\right)\mathbf{dx}^{\nu}, & (6.277)
\end{aligned}$$

and equation (6.272) we get the general relation between the Lie derivative and the exterior differentiation

$$\pounds_{\mathbf{X}}\alpha = \mathbf{d}(\alpha(\mathbf{X})) + \mathbf{d}\alpha(\mathbf{X}). \tag{6.278}$$

One may show that this relation holds for any p-form, thus one may write

$$\boxed{\pounds_{\mathbf{X}} = \mathbf{d}\circ\iota_{\mathbf{X}} + \iota_{\mathbf{X}}\circ\mathbf{d},} \tag{6.279}$$

where $\iota_{\mathbf{X}}$ is the interior product (or a contraction), for Lie derivatives on forms. This formula is called *H. Cartan's Formula*. Using $\mathbf{d}^2 = 0$, we note that

$$\begin{aligned}
\pounds_{\mathbf{X}}\circ\mathbf{d} &= \mathbf{d}\circ\iota_{\mathbf{X}}\circ\mathbf{d} + \iota_{\mathbf{X}}\circ\mathbf{d}^2 = \mathbf{d}\circ\iota_{\mathbf{X}}\circ\mathbf{d} & (6.280) \\
\mathbf{d}\circ\pounds_{\mathbf{X}} &= \mathbf{d}^2\circ\iota_{\mathbf{X}} + \mathbf{d}\circ\iota_{\mathbf{X}}\circ\mathbf{d} = \mathbf{d}\circ\iota_{\mathbf{X}}\circ\mathbf{d}. & (6.281)
\end{aligned}$$

Thus in general

$$\boxed{\pounds_{\mathbf{X}}\circ\mathbf{d} = \mathbf{d}\circ\pounds_{\mathbf{X}}.} \tag{6.282}$$

The Lie derivative commutes with the exterior derivative.

Example 6.11 (The divergence of a vector field)
Let ϵ be the volume form. The divergence of $\mathbf{X}$ is defined to be the scalar $\nabla \cdot \mathbf{X}$ given by the formula

$$\pounds_{\mathbf{X}}\epsilon = (\nabla \cdot \mathbf{X})\,\epsilon. \tag{6.283}$$

In local coordinates

$$\epsilon = \sqrt{|g|}\mathbf{dx}^1 \wedge ... \wedge \mathbf{dx}^n. \tag{6.284}$$

According to H. Cartan's formula

$$
\begin{aligned}
\pounds_{\mathbf{X}}\epsilon &= \mathbf{d}\,(\iota_{\mathbf{X}}\epsilon) = \mathbf{d}\sum_{\mu}(-1)^{\mu-1}\sqrt{|g|}\mathbf{dx}^1 \wedge ... \iota_{\mathbf{X}}\mathbf{dx}^\mu \wedge ... \wedge \mathbf{dx}^n \\
&= \mathbf{d}\sum_{\mu}(-1)^{\mu-1}\left(\sqrt{|g|}X^\mu\right)\mathbf{dx}^1 \wedge ...\widehat{\mathbf{dx}^\mu} \wedge ... \wedge \mathbf{dx}^n,
\end{aligned} \tag{6.285}
$$

where $\widehat{\mathbf{dx}^\mu}$ means that $\mathbf{dx}^\mu$ shall be omitted from the wedge product. Taking the exterior derivative we get

$$
\begin{aligned}
\pounds_{\mathbf{X}}\epsilon &= \sum_{\mu}(-1)^{\mu-1}\left[\frac{\partial}{\partial x^\mu}\left(\sqrt{|g|}X^\mu\right)\mathbf{dx}^\mu\right] \wedge \mathbf{dx}^1 \wedge ...\widehat{\mathbf{dx}^\mu} \wedge ... \wedge \mathbf{dx}^n \\
&= \sum_{\mu}\left[\frac{\partial}{\partial x^\mu}\left(\sqrt{|g|}X^\mu\right)\right]\mathbf{dx}^1 \wedge ...\mathbf{dx}^\mu \wedge ... \wedge \mathbf{dx}^n.
\end{aligned} \tag{6.286}
$$

Hence,

$$\boxed{\nabla \cdot \mathbf{X} = \frac{1}{\sqrt{|g|}}\frac{\partial}{\partial x^\mu}\left(\sqrt{|g|}X^\mu\right),} \tag{6.287}$$

which is valid in any metric space. Not surprisingly, this is the same as we got in eq. (6.25).

We can now, for instance, show that the Lie derivative of a tensor of rank $\{^0_2\}$ is

$$(\pounds_{\mathbf{X}}\mathbf{T})_{\mu\nu} = T_{\mu\nu;\alpha}X^\alpha + T_{\alpha\nu}X^\alpha_{;\mu} + T_{\mu\alpha}X^\alpha_{;\nu}. \tag{6.288}$$

Invariance and symmetry principles of tensor fields may be described by means of the Lie derivative. In this connection, the concept of *Lie transport* of a tensor field is applied. The tensors at different points of a tensor field $\mathbf{T}$ are connected by Lie transport along a curve if the Lie derivative of $\mathbf{T}$ along the curve vanishes. If $\mathbf{u}$ is the tangent vector along the curve, we say that $\mathbf{T}$ is *Lie transported* iff

$$\pounds_{\mathbf{u}}\mathbf{T} = 0. \tag{6.289}$$

More specifically, a scalar field connected by Lie transport along a curve is constant along it.

The vectors of a Lie transported vector field along a curve commutes with the tangent vectors of the curve. In this case the vector field is said to be *invariant* with respect to the transformation we denoted by ϕ_t. The geometrical interpretation is as follows. Assume that $\mathbf{u}$ is a tangent vector field of a congruence of curves $x^\mu(\lambda)$ and $\mathbf{v}$ a vector field. If the vectors $\mathbf{v}$ are connected by Lie transport along $\mathbf{u}$, they will connect points with the same value of λ on neighbouring curves of the congruence (see figure 6.7).

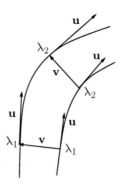

Figure 6.7: The Lie derivative.

6.11   Killing vectors and Symmetries

A important concept in almost all branches in physics is the concept of symmetry. In this section we will define what we mean by symmetries for spaces.

Killing vectors are useful when we are going to describe the symmetry properties of a space in an invariant way, independently of the choice of coordinates.

Consider a space with coordinate system $\{x^\mu\}$ and with a metric

$$\mathbf{g} = g_{\mu\nu}\mathbf{dx}^\mu \otimes \mathbf{dx}^\nu. \tag{6.290}$$

Let $\phi_t(x) \equiv (x'^\mu)$ be a one parameter group of diffeomorphisms, and define a new metric

$$\hat{\mathbf{g}}_t = \phi_t^*\mathbf{g} = \frac{\partial x'^\alpha}{\partial x^\nu}\frac{\partial x'^\beta}{\partial x^\mu}g_{\alpha\beta}\mathbf{dx}^\mu \otimes \mathbf{dx}^\nu. \tag{6.291}$$

This is just the metric at $\phi_t(x)$ instead of at x, pulled back to x. If now this metric happens to be equal to the original one, that is

$$\hat{\mathbf{g}}_t = \mathbf{g}, \tag{6.292}$$

then we say that ϕ_t is an *isometry*. The Killing vectors are related to the isometries as follows. Consider a vector field $\boldsymbol{\xi}$. If the one-parameter groups of diffeomorphisms generated by $\boldsymbol{\xi}$ is an isometry, then we call $\boldsymbol{\xi}$ a *Killing vector field*. This yields an equivalent definition of a Killing vector field. We define a Killing vector field by the relation

$$\boxed{\pounds_{\boldsymbol{\xi}}\mathbf{g} = 0.} \tag{6.293}$$

This can be seen as follows. If $\boldsymbol{\xi}$ is a Killing vector field, and ϕ_t its flow, then by definition

$$\phi_t^*\mathbf{g} = \mathbf{g}, \tag{6.294}$$

so that

$$\pounds_{\boldsymbol{\xi}}\mathbf{g} = \lim_{t \longrightarrow 0}\frac{1}{t}(\phi_t^*\mathbf{g} - \mathbf{g}) = 0. \tag{6.295}$$

The Lie derivative of the metric tensor along a Killing vector vanishes.
In component form this equation is

$$\left(\pounds_\xi g\right)_{\mu\nu} = g_{\mu\nu;\alpha}\xi^\alpha + g_{\mu\alpha}\xi^\alpha_{\ ;\nu} + g_{\alpha\nu}\xi^\alpha_{\ ;\mu} = 0. \tag{6.296}$$

Since the covariant derivative of the metric tensor vanishes, this equation can
be written

$$\boxed{\xi_{\mu;\nu} + \xi_{\nu;\mu} = 0.} \tag{6.297}$$

This is *Killing's equation*. In this form Killing's equation is valid in an arbitrary basis. In a coordinate basis the equation reduces to

$$\xi_{\mu,\nu} + \xi_{\nu,\mu} = 2\xi_\alpha \Gamma^\alpha_{\mu\nu}. \tag{6.298}$$

It is not difficult to show that if $\xi^{(1)}$ and $\xi^{(2)}$ are two Killing vectors, a
and b two constants, then $a\xi^{(1)} + b\xi^{(2)}$ is a Killing vector. Furthermore, the
commutator $[\xi^{(1)}, \xi^{(2)}]$ is also a Killing vector.

In an n-dimensional space there are maximally $\frac{n}{2}(n+1)$ linearly independent Killing vectors. In four-dimensional space-time there may be up to 10
such vectors. A metric and the corresponding space that admits the maximally number of Killing vectors is said to be maximally symmetric. These
spaces are classified as follows [Kob72].

Theorem   *Let M be an n-dimensional maximally symmetric Riemannian space.
Then M must be one of the following spaces.*

1. *The n-dimensional sphere, S^n.*

2. *The n-dimensional projective space, $\mathbb{P}^n$.*

3. *The n-dimensional Euclidean space, $\mathbb{E}^n$.*

4. *The n-dimensional hyperbolic space, $\mathbb{H}^n$.*

These spaces and their maximally symmetric metrics will be investigated
in the next chapter.

Example 6.12 (2-dimensional Symmetry surfaces)   Example
Let us consider a 2-dimensional plane with the Euclidean metric

$$ds^2 = dx^2 + dy^2. \tag{6.299}$$

In a Cartesian coordinate system the Christoffel symbols will vanish, and hence,
Killing's equation (6.297) reduces to

$$\xi_{x,x} = \xi_{y,y} = 0, \quad \xi_{x,y} + \xi_{y,x} = 0. \tag{6.300}$$

There are three independent solutions to this equation. These are

$$\xi_1 = \frac{\partial}{\partial x}, \quad \xi_2 = \frac{\partial}{\partial y}, \quad \xi_3 = x\frac{\partial}{\partial y} - y\frac{\partial}{\partial x}. \tag{6.301}$$

These Killing vectors are characteristic for plane symmetry. By calculating their commutators, we get

$$[\xi_1, \xi_2] = 0, \quad [\xi_1, \xi_3] = \xi_2, \quad [\xi_2, \xi_3] = -\xi_1. \tag{6.302}$$

If a space possesses 3 Killing vectors which span a 2-dimensional tangent space at every point, and, in addition, obey the commutator relations (6.302), we say that the space is *plane symmetric*. In this way we can characterize plane symmetry entirely on the basis of the Killing vectors.

Similarly, assume that a space possesses 3 Killing vectors which everywhere span a 2-dimensional tangent plane. Then we call the space

1. *Spherical symmetric* if the commutator relations are

$$[\boldsymbol{\xi}_1, \boldsymbol{\xi}_2] = \boldsymbol{\xi}_3, \quad [\boldsymbol{\xi}_2, \boldsymbol{\xi}_3] = \boldsymbol{\xi}_1, \quad [\boldsymbol{\xi}_3, \boldsymbol{\xi}_1] = \boldsymbol{\xi}_2.$$

2. *Hyperbolic-plane symmetric* if the commutator relations are

$$[\boldsymbol{\xi}_1, \boldsymbol{\xi}_2] = -\boldsymbol{\xi}_3, \quad [\boldsymbol{\xi}_2, \boldsymbol{\xi}_3] = \boldsymbol{\xi}_1, \quad [\boldsymbol{\xi}_3, \boldsymbol{\xi}_1] = \boldsymbol{\xi}_2.$$

Furthermore, a space is called *cylindrical symmetric* if the space possesses two commuting Killing vectors – i.e. $[\boldsymbol{\xi}_1, \boldsymbol{\xi}_2] = 0$ – where the corresponding one-parameter family of diffeomorphisms to $\boldsymbol{\xi}_1$, say, is periodic. The periodicity implies that $\phi_t = \phi_{t+\ell}$ for a constant ℓ. The ℓ corresponds to going around the circumference of the cylinder once.

An *invariant basis* is defined as a basis-field where the basis-vectors are connected by Lie transport along Killing vectors. Let $\{e_\mu\}$ be an invariant basis. Then

$$\pounds_{\boldsymbol{\xi}} e_\mu = [\boldsymbol{\xi}, e_\mu] = 0, \tag{6.303}$$

for an arbitrary Killing vector $\boldsymbol{\xi}$.

The components of a tensor are scalar functions. This means that the Lie derivative of for example the components of the metric tensor, $g_{\mu\nu}$, along a Killing vector field is equal to the directional derivative of $g_{\mu\nu}$ along $\boldsymbol{\xi}$. Thus

$$\boldsymbol{\xi}(g_{\mu\nu}) = \pounds_{\boldsymbol{\xi}} [\mathbf{g}(e_\mu, e_\nu)] = \mathbf{g}\left(\pounds_{\boldsymbol{\xi}} e_\mu, e_\nu\right) + \mathbf{g}\left(e_\mu, \pounds_{\boldsymbol{\xi}} e_\nu\right). \tag{6.304}$$

If $\{e_\mu\}$ is an invariant basis, then

$$\boldsymbol{\xi}(g_{\mu\nu}) = 0. \tag{6.305}$$

The components of the metric tensor are constants along Killing vectors in a space with an invariant basis field.

There is an interesting relation between particle motion and Killing vectors. In Lagrangian dynamics we have the notion of cyclic coordinates, and we will now find a relation between Killing vectors and cyclic coordinates. Assume that x^α is a cyclic coordinate, and consider the vector

$$\frac{\partial}{\partial x^\alpha} = \delta^\mu{}_\alpha \frac{\partial}{\partial x^\mu}. \tag{6.306}$$

The covariant derivative of the covariant components of this vector is

$$\begin{aligned} (g_{\mu\rho}\delta^\rho{}_\alpha)_{;\nu} &= g_{\mu\rho}\delta^\rho{}_{\alpha;\nu} = g_{\mu\rho}\Gamma^\rho{}_{\nu\alpha} = \Gamma_{\mu\nu\alpha} \\ &= \frac{1}{2}\left(g_{\mu\nu,\alpha} + g_{\mu\alpha,\nu} - g_{\nu\alpha,\mu}\right). \end{aligned} \tag{6.307}$$

Since x^α is a cyclic coordinate $g_{\mu\nu,\alpha} = 0$, so that

$$(g_{\mu\rho}\delta^\rho{}_\alpha)_{;\nu} = g_{\alpha[\mu,\nu]}. \tag{6.308}$$

The fact that $(g_{\mu\rho}\delta^\rho_\alpha)_{;\nu}$ is antisymmetric in μ and ν is a sufficient condition that the vector $\frac{\partial}{\partial x^\alpha}$ fulfills Killing's equations. We then have the following result. *The coordinate basis vector $\frac{\partial}{\partial x^\alpha}$ associated with a cyclic coordinate is a Killing vector.* Even if this does not give all the Killing vectors of a space, this is a useful result when one shall find the Killing vectors of the space.

We can also find the relation between the Killing vectors of a space and the constants of motion of a particle moving freely in that space.

A free particle moves along a geodesic curve, with equation

$$\nabla_{\mathbf{u}}\mathbf{u} = 0, \tag{6.309}$$

where $\mathbf{u}$ is the four-velocity of the particle. Consider the scalar product $\mathbf{u} \cdot \boldsymbol{\xi}$ where $\boldsymbol{\xi}$ is a Killing vector field. The covariant directional derivative of this product along the geodesic curve is

$$\nabla_{\mathbf{u}}(\mathbf{u} \cdot \boldsymbol{\xi}) = u^\alpha u^\mu_{;\alpha}\xi_\mu + u^\alpha u^\mu \xi_{\mu;\alpha}. \tag{6.310}$$

Here the first term vanishes because of eq. (6.309) and the second vanishes since $u^\alpha u^\mu$ is symmetric and $\xi_{\mu;\alpha}$ is antisymmetric in μ and α. Hence,

$$\nabla_{\mathbf{u}}(\mathbf{u} \cdot \boldsymbol{\xi}) = 0. \tag{6.311}$$

We then have the result that $\mathbf{u} \cdot \boldsymbol{\xi}$ is constant along a geodesic curve. For a particle with constant rest mass this may also be expressed as $\mathbf{p} \cdot \boldsymbol{\xi}$ where $\mathbf{p}$ is the four-momentum of the particle.

In the case where $\boldsymbol{\xi}_\alpha$ is associated with a cyclic coordinate x^α we get

$$\mathbf{p} \cdot \boldsymbol{\xi}_\alpha = p_\mu \xi^\mu_\alpha = p_\mu \delta^\mu_\alpha = p_\alpha. \tag{6.312}$$

Thus $\mathbf{p} \cdot \boldsymbol{\xi}_\alpha$ is equal to the covariant canonical momentum to a cyclic coordinate. As we have seen earlier, this is a constant of motion for a free particle in a gravitational field.

Problems

6.1. *Loop integral of a closed form*
By using complex analysis we will show that the integral

$$\oint \omega = \oint \frac{x\mathbf{dy} - y\mathbf{dx}}{x^2 + y^2}, \tag{6.313}$$

equals 2π for *any* loop that encircles the origin once in the anti-clockwise direction.

(a) Let us define the complex variable z by $z = x + iy$. Show that

$$\oint \frac{dz}{iz} = \oint \left[\frac{x dx + y dy}{i(x^2 + y^2)} + \frac{x dy - y dx}{(x^2 + y^2)}\right]. \tag{6.314}$$

(b) Using the residue theorem from complex analysis, show that

$$\oint \frac{x dx + y dy}{i(x^2 + y^2)} = 0. \tag{6.315}$$

(c) Show that for any loop c_1 that encircles the origin $z = 0$ once in the anti-clockwise direction, the integral is given by

$$\oint_{c_1} \omega = 2\pi, \qquad (6.316)$$

and show that in general for a loop c_n that encircles the origin n times, the integral is given by

$$\oint_{c_n} \omega = 2\pi n. \qquad (6.317)$$

Note also that the orientation is incorporated into this formula. If the loop goes in the anti-clockwise direction, the integer n is positive, while if it is clockwise then n is negative.

6.2. *The covariant derivative*

(a) Assume that $A^{\mu\nu}{}_\lambda$ are the components of a tensor. Show that $A^{\mu\nu}{}_\nu$ will transform as vector components, while $A^{\mu\mu}{}_\lambda$ will not (summation over repeated indices).

(b) Show, using the expression

$$\Gamma_{\mu\nu\lambda} = \frac{1}{2}\left(g_{\mu\nu,\lambda} + g_{\mu\lambda,\nu} - g_{\nu\lambda,\mu}\right),$$

that $\Gamma^\mu{}_{\nu\lambda}$ are not the components of a tensor.

(c) Assume that $A^\mu(x)$ is a vector field. Show that $A^\mu{}_{,\nu} \equiv \frac{\partial A^\mu}{\partial x^\nu}$ does not transform according to a tensor, but that the covariant derivative

$$A^\mu{}_{;\nu} = A^\mu{}_{,\nu} + \Gamma^\mu{}_{\lambda\nu} A^\lambda \qquad (6.318)$$

does.

(d) Show the following relations:

$$\begin{aligned} g_{\mu\nu;\lambda} &= 0, \\ (A^\mu B_\nu)_{;\lambda} &= A^\mu{}_{;\lambda} B_\nu + A^\mu B_{\nu;\lambda}. \end{aligned}$$

$$(6.319)$$

Show also that the covariant divergence can be expressed as

$$\nabla \cdot \mathbf{A} \equiv A^\mu{}_{;\mu} = \frac{1}{\sqrt{|g|}} \frac{\partial}{\partial x^\mu}\left(\sqrt{|g|} A^\mu\right). \qquad (6.320)$$

6.3. *The Poincaré half-plane*

The Poincaré half-plane is the upper half of $\mathbb{R}^2$ given by $\mathbb{R}^2_+ = \{(x,y) \in \mathbb{R}^2 | y > 0\}$, see Fig. 6.8, equipped with the metric

$$ds^2 = \frac{dx^2 + dy^2}{y^2}. \qquad (6.321)$$

(a) Use the orthonormal frame formalism and calculate the rotation forms.

(b) Using for instance the variational principle, show that the geodesics are semi-circles centered at $y = 0$ *or* lines of constant x.

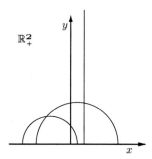

Figure 6.8: Geodesics in the Poincaré half-plane.

6.4. *The Christoffel symbols in a rotating reference frame with plane polar coordinates*
Show, using the transformation (5.6) on page 90, that the Christoffel symbols
in a rotating reference frame with plane polar coordinates are given by

$$\Gamma^r_{tt} = -\omega^2 r, \qquad \Gamma^r_{\theta\theta} = -r,$$
$$\Gamma^r_{\theta t} = \Gamma^r_{t\theta} = -\omega r, \qquad \Gamma^\theta_{rt} = \Gamma^\theta_{tr} = \frac{\omega}{r},$$
$$\Gamma^\theta_{\theta r} = \Gamma^\theta_{r\theta} = \frac{1}{r}. \tag{6.322}$$

All other components are zero.

7

Curvature

We have seen, for example in rotating reference frames, that the geometry in a space with non-vanishing acceleration of gravity, may be non-Euclidean. It is easy to visualize curves and surfaces in three-dimensional space but it is difficult to grasp visually what curvature means in three-dimensional space, or worse still, in four-dimensional space-time. However the curvature of such spaces may be discussed using the lower dimensional analogues of curves and surfaces. It is therefore important to have a good knowledge of the differential geometry of surfaces. Also the formalism used in describing surfaces may be taken over with minor modifications, when we are going to describe the geometric properties of curved space-time.

7.1 Curves

Let us consider curves in Euclidean three-dimensional space, $\mathbb{E}^3$, and let $\mathbf{r}(s)$ be the position vector of points parametrized by the *arc length* s. The unit tangent vector along the curve is

$$\mathbf{t} = \frac{d\mathbf{r}}{ds} = \frac{dx^i}{ds}\mathbf{e}_i. \qquad (7.1)$$

The faster the unit tangent vector changes the direction along the curve the more curved the curve is. The *curvature vector* of the curve is defined by

$$\mathbf{k} = \frac{d\mathbf{t}}{ds}. \qquad (7.2)$$

Since $\mathbf{t} \cdot \mathbf{t} = 1$ we have by differentiation $\mathbf{t} \cdot \mathbf{k} = 0$, thus the curvature vector is always orthogonal to the tangent vector. The length of the curvature vector is called the curvature of the curve, and is denoted by κ:

$$\kappa \equiv |\mathbf{k}|. \qquad (7.3)$$

A curve with vanishing curvature is a straight line.

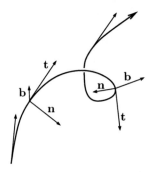

Figure 7.1: A curve in three-dimensional space.

Example 7.1 (The curvature of a circle)
Consider a circle of radius R. The tangent vector is

$$\mathbf{t} = \frac{d\mathbf{r}}{ds} = \frac{dr}{ds}\mathbf{e}_r + \frac{d\theta}{ds}\mathbf{e}_\theta = \frac{1}{R}\mathbf{e}_\theta \tag{7.4}$$

since $r = R$ and $s = R\theta$ along the circle. The curvature vector can now be found with

$$\mathbf{k} = \frac{d\mathbf{t}}{ds} = \frac{1}{R^2}\frac{d\mathbf{e}_\theta}{d\theta} = -\frac{1}{R}\mathbf{e}_r. \tag{7.5}$$

Thus the curvature of the circle is $\kappa = \frac{1}{R}$.

The unit vector $\mathbf{n}$ defined by

$$\boxed{\frac{d\mathbf{t}}{ds} = \kappa\mathbf{n},} \tag{7.6}$$

is called the *principal normal vector* of the curve. The vectors $\mathbf{t}$ and $\mathbf{n}$ span a plane which is called the *osculating plane* of the curve. This plane turns as we move along the curve. The unit normal vector $\mathbf{b}$ of this plane, defined by

$$\mathbf{b} = \mathbf{t} \times \mathbf{n}, \tag{7.7}$$

is called the *binormal vector* of the curve (see Fig. 7.1). The rate of turning of the osculating plane is given by $\frac{d\mathbf{b}}{ds}$. Since $\mathbf{t} \cdot \mathbf{b} = 0$, it follows by differentiation that

$$\frac{d\mathbf{t}}{ds} \cdot \mathbf{b} + \mathbf{t} \cdot \frac{d\mathbf{b}}{ds} = 0. \tag{7.8}$$

Combining this with equations (7.6) and (7.7) it follows that $\frac{d\mathbf{b}}{ds}$ is orthogonal to $\mathbf{t}$. Since $\mathbf{b}$ has constant length, $\frac{d\mathbf{b}}{ds}$ has no component along $\mathbf{b}$ either. Thus $\frac{d\mathbf{b}}{ds}$ points in the $\mathbf{n}$ direction. The *torsion* τ of a curve is defined by the equation

$$\boxed{\frac{d\mathbf{b}}{ds} = -\tau\mathbf{n}.} \tag{7.9}$$

The vectors $\{\mathbf{t}, \mathbf{n}, \mathbf{b}\}$ represent three orthonormal basis vector fields along the curve (see Fig. 7.1). They are related by eq. (7.7) together with

$$\mathbf{t} = \mathbf{n} \times \mathbf{b}, \qquad \mathbf{n} = \mathbf{b} \times \mathbf{t}. \tag{7.10}$$

The variation of **n** along the curve is now given by

$$\frac{d\mathbf{n}}{ds} = \frac{d\mathbf{b}}{ds} \times \mathbf{t} + \mathbf{b} \times \frac{d\mathbf{n}}{ds} = -\tau \mathbf{n} \times \mathbf{t} + \kappa \mathbf{b} \times \mathbf{n} \tag{7.11}$$

so

$$\boxed{\frac{d\mathbf{n}}{ds} = \tau \mathbf{b} - \kappa \mathbf{t}.} \tag{7.12}$$

Equations (7.6), (7.9) and (7.12) are called the *Serret-Frenet equations*.

7.2 Surfaces

Consider a two-dimensional surface embedded in three-dimensional Euclidean space. Let u and v be coordinates (or parameters) on the surface. Then, at every point on the surface the basis vectors

$$\mathbf{e}_u = \frac{\partial}{\partial u}, \qquad \mathbf{e}_v = \frac{\partial}{\partial v}, \tag{7.13}$$

define a tangent plane of the surface, see Fig. 7.2.

The line element on the surface is

$$ds^2 = g_{\mu\nu} dx^\mu dx^\nu, \tag{7.14}$$

where $x^1 = u$ and $x^2 = v$. This is often called *the first fundamental form* of the surface.

The directional derivative of $\mathbf{e}_\mu$ along $\mathbf{e}_\nu$ has generally one component in the tangent plane of the surface and one component orthogonal to the surface,

$$\boxed{\mathbf{e}_{\mu,\nu} = \Gamma^\alpha_{\mu\nu} \mathbf{e}_\alpha + K_{\mu\nu} \mathbf{n},} \tag{7.15}$$

where $\Gamma^\alpha_{\mu\nu}$ are the connection coefficients of the u, v system and **n** is the unit normal vector field on the surface. The coefficients $K_{\mu\nu}$ are defined by this equation. Since $\mathbf{e}_{\mu,\nu} = \frac{\partial^2}{\partial x^\mu \partial x^\nu} = \frac{\partial^2}{\partial x^\nu \partial x^\mu} = \mathbf{e}_{\nu,\mu}$ these coefficients are symmetric. Note also that the eq. (7.15) provides us with an interpretation of the covariant derivative and the connection coefficients $\Gamma^\alpha_{\mu\nu}$. The covariant derivative of a surface embedded in an Euclidean space is the ordinary derivative in Euclidean space, projected onto the tangent space of the surface.

The equation (7.15) is usually called *Gauss' equations*. Using

$$K_{\mu\nu} = \mathbf{e}_{\mu,\nu} \cdot \mathbf{n}, \tag{7.16}$$

we can calculate the coefficients $K_{\mu\nu}$. Let now **u** be the tangent vector of a curve in the surface, parametrized by λ. Using equation (7.15) we have

$$\frac{d\mathbf{u}}{d\lambda} = u^\mu_{;\nu} u^\nu \mathbf{e}_\mu + K_{\mu\nu} u^\mu u^\nu \mathbf{n}. \tag{7.17}$$

The λ-independent coefficient in front of **n**, $K_{\mu\nu} dx^\mu dx^\nu$, is called *the second fundamental form* of the surface. Whereas the first fundamental form determines the intrinsic geometry of the surface, the second fundamental form reflects the extrinsic geometry, i.e., how the surface curves in the ambient three-dimensional Euclidean space in which it is embedded.

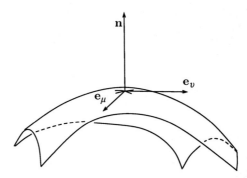

Figure 7.2: A two-dimensional surface embedded in three-dimensional Euclidean space.

If λ is the arc-length, we can write equation (7.17) as

$$\frac{d\mathbf{u}}{d\lambda} = \kappa_g \mathbf{e} + \kappa_n \mathbf{n}, \quad \kappa_g \equiv |u^\mu_{;\nu} u^\nu \mathbf{e}_\mu|, \quad \kappa_n \equiv K_{\mu\nu} u^\mu u^\nu, \tag{7.18}$$

where $\mathbf{e}$ is a unit vector. The quantities κ_g and κ_n are called *the geodesic curvature* and *the normal curvature* of the surface, respectively.

Since $\frac{d\mathbf{u}}{d\lambda}$ is orthogonal to $\mathbf{u}$ and $\mathbf{n}$, the unit vector $\mathbf{e}$ in the surface is given by

$$\mathbf{e} = \pm \mathbf{n} \times \mathbf{u}. \tag{7.19}$$

We see that the normal curvature can also be written

$$\kappa_n = \frac{d\mathbf{u}}{d\lambda} \cdot \mathbf{n}. \tag{7.20}$$

By differentiating $\mathbf{u} \cdot \mathbf{n}$ we get

$$\mathbf{u} \cdot \frac{d\mathbf{n}}{d\lambda} = -\kappa_n = -K_{\mu\nu} u^\mu u^\nu. \tag{7.21}$$

This is *Weingarten's equation*.

The geodesic and normal curvatures, taken separately, characterize the extrinsic geometry of the surface. We shall now find a quantity, called the Gaussian curvature, which is a measure of the intrinsic geometry of the surface.

At an arbitrary point on the surface we consider geodesic curves through the point with tangent vectors $\mathbf{u} = u^\mu \mathbf{e}_\mu$. In order to compare the normal curvature of the geodesics having different directions, we choose tangent vectors of unit length,

$$\mathbf{u} \cdot \mathbf{u} = g_{\mu\nu} u^\mu u^\nu = 1. \tag{7.22}$$

The directions with maximal and minimal values of the normal curvature are found by extremizing κ_n as given by eq. (7.18) with the constraint eq. (7.22). Thus we have to solve the variational problem δF for arbitrary u^μ, where

$$F = K_{\mu\nu} u^\mu u^\nu - k(g_{\mu\nu} u^\mu u^\nu - 1). \tag{7.23}$$

Here, k is a Lagrange multiplier. Variation with respect to u^μ gives

$$\delta F = 2(K_{\mu\nu} - k g_{\mu\nu}) u^\nu \delta u^\mu. \tag{7.24}$$

Thus we must have

$$(K_{\mu\nu} - kg_{\mu\nu})u^\nu = 0. \tag{7.25}$$

This set of equations has non-trivial solutions whenever

$$\det (K_{\mu\nu} - kg_{\mu\nu}) = 0, \tag{7.26}$$

which yields the following quadratic equation for k

$$k^2 \det(g_{\mu\nu}) - k(g_{11}K_{22} - 2g_{12}K_{12} + g_{22}K_{11}) + \det(K_{\mu\nu}) = 0, \tag{7.27}$$

with solutions k_1 and k_2. These are the extremal values of k.

To see the meaning of k we multiply eq. (7.25) by u^μ and use eq. (7.18). This gives

$$k = \kappa_n. \tag{7.28}$$

The extremal values of k are called the *principal curvatures* of the surface. The *Gaussian curvature* of the surface is defined as

$$K = k_1 k_2. \tag{7.29}$$

From eq. (7.27) it follows that

$$\boxed{K = \frac{\det(K_{\mu\nu})}{\det(g_{\mu\nu})}.} \tag{7.30}$$

Let the directions corresponding to the principal curvatures be characterized by the vectors $\mathbf{u}$ and $\mathbf{v}$. From equation (7.25) and the symmetry of $K_{\mu\nu}$ follows

$$(k_1 - k_2)(\mathbf{u} \cdot \mathbf{v}) = 0. \tag{7.31}$$

For $k_1 \neq k_2$ this implies that $\mathbf{u}$ and $\mathbf{v}$ are orthogonal. The principal curvatures are found in orthogonal directions.

A positive Gaussian curvature means that the principal curvatures are of the same sign. The surface is locally similar to a distorted sphere, and the geometry is locally elliptic. If the Gaussian curvature vanishes, one of the principal curvatures has to vanish and the geometry is locally planar and the geometry is locally Euclidean. Finally, if the Gaussian curvature is negative the surface is locally saddle-shaped, i.e. like a hyperbolic surface. In this case the geometry is said to be locally hyperbolic.

In the following sections we will show that the Gaussian curvature represents a measure of the intrinsic geometry of the surface. This will be achieved by expressing it in terms of the components of the Riemann tensor, which we will introduce in the next section. Note however that curves have no intrinsic curvature.

7.3 The Riemann Curvature Tensor

In this section we will not restrict ourselves to two-dimensional surfaces, but describe spaces of arbitrary number of dimensions. We will introduce the important concept of curvature in an invariant way, described by the Riemann curvature tensor.

Consider two nearby points Q and P connected by a vector $\delta\lambda\mathbf{v}$ of infinitesimal length. Let $\mathbf{A}$ be a vector field. Then the difference between the vector field at Q, denoted $\mathbf{A}_Q$, and the vector $\mathbf{A}_P$ parallel-transported from P to Q (which is denoted $\mathbf{A}_{PQ}$) is to first order in $\delta\lambda$ given by the directional derivative at Q of $\mathbf{A}$ in the v-direction

$$\delta\lambda\nabla_{\mathbf{v}}\mathbf{A} \approx \mathbf{A}_Q - \mathbf{A}_{PQ}. \tag{7.32}$$

Thus,

$$\mathbf{A}_{PQ} \approx (1 - \delta\lambda\nabla_{\mathbf{v}})\,\mathbf{A}_Q. \tag{7.33}$$

To second order $\mathbf{A}_{PQ}$ is given by the first terms of the Taylor expansion

$$\mathbf{A}_{PQ} \approx \left(1 - \delta\lambda\nabla_{\mathbf{v}} + \frac{1}{2}\delta\lambda^2\nabla_{\mathbf{v}}\nabla_{\mathbf{v}}\right)\mathbf{A}_Q. \tag{7.34}$$

We are now going to parallel-transport the vector $\mathbf{A}_P$ around the polygon shown in Fig. 7.3. Parallel-transporting $\mathbf{A}_{PQ}$ from Q to R gives

$$\mathbf{A}_{PQR} \approx \left(1 - \delta\lambda\nabla_{\mathbf{u}} + \frac{1}{2}\delta\lambda^2\nabla_{\mathbf{u}}\nabla_{\mathbf{u}}\right)\left(1 - \delta\lambda\nabla_{\mathbf{v}} + \frac{1}{2}\delta\lambda^2\nabla_{\mathbf{v}}\nabla_{\mathbf{v}}\right)\mathbf{A}_R. \tag{7.35}$$

Proceeding around the polygon gives

$$\begin{aligned}
\mathbf{A}_{PQRSTP} \approx &\left(1 + \delta\lambda\nabla_{\mathbf{u}} + \frac{1}{2}\delta\lambda^2\nabla_{\mathbf{u}}\nabla_{\mathbf{u}}\right)\left(1 + \delta\lambda\nabla_{\mathbf{v}} + \frac{1}{2}\delta\lambda^2\nabla_{\mathbf{v}}\nabla_{\mathbf{v}}\right) \\
&\times\ \left(1 - \delta\lambda^2\nabla_{[\mathbf{u},\mathbf{v}]}\right) \\
&\times\ \left(1 - \delta\lambda\nabla_{\mathbf{u}} + \frac{1}{2}\delta\lambda^2\nabla_{\mathbf{u}}\nabla_{\mathbf{u}}\right)\left(1 - \delta\lambda\nabla_{\mathbf{v}} + \frac{1}{2}\delta\lambda^2\nabla_{\mathbf{v}}\nabla_{\mathbf{v}}\right)\mathbf{A}_P.
\end{aligned} \tag{7.36}$$

Thus, to second order, the change of the vector after parallel-transport around the polygon is

$$\delta\mathbf{A} = \mathbf{A}_{PQRSTP} - \mathbf{A}_P = \left([\nabla_{\mathbf{u}}, \nabla_{\mathbf{v}}] - \nabla_{[\mathbf{u},\mathbf{v}]}\right)\delta\lambda^2\mathbf{A}_P. \tag{7.37}$$

In flat space there would be no such change of $\mathbf{A}$. This change is due to the curvature of the space. It can be shown that $\delta\mathbf{A}$ as given in eq. (7.37) is a vector which is linear in $\mathbf{A}$, $\mathbf{u}$ and $\mathbf{v}$ (see e.g. [vW81]). Thus $\delta\mathbf{A}$ may be expressed by a tensor of rank $\{^1_3\}$. This tensor $\mathbf{R}$ is called the *Riemann curvature tensor* and is defined by

$$\boxed{\mathbf{R}(\mathbf{u}, \mathbf{v})\mathbf{A} \equiv \left([\nabla_{\mathbf{u}}, \nabla_{\mathbf{v}}] - \nabla_{[\mathbf{u},\mathbf{v}]}\right)\mathbf{A}.} \tag{7.38}$$

The components of this tensor are given by

$$\mathbf{e}_\mu R^\mu{}_{\nu\alpha\beta} = \left([\nabla_\alpha, \nabla_\beta] - \nabla_{[\mathbf{e}_\alpha, \mathbf{e}_\beta]}\right)\mathbf{e}_\nu. \tag{7.39}$$

It follows that the Riemann curvature tensor is antisymmetric in $\mathbf{u}$ and $\mathbf{v}$, i.e., in α and β. We can therefore define a matrix of two-forms

$$\boxed{\mathbf{R}^\mu{}_\nu = \frac{1}{2}R^\mu{}_{\nu\alpha\beta}\boldsymbol{\omega}^\alpha \wedge \boldsymbol{\omega}^\beta,} \tag{7.40}$$

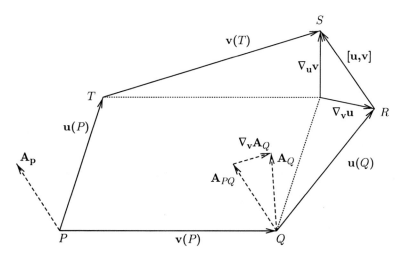

Figure 7.3: The vector $\mathbf{A}_P$ parallel transported around the polygon $PQRSTP$.

which are called the *curvature forms*.

Equation (7.37) may now be written

$$\delta\mathbf{A} = \mathbf{e}_\mu R^\mu{}_{\nu\alpha\beta} A^\nu u^\alpha v^\beta. \tag{7.41}$$

The infinitesimal area of the polygon is to lowest order in $\mathbf{u}$ and $\mathbf{v}$

$$\Delta S^{\alpha\beta} = u^\alpha v^\beta - u^\beta v^\alpha. \tag{7.42}$$

Using the antisymmetry of $R^\mu{}_{\nu\alpha\beta}$ we may then write

$$\delta\mathbf{A} = \frac{1}{2} A^\nu R^\mu{}_{\nu\alpha\beta} \Delta S^{\alpha\beta} \mathbf{e}_\mu. \tag{7.43}$$

This shows that the change of a vector by parallel transport around a closed curve is proportional to the curvature of the space and to the area enclosed by the curve.

We shall now show how $R^\mu{}_{\nu\alpha\beta}$ may be expressed by the connection- and structure coefficients of a given basis. We find

$$
\begin{aligned}
\mathbf{e}_\mu R^\mu{}_{\nu\alpha\beta} &= \left([\nabla_\alpha, \nabla_\beta] - \nabla_{[\mathbf{e}_\alpha, \mathbf{e}_\beta]}\right) \mathbf{e}_\nu \\
&= \left(\nabla_\alpha \nabla_\beta - \nabla_\beta \nabla_\alpha - c^\rho{}_{\alpha\beta} \nabla_\rho\right) \mathbf{e}_\nu \\
&= \left(\Gamma^\mu{}_{\nu\beta,\alpha} + \Gamma^\rho{}_{\nu\beta} \Gamma^\mu{}_{\rho\alpha} - \Gamma^\mu{}_{\nu\alpha,\beta} - \Gamma^\rho{}_{\nu\alpha} \Gamma^\mu{}_{\rho\beta} - c^\rho{}_{\alpha\beta} \Gamma^\mu{}_{\nu\rho}\right) \mathbf{e}_\mu,
\end{aligned} \tag{7.44}
$$

which implies that

$$\boxed{R^\mu{}_{\nu\alpha\beta} = \Gamma^\mu{}_{\nu\beta,\alpha} - \Gamma^\mu{}_{\nu\alpha,\beta} + \Gamma^\rho{}_{\nu\beta} \Gamma^\mu{}_{\rho\alpha} - \Gamma^\rho{}_{\nu\alpha} \Gamma^\mu{}_{\rho\beta} - c^\rho{}_{\alpha\beta} \Gamma^\mu{}_{\nu\rho}.} \tag{7.45}$$

The connection coefficients and structure coefficients are calculated from a basis vector field in the space considered, without reference to any higher dimensional space. Thus eq. (7.45) shows that the Riemann tensor describes the intrinsic geometry of space.

This is an expression of *Gauss' Theorema Egregium*, which says that an inhabitant of, say, a three-dimensional space, may perform measurements, within the three-dimensional space, which reveals to him the curvature of that space. It is not necessary to embed the space in a higher-dimensional one.

The curvature forms may now be written in component form as

$$\mathbf{R}^{\mu}_{\nu} = \left(\Gamma^{\mu}_{\nu\beta,\alpha} - \frac{1}{2} c^{\rho}_{\alpha\beta} \Gamma^{\mu}_{\nu\rho} + \Gamma^{\rho}_{\nu\beta} \Gamma^{\mu}_{\rho\alpha} \right) \boldsymbol{\omega}^{\alpha} \wedge \boldsymbol{\omega}^{\beta}. \tag{7.46}$$

Using equations (6.137) and (6.166) we find

$$\boxed{\mathbf{R}^{\mu}_{\nu} = \mathbf{d}\boldsymbol{\Omega}^{\mu}_{\nu} + \boldsymbol{\Omega}^{\mu}_{\lambda} \wedge \boldsymbol{\Omega}^{\lambda}_{\nu}.} \tag{7.47}$$

This is *Cartan's 2nd structural equation*.

We shall now deduce some identities fulfilled by the Riemann tensor. Using eqs. (6.119) and (6.145) we get

$$
\begin{aligned}
A^{\mu}_{;\beta\alpha} &= A^{\lambda} \Gamma^{\mu}_{\lambda\beta,\alpha} - A^{\lambda} \Gamma^{\tau}_{\lambda\alpha} \Gamma^{\mu}_{\tau\beta} - A^{\mu}_{;\lambda} \Gamma^{\lambda}_{\beta\alpha} + A^{\mu}_{,\beta\alpha} \\
&\quad + \left(A^{\lambda}_{;\alpha} \Gamma^{\mu}_{\lambda\beta} + A^{\lambda}_{;\beta} \Gamma^{\mu}_{\lambda\alpha} \right).
\end{aligned}
\tag{7.48}
$$

The two latter terms are symmetric in α and β and do not contribute to the antisymmetric combination $A^{\mu}_{;\beta\alpha} - A^{\mu}_{;\alpha\beta}$. Hence, by using eqs. (7.45) and (6.135), we get

$$A^{\mu}_{;\beta\alpha} - A^{\mu}_{;\alpha\beta} = A^{\lambda} R^{\mu}_{\lambda\alpha\beta} - A^{\mu}_{;\lambda} T^{\lambda}_{\alpha\beta}. \tag{7.49}$$

Since this is a tensor equation it is valid in an arbitrary basis. Eq. (7.49) is called the *Ricci identity*. In Riemannian geometry it reduces to

$$\boxed{A^{\mu}_{;\beta\alpha} - A^{\mu}_{;\alpha\beta} = R^{\mu}_{\lambda\alpha\beta} A^{\lambda}.} \tag{7.50}$$

Combining this with eq. (6.164) leads to

$$\mathbf{d}^2 \mathbf{A} = \frac{1}{2} R^{\mu}_{\nu\alpha\beta} A^{\nu} \mathbf{e}_{\mu} \otimes \boldsymbol{\omega}^{\alpha} \wedge \boldsymbol{\omega}^{\beta}. \tag{7.51}$$

By exterior differentiation of Cartan's first structural equation, eq. (6.181) combined with Cartan's second structural equation and Poincaré's Lemma, we find

$$\mathbf{R}^{\mu}_{\nu} \wedge \boldsymbol{\omega}^{\nu} = \mathbf{d}\mathbf{T}^{\mu} + \boldsymbol{\Omega}^{\mu}_{\nu} \wedge \mathbf{T}^{\nu}. \tag{7.52}$$

By means of eq. (6.217), this equation takes the form

$$\mathbf{R}^{\mu}_{\nu} \wedge \boldsymbol{\omega}^{\nu} = \mathbf{D}\mathbf{T}^{\mu}. \tag{7.53}$$

This is *Bianchi's first identity*. In Riemannian geometry it reduces to

$$\boxed{\mathbf{R}^{\mu}_{\nu} \wedge \boldsymbol{\omega}^{\nu} = 0.} \tag{7.54}$$

On component form this equation is

$$\boxed{R^{\mu}_{[\nu\alpha\beta]} = 0.} \tag{7.55}$$

By exterior differentiation of Cartan's second structural equation, and using Poincaré's Lemma, we get

$$\mathbf{dR}^{\mu}_{\ \nu} = \mathbf{d\Omega}^{\mu}_{\ \lambda} \wedge \mathbf{\Omega}^{\lambda}_{\ \nu} - \mathbf{\Omega}^{\mu}_{\ \lambda} \wedge \mathbf{d\Omega}^{\lambda}_{\ \nu} = \mathbf{R}^{\mu}_{\ \lambda} \wedge \mathbf{\Omega}^{\lambda}_{\ \nu} - \mathbf{\Omega}^{\mu}_{\ \lambda} \wedge \mathbf{R}^{\lambda}_{\ \nu}, \qquad (7.56)$$

which is more usually written

$$\mathbf{dR}^{\mu}_{\ \nu} + \mathbf{\Omega}^{\mu}_{\ \lambda} \wedge \mathbf{R}^{\lambda}_{\ \nu} - \mathbf{R}^{\mu}_{\ \lambda} \wedge \mathbf{\Omega}^{\lambda}_{\ \nu} = 0. \qquad (7.57)$$

Applying eq. (6.215) to a matrix of two-forms this equation may be written

$$\boxed{\mathbf{DR}^{\mu}_{\ \nu} = 0.} \qquad (7.58)$$

This is *Bianchi's second identity*. On component form this identity becomes

$$\boxed{R^{\mu}_{\ \nu[\alpha\beta;\gamma]} = 0.} \qquad (7.59)$$

An additional symmetry of the Riemann tensor is most easily found by decomposing it in an orthonormal basis field. Applying eq. (6.170) and Cartan's second structural equation (7.47), we get

$$\boxed{R_{\mu\nu\alpha\beta} = -R_{\nu\mu\alpha\beta}.} \qquad (7.60)$$

The fourth and last symmetry of the Riemann tensor is found by applying geodesic normal coordinates. From eqs. (7.45), (6.111) and (6.233) we then get

$$R_{\mu\nu\alpha\beta} = \frac{1}{2}(g_{\mu\beta,\nu\alpha} - g_{\mu\alpha,\nu\beta} + g_{\nu\alpha,\mu\beta} - g_{\nu\beta,\mu\alpha}). \qquad (7.61)$$

It follows that

$$\boxed{R_{\mu\nu\alpha\beta} = R_{\alpha\beta\mu\nu}.} \qquad (7.62)$$

The four symmetries of the Riemann tensor reduce its number of independent components in an n-dimensional space from n^4 to $\frac{1}{12}n^2(n^2-1)$, in four-dimensional space-time from 256 to 20.

We shall now construct a curvature tensor of rank $\{^0_2\}$ by contraction of the Riemann tensor. Note first that

$$R^{\alpha}_{\ \alpha\mu\nu} = 0, \qquad (7.63)$$

because of the antisymmetry in the first two indices. Furthermore

$$R^{\alpha}_{\ \mu\nu\alpha} = -R^{\alpha}_{\ \mu\alpha\nu}. \qquad (7.64)$$

Thus there exists only one independent non-vanishing contraction of the Riemann tensor. This is called the *Ricci tensor* and is usually written

$$\boxed{R_{\mu\nu} = R^{\alpha}_{\ \mu\alpha\nu}.} \qquad (7.65)$$

From eq. (7.62) follows that it is symmetrical. It has $\frac{1}{2}n(n+1)$ independent components in an n-dimensional space; 10 components in four-dimensional space-time.

Contraction of the Ricci tensor gives the *Ricci scalar*

$$\boxed{R = R^{\mu}_{\;\mu}.}$$ (7.66)

Let us calculate the divergence of the Ricci tensor. Contraction of Bianchi's second identity eq. (7.58) (μ with α), yields

$$\begin{aligned} R^{\mu}_{\;\nu\mu\beta;\gamma} + R^{\mu}_{\;\nu\gamma\mu;\beta} + R^{\mu}_{\;\nu\beta\gamma;\mu} &=\\ R_{\nu\beta;\gamma} - R_{\nu\gamma;\beta} + R^{\mu}_{\;\nu\beta\gamma;\mu} &= 0. \end{aligned}$$ (7.67)

Raising the index ν and contracting with γ leads to

$$R^{\nu}_{\;\beta;\nu} - R_{;\beta} + R^{\mu}_{\;\beta;\mu} = 0.$$ (7.68)

Hence the divergence of the Ricci tensor is

$$R^{\nu}_{\;\mu;\nu} = \frac{1}{2}\left(\delta^{\nu}_{\;\mu}R\right)_{;\nu}.$$ (7.69)

Thus the tensor

$$E^{\nu}_{\;\mu} = R^{\nu}_{\;\mu} - \frac{1}{2}\delta^{\nu}_{\;\mu}R,$$ (7.70)

is divergence free, $E^{\nu}_{\;\mu;\nu} = 0$. This is *Einstein's curvature tensor*. Its covariant components are

$$\boxed{E_{\mu\nu} = R_{\mu\nu} - \frac{1}{2}Rg_{\mu\nu}.}$$ (7.71)

It follows immediately that this tensor is symmetric.

In an n-dimensional space the vanishing of the divergence of the Einstein tensor represents n equations. Thus the Einstein tensor has $\frac{1}{2}n(n-1)$ differentially independent components in general, and 6 differentially independent components in four-dimensional space-time.

The Weyl Curvature Tensor

Let us now focus on the case where the dimension of the manifold is 4, like our four-dimensional spacetime. The symmetries of the Riemann tensor imply that the Riemann tensor has 20 independent components in four dimensions. The Ricci tensor, on the other hand, has only 10 independent components. The components of the Riemann tensor which is not captured in the Ricci part form what is called the *Weyl curvature tensor*.

In four dimensions the Weyl curvature tensor is defined by

$$C_{\alpha\beta\gamma\delta} = R_{\alpha\beta\gamma\delta} - g_{\alpha[\gamma}R_{\delta]\beta} + g_{\beta[\gamma}R_{\delta]\alpha} + \frac{1}{3}Rg_{\alpha[\gamma}g_{\delta]\beta}.$$ (7.72)

It possesses the same symmetries as the Riemann tensor,

$$C_{\alpha\beta\gamma\delta} = C_{\gamma\delta\alpha\beta}, \quad C_{\alpha\beta\gamma\delta} = -C_{\beta\alpha\gamma\delta}, \quad C_{\alpha[\beta\gamma\delta]} = 0,$$

and in addition, *contraction over any pair of indices yields zero*:

$$C^{\alpha}_{\;\beta\alpha\delta} = 0.$$ (7.73)

Hence, when contracting the Riemann tensor over two indices only the Ricci part of it will survive. This gives, as we will see later, a physical interpretation of the Weyl tensor. Einstein's equations, which will be introduced in the next chapter, will only involve the Ricci tensor and hence the Weyl tensor represents the *free gravitational field*. Thus even if the Ricci tensor is zero, there can be a free gravitational field encoded in the Weyl tensor. This property gives rise to many interesting phenomena; many of which will be discussed in this book. Two important examples are gravitational waves, and the Schwarzschild vacuum solution.

The definition of the Weyl tensor in any dimension is given in section 18.4. This section also discusses some further properties of the Weyl tensor.

7.4 Extrinsic and Intrinsic Curvature

From eq. (7.45) it is seen that the components of the Riemann tensor may be calculated from the components of the metric tensor. They are defined in terms of basis-vectors in a space which has a curvature specified by the Riemann tensor. Thus the specification of the metric does not presuppose any embedding of the curved space in a higher-dimensional flat space. The Riemann curvature tensor represents intrinsic geometric properties of space, which may be measured by inhabitants of that space. (Such inhabitants are always assumed to be creatures with the same number of dimensions as that of the space they inhabit.) Therefore one says that the Riemann tensor is a measure of the *intrinsic* curvature of space.

In the case of a two-dimensional surface, say a balloon, the intrinsic geometry is measured by two-dimensional creatures on the surface, "flatlanders". In general, if the Riemann tensor in a space vanishes, the space is flat. For a two-dimensional surface this means that the surface may be rolled onto an Euclidean plane without any local changes of the geometry. If the surface is an elastic membrane, no stresses or strains are introduced by this "rolling out" of it. The intrinsic geometry of a cylindrical surface for example, is Euclidean, and has no intrinsic curvature.

However, as seen from an external Euclidean three-dimensional space, the cylindrical surface looks curved. The surface has external or *extrinsic curvature*. We shall now introduce a tensor measuring the extrinsic curvature of a space, which is embedded in a space of one more dimension.

In the following we shall consider a curved n-dimensional space $\mathcal{M}^n$ embedded in an $(n + 1)$-dimensional space $\mathcal{M}^{n+1}$ which also may be curved. Such a space embedded in an space with one dimensional higher, is called a *hypersurface*. Furthermore, Greek indices are associated with $\mathcal{M}^{n+1}$ and Latin indices with $\mathcal{M}^n$.

A measure of the extrinsic curvature of a space is obtained by considering how the direction of the unit normal vector n to the hypersurface changes with position on the hypersurface. *The extrinsic curvature tensor* **K** is a tensor on $\mathcal{M}^n$ of rank $\{^0_2\}$ defined up to a sign ambiguity by

$$K_{ab} = -\mathbf{e}_b \cdot \nabla_a \mathbf{n}, \tag{7.74}$$

where the covariant derivative is taken in the ambient space $\mathcal{M}^{n+1}$. Since n is orthogonal to the basis vectors $\mathbf{e}_b$ on $\mathcal{M}^n$, so that $\nabla_a(\mathbf{e}_b \cdot \mathbf{n}) = 0$ we get

$$K_{ab} = \mathbf{n} \cdot \nabla_a \mathbf{e}_b = \mathbf{n} \cdot \mathbf{e}_\alpha \Gamma^\alpha_{ba} = n^\alpha g_{\alpha\beta} \Gamma^\beta_{ba} = n_\alpha \Gamma^\alpha_{ba}, \tag{7.75}$$

where $\mathbf{n} = n^\alpha \mathbf{e}_\alpha$. We introduce an orthonormal basis $\{\mathbf{e}_{\hat{a}}\}$ on $\mathcal{M}^n$ and a normal unit vector $\mathbf{n} = \mathbf{e}_{\hat{n}}$. By defining $\mathbf{n} \cdot \mathbf{n} = \epsilon = \pm 1$ we get

$$K_{\hat{a}\hat{b}} = \Gamma^{\hat{n}}_{\hat{a}\hat{b}}, \qquad g_{\hat{n}\hat{n}} = \epsilon. \tag{7.76}$$

Equation (7.75) shows that the extrinsic curvature is symmetric.

Let us now deduce the relation between the Riemann tensors of $\mathcal{M}^{n+1}$ and $\mathcal{M}^n$ and the extrinsic curvature of $\mathcal{M}^n$.

We will calculate the Riemann tensors using orthonormal frames. The Riemann tensor in $\mathcal{M}^{n+1}$ projected onto $\mathcal{M}^n$ is given by

$$\left(\mathbf{d}^2 \mathbf{e}_{\hat{b}} \right)_\perp = \left(\frac{1}{2} {}^{(n+1)} R^{\hat{\lambda}}_{\hat{b}\hat{\alpha}\hat{\beta}} \mathbf{e}_{\hat{\lambda}} \otimes \boldsymbol{\omega}^{\hat{\alpha}} \wedge \boldsymbol{\omega}^{\hat{\beta}} \right)_\perp = \frac{1}{2} {}^{(n+1)} R^{\hat{a}}_{\hat{b}\hat{c}\hat{d}} \mathbf{e}_{\hat{a}} \otimes \boldsymbol{\omega}^{\hat{c}} \wedge \boldsymbol{\omega}^{\hat{d}}. \tag{7.77}$$

We can also write this in another way, using instead the Riemann tensor in $\mathcal{M}^n$

$$\begin{aligned}
\left(\mathbf{d}^2 \mathbf{e}_{\hat{b}} \right)_\perp &= \left(\mathbf{d} \left[\mathbf{e}_{\hat{a}} \otimes \boldsymbol{\Omega}^{\hat{a}}_{\hat{b}} \right] \right)_\perp = \left(\mathbf{d}\mathbf{e}_{\hat{a}} \otimes \boldsymbol{\Omega}^{\hat{a}}_{\hat{b}} \right)_\perp + \left(\mathbf{e}_{\hat{a}} \otimes \mathbf{d}\boldsymbol{\Omega}^{\hat{a}}_{\hat{b}} \right)_\perp \\
&= \mathbf{e}_{\hat{a}} \otimes \left(\mathbf{d}\boldsymbol{\Omega}^{\hat{a}}_{\hat{b}} + \boldsymbol{\Omega}^{\hat{a}}_{\hat{\alpha}} \wedge \boldsymbol{\Omega}^{\hat{\alpha}}_{\hat{b}} \right)_\perp.
\end{aligned} \tag{7.78}$$

We can now use Cartan's second structural equation in $\mathcal{M}^n$ by decomposing the wedge product

$$\begin{aligned}
\left(\mathbf{d}^2 \mathbf{e}_{\hat{b}} \right)_\perp &= \mathbf{e}_{\hat{a}} \otimes \left(\mathbf{d}\boldsymbol{\Omega}^{\hat{a}}_{\hat{b}} + \boldsymbol{\Omega}^{\hat{a}}_{\hat{\lambda}} \wedge \boldsymbol{\Omega}^{\hat{\lambda}}_{\hat{b}} + \boldsymbol{\Omega}^{\hat{a}}_{\hat{n}} \wedge \boldsymbol{\Omega}^{\hat{n}}_{\hat{b}} \right)_\perp \\
&= \mathbf{e}_{\hat{a}} \otimes \left({}^{(n)} \mathbf{R}^{\hat{a}}_{\hat{b}} + \left[\boldsymbol{\Omega}^{\hat{a}}_{\hat{n}} \wedge \boldsymbol{\Omega}^{\hat{n}}_{\hat{b}} \right]_\perp \right) \\
&= \frac{1}{2} \left({}^{(n)} R^{\hat{a}}_{\hat{b}\hat{c}\hat{d}} + \Gamma^{\hat{a}}_{\hat{n}\hat{c}} \Gamma^{\hat{n}}_{\hat{b}\hat{d}} \right) \mathbf{e}_{\hat{a}} \otimes \boldsymbol{\omega}^{\hat{c}} \wedge \boldsymbol{\omega}^{\hat{d}} \\
&= \frac{1}{2} \left({}^{(n)} R^{\hat{a}}_{\hat{b}\hat{c}\hat{d}} \pm K^{\hat{a}}_{\hat{c}} K_{\hat{b}\hat{d}} \right) \mathbf{e}_{\hat{a}} \otimes \boldsymbol{\omega}^{\hat{c}} \wedge \boldsymbol{\omega}^{\hat{d}}.
\end{aligned} \tag{7.79}$$

From equations (7.77) and (7.79) it follows that

$$ {}^{(n)} R^{\hat{a}}_{\hat{b}\hat{c}\hat{d}} = {}^{(n+1)} R^{\hat{a}}_{\hat{b}\hat{c}\hat{d}} \pm 2 K^{\hat{a}}_{[\hat{c}} K_{\hat{d}]\hat{b}}. \tag{7.80}$$

Let us write equation (7.80) in a covariant manner. The Riemann tensor on the right side in eq. (7.80) is the projected Riemann tensor of the ambient space $\mathcal{M}^{n+1}$. If we split the metric tensor $g_{\alpha\beta}$ into

$$g_{\alpha\beta} = h_{\alpha\beta} + \epsilon n_\alpha n_\beta, \tag{7.81}$$

where n_α are the components of the unit normal vector $\mathbf{n}$, then the tensor $h_{\alpha\beta}$ will act as a projection tensor onto the space $\mathcal{M}^n$. In addition, the tensor $h_{\alpha\beta}$ will be the metric tensor of the space $\mathcal{M}^n$ for vectors on $\mathcal{M}^n$. Since the extrinsic curvature and the Riemann tensor in $\mathcal{M}^n$ already are projected we have

$$h^\lambda_\alpha K_{\lambda\beta} = K_{\alpha\beta}, \tag{7.82}$$

i.e., they are eigentensors to the projection map h^λ_α. The equation (7.80) can now be written covariantly as

$$\boxed{ {}^{(n)} R^\alpha_{\beta\mu\nu} = {}^{(n+1)} R^\lambda_{\gamma\sigma\rho} h^\alpha_\lambda h^\gamma_\beta h^\sigma_\mu h^\rho_\nu + \epsilon \left(K^\alpha_\mu K_{\beta\nu} - K^\alpha_\nu K_{\beta\mu} \right). } \tag{7.83}$$

Similarly, we can show that

$$\boxed{{}^{(n)}\nabla_\alpha K_{\beta\mu} - {}^{(n)}\nabla_\mu K_{\beta\alpha} = {}^{(n+1)}R^\lambda{}_{\sigma\rho\delta} n_\lambda h^\sigma{}_\beta h^\rho{}_\alpha h^\delta{}_\mu,} \tag{7.84}$$

where ${}^{(n)}\nabla_\alpha = h^\beta{}_\alpha \nabla_\beta$ is the n-dimensional connection. The derivation of this is left as a problem (see problem 7.4). This equation is called the *Codazzi equation*.

In the special case where $\mathcal{M}^{n+1}$ is flat, eq. (7.80) reduce to

$$ {}^{(n)}R_{\hat{a}\hat{b}\hat{c}\hat{d}} = \pm \left(K_{\hat{a}\hat{c}} K_{\hat{d}\hat{b}} - K_{\hat{a}\hat{d}} K_{\hat{c}\hat{b}} \right). \tag{7.85}$$

For a two dimensional surface $\mathcal{M}^2$ these equations reduce to the single equation

$$R_{\hat{1}\hat{2}\hat{1}\hat{2}} = K_{\hat{1}\hat{1}} K_{\hat{1}\hat{2}} - (K_{\hat{1}\hat{2}})^2 = \det(K_{\hat{a}\hat{b}}). \tag{7.86}$$

Comparing with eq. (7.30) of a surface (with $\det(g_{\hat{a}\hat{b}}) = 1$) we obtain

$$K = R_{\hat{1}\hat{2}\hat{1}\hat{2}}. \tag{7.87}$$

Thus, the Gaussian curvature represents the *intrinsic* geometry of the surface.

Example 7.2 (The curvature of a straight circular cone)   Example
From figure 7.4 it is seen that the line-element of the conical surface is

$$ds^2 = dl^2 + r^2 d\theta^2 = dl^2 + \left(\frac{R}{H} \right)^2 l^2 d\theta^2. \tag{7.88}$$

Here are l and θ coordinates on the surface, and r a coordinate normal to the axis of the cone.

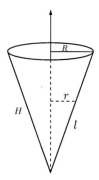

Figure 7.4: A cone is intrinsically flat.

We introduce an orthonormal basis on the surface

$$\omega^{\hat{l}} = dl, \quad \omega^{\hat\theta} = \left(\frac{R}{H} \right) d\theta. \tag{7.89}$$

Exterior differentiation yields

$$\mathbf{d}\omega^{\hat{l}} = 0, \quad \mathbf{d}\omega^{\hat\theta} = \frac{1}{l} \omega^{\hat{l}} \wedge \omega^{\hat\theta}. \tag{7.90}$$

Cartan's first structural equation now give the non-vanishing connection forms

$$\Omega^{\hat{\theta}}_{\hat{l}} = -\Omega^{\hat{l}}_{\hat{\theta}} = \frac{1}{l}\omega^{\hat{\theta}}. \tag{7.91}$$

Thus

$$\mathbf{d}\Omega^{\hat{\theta}}_{\hat{l}} = \left(\mathbf{d}\frac{1}{l}\right)\omega^{\hat{\theta}} + \frac{1}{l}\mathbf{d}\omega^{\hat{\theta}} = 0, \tag{7.92}$$

and Cartan's second structural equation gives

$$\mathbf{R}^{\hat{\theta}}_{\hat{l}} = \mathbf{R}^{\hat{l}}_{\hat{\theta}} = 0, \tag{7.93}$$

which shows that the intrinsic geometry of the conical surface is Euclidean.

The extrinsic curvature is given by eq. (7.75). In the present case the single non-vanishing component of this tensor is

$$K_{\hat{\theta}\hat{\theta}} = \Gamma^{\hat{n}}_{\hat{\theta}\hat{\theta}}. \tag{7.94}$$

In example 6.9 we found that $\Gamma^{\hat{r}}_{\hat{\theta}\hat{\theta}} = -\frac{1}{r}$, so

$$\mathbf{d}\mathbf{e}_{\hat{\theta}} = -\frac{1}{r}\mathbf{e}_{\hat{r}} \otimes \omega^{\hat{\theta}}. \tag{7.95}$$

From Fig. 7.4 we see that

$$\mathbf{e}_{\hat{r}} = \frac{H}{L}\mathbf{e}_{\hat{n}} + \frac{R}{L}\mathbf{e}_{\hat{l}}, \quad L = (H^2 + L^2)^{\frac{1}{2}}. \tag{7.96}$$

This yields

$$\mathbf{d}\mathbf{e}_{\hat{\theta}} = -\frac{H}{Lr}\mathbf{e}_{\hat{n}} \otimes \omega^{\hat{\theta}} - \frac{R}{Lr}\mathbf{e}_{\hat{l}} \otimes \omega^{\hat{\theta}}, \tag{7.97}$$

which shows that

$$\Gamma^{\hat{n}}_{\hat{\theta}\hat{\theta}} = -\frac{H}{Lr}. \tag{7.98}$$

Hence, the conical surface has a non-vanishing extrinsic curvature given by

$$K_{\hat{\theta}\hat{\theta}} = -\left(1 + \frac{L^2}{H^2}\right)^{-\frac{1}{2}} \cdot \frac{1}{r}. \tag{7.99}$$

The limit $H \to \infty, r \to R$ represents a straight cylinder with $K_{\hat{\theta}\hat{\theta}} = -\frac{1}{R}$. The limit $H \to 0$ represents a plane with $K_{\hat{\theta}\hat{\theta}} = 0$.

7.5 The equation of geodesic deviation

Consider two nearby geodesic curves, both parametrised by a parameter λ. Let s be a vector connecting points on the two curves with the same value of λ. The connecting vector s is said to measure the *geodesic deviation* of the curves.

In order to deduce an equation describing how the geodesic deviation varies along the curves, we consider the covariant directional derivative of s along the curve $\nabla_{\mathbf{u}}\mathbf{s}$ where u is the tangent vector to the curve (see Fig. 7.5).

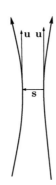

Figure 7.5: The two solid lines are neighbouring geodesics. They are connected by an infinitesimal vector s that obeys the equation of geodesic deviation.

Let **u** and **s** be coordinate basis vectors of a coordinate system. Then $[\mathbf{s}, \mathbf{u}] = 0$, so that

$$\nabla_\mathbf{u}\mathbf{s} = \nabla_\mathbf{s}\mathbf{u}, \tag{7.100}$$

giving

$$\nabla_\mathbf{u}\nabla_\mathbf{u}\mathbf{s} = \nabla_\mathbf{u}\nabla_\mathbf{s}\mathbf{u}. \tag{7.101}$$

Furthermore

$$
\begin{aligned}
\mathbf{R}(\mathbf{u}, \mathbf{s})\mathbf{u} &= \left([\nabla_\mathbf{u}, \nabla_\mathbf{s}] - \nabla_{[\mathbf{u}, \mathbf{s}]}\right)\mathbf{u} \\
&= [\nabla_\mathbf{u}, \nabla_\mathbf{s}]\mathbf{u}.
\end{aligned} \tag{7.102}
$$

Thus

$$\nabla_\mathbf{u}\nabla_\mathbf{u}\mathbf{s} = \nabla_\mathbf{s}\nabla_\mathbf{u}\mathbf{u} + \mathbf{R}(\mathbf{u}, \mathbf{s})\mathbf{u}. \tag{7.103}$$

Since the curves are geodesics $\nabla_\mathbf{u}\mathbf{u} = 0$, and the equation reduces to

$$\boxed{\nabla_\mathbf{u}\nabla_\mathbf{u}\mathbf{s} + \mathbf{R}(\mathbf{s}, \mathbf{u})\mathbf{u} = 0,} \tag{7.104}$$

where we have used the antisymmetry of the Riemann tensor. Equation (7.104) is called the *equation of geodesic deviation*. The component form of the equation is

$$\left(\frac{d^2\mathbf{s}}{d\lambda^2}\right)^\mu + R^\mu{}_{\alpha\nu\beta}u^\alpha s^\nu u^\beta = 0. \tag{7.105}$$

This equation shows that the Riemann tensor can be determined entirely from measurements of geodesic deviation.

In comoving geodesic normal coordinates with $\mathbf{u} = (1, 0, 0, 0)$ the equation reduces to

$$\left(\frac{d^2\mathbf{s}}{d\lambda^2}\right)^i + R^i{}_{0j0}s^j = 0. \tag{7.106}$$

7.6   Spaces of constant curvature

We have in section 6.11 seen how important Killing vectors may be for the
solvability of the equations that govern particle motion. Highly symmetric
spaces are therefore important both in mathematics and in physics. We will
study maximally symmetric spaces with a Riemannian metric, S^n, $\mathbb{P}^n$, $\mathbb{E}^n$ and
$\mathbb{H}^n$. The first 2 of these differ only at a global scale, that is they are locally the
same but their topology is different. Maybe the mathematically most interesting one is the hyperbolic space $\mathbb{H}^n$. We will start out with the most familiar
one, the Euclidean space $\mathbb{E}^n$.

The Euclidean space, $\mathbb{E}^n$

We are all familiar with this space. The metric can be written

$$ds^2 = g_{ab}\mathbf{dx}^a \otimes \mathbf{dx}^b, \tag{7.107}$$

where $g_{ab} = \delta_{ab}$. We can easily generalize this to the Minkowskian case by
letting the components of the metric tensor be of either sign, $g_{ab} = \pm\delta_{ab}$. As
claimed, this is a maximally symmetric space, thus having $\frac{1}{2}n(n+1)$ linearly
independent Killing vectors. We should stress that when we say linearly independent in this context, we mean linearly independent *solutions* of Killing's
equations. The Killing vectors are not linearly independent in the space itself.
Killing's equations do in this case reduce to

$$\xi_{i,j} + \xi_{j,i} = 0, \tag{7.108}$$

because the connection coefficients all vanish identically. We can easily find n
linearly independent solutions to this equation,

$$\boldsymbol{\xi}_{(a)} = \frac{\partial}{\partial x^b}. \tag{7.109}$$

Another set of solutions are[1]

$$\boldsymbol{\xi}_{(ab)} = x^b\frac{\partial}{\partial x^a} - x^a\frac{\partial}{\partial x^b}. \tag{7.110}$$

The solutions $\boldsymbol{\xi}_{(ab)}$ are antisymmetric in the indices so they represent $\frac{1}{2}n(n-1)$
linearly independent solutions. Thus all in all we have found $n + \frac{1}{2}n(n-1)$
$= \frac{1}{2}n(n+1)$ linearly independent solutions of Killing's equation. Hence, as
claimed, the Euclidean spaces admit $\frac{1}{2}n(n+1)$ linearly independent Killing
vectors and is therefore maximally symmetric.

The one-parameter group of diffeomorphisms associated to these Killing
vectors can be found by solving the equations

$$\frac{d}{dt}\phi_t = \boldsymbol{\xi}. \tag{7.111}$$

Representing ϕ_t by a vector $\mathbf{V}$, the equations for $\boldsymbol{\xi}_{(a)}$ can be written

$$\frac{d}{dt}\mathbf{V}_{(a)} = \mathbf{e}_a. \tag{7.112}$$

[1]Since we use Cartesian coordinates we have $x_i \equiv g_{ij}x^j = x^i$. Hence, the position of the
index does not matter.

It has solutions

$$\mathbf{V}_{(a)} = \mathbf{V}_0 + t e_a, \tag{7.113}$$

where $\mathbf{V}_0$ is a constant vector corresponding to the initial condition. These mappings are mere translations a distance t in the a-direction. The Euclidean spaces have translation invariance.

Defining $J_{(ab)}$ as the antisymmetric matrix with indices

$$\left(J_{(ab)}\right)_{ij} = \delta_{ai}\delta_{bj} - \delta_{bi}\delta_{aj}, \tag{7.114}$$

we can write for the Killing vectors $\xi_{(ab)}$

$$\frac{d}{dt}\mathbf{V}_{(ab)} = J_{(ab)}\mathbf{V}_{(ab)}. \tag{7.115}$$

This equation have solutions

$$\mathbf{V}_{(ab)} = e^{tJ_{(ab)}} \cdot \mathbf{V}_0, \tag{7.116}$$

where

$$e^{tJ_{(ab)}} = 1 + tJ_{(ab)} + \frac{1}{2}t^2 J_{(ab)}^2 + \ldots = R_{(ab)}(t), \tag{7.117}$$

is the rotation matrix through an angle t with respect to the (ab)-plane. Showing this is left as an exercise, see the problems in the end of this chapter. Thus, the Euclidean spaces are also rotationally symmetric. The three-dimensional Euclidean space has 6 linearly independent Killing vectors, representing translation and rotational invariance. There is one significant difference between these operations. While the translations move "the whole space", the rotations leave an axis fixed. For a point p, the subgroup of the isometry group that leaves p fixed, is called *the isotropy subgroup*. In the case of $\mathbb{E}^3$, the isotropy subgroup of any point is the group of rotations with respect to this point. We will come back to these concepts in a later chapter, where we will define these groups more rigorously.

The elliptic spaces, S^n and $\mathbb{P}^n$

Let us first start out by defining the spheres S^n. Consider the $(n+1)$-dimensional Euclidean space and look at the hypersurface

$$(x^1)^2 + (x^2)^2 + \ldots + (x^{n+1})^2 = 1. \tag{7.118}$$

This hypersurface with the induced metric is the sphere S^n. The two-dimensional sphere is illustrated in Fig. 7.6. We claim now that the Killing vectors of the sphere is the Killing vectors from the Euclidean space that generate maps that map the sphere onto itself. These are the rotations with respect to the origin, with corresponding Killing vectors

$$\xi_{(\alpha\beta)} = x^\beta \frac{\partial}{\partial x^\alpha} - x^\alpha \frac{\partial}{\partial x^\beta}, \tag{7.119}$$

where Latin indices run from 1 to $(n+1)$. Here we have $\frac{1}{2}n(n+1)$ of them, and when constrained to the sphere, they form the Killing vectors of the sphere. Thus this space is maximally symmetric. Let us use eq. (7.83) to calculate its

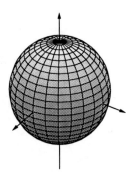

Figure 7.6: The two-dimensional sphere embedded in a Euclidean space.

curvature properties. Given a point on the sphere, then for one of the coordinates $x^a \neq 0$. Assume that $x^{n+1} \neq 0$. Then we can define the following basis vectors

$$\mathbf{e}_a = \boldsymbol{\xi}_{(a(n+1))} = x^{n+1}\frac{\partial}{\partial x^a} - x^a\frac{\partial}{\partial x^{n+1}}, \tag{7.120}$$

where Greek indices run from 1 to n. The metric in this basis is

$$\begin{aligned}
g_{ab} &= \mathbf{e}_a \cdot \mathbf{e}_b = \left(x^{n+1}\frac{\partial}{\partial x^a} - x^a\frac{\partial}{\partial x^{n+1}}\right) \cdot \left(x^{n+1}\frac{\partial}{\partial x^b} - x^b\frac{\partial}{\partial x^{n+1}}\right) \\
&= x^a x^b + \left(x^{n+1}\right)^2 \delta_{ab}.
\end{aligned} \tag{7.121}$$

The radial vector

$$\mathbf{e}_r = x^\alpha\frac{\partial}{\partial x^\alpha}, \tag{7.122}$$

has unit length on the sphere. Using this we can calculate the components of the extrinsic curvature

$$\begin{aligned}
K_{ab} &= \mathbf{e}_r \cdot (\mathbf{de}_a(\mathbf{e}_b)) = \mathbf{e}_r \cdot \left[\frac{\partial}{\partial x^a}\left(\mathbf{dx}^{n+1}(\mathbf{e}_b)\right) - \frac{\partial}{\partial x^{n+1}}\left(\mathbf{dx}^a(\mathbf{e}_b)\right)\right] \\
&= \mathbf{e}_r \cdot \left[x^b\frac{\partial}{\partial x^a} + \delta_{ab}(x^{n+1})\frac{\partial}{\partial x^{n+1}}\right] = x^a x^b + \left(x^{n+1}\right)^2 \delta_{ab}.
\end{aligned} \tag{7.123}$$

Thus we have

$$K_{ab} = g_{ab}. \tag{7.124}$$

Eq. (7.83) now gives us the Riemann tensor

$$R_{abcd} = g_{ac}g_{bd} - g_{ad}g_{bc}. \tag{7.125}$$

Contracting once,

$$R_{bd} = (n-1)g_{bd}, \tag{7.126}$$

we see that the Ricci tensor is proportional to the metric[2]. Thus in an orthonormal frame, the Ricci tensor will be everywhere positive and constant. Contracting once more to obtain the Ricci scalar

$$R = n(n-1). \tag{7.127}$$

---

[2]More generally, spaces for which the Ricci tensor can be written $R_{ab} = \lambda g_{ab}$ where λ is a constant, are called *Einstein spaces*. All the constant curvature spaces are Einstein spaces.

The projective space $\mathbb{P}^n$ can now be obtained by identifying opposite points on the sphere S^n. The projective space is therefore basically half of the sphere. There is a pathology however, for n even, the projective space is *non-orientable*. That means there is no globally defined unit normal vector field on the space. However, for n odd (including therefore dimension 3), the space is orientable, and no problem of such kind exists.

Let us derive a useful form of the metric of the sphere. We introduce a radial coordinate r by

$$r^2 = \sum_a (x^a)^2,\qquad(7.128)$$

so that

$$x^{n+1} = \pm\sqrt{1 - r^2}.\qquad(7.129)$$

Hence,

$$\left(dx^{n+1}\right)^2 = \frac{r^2 dr^2}{1 - r^2}.\qquad(7.130)$$

Assuming now that $d\Omega_{n-1}$ is the metric on the $(n - 1)$-dimensional sphere, S^{n-1}, then

$$\sum_a (dx^a)^2 = dr^2 + r^2 d\Omega_{n-1}^2,\qquad(7.131)$$

which is just the expression for the Euclidean metric in spherical coordinates. The metric on the sphere, S^n, is now

$$ds^2 = \left(dx^{n+1}\right)^2 + \sum_a (dx^a)^2 = \frac{dr^2}{1 - r^2} + r^2 d\Omega_{n-1}^2.\qquad(7.132)$$

Note that this metric only covers half of the sphere. For the case $\mathbb{P}^n$ it covers the whole space, except for the points on the equator which forms a set of measure zero.

The Hyperbolic spaces, $\mathbb{H}^n$

As for the sphere, the hyperbolic space can be viewed upon as a hypersurface in a flat $(n + 1)$-dimensional space. A two dimensional hyperbolic space is illustrated in Fig. 7.7. But now we have to use the flat $(n + 1)$-dimensional Minkowski space with metric

$$ds^2 = \eta_{\alpha\beta}\mathbf{dx}^\alpha \otimes \mathbf{dx}^\beta = -\mathbf{dx}^{n+1} \otimes \mathbf{dx}^{n+1} + \sum_a \mathbf{dx}^a \otimes \mathbf{dx}^a.\qquad(7.133)$$

The hyperbolic space is defined as the hyperboloid

$$-(x^{n+1})^2 + (x^1)^2 + ... + (x^n)^2 = -1,\quad x^{n+1} > 0.\qquad(7.134)$$

This surface is space-like and has a Riemannian metric. Its symmetries can be found by using the same argument as for the sphere. The symmetries for the Minkowski space that leave the hyperboloid invariant, are the Lorentz transformations in $(n + 1)$ dimensions. The Killing vectors for the Lorentz transformations come in two classes, boosts

$$\boldsymbol{\xi}_{(a)} = x^{n+1}\frac{\partial}{\partial x^a} + x^a\frac{\partial}{\partial x^{n+1}},\qquad(7.135)$$

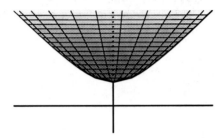

Figure 7.7: The Hyperbolic plane embedded in three-dimensional Minkowski space.

and rotations

$$\boldsymbol{\xi}_{(ab)} = x^b \frac{\partial}{\partial x^a} - x^a \frac{\partial}{\partial x^b}. \tag{7.136}$$

As for the sphere we can now choose a set of basis vectors by

$$\mathbf{e}_a = \boldsymbol{\xi}_{(a)}, \tag{7.137}$$

but these will now be globally defined since $x^{n+1} \geq 1$. The metric in this basis is

$$
\begin{aligned}
g_{ab} &= \mathbf{e}_a \cdot \mathbf{e}_b = \left(x^{n+1} \frac{\partial}{\partial x^a} + x^a \frac{\partial}{\partial x^{n+1}} \right) \cdot \left(x^{n+1} \frac{\partial}{\partial x^b} + x^b \frac{\partial}{\partial x^{n+1}} \right) \\
&= -x^a x^b + \left(x^{n+1} \right)^2 \delta_{ab},
\end{aligned} \tag{7.138}
$$

since $\frac{\partial}{\partial x^{n+1}} \cdot \frac{\partial}{\partial x^{n+1}} = -1$. The radial vector

$$\mathbf{e}_r = x^\alpha \frac{\partial}{\partial x^\alpha} \tag{7.139}$$

has on the hyperboloid a length

$$\mathbf{e}_r \cdot \mathbf{e}_r = -(x^{n+1})^2 + \sum_a (x^a)^2 = -1. \tag{7.140}$$

Following an almost identical procedure to the case of the sphere, we can calculate the extrinsic curvature

$$
\begin{aligned}
K_{ab} &= \mathbf{e}_r \cdot (\mathbf{d}\mathbf{e}_a(\mathbf{e}_b)) \\
&= g_{ab}.
\end{aligned} \tag{7.141}
$$

Eq. (7.83) now turns into (we have to choose the negative sign due to the negative length of the normal unit vector)

$$R_{abcd} = -(g_{ac}g_{bd} - g_{ad}g_{bc}). \tag{7.142}$$

Contracting once,

$$R_{bd} = -(n-1)g_{bd}. \tag{7.143}$$

This space has negative curvature, and basically just the same curvature properties as for the sphere, just with an opposite sign of the curvature. A metric
on the hyperbolic space can written by the following. Introduce spherical coordinates,

$$r^2 = \sum_a (x^a)^2, \tag{7.144}$$

and following the procedure as for the sphere, the metric can be written

$$ds^2 = \frac{dr^2}{1 + r^2} + r^2 d\Omega_{n-1}^2, \tag{7.145}$$

where $d\Omega_{n-1}^2$ is the metric on the $(n-1)$-dimensional sphere. In this case the
metric covers the whole of $\mathbb{H}^n$.

As we see, this space has infinite volume in contrast to the sphere. Note
that there exist (and a lot of them) compact hyperbolic spaces as well, but these
breaks the isometries for the $\mathbb{H}^n$ at a global scale. Locally, however, they have
a hyperbolic metric and are locally isometric to $\mathbb{H}^n$.

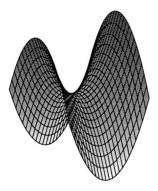

Figure 7.8: The hyperbolic plane can be seen upon as a saddle-surface.

A more intuitive view of the hyperbolic plane is obtained by considering it
as the surface of a saddle, Fig. 7.8. A saddle surface in Euclidean three-space
has a negative curvature and has therefore much of the same properties as
the constantly curved hyperbolic plane. We should emphasize that the whole
hyperbolic plane given by eq. (7.134) for $n = 2$ cannot be embedded in the
three-dimensional Euclidean space. So the surface in Fig. 7.8 has not a constant curvature.

We can find a constant negatively curved space in Euclidean space. Consider the surface generated by the rotation around the z-axis of the *tractrix*

$$z = \ln \left| \frac{1 \pm \sqrt{1 - r^2}}{r} \right| \mp \sqrt{1 - r^2}, \tag{7.146}$$

where $r^2 = x^2 + y^2$. This space will have constant negative curvature and
are sometimes called a *pseudo-sphere*. The tractrix and the pseudo-sphere are
depicted in Fig. 7.9.

The name tractrix is due to the fact that this is the curve which is traced
out by an object on the end of a rope of unit length held by a running child

along the z-axis. If the object and the child start in the xz-plane at $(1,0)$ and $(0,0)$, respectively, then the curve is precisely the curve eq. (7.146) with $y = 0$. The points along the circle $r = 1$ correspond to a singular line, and hence, this cannot be the whole hyperbolic space.

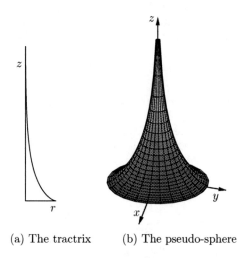

(a) The tractrix   (b) The pseudo-sphere

Figure 7.9: The tractrix and the pseudo-sphere. The pseudo-sphere is obtained by rotating the tractrix around the z-axis.

Problems

7.1. Rotation matrices
Show that

$$e^{tJ_{(ab)}} = 1 + tJ_{(ab)} + \frac{1}{2}t^2 J_{(ab)}^2 + \dots = R_{(ab)}(t), \tag{7.147}$$

i.e. verify eq. (7.117).

7.2. Inverse metric on S^n
In this problem we will use the basis vectors e_a defined in eq. (7.120):

$$e_a = \xi_{(a(n+1))} = x^{n+1} \frac{\partial}{\partial x^a} - x^a \frac{\partial}{\partial x^{n+1}}. \tag{7.148}$$

(a) Assume that the basis forms on the sphere S^n has the form

$$\omega^a = C^a_{\ \mu} dx^\mu. \tag{7.149}$$

Find the coefficients $C^a_{\ \mu}$.

(b) Find g^{ab}, and verify that $g^{ab} g_{bc} = \delta^a_{\ c}$.

7.3. The curvature of a curve
Consider a curve $y = y(x)$ in a two-dimensional plane. Utilize that the differential of the arc length is $ds = (1 + y'^2)^{1/2} dx$ and show that the tangent

vector, the curvature vector and the curvature of the curve are given by

$$\begin{aligned}
\mathbf{t} &= \left(1+y'^2\right)^{-1/2}\left(\mathbf{e}_x + y'\mathbf{e}_y\right), \\
\mathbf{k} &= \left(1+y'^2\right)^{-2} y'' \left(y'\mathbf{e}_x + \mathbf{e}_y\right), \\
\kappa &= \left(1+y'^2\right)^{-3/2} y'',
\end{aligned} \tag{7.150}$$

respectively.

7.4. The Gauss-Codazzi equations

The Gauss-Codazzi equations are of great importance and tell us much about the geometry of hypersurfaces. We will need these equations for a later chapter, and in this problem we will show an important relation between the Ricci curvatures in the ambient space and on the hypersurface.

(a) We will first show that the projection operator $h^\alpha{}_\beta$, defined by eq. (7.81) is what it is claimed to be. We define $h_{\alpha\beta}$ with eq. (7.81):

$$g_{\alpha\beta} = h_{\alpha\beta} + \epsilon n_\alpha n_\beta, \tag{7.151}$$

where $g_{\alpha\beta}$ is the metric in the ambient space, and n_α are the components of the unit normal vector. The operator $h^\alpha{}_\beta$ is now defined by $h^\alpha{}_\beta = g^{\alpha\lambda} h_{\lambda\beta}$. Show the following:

$$\begin{aligned}
h_{\alpha\beta} &= h_{\beta\alpha} \\
h^\alpha{}_\lambda h^\lambda{}_\beta &= h^\alpha{}_\beta.
\end{aligned}$$

These properties define a projection map. It must be shown that it projects onto the space that is orthogonal to n^α. Hence, show that

$$\begin{aligned}
h^\alpha{}_\beta v^\beta &= 0, &&\text{iff } v^\beta \text{ is parallel to } n^\beta. \\
h^\alpha{}_\beta v^\beta &= v^\alpha, &&\text{iff } v^\beta \text{ is orthogonal to } n^\beta.
\end{aligned}$$

Explain now we can define the projection of an arbitrary tensor by projecting the tensor, index by index.

(b) Verify the Codazzi equation, eq. (7.84). (Hint: Calculate $\left(\mathbf{d}^2\mathbf{e}_{\hat{b}}\right) \cdot \mathbf{n}$.)

(c) Show that

$$^{(n+1)}R = {}^{(n)}R - \epsilon\left(K^2 - K^{\alpha\beta}K_{\alpha\beta}\right) + 2\epsilon {}^{(n+1)}R_{\alpha\beta}n^\alpha n^\beta, \tag{7.152}$$

which can be written

$$-2\epsilon {}^{(n+1)}E_{\alpha\beta}n^\alpha n^\beta = {}^{(n)}R - \epsilon\left(K^2 - K^{\alpha\beta}K_{\alpha\beta}\right). \tag{7.153}$$

7.5. The Poincaré half-space

Consider half of $\mathbb{R}^3$, $z > 0$ with metric

$$ds^2 = \frac{1}{z^2}\left(dx^2 + dy^2 + dz^2\right). \tag{7.154}$$

(a) Calculate the connection forms and the curvature forms using the structural equations of Cartan.

(b) Calculate the Riemann tensor, the Ricci tensor and the Ricci scalar.

(c) Show that

$$R_{abcd} = -(g_{ac}g_{bd} - g_{ad}g_{bc}).$$ (7.155)

Compare this with the three-dimensional hyperbolic space. Are there any way we can differentiate between these two cases? Are they different manifestations of the same space?

7.6. The pseudo-sphere
Show that the tractrix, eq. (7.146), obeys the differential equation

$$\left(\frac{dr}{dz}\right)^2 = \frac{r^2}{1-r^2}.$$ (7.156)

Substitute this into the line-element for flat space, make the substitution $r = \sin\theta$, and show that the metric on the pseudo-sphere can be written

$$ds^2 = d\theta^2 + \sinh^2\theta d\phi^2.$$ (7.157)

Show that this metric has constant negative curvature.

7.7. A non-Cartesian coordinate system in two dimensions
Consider the following metric on a two-dimensional surface:

$$ds^2 = v^2 du^2 + u^2 dv^2.$$ (7.158)

You are going to show, in two different ways, that this is only the flat Euclidean plane in disguise.

(a) Use the orthonormal frame approach and find the connection one-forms $\Omega^{\hat{a}}_{\hat{b}}$. Find also the curvature two-forms $\mathbf{R}^{\hat{a}}_{\hat{b}}$ and show that they are identically zero.

(b) Show that the metric can be put onto the form

$$ds^2 = dx^2 + dy^2$$

by finding a transformation matrix $\mathbf{M} = (M^i{}_a)$ connecting the basis vectors $\mathbf{e}_u$ and $\mathbf{e}_v$, and $\mathbf{e}_x$ and $\mathbf{e}_y$. This can be done using the following relations

$$\begin{aligned} g_{ab} &= g_{ij} M^i{}_a M^j{}_b \\ \frac{\partial M^i{}_a}{\partial x^b} &= \frac{\partial M^i{}_b}{\partial x^a}. \end{aligned}$$ (7.159)

Where do these relations come from?

7.8. The curvature tensor of a sphere
Introduce an orthonormal basis on the sphere, S^2, and use Cartan's structural equations to find the physical components of the Riemann curvature tensor.

7.9. The curvature scalar of a surface of simultaneity
The spatial line-element of a rotating disc is

$$d\ell^2 = dr^2 + \frac{r^2}{1-\frac{\omega^2 r^2}{c^2}}d\phi^2.$$ (7.160)

Introduce an orthonormal basis on this surface and use Cartan's structural equations to find the Ricci scalar.

7.10. *The tidal force pendulum and the curvature of space*
We will again consider the tidal force pendulum from Example 1.3. Here we shall use the equation for geodesic deviation, eq. (7.104), to find the period of the pendulum.

(a) Why can the equation for geodesic equation be used to find the period of the pendulum in spite of the fact that the particles do not move along geodesics? Explain also why the equation can be used even though the centre of the pendulum does not follow a geodesic.

(b) Assume that the centre of the pendulum is fixed at a distance R from the centre of mass of the Earth. Introduce an orthonormal basis $\{e_{\hat{a}}\}$ with the origin at the centre of the pendulum (see Fig. 7.10).

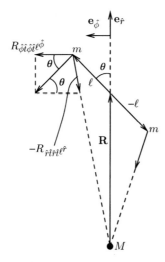

Figure 7.10: The tidal force pendulum.

Show that, to first order in v/c and ϕ/c^2, where v is the three-velocity of the masses and ϕ the gravitational potential at the position of the pendulum, that the equation of geodesic equation takes the form

$$\frac{d^2 \ell_{\hat{i}}}{dt^2} + R_{\hat{i}\hat{0}\hat{j}\hat{0}}\ell^{\hat{j}} = 0. \tag{7.161}$$

(c) Find the period of the pendulum expressed in terms of the components of Riemann's curvature tensor.

7.11. *The Weyl tensor vanishes for spaces of constant curvature*
Use the definition eq. (7.72) to show that the constant curvature spaces S^4, $\mathbb{E}^4$, and $\mathbb{H}^4$ all have zero Weyl tensor.

7.12. *Frobenius' Theorem*
In this problem we will consider integrability conditions for vector fields. In particular, we shall obtain a necessary and sufficent criterion for when a vectorfield is hypersurface orthogonal. Assume that we have an m-dimensional manifold M and a smooth collection of r-dimensional subspaces $\mathcal{D}_p \in T_pM$,

one for each $p \in M$. A submanifold N of M is called an *integral submanifold* if $\mathcal{D}_p = T_p N$ for each $p \in N$. Frobenius' theorem can now be stated:

Frobenius' Theorem: Given a smooth collection of spaces $\mathcal{D}_p$ as above. Then there exists an integral submanifold N at every p if and only if $\{\mathcal{D}_p\}$ is involute; i.e., for all $p \in M$,

$$[\mathbf{X}_i, \mathbf{X}_j] \in \mathcal{D}_p, \quad \forall \mathbf{X}_k \in \mathcal{D}_p.$$

(a) Assume that $\boldsymbol{\omega}^a$ is a set of one-forms that span the orthogonal complement of $\mathcal{D}_p$; i.e. $\boldsymbol{\omega}^a(\mathbf{X}_i) = 0$ for all $\mathbf{X}_i \in \mathcal{D}_p$. Show that $\mathbf{d}\boldsymbol{\omega}^a$ must be on the form

$$\mathbf{d}\boldsymbol{\omega}^a = \sum_b \boldsymbol{\omega}^b \wedge \boldsymbol{\rho}^b,$$

where each $\boldsymbol{\rho}^b$ is in $T_p^* M$. (Hint: Show first that $X^\mu Y^\nu \nabla_{[\mu} \omega^a_{\nu]} = 0$ for all $X^\mu, Y^\nu \in T_p N$.)

(b) In particular, this means if M is an m-dimensional manifold and ξ_μ is a covariant vector field, then ξ_μ is hypersurface orthogonal to some $(m-1)$-dimensional submanifold if and only if $\nabla_{[\mu} \xi_{\nu]} = \xi_{[\mu} \rho_{\nu]}$. Show that a sufficent condition for ξ_μ to be hypersurface orthogonal is

$$\xi_{[\mu} \nabla_\nu \xi_{\sigma]} = 0. \tag{7.162}$$

Part III

EINSTEIN'S FIELD EQUATIONS

8

Einstein's Field Equations

Einstein's field equations are the relativistic generalization of Newton's law of gravitation. Einstein's vision, based on the equality of inertial and gravitational masses, was that there is no gravitational force at all. What is said to be "particle motion under the influence of the gravitational force" in Newtonian theory, is according to the general theory of relativity, free motion along geodesic curves in a curved space-time.

First we will use a heuristic argument to derive Einstein's field equations. Then we shall use Hilbert's variational principle to deduce Einstein's field equations.

8.1 From Newton's law of gravitation to Einstein's field equations

We will first motivate Einstein's field equations by trying to replace Newton's law of gravitation by a covariant tensor equation. Newton's gravitational law tells how mass generates gravitational force. Einstein demanded from his field equations that they should tell how matter and energy curve space-time and that energy-momentum conservation should follow from them. He knew that the energy-momentum conservation of a continuum of matter and energy could be described covariantly by the vanishing of the divergence of a symmetric energy-momentum tensor of rank 2. Thus the field equations must be of the same form: A symmetric and divergence-free curvature tensor of rank 2 is proportional to the energy-momentum tensor.

We will start with the local form of Newton's law of gravitation. As explained in Chapter 1, the local form is given in terms of Poisson's equation, eq. (1.32):

$$\nabla^2 \phi(\mathbf{r}) = 4\pi G \rho(\mathbf{r}). \tag{8.1}$$

The right hand side of this equation involves the matter density and therefore we would like to replace this side with the energy-momentum tensor. The left hand side involves a gravitational potential. In the new theory motion under the influence of gravity could be seen as free motion in a curved space. The potential should therefore somehow be represented by the metric of spacetime. Since the left hand side of Poisson's equation involves the second derivative of the potential, we should replace the left hand side of Poisson's equation with a tensor involving second derivatives of the metric.

An alternative way to see this is if we keep in mind eq. (1.33) which tells us how acceleration of gravity can be deduced in Newton's theory: acceleration of gravity is generated by the gradient of the potential ϕ. Recall from chapter 6 that for a free particle instantaneously at rest, the acceleration is given by $\ddot{x}^i = -\Gamma^i{}_{00}$, see eq. (6.112); hence, the acceleration of gravity can be represented by the Christoffel symbols in an appropriate frame. According to eq. (1.33), we therefore seek a tensor which contains first derivatives of the Christoffel symbols.

The Christoffel symbols contain first derivatives of the metric, and the Riemann curvature tensor contains first derivatives of the Christoffel symbols. Hence, the Riemann curvature tensor contains the right order of derivatives to represent the left hand side of Poisson's equation. We have already argued that the right hand side is proportional to the energy-momentum tensor $T_{\mu\nu}$. This is a symmetric tensor of rank 2. The first natural choice is therefore to consider the Ricci tensor, which is obtained by contracting the Riemann tensor once. Therefore Einstein initially tried:

$$R_{\mu\nu} \propto T_{\mu\nu}.$$

He discovered, however, that this is not quite satisfactory; the Ricci tensor is in general not divergence-free. With the help of Marcel Grossmann Einstein then discovered that the combination $R_{\mu\nu} - (1/2)Rg_{\mu\nu}$ is divergence-free. This is the Einstein tensor, see eq. (7.71), and is the simplest divergence-free combination of the Ricci tensor. The Einstein tensor has all the required properties: it is a divergence-free symmetric tensor of rank two. Hence, *the Einstein tensor has the right properties to represent the geometrical part of Einstein's equations.*

We therefore arrive at the relation

$$R_{\mu\nu} - \frac{1}{2}Rg_{\mu\nu} \propto T_{\mu\nu}, \tag{8.2}$$

by "covariantising" the local form of Newton's gravitational law. Einstein also realised that since the metric $g_{\mu\nu}$ is also symmetric and divergence-free (in fact, $g_{\mu\nu;\rho} = 0$) we can even add a term $\Lambda g_{\mu\nu}$, where Λ is a constant. The result is *Einstein's Field equations*

$$\boxed{R_{\mu\nu} - \frac{1}{2}Rg_{\mu\nu} + \Lambda g_{\mu\nu} = \kappa T_{\mu\nu},} \tag{8.3}$$

where κ is a constant which needs to be determined.

8.2 Deduction of Einstein's vacuum field equations from Hilbert's variational principle

Since the previous argument was somewhat heuristic, we will derive the Einstein field equations in a more formal way using a variational principle,

$$\delta S_G = 0, \tag{8.4}$$

where S_G is the action integral for gravitation. S_G is of a geometrical nature, and is of the form

$$S_G = \frac{1}{2\kappa} \int_M \mathcal{L}[g_{\mu\nu}]\sqrt{-g}d^4x, \tag{8.5}$$

where κ is a constant. The constant κ will be determined under the requirement that the field equations reduce to Newton's law in the weak field limit. The function $\mathcal{L}[g_{\mu\nu}]$ has to be a scalar for the integral to transform in an invariant manner. Since the simplest scalar involving curvature is the Ricci curvature scalar, we will use

$$\mathcal{L}[g_{\mu\nu}] = R - 2\Lambda, \tag{8.6}$$

where we have also allowed for a pure constant in the action, Λ. This constant is termed *the cosmological constant* and this name will be clear later on. The action therefore reads

$$\boxed{S_G = \frac{1}{2\kappa} \int (R - 2\Lambda)\sqrt{-g}d^4x.} \tag{8.7}$$

We are going to vary the action inside an infinitesimal region V, letting the variation of the metric and its derivative vanish on the boundary of the region. Then we calculate the variation of the action integrals, and deduce Einstein's field equations from the requirement that $\delta S_G = 0$ for arbitrary variations of the metric.

Writing

$$S_G = \frac{1}{2\kappa} \int \left(R_{\mu\nu}g^{\mu\nu}\sqrt{-g} - 2\Lambda\sqrt{-g}\right)d^4x, \tag{8.8}$$

we get

$$\delta S_G = \frac{1}{2\kappa} \int \left(g^{\mu\nu}\sqrt{-g}\delta R_{\mu\nu} + R_{\mu\nu}\delta\left[g^{\mu\nu}\sqrt{-g}\right] - 2\Lambda\delta\sqrt{-g}\right)d^4x. \tag{8.9}$$

Introducing a local coordinate system with vanishing Christoffel symbols in V, the components of the Ricci tensor reduce to

$$R_{\mu\nu} = \Gamma^{\lambda}_{\mu\nu,\lambda} - \Gamma^{\lambda}_{\mu\lambda,\nu}. \tag{8.10}$$

Thus,

$$\delta R_{\mu\nu} = \delta\Gamma^{\lambda}_{\mu\nu,\lambda} - \delta\Gamma^{\lambda}_{\mu\lambda,\nu}. \tag{8.11}$$

The variation commutes with the partial derivatives, so

$$\delta R_{\mu\nu} = \left(\delta\Gamma^{\lambda}_{\mu\nu}\right)_{,\lambda} - \left(\delta\Gamma^{\lambda}_{\mu\lambda}\right)_{,\nu}. \tag{8.12}$$

Since the partial derivatives of the metric vanish in V this equation may be written

$$g^{\mu\nu}\delta R_{\mu\nu} = \left(g^{\mu\nu}\delta\Gamma^{\lambda}_{\mu\nu} - g^{\mu\lambda}\delta\Gamma^{\nu}_{\mu\nu}\right)_{,\lambda}. \tag{8.13}$$

According to eq. (6.96) the contravariant index of the Christoffel symbols transform as a tensor index. Thus we may define a vector **A** by

$$A^{\lambda} = g^{\mu\nu}\delta\Gamma^{\lambda}_{\mu\nu} - g^{\mu\lambda}\delta\Gamma^{\nu}_{\mu\nu}. \tag{8.14}$$

Equation (8.13) now takes the form

$$g^{\mu\nu}\delta R_{\mu\nu} = A^{\mu}{}_{,\mu}. \tag{8.15}$$

This is a total divergence, and hence, according to Stoke's Theorem (or the Gauss' integral theorem), the integral of this term only contributes with a boundary term. Since the metric and its derivative vanishes on the boundary of V, it follows that

$$\int \left(g^{\mu\nu}\sqrt{-g}\delta R_{\mu\nu} \right) d^4x = 0; \tag{8.16}$$

thus the first term of equation (8.9) does not contribute to δS_G.

We shall now consider the last term in eq. (8.9). The variation of $\sqrt{-g}$ is

$$\delta\sqrt{-g} = \left[\frac{\partial\sqrt{-g}}{\partial g_{\alpha\beta}} \right]\delta g_{\alpha\beta} = -\frac{1}{2\sqrt{-g}}\left(\frac{\partial g}{\partial g_{\alpha\beta}} \right)\delta g_{\alpha\beta}. \tag{8.17}$$

To calculate $\frac{\partial g}{\partial g_{\alpha\beta}}$ we use the formula

$$g = \sum_{\alpha} g_{\alpha\beta}\mathsf{Cof}^{\alpha\beta} = \frac{\mathsf{Cof}^{\alpha\beta}}{g^{\alpha\beta}}, \tag{8.18}$$

where $\mathsf{Cof}^{\alpha\beta}$ is the cofactor matrix of the element $g_{\alpha\beta}$ in the matrix made of the components of the metric tensor. This gives

$$\frac{\partial g}{\partial g_{\alpha\beta}} = \mathsf{Cof}^{\alpha\beta} = gg^{\alpha\beta}, \tag{8.19}$$

and therefore

$$\delta\sqrt{-g} = \frac{1}{2}\sqrt{-g}g^{\alpha\beta}\delta g_{\alpha\beta}. \tag{8.20}$$

It remains to calculate the second term in eq. (8.9). From

$$\delta\left[g^{\mu\nu}\sqrt{-g} \right] = \sqrt{-g}\delta g^{\mu\nu} + g^{\mu\nu}\delta\sqrt{-g}, \tag{8.21}$$

we see that it suffices to calculate $\delta g_{\alpha\beta}$. Since

$$g^{\mu\alpha}g_{\alpha\beta} = \delta^{\mu}{}_{\beta}, \tag{8.22}$$

we get

$$\delta\left(g^{\mu\alpha}g_{\alpha\beta} \right) = 0, \tag{8.23}$$

which leads to

$$\delta g_{\alpha\beta} = -g_{\alpha\mu}g_{\beta\nu}\delta g^{\mu\nu}. \tag{8.24}$$

Thus we get

$$\begin{aligned} \delta\left[g^{\mu\nu}\sqrt{-g} \right] &= \sqrt{-g}\left(\delta g^{\mu\nu} + \frac{1}{2}g^{\mu\nu}g^{\alpha\beta}\delta g_{\alpha\beta} \right) \\ &= \sqrt{-g}\left(\delta g^{\mu\nu} - \frac{1}{2}g^{\mu\nu}g_{\alpha\beta}\delta g^{\alpha\beta} \right). \end{aligned} \tag{8.25}$$

Inserting equations (8.20) and (8.25) into eq. (8.9) gives

$$\delta S_G = \frac{1}{2\kappa} \int \sqrt{-g} \left(R_{\alpha\beta} - \frac{1}{2} R g_{\alpha\beta} + \Lambda g_{\alpha\beta} \right) \delta g^{\alpha\beta} d^4 x. \tag{8.26}$$

The vacuum field equations of the general theory of relativity result from the requirement $\delta S_G = 0$ for any variation of the metric. This leads to

$$\boxed{ R_{\alpha\beta} - \frac{1}{2} R g_{\alpha\beta} + \Lambda g_{\alpha\beta} = 0. } \tag{8.27}$$

As noted in chapter 7 the Einstein tensor has only six independent components. So there are only six field equations. However, the metric tensor has 10 independent components in four-dimensional space-time. This leaves us with four degrees of freedom in the metric tensor; just the right number to permit a free choice of coordinate system.

8.3   The field equations in the presence of matter and energy

The field equations at a point with non-vanishing energy-momentum tensor is obtained from the variational principle

$$\delta(S_G + S_M) = 0, \tag{8.28}$$

where S_M is the action integral for matter and energy, which can be written as

$$S_M = \int \mathcal{L}_M \sqrt{-g} d^4 x, \tag{8.29}$$

where $\mathcal{L}_M$ is the Lagrangian density of the matter and energy.
Variation of the argument in eq. (8.29) gives

$$\delta \left[\sqrt{-g} \mathcal{L}_M \right] = \frac{\partial \left[\sqrt{-g} \mathcal{L}_M \right]}{\partial g^{\mu\nu}} \delta g^{\mu\nu} + \frac{\partial \left[\sqrt{-g} \mathcal{L}_M \right]}{\partial g^{\mu\nu}{}_{,\lambda}} \delta g^{\mu\nu}{}_{,\lambda}, \tag{8.30}$$

since the Lagrangian in general depends on both the metric and on the derivatives of the metric. This is the case because the covariant expression for $\mathcal{L}_M$ may be found from the special relativistic expressing by replacing partial derivatives by their covariant derivatives. This introduces Christoffel symbols, i.e. derivatives of the metric, into the expression.
We define a vector **B** by

$$B^\lambda = \frac{\partial \left[\sqrt{-g} \mathcal{L}_M \right]}{\partial g^{\mu\nu}{}_{,\lambda}} \delta g^{\mu\nu}. \tag{8.31}$$

The ordinary (not covariant) divergence of **B** is

$$B^\lambda{}_{,\lambda} = \left\{ \frac{\partial \left[\sqrt{-g} \mathcal{L}_M \right]}{\partial g^{\mu\nu}{}_{,\lambda}} \right\}_{,\lambda} \delta g^{\mu\nu} + \frac{\partial \left[\sqrt{-g} \mathcal{L}_M \right]}{\partial g^{\mu\nu}{}_{,\lambda}} \delta g^{\mu\nu}{}_{,\lambda}. \tag{8.32}$$

Inserting this into (8.30) gives

$$\delta \left[\sqrt{-g} \mathcal{L}_M \right] = \frac{\partial \left[\sqrt{-g} \mathcal{L}_M \right]}{\partial g^{\mu\nu}} \delta g^{\mu\nu} - \left\{ \frac{\partial \left[\sqrt{-g} \mathcal{L}_M \right]}{\partial g^{\mu\nu}{}_{,\lambda}} \right\}_{,\lambda} \delta g^{\mu\nu} + B^\lambda{}_{,\lambda}. \tag{8.33}$$

Thus the term $\int B^\lambda_{,\lambda} d^4x$ contributes only with a boundary term, due to Gauss' integral theorem. This boundary term vanishes because we have assumed that the variation vanishes on the boundary. Finally this yields

$$\delta S_M = \int \left(\frac{\partial [\sqrt{-g}\mathcal{L}_M]}{\partial g^{\mu\nu}} - \left\{ \frac{\partial [\sqrt{-g}\mathcal{L}_M]}{\partial g^{\mu\nu}_{,\lambda}} \right\}_{,\lambda} \right) \delta g^{\mu\nu} d^4x. \tag{8.34}$$

The energy-momentum tensor $T_{\mu\nu}$ of a system with Lagrangian density $\mathcal{L}_M$ is a symmetric tensor defined by

$$T_{\mu\nu} = -\frac{2}{\sqrt{-g}} \left(\frac{\partial [\sqrt{-g}\mathcal{L}_M]}{\partial g^{\mu\nu}} - \left\{ \frac{\partial [\sqrt{-g}\mathcal{L}_M]}{\partial g^{\mu\nu}_{,\lambda}} \right\}_{,\lambda} \right). \tag{8.35}$$

This gives

$$\delta S_M = -\frac{1}{2} \int T_{\mu\nu} \sqrt{-g} \delta g^{\mu\nu} d^4x. \tag{8.36}$$

Using equations (8.26) and (8.36) the variational principle then yields the gravitational field equations for the general theory of relativity

$$\boxed{R_{\mu\nu} - \frac{1}{2}R g_{\mu\nu} + \Lambda g_{\mu\nu} = \kappa T_{\mu\nu}.} \tag{8.37}$$

These are the famous Einstein's field equations. We will later see (Chapter 9) that the constant κ can be determined to be

$$\kappa = \frac{8\pi G}{c^4}.$$

Contracting eq. (8.37) we get

$$R = -\kappa T + 4\Lambda, \tag{8.38}$$

where T is the contracted energy-momentum tensor, $T = T^\mu_\mu$. Inserting eq. (8.38) into eq. (8.37) leads to

$$R_{\mu\nu} = \Lambda g_{\mu\nu} + \kappa \left(T_{\mu\nu} - \frac{1}{2}T g_{\mu\nu} \right). \tag{8.39}$$

This equation reflects a symmetry in the equations, the Ricci tensor and the energy-momentum tensor is invariant under a permutation between the two tensors. The vacuum equations with a cosmological constant are

$$R_{\mu\nu} = \Lambda g_{\mu\nu}. \tag{8.40}$$

With $\Lambda = 0$ this equation says that the Ricci tensor must vanish for a vacuum space-time without a cosmological constant:

$$R_{\mu\nu} = 0. \tag{8.41}$$

Note, however, that this does not mean that such a space-time is flat. Already in the next chapter we will see this. The reason for this is that the Riemann tensor consists of basically two parts, one gives the contribution to the Ricci tensor under contraction, while the other part, the trace-free part of the Riemann tensor, will not give any contribution to the Ricci tensor; hence, it is not determined directly by the Einstein equations.

8.4   Energy-momentum conservation

The law of conservation of energy-momentum asserts that the total flux of energy-momentum into a four-dimensional region Ω is equal to zero,

$$\int_{\partial\Omega} T^{\mu\nu} n_\nu d\sigma = 0, \tag{8.42}$$

where $\partial\Omega$ is the boundary of Ω and n^ν is the outward normal vector of $\partial\Omega$. From Gauss' integral theorem we obtain

$$\int_{\Omega} T^{\mu\nu}{}_{;\nu} \sqrt{-g} d^4x = 0, \tag{8.43}$$

for an arbitrary region Ω. Hence, the local formulation of the law of energy-momentum conservation has the form

$$\boxed{T^{\mu\nu}{}_{;\nu} = 0.} \tag{8.44}$$

The energy-momentum tensor is divergence-free.

   The time component describes conservation of energy, and the space components conservation of momentum. Note that the energy-momentum conservation follows from the Einstein's field equations since the Einstein tensor is divergence-free.

8.5   Some energy-momentum tensors

We will in this section give a few examples of different energy-momentum tensors that occur in general relativity. From now on, unless stated otherwise, units where the velocity of light is set to unity, i.e., $c = 1$, will be used.

Electromagnetic fields

The Lagrangian density of an electromagnetic field is the *energy-scalar* representing the energy-density of the field in a local frame moving so that the magnetic field vanishes

$$\mathcal{L} = -\frac{1}{4} F_{\alpha\beta} F^{\alpha\beta} = -\frac{1}{4} g^{\alpha\beta} g^{\mu\nu} F_{\mu\alpha} F_{\mu\beta}. \tag{8.45}$$

Since this Lagrangian does not contain any derivatives of the metric, we have

$$T_{\mu\nu} = -\frac{2}{\sqrt{-g}} \frac{\partial [\sqrt{-g}\mathcal{L}_M]}{\partial g^{\mu\nu}} = -2 \frac{\partial\mathcal{L}}{\partial g^{\mu\nu}} - \frac{\mathcal{L}}{g} \cdot \frac{\partial g}{\partial g^{\mu\nu}}. \tag{8.46}$$

Using that

$$\frac{\partial g}{\partial g^{\mu\nu}} = -g_{\alpha\mu} g_{\beta\nu} \frac{\partial g}{\partial g_{\alpha\beta}} = -g g_{\mu\nu}, \tag{8.47}$$

we find $T_{\mu\nu}$ to be

$$T_{\mu\nu} = -2 \frac{\partial\mathcal{L}}{\partial g^{\mu\nu}} + g_{\mu\nu}\mathcal{L}. \tag{8.48}$$

Inserting the Lagrangian density of an electromagnetic field leads to

$$T_{\mu\nu} = F^\alpha_\mu F_{\alpha\nu} - \frac{1}{4} g_{\mu\nu} F_{\alpha\beta} F^{\alpha\beta}. \qquad (8.49)$$

We note that the energy-momentum tensor of an electromagnetic field is trace-free:

$$T^\mu_\mu = 0. \qquad (8.50)$$

Perfect Fluids

In the theory of relativity the word *fluid* has a wide meaning, encompassing not only what is called ordinary fluids, but also gases, radiation and even vacuum energy. A fluid is said to be *perfect* when it has no viscosity and no heat conduction. It can be characterised by a four-velocity u and by two of the following scalar quantities: the proper density ρ, the isotropic pressure p, the temperature T, the specific entropy s, or the specific enthalpy $w = \frac{\rho+p}{n}$, where n is the baryon number density. These quantities are defined in a comoving orthonormal basis field in the fluid. Here, n is given in terms of a baryon number flux vector density

$$n^\mu = n\sqrt{-g}u^\mu, \qquad (8.51)$$

so that

$$n = \sqrt{\frac{g_{\mu\nu}n^\mu n^\nu}{g}}. \qquad (8.52)$$

We shall now deduce the form of the energy-momentum tensor of a perfect fluid from eq. (8.35) under the constraint that the rates of entropy and particle production are conserved under variation of the metric. The Lagrangian density of a perfect fluid is the energy scalar representing the energy in a local rest frame of the fluid, i.e., the proper density ρ,

$$\mathcal{L} = -\rho. \qquad (8.53)$$

Again the last term of eq. (8.35) vanishes and the energy-momentum tensor is given by eq. (8.48) under the constraints

$$\delta s = 0, \qquad (8.54)$$
$$\delta n^\mu = 0. \qquad (8.55)$$

From the thermodynamical relation

$$\left(\frac{\partial\rho}{\partial n}\right)_s = w, \qquad (8.56)$$

we have

$$\delta\rho = w\delta n. \qquad (8.57)$$

Using equations (8.51), (8.52) and (8.55) we obtain

$$\delta n = \frac{1}{2n}\left(\frac{n^\mu n^\nu}{g}\delta g_{\mu\nu} - n^\mu n^\nu \frac{g_{\mu\nu}}{g}\delta g\right) = \frac{n}{2}\left(-u^\mu u^\nu \delta g_{\mu\nu} + \frac{u^\mu u_\mu}{g}\delta g\right). \qquad (8.58)$$

Substituting from (8.19), (8.24), (8.47) and using that $u^\mu u_\mu = -1$, we get

$$\delta n = \frac{n}{2}\left(u_\mu u_\nu + g_{\mu\nu}\right)\delta g^{\mu\nu}. \tag{8.59}$$

From equations (8.53), (8.57) and (8.59) follows

$$\frac{\partial \mathcal{L}}{\partial g^{\mu\nu}} = -\frac{nw}{2}\left(u_\mu u_\nu + g_{\mu\nu}\right). \tag{8.60}$$

Inserting the expression for w we get

$$\frac{\partial \mathcal{L}}{\partial g^{\mu\nu}} = -\frac{1}{2}(\rho + p)\left(u_\mu u_\nu + g_{\mu\nu}\right). \tag{8.61}$$

Equations (8.48), (8.53) and (8.61) give the following expression for the energy-momentum tensor of a perfect fluid

$$\boxed{T_{\mu\nu} = (\rho + p)u_\mu u_\nu + p g_{\mu\nu}.} \tag{8.62}$$

It may happen that one knows the components of $T_{\mu\nu}$ for a material system from information or calculations not involving eq. (8.62). Then one needs a general physical interpretation of the components $T_{\mu\nu}$ without using this expression. This is provided as follows.

The eigenvalues $\lambda_{(\alpha)}$ and eigenvectors $\mathbf{u}_{(\alpha)}$ of the energy-momentum tensor are given by

$$\det |T^\mu_{\ \nu} - \lambda \delta^\mu_{\ \nu}| = 0, \tag{8.63}$$

and

$$T^\mu_{\ \nu} u^\nu_{(\alpha)} = \lambda_{(\alpha)} u^\mu_{(\alpha)}, \tag{8.64}$$

respectively.

Equation (8.63) is an equation of fourth degree with four roots $\lambda_{(t)}$, $\lambda_{(i)}$, $i = 1, 2, 3$. Equation (8.64) gives the four corresponding eigenvectors $\mathbf{u}_{(t)}$, $\mathbf{u}_{(i)}$. It follows from the symmetry of $T^{\mu\nu}$ that they are orthogonal and they are fixed by choosing them to be unit vectors. These vectors can then represent a comoving orthonormal basis field of the fluid, and $\mathbf{u}_{(t)}$ is its four-velocity (assuming $\mathbf{u}_{(t)}$ is not null). Furthermore, $\lambda_{(t)}$ is the energy (or mass) density as measured by an observer comoving with the fluid, and $\lambda_{(i)}$ are the scalar stresses he measures. In the case of a fluid with with isotropic pressure $\lambda_{(1)} = \lambda_{(2)} = \lambda_{(3)} = p$. Note that $\lambda_{(i)}$ need not to be positive. A negative $\lambda_{(i)}$ means strain. For the tensor in eq. (8.62) we have $\lambda_{(t)} = \rho$, $\lambda_{(i)} = p$ and $\mathbf{u}_{(t)} = \mathbf{u}$.

8.6   Some particular fluids

In this section we shall deduce the equation of state for vacuum energy, electromagnetic radiation and dust, described as perfect fluids. We shall also look at a cosmic magnetic field.

Lorentz invariant vacuum energy, LIVE

The energy-momentum tensor for a LIVE can be deduced from the requirement that its components must be Lorentz invariant. Thus

$$T_{\hat\mu\hat\nu} = T_{\hat\mu'\hat\nu'} = \Lambda^{\hat\alpha}_{\ \hat\mu'}\Lambda^{\hat\beta}_{\ \hat\nu'}T_{\hat\alpha\hat\beta}, \tag{8.65}$$

for arbitrary Lorentz transformations $\Lambda^{\hat\mu}_{\ \hat\mu'}$. Consider first a boost in the $x^{\hat 1}$-direction:

$$\Lambda^{\hat\mu}_{\ \hat\mu'} = \begin{bmatrix} \gamma & v\gamma & 0 & 0 \\ v\gamma & \gamma & 0 & 0 \\ 0 & 0 & 1 & 0 \\ 0 & 0 & 0 & 1 \end{bmatrix}, \quad \gamma = \frac{1}{\sqrt{1-v^2}}. \tag{8.66}$$

Equations (8.65) and (8.66) give

$$v(T_{\hat 0\hat 0} + T_{\hat 1\hat 1}) + T_{\hat 0\hat 1} + T_{\hat 1\hat 0} = 0. \tag{8.67}$$

Transformation of $T_{\hat 1\hat 1}$ gives the same equation. In a similar way transformation of $T_{\hat 0\hat 1}$ and $T_{\hat 1\hat 0}$ leads to

$$T_{\hat 0\hat 0} + T_{\hat 1\hat 1} + v(T_{\hat 0\hat 1} + T_{\hat 1\hat 0}) = 0. \tag{8.68}$$

From these equations follow that

$$T_{\hat 0\hat 0} = -T_{\hat 1\hat 1}, \quad T_{\hat 0\hat 1} = -T_{\hat 1\hat 0}. \tag{8.69}$$

Transformations on $T_{\hat 0\hat 2}$ and $T_{\hat 1\hat 2}$ give, respectively

$$T_{\hat 0\hat 2} = \gamma(T_{\hat 0\hat 2} + vT_{\hat 1\hat 2}), \tag{8.70}$$
$$T_{\hat 1\hat 2} = \gamma(vT_{\hat 0\hat 2} + T_{\hat 1\hat 2}), \tag{8.71}$$

which demands that

$$T_{\hat 0\hat 2} = T_{\hat 1\hat 2} = 0. \tag{8.72}$$

In the same way one finds

$$T_{\hat 2\hat 0} = T_{\hat 2\hat 1} = T_{\hat 0\hat 3} = T_{\hat 1\hat 3} = T_{\hat 3\hat 0} = T_{\hat 3\hat 1} = 0. \tag{8.73}$$

Thus as a result of Lorentz invariance of the components $T_{\hat\mu\hat\nu}$ under a boost in the $x^{\hat 1}$-direction, we have managed to reduce the energy momentum tensor to the following for the vacuum fluid

$$T_{\hat\mu\hat\nu} = \begin{bmatrix} T_{\hat 0\hat 0} & T_{\hat 0\hat 1} & 0 & 0 \\ -T_{\hat 0\hat 1} & -T_{\hat 0\hat 0} & 0 & 0 \\ 0 & 0 & T_{\hat 2\hat 2} & T_{\hat 2\hat 3} \\ 0 & 0 & T_{\hat 3\hat 2} & T_{\hat 3\hat 3} \end{bmatrix}. \tag{8.74}$$

Demanding Lorentz invariance under a boost in the $x^{\hat 2}$ direction gives the additional equations

$$T_{\hat 0\hat 1} = T_{\hat 1\hat 0} = T_{\hat 2\hat 3} = T_{\hat 3\hat 2} = 0, \quad T_{\hat 2\hat 2} = T_{\hat 0\hat 0}. \tag{8.75}$$

Lastly, Lorentz invariance under a boost in the $x^{\hat{1}}$-direction gives the additional equation

$$T_{\hat{3}\hat{3}} = T_{\hat{0}\hat{0}}. \tag{8.76}$$

It follows that the energy-momentum tensor for the vacuum fluid has to be

$$T_{\hat{\mu}\hat{\nu}} = T_{\hat{0}\hat{0}}\mathrm{diag}(-1,1,1,1) = T_{\hat{0}\hat{0}}\eta_{\hat{\mu}\hat{\nu}}, \tag{8.77}$$

where $\eta_{\hat{\mu}\hat{\nu}}$ are the components of the Minkowski metric. Transforming to an arbitrary basis the Minkowski metric can be replaced by a general metric $g_{\mu\nu}$. From the physical interpretation of the components of the energy-momentum tensor, it follows that $T_{\hat{0}\hat{0}} = -\rho$, where ρ is the energy-density of the vacuum. Thus,

$$\boxed{T_{\mu\nu} = -\rho g_{\mu\nu}.} \tag{8.78}$$

Comparing with equation (8.62) shows that this is the energy-momentum tensor of a perfect fluid with equation of state

$$p = -\rho. \tag{8.79}$$

Hence, the vacuum is in a state of extreme stress.

Generally the density of vacuum is a scalar function of the four spacetime coordinates. If vacuum is homogeneous, the density depends upon time only. Due to the relativity of simultaneity this condition is Lorentz invariant only if $\rho = $ constant. In this case the energy-density of the LIVE appears as a cosmological constant.

Quintessence

There are more general forms of "vacuum energies" than LIVE. They are represented by different vacuum fields and have been called "quintessence energy". We shall here consider the simple case where the energy is given by a real scalar field ϕ with Lagrange density

$$\mathcal{L} = -\frac{1}{2}\frac{\partial\phi}{\partial x^\mu}\frac{\partial\phi}{\partial x_\mu} - V(\phi), \tag{8.80}$$

where $V(\phi)$ is the potential of the field. The Lagrange density of eq. (8.80) does not contain any derivatives of the metric. Hence we can use the expression eq. (8.48) for the energy-momentum tensor. This leads to

$$T_{\mu\nu} = \frac{\partial\phi}{\partial x^\mu}\frac{\partial\phi}{\partial x^\nu} - g_{\mu\nu}\left(\frac{1}{2}\frac{\partial\phi}{\partial x^\mu}\frac{\partial\phi}{\partial x_\mu} + V(\phi)\right). \tag{8.81}$$

In the comoving expanding frame of a homogeneous and isotropic universe model this energy-momentum tensor reduces to

$$T_{\mu\nu} = \mathrm{diag}\left(\frac{1}{2}\dot{\phi}^2 + V(\phi), \frac{1}{2}\dot{\phi}^2 - V(\phi), \frac{1}{2}\dot{\phi}^2 - V(\phi), \frac{1}{2}\dot{\phi}^2 - V(\phi)\right). \tag{8.82}$$

Let us consider the vacuum energy as a perfect fluid. In an orthonormal basis comoving with the fluid the non-vanishing components of the energy-momentum tensor are

$$T_{\mu\nu} = \mathrm{diag}(\rho, p, p, p). \tag{8.83}$$

Comparing with eq. (8.82) gives the density and pressure (or stress) of a homogeneous scalar field

$$\boxed{\rho = \frac{1}{2}\dot{\phi}^2 + V(\phi), \quad p = \frac{1}{2}\dot{\phi}^2 - V(\phi).}$$
(8.84)

Hence, the equation of state of this energy is

$$p = \frac{\frac{1}{2}\dot{\phi}^2 + V(\phi)}{\frac{1}{2}\dot{\phi}^2 - V(\phi)}\rho.$$
(8.85)

Gas consisting of ultra-relativistic particles. Radiation

If the velocities of the gas particles approach that of light their rest energy becomes negligible compared to their total energy. In this limit their rest masses can be neglected and the fluid behaves like a gas of photons, i.e. like electromagnetic radiation. From eq. (8.50) we know that the trace of the mixed components of the energy-momentum tensor vanishes. Taking the trace of the energy-momentum tensor eq. (8.62) for a perfect fluid we get

$$T^{\mu}_{\mu} = 3p - \rho.$$
(8.86)

This shows that the equation of state for a gas of ultra-relativistic particles, and for electromagnetic radiation is

$$\boxed{p = \frac{1}{3}\rho.}$$
(8.87)

Dust

For a gas of slowly moving particles the energy will be dominated by the rest energy of the particles. Even if the pressure gradient will be important for the motion of the fluid elements in inhomogeneous regions, the gravitational effects of the pressure can be neglected in the non-relativistic limit.

A gas of particles with vanishing pressure is called dust. Thus the equation of state of dust is

$$\boxed{p_{\text{dust}} = 0}$$
(8.88)

and the energy-momentum tensor reduces to

$$T_{\mu\nu} = \rho u_{\mu} u_{\nu}.$$
(8.89)

A cosmic magnetic field

Observations indicate that galaxies are surrounded by huge magnetic fields. Whether or not there exist magnetic fields at a cosmic scale is still unsettled, but it is by no means ruled out that the universe has a such a field.

Consider a pure magnetic field in an orthonormal frame. Note that in this case the character of the electromagnetic field is dependent of the frame chosen. We will choose a frame where there is only a magnetic field present, i.e., $E_i = 0$.

Using the electromagnetic field tensor in eq. (6.31) we find that the energy-momentum tensor eq. (8.49) can be written as

$$T_{\hat{\mu}\hat{\nu}} = (\rho + p)u_{\hat{\mu}}u_{\hat{\nu}} + pg_{\hat{\mu}\hat{\nu}} + \pi_{\hat{\mu}\hat{\nu}}, \tag{8.90}$$

where

$$\boxed{\rho = 3p = \frac{1}{2}B^2,} \tag{8.91}$$

and $\pi_{\hat{\mu}\hat{\nu}}$ is given by

$$\pi_{ij} = -B_iB_j + \frac{1}{3}B^2\delta_{ij}, \tag{8.92}$$

$$\pi_{0i} = \pi_{i0} = \pi_{00} = 0. \tag{8.93}$$

The tensor $\pi_{\mu\nu}$ is called the *anisotropic stress tensor* and it is in general symmetric and trace-free; i.e.,

$$\pi_{\mu\nu} = \pi_{\nu\mu}, \quad \pi^{\mu}_{\ \mu} = 0. \tag{8.94}$$

We note that the magnetic field has a perfect fluid part which behaves like radiation fluid, but it is not a perfect fluid because of this anisotropic stress tensor.

8.7 The paths of free point particles

Consider a system of free point particles in curved space-time and assume that the particles do not collide with each other. This system will be described as a pressure-free perfect fluid, i.e., as dust. From Einstein's field equations, as applied to a dust-filled region, follow

$$(\rho u^{\mu}u^{\nu})_{;\nu} = 0, \tag{8.95}$$

or

$$(\rho u^{\nu})_{;\nu}\, u^{\mu} + \rho u^{\nu}u^{\mu}_{\ ;\nu} = 0. \tag{8.96}$$

The four-velocity identity $u^{\mu}u_{\mu} = -1$ gives

$$u^{\mu}_{\ ;\nu}u_{\mu} = 0. \tag{8.97}$$

In order to utilize this equation we multiply eq. (8.96) by u_{μ}. This leads to

$$(\rho u^{\nu})_{;\nu} = 0. \tag{8.98}$$

Inserting this into eq. (8.96) we obtain

$$\boxed{u^{\nu}u^{\mu}_{\ ;\nu} = 0,} \tag{8.99}$$

which is just the geodesic equation as given in eq. (6.104). Thus *it follows from Einstein's field equations that free particles move along geodesic curves in space-time.*

Note that eq. (8.99), or (6.104), is equivalent to eq. (6.105), which may be written

$$\frac{d\mathbf{u}}{d\tau} = 0, \tag{8.100}$$

where $\mathbf{u} = (dx^\mu/d\tau)\mathbf{e}_\mu = u^\mu\mathbf{e}_\mu$ is the four-velocity of the particle. According to eq. (3.16) this equation takes the form

$$\mathbf{A} = 0, \qquad (8.101)$$

saying that *a free particle has vanishing four-acceleration.*

A non-vanishing four-acceleration of a particle means that it is acted upon by non-gravitational forces and moves non-geodesically. This is an invariant or intrinsic property of the particle. In the language of Kant the four-acceleration represents "Das Ding an Sich". On the other hand, a non-vanishing three-acceleration means that a particle accelerates relative to a reference frame or an observer. This is a relational property of the particle and the observer. The three-acceleration represents "Das Ding an Mich".

Problems

8.1. *Lorentz transformation of a perfect fluid*
Consider a homogeneous perfect fluid. In the rest frame of the fluid the equation of state is $p = w\rho$ (with $c = 1$), and the energy-momentum tensor has the form

$$T_{\mu\nu} = \rho\,\mathrm{diag}(1, w, w, w). \qquad (8.102)$$

(a) Make a Lorentz transformation in the $1-$direction with velocity v and show that the transformed energy-momentum tensor has the form

$$T_{\mu'\nu'} = \rho \begin{bmatrix} \gamma^2(1+v^2w) & \gamma^2 v(1+w) & 0 & 0 \\ \gamma^2 v(1+w) & \gamma^2(v^2+w) & 0 & 0 \\ 0 & 0 & w & 0 \\ 0 & 0 & 0 & w \end{bmatrix}, \gamma = \frac{1}{\sqrt{1-v^2}}. \quad (8.103)$$

(b) The weak energy condition requires that the energy-density is positive. What restriction does this put on w?

(c) Which value of w makes the components of the energy-momentum tensor Lorentz invariant?

8.2. *Geodesic equation and constants of motion*
Show that the covariant components of the geodesic equation have the form

$$\dot{u}_\mu = \frac{1}{2}g_{\alpha\beta,\mu}u^\alpha u^\beta.$$

What does this equation tell about constants of motion for free particles?

8.3. *The electromagnetic energy-momentum tensor*
Consider a general energy-momentum tensor $T_{\mu\nu}$ and a time-like vector u^μ. We can always decompose $T_{\mu\nu}$ as follows:

$$T_{\mu\nu} = \rho u_\mu u_\nu + p h_{\mu\nu} + 2u_{(\mu}q_{\nu)} + \pi_{\mu\nu}, \qquad (8.104)$$

where ρ is the energy density; p is the isotropic pressure; q_μ is the energy flux; $\pi_{\mu\nu}$ is the anisotropic stress tensor; and $h_{\mu\nu} = g_{\mu\nu} + u_\mu u_\nu$ is the 3-metric tensor on the hypersurfaces orthogonal to u_μ. These fulfill the following relations:

$$u^\mu q_\mu = u^\mu \pi_{\mu\nu} = u^\mu h_{\mu\nu} = \pi^\mu{}_\mu = 0, \quad \pi_{\nu\mu} = \pi_{\mu\nu}.$$

(a) Show that

$$\rho = T_{\mu\nu}u^{\mu}u^{\nu}, \qquad p = \frac{1}{3}T_{\mu\nu}h^{\mu\nu},$$

$$q_{\nu} = -u^{\mu}h^{\rho}{}_{\nu}T_{\mu\rho}, \qquad \pi_{\mu\nu} = T_{\rho\sigma}\left(h^{\rho}{}_{\mu}h^{\sigma}{}_{\nu} - \frac{1}{3}h^{\rho\sigma}h_{\mu\nu}\right). \quad (8.105)$$

(b) Consider the electromagnetic field tensor, eq. (6.31), in an orthonormal basis $\{e_t, e_i\}$. Assume that u^{μ} is aligned with the time-like basis vector e_t. Use the electromagnetic energy-momentum tensor, eq. (8.49), to show that

$$\rho = 3p = \frac{1}{2}(E^k E_k + B^k B_k),$$

$$q_i = -\varepsilon_{ijk}E^j B^k,$$

$$\pi_{ij} = -E_i E_j - B_i B_j + \frac{1}{3}\delta_{ij}(E^k E_k + B^k B_k). \quad (8.106)$$

What is the physical interpretation of q_i?

8.4. *Lorentz-invariant radiation*

Consider a region filled with photons of all frequencies and moving in all directions. Let $n(\nu, e)$ be the number of photons per unit volume, per frequency interval, and per unit solid angle moving in the direction e, as referred to an orthonormal basis Σ. Let primed quantities be measured in a basis Σ' moving with a velocity v relative to Σ. A comoving volume element dV has a velocity u in Σ. The corresponding rest volume is dV_0.

The quantity $n(\nu, e)dV d\nu d\Omega$ represents the number of photons occupying a volume dV, with frequencies between ν and $\nu + d\nu$, and moving with directions within a solid angle $d\Omega = \sin\theta d\theta d\phi$. It is an invariant quantity. Hence,

$$n(\nu, e)dV d\nu d\Omega = n'(\nu', e')dV' d\nu' d\Omega'.$$

(a) Use that dV_0 is invariant to show that $dV' = \gamma_{u'}^{-1}\gamma_u dV$ where $\gamma_{u'} = (1 - |u'|^2/c^2)^{-1/2}$ and $\gamma_u = (1 - |u|^2/c^2)^{-1/2}$.

(b) Choose the x'- and x-axes to be directed along v (so that $v = v e_x$) and use the transformation formulae of the velocity components,

$$u'_x = \frac{u_x - v}{1 - \frac{u_x v}{c^2}}, \quad u'_y = \frac{u_y}{\gamma_v\left(1 - \frac{u_x v}{c^2}\right)}, \quad \gamma_v = \left(1 - \frac{v^2}{c^2}\right)^{-\frac{1}{2}},$$

to show that

$$\gamma_{u'} = \gamma_u \gamma_v \left(1 - \frac{u \cdot v}{c^2}\right)^{-1};$$

and hence that

$$dV' = \gamma_v^{-1}\left(1 - \frac{u \cdot v}{c^2}\right)^{-1} dV.$$

(c) Let θ be the angle between the transformation velocity v and the velocity u of the volume. Since the volume is comoving with photons moving in the u-direction, we now set $|u| = c$. Show that this leads to $dV' = \kappa^{-1}dV$, where $\kappa = \gamma_v[1 - (v/c)\cos\theta]$.

(d) Use the relativistic equations for the Doppler effect and abberation (see problem 2.8), and show that the transformation equations for the differentials of frequency and solid angle are

$$d\nu' = \kappa d\nu, \quad d\Omega' = \kappa^{-2}d\Omega.$$

(e) Deduce that $n'(\nu', \mathbf{e}') = \kappa^2 n(\nu, \mathbf{e})$ and use the transformation equation for the frequency to show that $n(\nu, \mathbf{e})/\nu^2$ is a Lorentz-invariant quantity.

(f) Since ν is not Lorentz invariant it follows that $n(\nu, \mathbf{e})/\nu^2$ must be independent of ν. Use this, together with the fact that the energy of a photon is given by $h\nu$, to find how the energy-density per frequency interval and solid angle of a Lorentz-invariant radiation depends upon the frequency.

The Linear Field Approximation

Einstein's theory of general relativity leads to Newtonian gravity in the limit when the gravitational field is weak and static and the particles in the gravitational field moves slowly compared to the velocity of light. In the case of mass distributions of limited extension the field is weak at distances much larger than the Schwarzschild radius of the mass (see Chapter 10). At such distances the absolute value of the gravitational potential is much less than 1, and there is approximately Minkowski spacetime.

In the linear field approximation the field is weak, but it need not be static, and particles are allowed to move with relativistic velocities.

9.1 The linearised field equations

We shall describe small deviations from Minkowski spacetime. Then the properties of the spacetime are given by the metric tensor with components

$$g_{\mu\nu} = \eta_{\mu\nu} + h_{\mu\nu}, \qquad |h_{\mu\nu}| \ll 1. \tag{9.1}$$

Let us consider the transformation of these components

$$g_{\rho'\sigma'} = g_{\mu\nu} \frac{\partial x^\mu}{\partial x^{\rho'}} \frac{\partial x^\nu}{\partial x^{\sigma'}}, \tag{9.2}$$

under an infinitesimal coordinate transformation at a point P,

$$x^{\mu'}(P) = x^\mu(P) + \xi^\mu(P), \qquad |\xi^\mu| \ll |x^\mu|. \tag{9.3}$$

This gives

$$g_{\rho'\sigma'}\Big|_{x^\mu} = \frac{\partial x^\mu}{\partial x^{\rho'}} \frac{\partial x^\nu}{\partial x^{\sigma'}} g_{\mu\nu}\Big|_{(x^{\mu'} - \xi^\mu)}. \tag{9.4}$$

All calculations will be performed only to first order in $h_{\mu\nu}$, ξ^{μ} and their derivatives. Hence,

$$\frac{\partial x^{\mu}}{\partial x^{\rho'}} = \delta^{\mu}_{\rho} - \frac{\partial \xi^{\mu}}{\partial x^{\rho}} \equiv \delta^{\mu}_{\rho} - \xi^{\mu}_{,\rho}, \tag{9.5}$$

$$g_{\mu\nu}\Big|_{(x^{\mu'} - \xi^{\mu})} = \eta_{\mu\nu} + h_{\mu\nu}, \tag{9.6}$$

which gives to first order

$$\begin{aligned} g_{\rho'\sigma'} &= (\delta^{\mu}_{\rho} - \xi^{\mu}_{,\rho})(\delta^{\nu}_{\sigma} - \xi^{\nu}_{,\sigma})(\eta_{\mu\nu} + h_{\mu\nu}) \\ &\approx \eta_{\rho\sigma} + h_{\rho\sigma} - \xi_{\sigma,\rho} - \xi_{\rho,\sigma}. \end{aligned} \tag{9.7}$$

Since

$$g_{\rho'\sigma'} = \eta_{\rho\sigma} + h_{\rho'\sigma'}, \tag{9.8}$$

we get

$$h_{\rho'\sigma'} = h_{\rho\sigma} - \xi_{\sigma,\rho} - \xi_{\rho,\sigma}. \tag{9.9}$$

Because this transformation was induced by a coordinate transformation, such a transformation in the linear field approximation is called a *gauge transformation*. In this approximation we see that generally the components of the metric tensor are not gauge invariant. In the case that the components of the metric tensor are gauge invariant, the transformation is called an isometry, and the vector $\boldsymbol{\xi}$ is a Killing vector. Then $\xi_{\sigma;\rho} + \xi_{\rho;\sigma} = 0$ which are the Killing equations.

To 1st order in $h_{\mu\nu}$ we may neglect products of the Christoffel symbols in eq. (7.45) and the Riemann curvature tensor is

$$R_{\alpha\mu\beta\nu} = \Gamma_{\alpha\mu\nu,\beta} - \Gamma_{\alpha\mu\beta,\nu}, \tag{9.10}$$

where

$$\Gamma_{\alpha\mu\nu} = \frac{1}{2}\left(h_{\mu\alpha,\nu} + h_{\nu\alpha,\mu} - h_{\mu\nu,\alpha}\right). \tag{9.11}$$

Hence,

$$R_{\alpha\mu\beta\nu} = \frac{1}{2}\left(h_{\nu\alpha,\mu\beta} + h_{\mu\beta,\alpha\nu} - h_{\mu\nu,\alpha\beta} - h_{\alpha\beta,\mu\nu}\right). \tag{9.12}$$

The Ricci tensor is thus to 1st order

$$R_{\mu\nu} = \frac{1}{2}\left(h^{\alpha}_{\nu,\alpha\mu} + h^{\alpha}_{\mu,\alpha\nu} - h_{,\mu\nu} - \Box h_{\mu\nu}\right), \tag{9.13}$$

where $\Box \equiv \eta^{\alpha\beta}\partial_{\alpha}\partial_{\beta} = -\partial^2/\partial t^2 + \nabla^2$ is the d'Alembert wave operator in Minkowski spacetime. Contracting once more with $\eta^{\mu\nu}$ the Ricci scalar is obtained as

$$R = h^{\mu\nu}_{,\mu\nu} - \Box h, \quad h \equiv h^{\alpha}_{\alpha}. \tag{9.14}$$

The linearised Einstein tensor is

$$E_{\mu\nu} = \frac{1}{2}\left[h^{\alpha}_{\nu,\alpha\mu} + h^{\alpha}_{\mu,\alpha\nu} - h_{,\mu\nu} - \Box h_{\mu\nu} - \eta_{\mu\nu}(h^{\alpha\beta}_{,\alpha\beta} - \Box h)\right]. \tag{9.15}$$

Hence, the linearised field equations take the form

$$h^\alpha{}_{\nu,\alpha\mu} + h^\alpha{}_{\mu,\alpha\nu} - h_{,\mu\nu} - \Box h_{\mu\nu} - \eta_{\mu\nu}(h^{\alpha\beta}{}_{,\alpha\beta} - \Box h) = 2\kappa T_{\mu\nu}. \qquad (9.16)$$

It proves useful to introduce

$$\bar{h}_{\mu\nu} = h_{\mu\nu} - \frac{1}{2}\eta_{\mu\nu}h, \qquad (9.17)$$

which simplifies the field equations to

$$\bar{h}^\alpha{}_{\nu,\alpha\mu} + \bar{h}^\alpha{}_{\mu,\alpha\nu} - \Box \bar{h}_{\mu\nu} - \eta_{\mu\nu}\bar{h}^{\alpha\beta}{}_{,\alpha\beta} = 2\kappa T_{\mu\nu}. \qquad (9.18)$$

In order to simplify the equations still more, we perform a gauge transformation (9.9). The transformed metric $\bar{h}'_{\alpha\beta}$ then becomes

$$\bar{h}'_{\alpha\beta} = \bar{h}_{\alpha\beta} - \xi_{\alpha,\beta} - \xi_{\beta,\alpha} + \eta_{\alpha\beta}\eta^\sigma{}_{,\sigma}. \qquad (9.19)$$

The transformed divergence of $\bar{h}_{\alpha\beta}$ becomes

$$\bar{h}'^\beta{}_{\alpha,\beta} = \bar{h}^\beta{}_{\alpha,\beta} - \Box\xi_\alpha. \qquad (9.20)$$

Choosing gauge functions ξ_α fulfilling $\Box\xi_\alpha = \bar{h}^\beta{}_{\alpha,\beta}$, one obtains (dropping the prime from now on)

$$\bar{h}^\beta{}_{\alpha,\beta} = 0. \qquad (9.21)$$

This is called the *Lorenz condition*, or *Lorenz gauge*. In this gauge the field equations reduce to

$$\boxed{\Box h_{\mu\nu} = -2\kappa T_{\mu\nu}.} \qquad (9.22)$$

Coordinates that obey the Lorenz condition are called *harmonic*.

In the time-dependent case the field equations of empty space are

$$\Box h_{\mu\nu} = 0, \qquad (9.23)$$

which is d'Alemberts wave equation. The corresponding equation for the Riemann tensor is

$$\Box R_{\alpha\mu\beta\nu} = 0. \qquad (9.24)$$

This equation means that gravitational waves move in empty space with the speed of light.

We have seen that an infinitesimal coordinate transformation $x^{\mu'} = x^\mu + \xi^\mu$ causes a change in the metric tensor so that the metric perturbation takes the form (9.9). Both $h'_{\mu\nu}$ and $h_{\mu\nu}$ are solutions of the field equations. Hence, $\xi_{\mu,\nu} + \xi_{\nu,\mu}$ are also solutions of the field equations. Solutions of the linearised field equations of the form $\xi_{\mu,\nu} + \xi_{\nu,\mu}$ are called *Weyl solutions*. Calculating the Riemann tensor associated with a Weyl solution one finds

$$R^{\text{Weyl}}_{\alpha\mu\beta\nu} = 0. \qquad (9.25)$$

This means that the Weyl solutions do not represent properties of the spacetime. They only represent coordinate effects that may be transformed away.

9.2   The Newtonian limit of general relativity

The motion of a free particle is given by the geodesic equation, eq. (6.105). Using the proper time τ of the particle as parameter, it takes the form

$$\frac{d^2x^\mu}{d\tau^2} + \Gamma^\mu{}_{\alpha\beta}\frac{dx^\alpha}{d\tau}\frac{dx^\beta}{d\tau} = 0. \tag{9.26}$$

Taking the Newtonian limit, $\frac{dx^i}{d\tau} \ll c$ (in this section we shall retain the speed of light in the expressions) and keeping only terms to first order in the velocity, we have $d\tau \approx dt$ where dt is the usual Newtonian time. Assuming in addition that the metric is diagonal and time independent, the i-component of the acceleration of gravity is found to be

$$g^i = \frac{d^2x^i}{d\tau^2} = -\Gamma^i{}_{00}\frac{dx(ct)}{dt}\frac{d(ct)}{dt} = -c^2\Gamma^i{}_{00}. \tag{9.27}$$

We have hereby obtained a simple weak-field interpretation of the Christoffel symbols $\Gamma^i{}_{00}$. They represent the components of the acceleration of gravity. Using eq. (6.96) and remembering that the metric tensor is assumed to be diagonal, we get

$$\Gamma^i{}_{00} = -\frac{1}{2}g^{\alpha i}\frac{\partial g_{00}}{\partial x^\alpha} \approx \frac{1}{2}\eta^{ii}\frac{\partial h_{00}}{\partial x^i} = \frac{1}{2}\frac{\partial h_{00}}{\partial x^i}. \tag{9.28}$$

Inserting this into eq. (9.27) we have

$$g^i = \frac{c^2}{2}\frac{\partial h_{00}}{\partial x^i}. \tag{9.29}$$

This equation shows explicitly how, in the Newtonian limit, the time component of the metric tensor determines the acceleration of gravity.

We shall now take the Newtonian limit of Einstein's field equations. With the assumptions above the line element of space-time can be written

$$ds^2 = -(1 - h_{00})c^2dt^2 + (\eta_{ii} + h_{ii})dx^i dx^i. \tag{9.30}$$

In this case we need only one independent equation, which can be taken as the 00-component of eq. (8.39) with $\Lambda = 0$

$$R_{00} = \kappa\left(T_{00} - \frac{1}{2}g_{00}T\right). \tag{9.31}$$

From eq. (9.12) we have

$$R_{\mu 0 \alpha 0} = \frac{1}{2}\left(h_{\mu 0,0\alpha} - h_{\mu\alpha,00} - h_{00,\mu\alpha} + h_{0\alpha,\mu 0}\right). \tag{9.32}$$

Considering a static field, all terms with time derivatives are equal to zero. In this case we get

$$R_{\mu 0 \alpha 0} = -\frac{1}{2}h_{00,\mu\alpha}. \tag{9.33}$$

Contracting μ with α leads to

$$R_{00} = R^\alpha{}_{0\alpha 0} = -\frac{1}{2}h_{00,}{}^\alpha{}_\alpha = -\frac{1}{2}\frac{\partial}{\partial x^i}\left(\frac{\partial h_{00}}{\partial x^i}\right), \tag{9.34}$$

since derivatives with respect to time vanish.

Using (9.29) this can be written

$$R_{00} = -\frac{1}{c^2}\frac{\partial g^i}{\partial x^i}.$$

(9.35)

In the limit with $|h_{\mu\nu}| \ll 1$ we can use the Cartesian expression for the divergence, so that

$$R_{00} = -\frac{1}{c^2}\nabla \cdot \mathbf{g}.$$

(9.36)

Considering the components of the energy-momentum tensor of a perfect fluid as given in eq. (8.62), we see that in the Newtonian limit the term $T_{00} = \rho c^2$ is dominating; all the other terms can be neglected compared to T_{00}. Regarding the trace of the energy-momentum tensor in the Newtonian limit, we find

$$T = T^0_0 = \eta^{0\alpha}T_{\alpha 0} = -T_{00}.$$

(9.37)

This gives

$$T_{00} - \frac{1}{2}g_{00}T \approx T_{00} - \frac{1}{2}\eta_{00}T = \frac{1}{2}T_{00}.$$

(9.38)

Hence, equation (9.31) can be written

$$R_{00} = \frac{1}{2}\kappa T_{00} = \frac{1}{2}\kappa\rho c^2.$$

(9.39)

Equations (9.36) and (9.39) then give

$$\nabla \cdot \mathbf{g} = -\frac{1}{2}\kappa\rho c^4.$$

(9.40)

This represents the Newtonian limit of Einstein's gravitational field equations in the case of static fields.

Comparing equation (9.40) with equations (1.32) and (1.33) we see that the relativistic equation reduces to the "Newtonian" gravitational field equations if

$$\kappa = \frac{8\pi G}{c^4}.$$

(9.41)

Thus we have to conclude that the Einstein field equations with the correct constant is

$$\boxed{R_{\mu\nu} - \frac{1}{2}Rg_{\mu\nu} + \Lambda g_{\mu\nu} = \frac{8\pi G}{c^4}T_{\mu\nu}.}$$

(9.42)

9.3 Solutions of the linearised field equations

We shall now consider solutions of the linearised field equations with a non-relativistic mass-distribution as a source. "Non-relativistic" means that the pressure is so small that it may be neglected compared to the mass density, and the fluid moves so slowly that it is sufficient to include terms of 1st order in the velocity in the energy-momentum tensor.

Einstein's field equations may be written

$$\Box h_{\mu\nu} = -2\kappa \left(T_{\mu\nu} - \frac{1}{2}\eta_{\mu\nu}T \right). \tag{9.43}$$

The solution to this equation can be written as the retarded potential

$$h_{\mu\nu} = \frac{\kappa}{2\pi} \int \frac{\left[T_{\mu\nu} - \frac{1}{2}\eta_{\mu\nu}T \right](t', \mathbf{x}')}{|\mathbf{x} - \mathbf{x}'|} d^3x', \tag{9.44}$$

where the retarded time, t', is given by $t' = t - |\mathbf{x} - \mathbf{x}'|/c$. In the following we shall, however, assume that the distances are so small, and the variation of the source so slow, that we may put $t' = t$.

We shall solve the field equations in the presence of a perfect fluid with energy-momentum tensor

$$T^{\mu\nu} = \frac{p}{c^2}g^{\mu\nu} + \left(\frac{p}{c^2} + \rho \right) \frac{dx^\mu}{d\tau}\frac{dx^\nu}{d\tau}. \tag{9.45}$$

With $p = 0$ and small velocities we get

$$T^{\mu\nu} = \rho\frac{dx^\mu}{dt}\frac{dx^\nu}{dt}, \qquad T = -\rho, \tag{9.46}$$

so that

$$T_{00} - \frac{1}{2}\eta_{00}T = \frac{1}{2}\rho, \quad T_{i0} - \frac{1}{2}\eta_{i0}T = -\rho v^i,$$

$$T_{ij} - \frac{1}{2}\eta_{ij}T = \frac{1}{2}\rho\delta_{ij}. \tag{9.47}$$

In this case the field equations take the form

$$h_{00} = h_{ii} = \frac{2G}{c^2}\int\frac{\rho}{r}d^3\mathbf{r} = -2\frac{\phi}{c^2}, \tag{9.48}$$

there ϕ is the Newtonian gravitational potential of eq.(1.17), and

$$h_{i0} = -\frac{4G}{c^2}\int\frac{\rho v_i}{r}d^3\mathbf{r} \equiv A_i, \qquad h_{ij} = 0, \text{ for } i \neq j. \tag{9.49}$$

Here, A_i is the i-component of a vector potential.

Assume that the source is non-rotating and spherically symmetric. Then

$$A_i = -\frac{4Gv_i}{c^2}\int\frac{\rho}{r}d^3\mathbf{r}. \tag{9.50}$$

Outside the mass-distribution this gives

$$A_i = -4\frac{Gm}{c^2r}v_i = -2\frac{R_S}{r}v_i, \tag{9.51}$$

where $R_S \equiv 2Gm/c^2$ is called the *Schwarzschild radius* of the source, and r is the distance from its centre to the field point. Hence, the external metric is

$$ds^2 = -\left(1 - \frac{R_S}{r} \right)c^2dt^2 + \left(1 + \frac{R_S}{r} \right)(dx^2 + dy^2 + dz^2)$$

$$-\frac{2R_S}{r}(v_xdx + v_ydy + v_zdz)\,dt. \tag{9.52}$$

In the static case the metric reduces to

$$ds^2 = -\left(1 - \frac{R_S}{r}\right)c^2 dt^2 + \left(1 + \frac{R_S}{r}\right)(dx^2 + dy^2 + dz^2). \quad (9.53)$$

In general one must have $h_{\mu\nu} \to 0$ infinitely far from the mass distribution in order for the integrals to converge. This is the reason for using isotropic coordinates in the linear field approximation.

The internal metric can easily be found for the special case that the density is constant for $r < R$ and vanishes for $r > R$. Let m be the total mass of the system. In problem 1.3 it was shown that the Newtonian gravitational potential at a distance r from the centre of the mass distribution is

$$\phi = -\frac{Gm}{2c^2 R}\left(3 - \frac{r^2}{R^2}\right). \quad (9.54)$$

Hence, the internal metric is

$$\begin{aligned} ds^2 &= -\left[1 - \frac{Gm}{2c^2 R}\left(3 - \frac{r^2}{R^2}\right)\right]c^2 dt^2 \\ &+ \left[1 + \frac{Gm}{2c^2 R}\left(3 - \frac{r^2}{R^2}\right)\right](dx^2 + dy^2 + dz^2). \end{aligned} \quad (9.55)$$

The generalisation of the solutions (9.53) and (9.55) to gravitational fields of arbitrary strength are the external and internal Schwarzschild solutions of the full field equations and will be derived in Chapter 10.

9.4 Gravitoelectromagnetism

The weak field approximation of Einstein's equations is valid to great accuracy in, for example, the Solar system. The resemblance between the electromagnetic wave equation, eq. (6.54), and eq. (9.22) is evident. The similarity between electromagnetism and the linearised Einstein equations goes even further.

The solution of eq. (9.22) may be written in terms of retarded potentials as

$$\bar{h}_{\mu\nu} = \frac{\kappa}{2\pi}\int \frac{T_{\mu\nu}(t - |\mathbf{x} - \mathbf{x}'|/c, \mathbf{x}')}{|\mathbf{x} - \mathbf{x}'|}d^3x', \quad (9.56)$$

where $\mathbf{x}$ is a spatial vector and $T_{\mu\nu} = T_{\mu\nu}(t, \mathbf{x})$. The energy-momentum tensor $T_{\mu\nu}$ mimics the behaviour of a electromagnetic four-current J_μ and the tensor potential $\bar{h}_{\mu\nu}$ mimics a field potential A_μ.

We will assume that the energy-momentum tensor obeys $|T_{00}| \gg |T_{ij}|$ and $|T_{0i}| \gg |T_{ij}|$ in the weak field approximation. Hence from eq. (9.56) $|\bar{h}_{00}| \gg |\bar{h}_{ij}|$ and $|\bar{h}_{0i}| \gg |\bar{h}_{ij}|$. Then we can write

$$\bar{h}_{00} = -\frac{4\phi}{c^2}, \quad (9.57)$$

$$\bar{h}_{0i} = \frac{2A_i}{c^2}. \quad (9.58)$$

Here, ϕ is the Newtonian or "gravitoelectric" potential

$$\phi = -\frac{Gm}{r}, \quad (9.59)$$

and A_i is the "gravitomagnetic" vector potential given in terms of the total angular momentum **S** of the system

$$A_i = \frac{G}{c} \frac{S^j x^k}{r^3} \varepsilon_{ijk}.$$ (9.60)

The mass m is related to the mass-density $\rho = T_{00}/c^2$ by

$$\int \rho d^3 x = m,$$ (9.61)

and the angular momentum **S** to the mass-current density $j^i = T^{0i}/c$ by

$$S^i = 2 \int \varepsilon^i_{jk} x^j j^k d^3 x.$$ (9.62)

The Lorenz gauge condition $\bar{h}^{\mu\alpha}{}_{,\alpha} = 0$ can be written in terms of the potentials ϕ and **A**

$$\frac{1}{c} \frac{\partial \phi}{\partial t} + \frac{1}{2} \nabla \cdot \mathbf{A} = 0.$$ (9.63)

This is, apart from a factor $1/2$, the Lorenz gauge condition in electromagnetism. This factor relates to the fact that the electromagnetic field is a spin-1 field, while the geometrodynamical field involves a spin-2 field.

Defining the gravitoelectric and gravitomagnetic fields $\mathbf{E}_G$ and $\mathbf{B}_G$ by

$$\mathbf{E}_G = -\nabla\phi - \frac{1}{2c} \frac{\partial \mathbf{A}}{\partial t}$$ (9.64)

$$\mathbf{B}_G = \nabla \times \mathbf{A},$$ (9.65)

the equations (9.22) – using eqs. (9.56), (9.57), (9.58), (9.63), (9.64) and (9.65) – reduces to

$$\nabla \cdot \mathbf{E}_G = -4\pi G\rho$$ (9.66)

$$\nabla \cdot \mathbf{B}_G = 0$$ (9.67)

$$\nabla \times \mathbf{E}_G = -\frac{1}{2c} \frac{\partial \mathbf{B}_G}{\partial t}$$ (9.68)

$$\nabla \times \frac{1}{2} \mathbf{B}_G = -\frac{4\pi G}{c} \mathbf{j} + \frac{1}{c} \frac{\partial \mathbf{E}_G}{\partial t}.$$ (9.69)

These are the *Maxwell equations for the gravitoelectromagnetic (GEM) fields*.

These fields describes the spacetime outside a rotating object in terms of the gravitoelectric and gravitomagnetic fields. The metric tensor can be written in terms of the gravitoelectric and gravitomagnetic potential as

$$ds^2 = -\left(1 + \frac{2\phi}{c^2}\right) c^2 dt^2 - \frac{4}{c} A_i dx^i dt + \left(1 - \frac{2\phi}{c^2}\right) \delta_{ij} dx^i dx^j.$$ (9.70)

In the weak field approximation gravity can be considered analogous to electromagnetism. Furthermore, for a weakly gravitating rotating body, the gravitomagnetic field can be written as a dipole field

$$\mathbf{B}_G = -\frac{4G}{c} \frac{3\mathbf{r}\,(\mathbf{r} \cdot \mathbf{S}) - \mathbf{S}r^2}{2r^5}.$$ (9.71)

In the Newtonian theory there will not be any gravitomagnetic effects; the Newtonian potential is the same irrespective of whether or not the body is rotating. Hence the gravitomagnetic field is a purely relativistic effect. The gravitoelectric field is the Newtonian part of the gravitational field, while the gravitomagnetic field is the non-Newtonian part.

This can also be seen if we note a further analogy between the weak field approximation and electromagnetic fields. The geodesic equation for a test particle is

$$\frac{d^2 x^\mu}{d\tau^2} + \Gamma^\mu_{\alpha\beta} \frac{dx^\alpha}{d\tau} \frac{dx^\beta}{d\tau} = 0, \tag{9.72}$$

where τ is the proper time of the particle. For a non-relativistic particle, we have $\frac{dx^0}{d\tau} \approx 1$ and $\frac{dx^i}{d\tau} \approx v^i/c$. Considering only linear terms in v^i/c, and restricting ourselves to static fields were $g_{\alpha\beta,0} = 0$, we obtain the expression

$$\frac{d\mathbf{v}}{dt} = \mathbf{E}_G + \frac{\mathbf{v}}{c} \times \mathbf{B}_G. \tag{9.73}$$

This is the *Lorentz's force-law for GEM fields*.

Particles orbiting a rotating body (like the Earth), will experience a gravitomagnetic field which will make their orbit precess. This precession is called the *Lense-Thirring effect* in honour of the physicists Josef Lense and Hans Thirring who first predicted this effect in 1918 [LT18].

An orbiting body has an orbital angular momentum **L**. The gravitomagnetic field interacts with this angular momentum and causes a torque given by

$$\boldsymbol{\tau} = \frac{1}{2c} \mathbf{L} \times \mathbf{B}_G. \tag{9.74}$$

The torque is, as usual, equal to the time derivative of the angular momentum, and hence,

$$\frac{d\mathbf{L}}{dt} = -G \frac{\mathbf{L} \times \left[3\mathbf{r} \left(\mathbf{r} \cdot \mathbf{S} \right) - \mathbf{S}r^2 \right]}{c^2 r^5}, \tag{9.75}$$

from which – using the formula $\frac{d\mathbf{L}}{dt} = \boldsymbol{\Omega} \times \mathbf{L}$ – we can read off the precession angular velocity

$$\boldsymbol{\Omega} = G \frac{3\mathbf{r} \left(\mathbf{r} \cdot \mathbf{S} \right) - \mathbf{S}r^2}{c^2 r^5}. \tag{9.76}$$

The Lense-Thirring effect will be taken up again in the next chapter were we will derive it from an exact solution of Einstein's field equations.

9.5 Gravitational waves

We shall now consider plane-wave solutions of the linearised field equations (9.22) for empty space. In this and the next section we use units so that $c = 1$. These equations admit the solutions

$$\bar{h}_{\mu\nu} = A_{\mu\nu} \cos(k_\alpha x^\alpha), \tag{9.77}$$

where $A_{\mu\nu}$ is a constant symmetric tensor of rank 2 and k_α is a constant *wave-vector*. Inserting this into eq. (9.22) gives

$$k_\alpha k^\alpha = 0. \tag{9.78}$$

Hence, k_α is a null-vector, which means that the gravitational waves propagate with the velocity of light. An observer with four-velocity U^μ would observe the wave to have a frequency

$$\omega = -k_\mu U^\mu. \tag{9.79}$$

The components of the wave-vector may therefore be written

$$k^\mu = (\omega, k^1, k^2, k^3), \qquad \omega^2 = k_i k^i. \tag{9.80}$$

A general solution of eq. (9.22) can be written as a superposition of such plane waves.

The solution (9.77) contains 13 parameters to specify the wave: ten for the coefficients $A_{\mu\nu}$ and three for the null vector k^μ. However, most of these are the result of coordinate freedom and gauge freedom.

Assume that the vector k_α is given. Then we will show that there are physically only two polarisations left, when the gauge freedom is eliminated. Using the Lorenz condition we have

$$k^\alpha A_{\alpha\beta} = 0. \tag{9.81}$$

This means that the wave is orthogonal or *transverse* to $A_{\alpha\beta}$. The Lorenz condition does not completely specify the gauge. We still have the freedom of choosing ξ_μ such that $\Box \xi_\mu = 0$. This gauge transformation preserves the Lorenz condition so we can use this ξ_μ to simplify $A_{\mu\nu}$ further. By a clever choice of ξ_μ we can require that

$$U^\alpha A_{\alpha\beta} = A^\alpha{}_\alpha = 0. \tag{9.82}$$

The two remaining free components of $A_{\alpha\beta}$ represent the two degrees of freedom – the two polarisations – in the plane gravitational wave.

In the comoving frame of the observer, where $U^\mu = (1, 0, 0, 0)$, the *transverse traceless gauge* conditions take the form

$$h^{\text{TT}}_{\mu 0} = h^{\text{TT}}_{kj,j} = h^{\text{TT}}_{ii} = 0. \tag{9.83}$$

The first of these equations tells that only the spatial components of the metric perturbation is non-zero. The second says that the spatial components are divergence-free, and the third says they are trace-free. Note also that since $h = h^\mu{}_\mu = 0$ there is no distinction between $\bar{h}_{\mu\nu}$ and $h_{\mu\nu}$ in this gauge.

If we choose the orientation of the coordinates such that the gravitational wave is travelling along the z-axis, the components of the metric perturbation can be written

$$h^{\text{TT}}_{\mu\nu} = \begin{bmatrix} 0 & 0 & 0 & 0 \\ 0 & h_{xx} & h_{xy} & 0 \\ 0 & h_{xy} & -h_{yy} & 0 \\ 0 & 0 & 0 & 0 \end{bmatrix}. \tag{9.84}$$

We shall now describe physical effects of gravitational waves. Since this is a 'curvature wave' we consider the relative motion of nearby particles as described by the equation of geodesic deviation, eq. (7.104), in the comoving geodesic normal coordinates of an observer,

$$\frac{d^2 s^i}{dt^2} = -R_{i0j0}s^j. \tag{9.85}$$

Using eq. (9.12) we find in the transverse traceless gauge

$$R_{i0j0} = \frac{1}{2}h^{TT}_{ij,00}. \tag{9.86}$$

Hence, eq. (9.85) takes the form

$$s^i_{,00} = \frac{1}{2}s^j h^{TT}_{ij,00}. \tag{9.87}$$

Inserting the components h_{ij} from eq. (9.84) we obtain the equations

$$s^x_{,tt} = \frac{1}{2}s^x h_{xx,tt} + \frac{1}{2}s^y h_{xy,tt} \tag{9.88}$$

$$s^y_{,tt} = \frac{1}{2}s^x h_{xy,tt} - \frac{1}{2}s^y h_{xx,tt} \tag{9.89}$$

$$s^z_{,tt} = 0. \tag{9.90}$$

These equations show that only the s^x and s^y components of the separation vector between two nearby, free particles will be disturbed by a gravitational wave travelling in the z-direction. Hence, test particles are only disturbed in directions perpendicular to the wave propagation.

We can use the above equation to describe what happens to a ring of free, stationary test particles in the xy-plane as a gravitational wave passes in the z-direction. Consider first two particles separated in the x-direction. To lowest order we can then neglect the terms with s^y at the right hand side of eqs. (9.88) and (9.89), so that

$$s^x_{,tt} = \frac{1}{2}s^x h_{xx,tt}, \qquad s^y_{,tt} = \frac{1}{2}s^x h_{xy,tt}. \tag{9.91}$$

Similarly, for two particles initially separated in the y-direction,

$$s^x_{,tt} = \frac{1}{2}s^y h_{xy,tt}, \qquad s^y_{,tt} = -\frac{1}{2}s^y h_{xx,tt}. \tag{9.92}$$

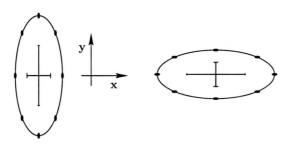

Figure 9.1: Displacement of test particles caused by a travelling gravitational wave with $+$ polarisation. The states are separated by a phase difference of π.

Suppose a wave with $h_{xx} \neq 0$, $h_{xy} = 0$ hits the particles. First the particles along the x-direction come towards each other and then they move away from each other as h_{xx} reverses sign. This is called the $+$ polarisation and is shown in Fig. 9.1. If the wave had $h_{xy} \neq 0$, $h_{xx} = h_{yy} = 0$ the particles respond as shown in Fig. 9.2. This is called the $\times$ polarisation.

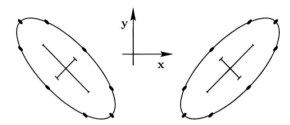

Figure 9.2: Displacement of test particles caused by a travelling gravitational wave with $\times$ polarisation. The states are separated by a phase difference of π.

Since h_{xy} and h_{xx} are independent, the figures 9.1 and 9.2 demonstrate the existence of two different states of polarisations, which are oriented at an angle $45°$ to each other.

9.6   Gravitational radiation from sources

We shall now consider the relation between the gravitational radiation, represented by $\bar{h}_{\mu\nu}$, and its source, represented by $T_{\mu\nu}$.

Let the source be a matter distribution localised near the origin O with source particles moving slowly compared to the speed of light. We calculate the field at a distance r from O which is large compared to the extension of the matter distribution. Then eq. (9.56) may be approximated by $(c = 1)$

$$\bar{h}_{\mu\nu}(t, \mathbf{r}) = \frac{4G}{r} \int T_{\mu\nu}(t - r, \mathbf{r})dV. \tag{9.93}$$

This means that we consider the gravitational radiation in the wave zone far from the source. In this zone the radiation looks like a plane wave, in which the radiative part of $\bar{h}_{\mu\nu}$ is determined by its spatial part $\bar{h}_{ij}$. Hence, we need only consider $\int T^{ij}dV$, which will be calculated following Foster and Nightingale [FN94].

The energy-momentum conservation equation $T^{\mu\nu}{}_{;\nu} = 0$ is equivalent to the component equations

$$T^{00}{}_{,0} + T^{0k}{}_{,k} = 0, \tag{9.94}$$

$$T^{i0}{}_{,0} + T^{ik}{}_{,k} = 0. \tag{9.95}$$

Furthermore, we are also going to use the integral identity

$$\int \left(T^{ik}x^j\right)_{,k} dV = \int T^{ik}{}_{,k}x^j dV + \int T^{ij}dV, \tag{9.96}$$

where the integrals are taken over a region of space enclosing the source, so that $T^{\mu\nu} = 0$ on the boundary of the region. Hence, transforming the integral on the left hand side to a surface integral by means of Gauss' integral theorem, eq. (6.87), we see that the left hand side vanishes. Therefore,

$$\int T^{ij} dV = -\int T^{ik}{}_{,k} x^j dV = \int T^{i0}{}_{,0} x^j dV = \frac{d}{dt} \int T^{i0} x^j dV. \tag{9.97}$$

Interchanging i and j and adding gives

$$\int T^{ij} dV = \frac{1}{2} \frac{d}{dt} \int \left(T^{i0} x^j + T^{j0} x^i \right) dV. \tag{9.98}$$

Furthermore,

$$\int \left(T^{0k} x^i x^j \right)_{,k} dV = \int T^{0k}{}_{,k} x^i x^j dV + \int \left(T^{i0} x^j + T^{j0} x^i \right) dV. \tag{9.99}$$

Again, using Gauss' integral theorem, the left hand side vanishes. Hence, using eq. (9.94) we have

$$\int \left(T^{i0} x^j + T^{j0} x^i \right) dV = \frac{d}{dt} \int T^{00} x^i x^j dV. \tag{9.100}$$

For slowly moving source particles $T^{00} \approx \rho$, where ρ is the proper density. Eqs. (9.93), (9.98) and (9.100) then yield the approximate expression

$$\bar{h}^{ij} = \frac{2G}{r} \frac{d^2}{dt^2} \left[\int \rho x^i x^j dV \right]_{t'=t-r}. \tag{9.101}$$

The *quadrupole moment* of the source is defined by

$$q^{ij} = \int \rho x^i x^j dV. \tag{9.102}$$

The solution then finally takes the form

$$\boxed{\bar{h}_{ij}(t, \mathbf{r}) = \frac{2G}{r} \ddot{q}_{ij}.} \tag{9.103}$$

This equation tells us that the gravitational radiation produced by an isolated non-relativistic object is proportional to the second derivative of the quadrupole moment of the mass distribution at the emission time.

Example 9.1 (Gravitational radiation emitted by a binary star)   Example
We consider two stars of mass M in a circular orbit with radius R in the xy-plane, at a distance r from their common centre of mass, as shown in Fig. 9.3. It is sufficient to treat the motion of the stars in the Newtonian approximation. Then, according to Newton's law of gravitation and Newton's 2nd law,

$$\frac{GM^2}{(2R)^2} = \frac{Mv^2}{R}, \tag{9.104}$$

which gives

$$v = \sqrt{\frac{GM}{4R}}. \tag{9.105}$$

The time it takes to complete a single orbit is $T = 2\pi R/v$. Hence, the angular velocity

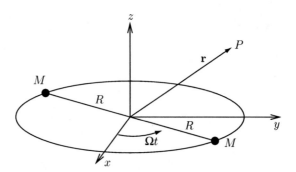

Figure 9.3: Two stars of equal mass M are in a circular orbit around their mass centre. The radius is R, and the orbital angular velocity is Ω. The observers are at a point P at a large distance compared to the radius R.

of the orbit is

$$\Omega = \frac{2\pi}{T} = \sqrt{\frac{GM}{4R^3}}. \tag{9.106}$$

The paths of the stars are given parametrically as

$$(x_A, y_A) = (R\cos\Omega t, R\sin\Omega t), \qquad (x_B, y_B) = (-R\cos\Omega t, -R\sin\Omega t), \tag{9.107}$$

for star A and B, respectively.

The mass density of the system is

$$\begin{aligned} \rho(t, \mathbf{r}) &= M\delta(z) \\ &\times [\delta(x - R\cos\Omega t)\delta(y - R\sin\Omega t) + \delta(x + R\cos\Omega t)\delta(y + R\sin\Omega t)]. \end{aligned} \tag{9.108}$$

Calculating the non-vanishing components of the quadrupole moment from eq. (9.102) now leads to

$$\begin{aligned} q_{xx} &= 2MR^2\cos^2\Omega t = MR^2(1 + \cos 2\Omega t) \\ q_{yy} &= 2MR^2\sin^2\Omega t = MR^2(1 - \cos 2\Omega t) \\ q_{yx} = q_{xy} &= 2MR^2\cos\Omega t \sin\Omega t = MR^2\sin 2\Omega t. \end{aligned} \tag{9.109}$$

Inserting this into eq.(9.103) gives the components of the metric perturbation

$$\bar{h}_{ij}(t, \mathbf{r}) = \frac{8GM\Omega^2 R^2}{r} \begin{bmatrix} -\cos[2\Omega(t - r)] & -\sin[2\Omega(t - r)] & 0 \\ -\sin[2\Omega(t - r)] & \cos[2\Omega(t - r)] & 0 \\ 0 & 0 & 0 \end{bmatrix}. \tag{9.110}$$

The frequency of the emitted radiation is thus twice the orbital frequency.

We shall finally set up the expression for the total power radiated gravitationally by a slowly moving source. Let us start by expanding the Newtonian potential ϕ in powers of r,

$$\phi = -\left(\frac{M}{r} + \frac{d_j n^j}{r^2} + \frac{3Q_{ij}n^i n^j}{2r^3} + \cdots\right), \qquad n^i = \frac{x^i}{r}. \tag{9.111}$$

Here d_j is the dipole moment of the source,

$$d_j \equiv \int \rho x^j \, dV, \tag{9.112}$$

and

$$Q_{ij} \equiv \int \rho \left(x_i x_j - \frac{1}{3} \delta_{ij} r^2 \right) = q_{ij} - \frac{1}{3} \delta_{ij} q^k_k \qquad (9.113)$$

is the trace-free part of the quadrupole moment of the mass distribution.

In the transverse traceless gauge one can introduce an effective energy-momentum tensor for gravitational waves by

$$T^{GW}_{\mu\nu} = \frac{1}{32\pi} \langle h_{ik,\mu} h_{ik,\nu} \rangle, \qquad (9.114)$$

where $\langle \ \rangle$ denotes the average over wavelengths. The total power crossing a sphere of radius r at a time t is

$$P(t, r) = \int T^{GW}_{0r} r^2 d\Omega. \qquad (9.115)$$

Using eq. (9.103) we have

$$
\begin{aligned}
T^{GW}_{0r} &= \frac{1}{32\pi} \langle \dot{h}_{ik} \dot{h}_{ik} \rangle \\
&= \frac{1}{8\pi r^2} \left\langle \dddot{Q}_{jk} \dddot{Q}_{jk} - 2 n_i \dddot{Q}_{ij} \dddot{Q}_{jk} n_k + \frac{1}{2} \left(n_j \dddot{Q}_{jk} n_k \right)^2 \right\rangle. \quad (9.116)
\end{aligned}
$$

The total power can be found by averaging the flux over all directions and multiplying the result by 4π. One then needs

$$\langle n_i \rangle = \langle n_i n_j n_k \rangle = 0, \quad \langle n_i n_j \rangle = \frac{1}{3} \delta_{ij},$$

$$\langle n_i n_j n_k n_l \rangle = \frac{1}{15} \left(\delta_{ij} \delta_{kl} + \delta_{ik} \delta_{jl} + \delta_{il} \delta_{jk} \right). \qquad (9.117)$$

Inserting the expression (9.116) into eq. (9.115) and performing the integration using the formulae (9.117) one finally arrives at the emitted power of gravitational radiation from a slowly moving source

$$\boxed{P(t, r) = \frac{G}{5} \left[\dddot{Q}_{ij} \dddot{Q}^{ij} \right]_{t'=t-r}.} \qquad (9.118)$$

Let us apply this formula to the gravitational radiation emitted by a binary star, as considered in Example 9.1. The components of the quadrupole is given in eq. (9.109). The traceless part of the quadrupole, as defined in eq. (9.113), is

$$Q_{ij} = \frac{1}{3} M R^2 \begin{bmatrix} 1 + 3\cos 2\Omega t & 3\sin 2\Omega t & 0 \\ 3\sin 2\Omega t & 1 - 3\cos 2\Omega t & 0 \\ 0 & 0 & -2 \end{bmatrix}. \qquad (9.119)$$

Its third derivative is

$$\dddot{Q}_{ij} = 8 M R^2 \Omega^3 \begin{bmatrix} \sin 2\Omega t & -\cos 2\Omega t & 0 \\ -\cos 2\Omega t & -\sin 2\Omega t & 0 \\ 0 & 0 & 0 \end{bmatrix}. \qquad (9.120)$$

Hence, the power radiated by the binary star is

$$P = \frac{128}{5} G M^2 R^4 \Omega^6. \qquad (9.121)$$

Using eq. (9.106) for the angular velocity this can be written

$$P = \frac{2}{5} \frac{G^4 M^5}{R^5}.$$

(9.122)

As expressed by the period T of the orbital motion the formula takes the form

$$P = \frac{128}{5} 4^{\frac{1}{3}} \frac{1}{G} \left(\frac{\pi G M}{T} \right)^{\frac{10}{3}}.$$

(9.123)

Inserting numerical values

$$P = 1.9 \cdot 10^{26} \left(\frac{M}{M_{\text{Sun}}} \frac{T_0}{T} \right)^{\frac{10}{3}} \frac{\text{J}}{\text{s}},$$

(9.124)

where M_{Sun} is the mass of the Sun and $T_0 = 1\text{h}$.

The effect of emitting gravitational radiation upon the period has been observed for the binary pulsar PSR B1913+16 [TW89]. The emission of radiation extracts energy from the system and hence decreases its period. The rate of decrease of the period can be calculated by applying the Newtonian approximation to this non-relativistic system. Its energy is

$$E = 2 \left(\frac{1}{2} M v^2 \right)^2 - \frac{M^2}{2R}.$$

(9.125)

Using eq. (9.105) to relate v to R and eq. (9.106) to relate R to the orbital period T gives

$$E = -\frac{M^2}{4R} = -\frac{M}{4} \left(\frac{4\pi M}{T} \right)^{\frac{2}{3}}.$$

(9.126)

Differentiating E with respect to t and equating dE/dt to $-P$ in eq.(9.123) leads to

$$\frac{dT}{dt} = -\frac{96}{5} \pi 4^{\frac{1}{3}} \left(\frac{2\pi M}{T} \right)^{\frac{5}{3}}.$$

(9.127)

Inserting numerical values gives

$$\frac{dT}{dt} = -3.4 \cdot 10^{-12} \left(\frac{M}{M_{\text{Sun}}} \frac{T_0}{T} \right)^{\frac{5}{3}}.$$

(9.128)

The mass of both the pulsar and its unseen companion is about $1.4 M_{\text{Sun}}$, and the orbital period is 7.75h. Eq. (9.128) then gives a predicted value of the rate of decrease of the period equal to about $10 \mu s$ per year. This slow decrease in the orbital period has been detected. Timing measurements over an epoch of many years gave $dT/dt = -(2.422 \pm 0.006) \cdot 10^{-12}$ in good agreement with more accurate calculations taking into account several observed parameters of the system.

Problems

9.1. The Linearised Einstein Field Equations
In this problem we will do a more careful analysis of the linearised Einstein field equations. We will assume that the metric is

$$g_{\mu\nu} = \eta_{\mu\nu} + h_{\mu\nu}$$

(9.129)

where $\eta_{\mu\nu}$ is the Minkowski metric and $|h_{\mu\nu}| \ll 1$. The linearised Einstein field equations are the Einstein field equations where we have only kept the terms linear in $h_{\mu\nu}$. In all the calculations in this problem we will therefore ignore the terms of higher order in $h_{\mu\nu}$ and will assume that the derivative operator $\frac{\partial}{\partial x^{\mu}}$ is the flat derivative operator with respect to $\eta_{\mu\nu}$.

(a) Show that the inverse metric is $g^{\mu\nu} = \eta^{\mu\nu} - h^{\mu\nu}$. Argue that the Riemann tensor can be calculated using eq. (7.61) on page 159. Show that the Ricci tensor can be written

$$R_{\alpha\beta} = h_{\mu(\alpha,\beta)}{}^{\mu} - \frac{1}{2}h_{\alpha\beta,\mu}{}^{\mu} - \frac{1}{2}h_{,\alpha\beta} \qquad (9.130)$$

where $h = h^{\mu}{}_{\mu}$.

(b) Write down the expression for the Einstein tensor $E_{\mu\nu}$ and show that it can be simplified with the introduction of

$$\bar{h}_{\alpha\beta} = h_{\alpha\beta} - \frac{1}{2}\eta_{\alpha\beta}h. \qquad (9.131)$$

Write down the Einstein field equations in terms of $\bar{h}_{\mu\nu}$.

(c) In section 6.9 we learned how we could make a change of coordinates with the aid of a vector field $\mathbf{X} = X^{\mu}\mathbf{e}_{\mu}$. The vector field generates a one-parameter group of diffeomorphisms which can be seen upon as a change of coordinates. This freedom of choosing coordinates is called a *gauge freedom*, similarly as for the electromagnetic field. The vector field induces an infinitesimal gauge transformation which transforms the metric as

$$h_{\mu\nu} \longmapsto h_{\mu\nu} + \mathscr{L}_{\mathbf{X}}\eta_{\mu\nu}. \qquad (9.132)$$

Assume that $\mathbf{X}$ is an infinitesimal vector field. Show, with the aid of eq. (6.288) on page 143 that the vector field $\mathbf{X}$ induces the gauge transformation

$$h_{\mu\nu} \longmapsto h_{\mu\nu} + X_{\alpha,\beta} + X_{\beta,\alpha}. \qquad (9.133)$$

These gauge transformations only change the coordinates and should not change the physical interpretation of the perturbation. This gauge transformation can be used to simplify the expression for the linearised Einstein field equations. Choose a vector field $\mathbf{X}$ where the components satisfy the equations

$$X_{\alpha,\mu}{}^{\mu} = -\bar{h}_{\alpha\mu}{}^{,\mu}. \qquad (9.134)$$

Show that we can perform a gauge transformation such that

$$\bar{h}_{\alpha\mu}{}^{,\mu} = 0. \qquad (9.135)$$

This is similar to the Lorenz gauge condition in electromagnetism. Write down the Einstein field equations in the new gauge and show that they are equal to

$$\bar{h}_{\alpha\beta,\mu}{}^{\mu} = -2\kappa T_{\alpha\beta}. \qquad (9.136)$$

(d) Assume now that the energy-momentum tensor has the 00-component $T_{00} = \rho$, while all other components are zero. Assume also that the metric is time-independent. The linearised Einstein field equations now reduce to

$$
\begin{aligned}
\nabla^2 \bar{h}_{00} &= -2\kappa\rho, \\
\nabla^2 \bar{h}_{ij} &= 0.
\end{aligned}
\tag{9.137}
$$

The only solutions of $\bar{h}_{ij}$ that are zero and well behaved at infinity are $\bar{h}_{ij} = 0$. Assume that this is the case. Define ϕ by

$$
\phi = -\frac{1}{4}h.
\tag{9.138}
$$

Show that

$$
\begin{aligned}
h_{ij} &= -2\eta_{ij}\phi, \\
h_{00} &= -2\phi,
\end{aligned}
\tag{9.139}
$$

where ϕ satisfies Poisson's equation

$$
\nabla^2 \phi = \frac{\kappa}{2}\rho.
\tag{9.140}
$$

Compare with eq. (1.32) on page 9 and find κ (when $c = 1$).
Finally write down the metric in terms of ϕ.

9.2. Gravitational waves

We will here consider gravitational waves in the weak field approximation of Einstein's equations using the Maxwell equations for the gravitoelectromagnetic fields.

(a) Use these equations for vacuum, and the Lorenz gauge condition, eq. (9.63), to show that ϕ and $\mathbf{A}$ satisfy the wave equations

$$
\begin{aligned}
\Box\phi &= 0, \\
\Box\mathbf{A} &= 0,
\end{aligned}
\tag{9.141}
$$

where $\Box$ is the d'Alembert operator in Minkowski space. Hence, not surprisingly, gravitational waves travel with the speed of light. We can assume that $\mathbf{A}$ has in general complex components. The physical vector potential is the real part of $\mathbf{A}$.

(b) Consider waves far from any sources, so that $\phi = 0$. Find particular solutions where the wave describes a plane-wave with wave-vector $\mathbf{k}$. What does the Lorenz gauge condition tell us about the nature of these gravitational waves?

(c) A test particle is initially at rest as one of the plane waves with wave-vector $\mathbf{k} = k\mathbf{e}_x$ passes by. The wave is plane-polarised so that $\mathbf{A}$ can be written

$$
\mathbf{A} = \mathbf{A}_0 e^{ik(x-ct)},
\tag{9.142}
$$

where $\mathbf{A}_0 = A_0\mathbf{e}_y$, and A_0 real. Assume that the test particle is placed at the origin and that the deviation from the origin as the wave passes by is very small compared to the wavelength of the wave. Hence, we can assume that $e^{ik(x-ct)} \approx e^{-ikct}$. Assume also that the speed of the particle v is non-relativistic: $v/c \ll 1$ (thus A_0 has to be sufficiently small). Use the Lorentz force law for GEM fields and derive the the position of the particle to lowest order in A_0/c^2 as the wave passes by.

(d) Explain why gravitational waves cannot have only a Newtonian part, and thus that there are no gravitational waves in the Newtonian theory (or that they move with infinite speed).

9.3. *The spacetime inside and outside a rotating spherical shell*
A spherical shell with mass M and radius R is rotating with a constant angular velocity ω. In this problem the metric inside and outside the shell shall be found using the linearised Einstein's field equations

$$\Box \bar{h}_{\alpha\beta} = -2\kappa T_{\mu\nu}, \tag{9.143}$$

where $\bar{h}_{\alpha\beta}$ is the metric perturbation with respect to the Minkowski metric (see section 9.4). The rotation is assumed to be non-relativistic, thus the calculations should be made to first order in $R\omega$.

Assume that the shell is composed of dust, so that the energy-momentum tensor can be expressed as

$$
\begin{aligned}
T_{\alpha\beta} &= \rho u_\alpha u_\beta, \quad \rho = \frac{M}{4\pi R^2}\delta(r-R) \\
u_\alpha &\approx (-1, -R\omega \sin\theta \sin\phi, R\omega \sin\theta \cos\phi, 0)
\end{aligned} \tag{9.144}
$$

where (r, θ, ϕ) are spherical coordinates:

$$x = r\sin\theta\cos\phi, \quad y = r\sin\theta\sin\phi, \quad r = \cos\theta. \tag{9.145}$$

Find the metric inside and outside the rotating shell and show that

$$
\begin{aligned}
ds^2 &= -\left(1 - \frac{2M}{R}\right)dt^2 + \left(1 + \frac{2M}{R}\right)(dx^2 + dy^2 + dz^2) \\
&\quad - \frac{8M\omega}{3R}r^2 \sin^2\theta d\phi dt, \qquad r < R \\
ds^2 &= -\left(1 - \frac{2M}{r}\right)dt^2 + \left(1 + \frac{2M}{r}\right)(dx^2 + dy^2 + dz^2) \\
&\quad - \frac{4J}{r}\sin^2\theta d\phi dt, \qquad r > R
\end{aligned} \tag{9.146}
$$

where $J = (2/3)MR^2\omega$ is the angular momentum of the shell.

9.4. *Plane-wave spacetimes*
We will in this problem consider the *plane wave* metric

$$ds^2 = 2du(dv + Hdu) + dx^2 + dy^2, \tag{9.147}$$

where the function $H = H(u, x, y)$ does not depend on v (but is otherwise arbitrary).

(a) From the above we have the metric components given by $g_{uv} = 1$, $g_{uu} = 2H$ and $g_{xx} = g_{yy} = 0$. What is $g^{\mu\nu}$? Calculate also the Christoffel symbols and show that the vector $\mathbf{k} = \frac{\partial}{\partial v}$ is covariantly constant; i.e., $k_{\mu;\nu} = 0$. Is $\mathbf{k}$ time-like, null or space-like?

(b) Use the Christoffel symbols to calculate the Einstein tensor and show that in vacuum ($\Lambda = 0$) the Einstein field equations reduce to

$$\left(\frac{\partial^2}{\partial x^2} + \frac{\partial^2}{\partial y^2} \right) H = 0. \tag{9.148}$$

What are the linearised field equations in this case?

(c) Show that

$$H = C_1(u)(x^2 - y^2) + 2C_2(u)xy,$$

where $C_1(u)$ and $C_2(u)$ are arbitrary functions of u, is a solution to Einstein's field equations.

Consider the complex coordinate $z = x + iy$, and an arbitrary analytic complex function $f(u, z)$. Show that

$$H = f(u, z) + \bar{f}(u, \bar{z}),$$

where a bar means complex conjugate, is a solution to Einstein's field equations. Is it also a solution to the linear field equations?

10

The Schwarzschild Solution
and Black Holes

We have now established the Einstein field equations and explained their contents. In this chapter we will explore the first known non-trivial solution to these equations. The solution is due to the astronomer Karl Schwarzschild, and in his honour the solution is referred to as the *Schwarzschild solution for empty space*. This solution represents a spacetime outside a non-rotating black hole. The Kerr solution representing spacetime outside a rotating black hole will also be deduced. Finally, interior solutions will be investigated.

10.1   The Schwarzschild solution for empty space

The Newtonian potential around a static point object is spherically symmetric. Also for objects like stars and planets the same is true to lowest order. Exterior to such objects there is a static, spherically symmetric empty space. Motivated by this we will study spherically symmetric solutions to the Einstein field equations for empty space.

From Example 4.6 follows that the line-element of Minkowski spacetime as expressed in spherical coordinates has the form (in units with $c = 1$)

$$ds^2 = -dt^2 + d\tilde{r}^2 + \tilde{r}^2(d\theta^2 + \sin^2\theta d\phi^2). \tag{10.1}$$

We shall solve the field equations for empty spacetime with static and spherically symmetric 3-space. Then it is reasonable to assume that the line-element can be written

$$ds^2 = -f(\tilde{r})dt^2 + g(\tilde{r})d\tilde{r}^2 + h(\tilde{r})\tilde{r}^2(d\theta^2 + \sin^2\theta d\phi^2). \tag{10.2}$$

Introducing a new radial coordinate $r = \tilde{r}\sqrt{h(\tilde{r})}$, the line-element becomes

$$ds^2 = -A(r)dt^2 + B(r)dr^2 + r^2(d\theta^2 + \sin^2\theta d\phi^2). \tag{10.3}$$

It has been customary to replace the functions $A(r)$ and $B(r)$ by exponential functions in order to obtain somewhat simpler expressions for the components of the Einstein tensor. Hence, we introduce the functions $\alpha(r)$ and $\beta(r)$ by $e^{2\alpha(r)} = A(r)$ and $e^{2\beta(r)} = B(r)$, obtaining

$$ds^2 = -e^{2\alpha}dt^2 + e^{2\beta}dr^2 + r^2(d\theta^2 + \sin^2\theta d\phi^2). \tag{10.4}$$

These coordinates are called *Schwarzschild coordinates*. Obviously, this is not the only choice we have. For instance, we could choose *isotropic coordinates*

$$ds^2 = -e^{2A}dt^2 + e^{2B}(dr^2 + r^2(d\theta^2 + \sin^2\theta d\phi^2)) \tag{10.5}$$

or coordinates where the metric has off-diagonal components. We will however, use the coordinates where the metric takes the form (10.4).

These coordinates are particularly convenient because the spatial surface of constant r and t has area $4\pi r^2$. We will use the Cartan formalism to derive the static solution subject to the vacuum condition $T_{\mu\nu} = 0$.

Introducing an orthonormal basis,

$$\begin{aligned}
\boldsymbol{\omega}^{\hat{t}} &= e^{\alpha}\mathbf{dt} \\
\boldsymbol{\omega}^{\hat{r}} &= e^{\beta}\mathbf{dr} \\
\boldsymbol{\omega}^{\hat{\theta}} &= r\mathbf{d\theta} \\
\boldsymbol{\omega}^{\hat{\phi}} &= r\sin\theta\mathbf{d\phi},
\end{aligned} \tag{10.6}$$

and taking the exterior derivatives, we get

$$\begin{aligned}
\mathbf{d}\boldsymbol{\omega}^{\hat{t}} &= \alpha'e^{-\beta}\boldsymbol{\omega}^{\hat{r}}\wedge\boldsymbol{\omega}^{\hat{t}} \\
\mathbf{d}\boldsymbol{\omega}^{\hat{r}} &= 0 \\
\mathbf{d}\boldsymbol{\omega}^{\hat{\theta}} &= \frac{e^{-\beta}}{r}\boldsymbol{\omega}^{\hat{r}}\wedge\boldsymbol{\omega}^{\hat{\theta}} \\
\mathbf{d}\boldsymbol{\omega}^{\hat{\phi}} &= \frac{e^{-\beta}}{r}\boldsymbol{\omega}^{\hat{r}}\wedge\boldsymbol{\omega}^{\hat{\phi}} + \frac{1}{r}\cot\theta\boldsymbol{\omega}^{\hat{\theta}}\wedge\boldsymbol{\omega}^{\hat{\phi}},
\end{aligned} \tag{10.7}$$

where a prime denotes derivative with respect to r. The next step is to use Cartan's first structural equation, eq. (6.182),

$$\mathbf{d}\boldsymbol{\omega}^{\rho} = -\boldsymbol{\Omega}^{\rho}{}_{\nu}\wedge\boldsymbol{\omega}^{\nu}, \tag{10.8}$$

and the antisymmetry of the connection forms, $\Omega_{\hat{\mu}\hat{\nu}} = -\Omega_{\hat{\nu}\hat{\mu}}$, to find the non-zero connection forms. From eq. (10.7) we see from the expression for $\mathbf{d}\boldsymbol{\omega}^{\hat{t}}$ that $\Omega^{\hat{t}}{}_{\hat{r}}$ must have the form

$$\boldsymbol{\Omega}^{\hat{t}}{}_{\hat{r}} = \alpha'e^{-\beta}\boldsymbol{\omega}^{\hat{t}} + F(r)\boldsymbol{\omega}^{\hat{r}}. \tag{10.9}$$

To determine the function $F(r)$ we utilize the antisymmetry of the connection forms which implies $\Omega^{\hat{t}}{}_{\hat{r}} = \Omega^{\hat{r}}{}_{\hat{t}}$. Using the expression for $\mathbf{d}\boldsymbol{\omega}^{\hat{r}}$, we get

$$\boldsymbol{\Omega}^{\hat{r}}{}_{\hat{t}} = G(r)\boldsymbol{\omega}^{\hat{t}}. \tag{10.10}$$

Comparing eqs. (10.9) and (10.10) yields $F(r) = 0$ and $G(r) = \alpha'e^{-\beta}$. The other connection forms are determined analogously. The calculations give the following expressions:

$$\begin{aligned}
\boldsymbol{\Omega}^{\hat{t}}{}_{\hat{r}} &= \boldsymbol{\Omega}^{\hat{r}}{}_{\hat{t}} &= \alpha'e^{-\beta}\boldsymbol{\omega}^{\hat{t}} \\
\boldsymbol{\Omega}^{\hat{\theta}}{}_{\hat{r}} &= -\boldsymbol{\Omega}^{\hat{r}}{}_{\hat{\theta}} &= \frac{e^{-\beta}}{r}\boldsymbol{\omega}^{\hat{\theta}} \\
\boldsymbol{\Omega}^{\hat{\phi}}{}_{\hat{r}} &= -\boldsymbol{\Omega}^{\hat{r}}{}_{\hat{\phi}} &= \frac{e^{-\beta}}{r}\boldsymbol{\omega}^{\hat{\phi}} \\
\boldsymbol{\Omega}^{\hat{\phi}}{}_{\hat{\theta}} &= -\boldsymbol{\Omega}^{\hat{\theta}}{}_{\hat{\phi}} &= \frac{1}{r}\cot\theta\boldsymbol{\omega}^{\hat{\phi}}.
\end{aligned} \tag{10.11}$$

From Cartan's second structural equation, eq. (7.47),

$$\mathbf{R}^{\hat{\mu}}_{\ \hat{\nu}} = d\mathbf{\Omega}^{\hat{\mu}}_{\ \hat{\nu}} + \mathbf{\Omega}^{\hat{\mu}}_{\ \hat{\lambda}} \wedge \mathbf{\Omega}^{\hat{\lambda}}_{\ \hat{\nu}}, \tag{10.12}$$

we can calculate the curvature matrix. The non-zero components are

$$
\begin{aligned}
\mathbf{R}^{\hat{t}}_{\ \hat{r}} &= -e^{-2\beta}(\alpha'' + \alpha'^2 - \alpha'\beta')\omega^{\hat{t}} \wedge \omega^{\hat{r}} \\
\mathbf{R}^{\hat{t}}_{\ \hat{\theta}} &= -\frac{1}{r}\alpha' e^{-2\beta}\omega^{\hat{t}} \wedge \omega^{\hat{\theta}} \\
\mathbf{R}^{\hat{t}}_{\ \hat{\phi}} &= -\frac{1}{r}\alpha' e^{-2\beta}\omega^{\hat{t}} \wedge \omega^{\hat{\phi}} \\
\mathbf{R}^{\hat{r}}_{\ \hat{\theta}} &= \frac{1}{r}\beta' e^{-2\beta}\omega^{\hat{r}} \wedge \omega^{\hat{\theta}} \\
\mathbf{R}^{\hat{r}}_{\ \hat{\phi}} &= \frac{1}{r}\beta' e^{-2\beta}\omega^{\hat{r}} \wedge \omega^{\hat{\phi}} \\
\mathbf{R}^{\hat{\theta}}_{\ \hat{\phi}} &= \frac{1}{r^2}(1 - e^{-2\beta})\omega^{\hat{\theta}} \wedge \omega^{\hat{\phi}}. \tag{10.13}
\end{aligned}
$$

By means of the formula $\mathbf{R}^{\hat{\mu}}_{\ \hat{\nu}} = \frac{1}{2}R^{\hat{\mu}}_{\ \hat{\nu}\hat{\alpha}\hat{\beta}}\omega^{\hat{\alpha}} \wedge \omega^{\hat{\beta}}$ we can now extract the components of the Riemann curvature tensor. Contracting once yields the Ricci tensor

$$R_{\hat{\alpha}\hat{\beta}} = R^{\hat{\mu}}_{\ \hat{\alpha}\hat{\mu}\hat{\beta}}. \tag{10.14}$$

One more contraction yields the curvature scalar

$$R = R^{\hat{\alpha}}_{\ \hat{\alpha}}. \tag{10.15}$$

Using the definition of the Einstein tensor,

$$E_{\hat{\mu}\hat{\nu}} = R_{\hat{\mu}\hat{\nu}} - \frac{1}{2}\eta_{\hat{\mu}\hat{\nu}}R, \tag{10.16}$$

we find

$$
\begin{aligned}
E_{\hat{t}\hat{t}} &= \frac{2}{r}\beta' e^{-2\beta} + \frac{1}{r^2}\left(1 - e^{-2\beta}\right) \tag{10.17} \\
E_{\hat{r}\hat{r}} &= \frac{2}{r}\alpha' e^{-2\beta} - \frac{1}{r^2}\left(1 - e^{-2\beta}\right) \tag{10.18} \\
E_{\hat{\theta}\hat{\theta}} = E_{\hat{\phi}\hat{\phi}} &= \frac{1}{r}e^{-2\beta}\left(r\alpha'' + r\alpha'^2 - r\alpha'\beta' + \alpha' - \beta'\right). \tag{10.19}
\end{aligned}
$$

The condition $E_{\mu\nu} = 0$ for empty space implies that the expressions (10.17), (10.18) and (10.19) equal zero. Adding equations (10.17) and (10.18) we get simply

$$\frac{2}{r}e^{-2\beta}(\alpha' + \beta') = 0. \tag{10.20}$$

This equation can be integrated to give

$$\alpha(r) + \beta(r) = K, \tag{10.21}$$

where K is a constant. We note that by a rescaling of the time-coordinate we can shift this constant to any value we like. It is therefore without loss of generality to choose $K = 0$ so we can set $\alpha(r) = -\beta(r)$. Equation $E_{\hat{t}\hat{t}} = 0$ can be written

$$\frac{1}{r^2}\left[r\left(1 - e^{-2\beta}\right)\right]' = 0. \tag{10.22}$$

This equation can be integrated to give

$$e^{-2\beta} = 1 - \frac{2M}{r}, \tag{10.23}$$

where M is an arbitrary constant. We can now easily check that this solution also solves equation (10.19). The *Schwarzschild solution for empty space* is therefore:

$$ds^2 = -\left(1 - \frac{2M}{r}\right)dt^2 + \frac{dr^2}{1 - \frac{2M}{r}} + r^2(d\theta^2 + \sin^2\theta d\phi^2). \tag{10.24}$$

There are a couple of things worth noting. First, for large r, the metric is approximately that of flat Minkowski spacetime. Second, the metric appears singular when $r = 0$ and when $r = 2M$. These two values for r have special physical importance as we will see later on. However, their nature is different; at $r = 0$ we have a physical singularity where the curvature tensors diverge; at $r = 2M$ the curvature tensors are well-behaved and finite, but the spacetime has a *horizon* at $r = 2M$ in these coordinates.

The physical interpretation of M can be understood by considering a free particle instantaneously at rest outside a spherical body and comparing with the Newtonian limit. In a Newtonian gravitational field the acceleration of a free particle is

$$g = -\frac{Gm}{r^2}, \tag{10.25}$$

where m is the mass of the attracting body, and G is the Newtonian gravitational constant. According to the theory of relativity the acceleration of a test particle is given by the geodesic equation, eq. (6.105):

$$\frac{d^2x^\mu}{d\tau^2} + \Gamma^\mu{}_{\alpha\beta}\frac{dx^\alpha}{d\tau}\frac{dx^\beta}{d\tau} = 0. \tag{10.26}$$

Assuming that the particle is instantaneously at rest in a weak gravitational field we can approximate the proper time $d\tau$ with dt and set $\frac{dx^\alpha}{d\tau} = (1,0,0,0)$ at that particular moment of time. The geodesic equation now simplifies to

$$g = \frac{d^2x^\mu}{dt^2} \approx -\Gamma^r{}_{tt}. \tag{10.27}$$

Since we use a coordinate basis, the connection coefficients are Christoffel symbols and $\Gamma^r{}_{tt}$ is given by equation (6.111):

$$\begin{aligned}\Gamma^r{}_{tt} &= \frac{1}{2}g^{r\alpha}\left(\frac{\partial g_{\alpha t}}{\partial t} + \frac{\partial g_{\alpha t}}{\partial t} - \frac{\partial g_{tt}}{\partial x^\alpha}\right) \\ &= -\frac{1}{2}(g_{rr})^{-1}\frac{\partial g_{tt}}{\partial r}. \end{aligned} \tag{10.28}$$

Inserting the found solution into the above equation we find to lowest order

$$g = -\Gamma^r{}_{tt} = -\frac{M}{r^2}. \tag{10.29}$$

Comparing with the classical case we see that the constant M must be interpreted as the mass of the gravitating body, m, times the Newtonian gravitational constant: $M = Gm$. If we include the speed of light c, we get $g = -Mc^2/r^2$, and hence,

$$M = \frac{Gm}{c^2}. \tag{10.30}$$

For a mass m the radius

$$R_S = \frac{2Gm}{c^2}, \tag{10.31}$$

is called the *Schwarzschild radius*. As we see, the metric apparently has a terrible flaw; it is singular at the Schwarzschild radius. However, for a relatively small gravitating body like the Earth, the Schwarzschild radius is so small that we do not have to worry that our metric breaks down at R_S. For the Earth $R_S \approx 9 \cdot 10^{-3}$m, while for an object at the size of a solar mass $R_S \approx 3 \cdot 10^3$m, i.e., R_S is well inside the surface of these bodies. Inside the surface of planets and stars the condition $T_{\mu\nu} = 0$ for empty space is no longer valid so the Schwarzschild solution is not applicable in these regions. Outside the surfaces of the Earth and the Sun we will have $r \gg R_S$, and the Schwarzschild solution can be used. In fact, for $r \gg R_S$ the *weak field approximation* is valid to great accuracy.

For a static observer at a radius r outside a gravitating body the proper time $d\tau$ will have a time dilatation given by

$$d\tau = \sqrt{1 - \frac{R_S}{r}}\, dt. \tag{10.32}$$

Since the metric is inhomogeneous and static the coordinate clocks showing the time t must flow at equal pace compared to standard clocks at infinity, $r \longrightarrow \infty$. As we descend deeper and deeper into the gravitational field, the standard clocks showing proper time tick slower and slower compared to the coordinate time clocks. At the Schwarzschild radius the standard clocks are apparently standing still; the time does not flow at all compared to the proper time of the observer at infinity.

The singular behaviour at the Schwarzschild radius is only a coordinate singularity. If we for instance calculate *Kretschmann's curvature scalar* defined as the "square" of the Riemann tensor we get

$$R^{\alpha\beta\gamma\delta}R_{\alpha\beta\gamma\delta} = \frac{48M^2}{r^6}. \tag{10.33}$$

This scalar diverges only at the origin; there is nothing special happening at the Schwarzschild radius. This indicates that the origin, $r = 0$, is a physical singularity, but the Schwarzschild radius is not. We should therefore be able to find a new set of coordinates where the Schwarzschild radius is perfectly regular in the metric.

Let us assume that we are near the Schwarzschild radius, but still outside. We introduce the variable x by $x^2 = 2r - 4M$. At the Schwarzschild radius, $x = 0$, so we can approximate the metric with

$$ds^2 = \frac{1}{4M}(-x^2 dt^2 + (4M)^2 dx^2) + (4M)^2(d\theta^2 + \sin^2\theta d\phi^2) \tag{10.34}$$

close to the Schwarzschild radius. The last two variables only form a two-sphere S^2 which is perfectly regular everywhere. The t and x coordinates form a *Rindler space*. By the transformation

$$
\begin{aligned}
T &= x \sinh[(4M)^{-1}t] & (10.35) \\
X &= x \cosh[(4M)^{-1}t], & (10.36)
\end{aligned}
$$

the metric simply turns into

$$
ds^2 = 4M(-dT^2 + dX^2) + (4M)^2(d\theta^2 + \sin^2\theta d\phi^2). \qquad (10.37)
$$

The Schwarzschild radius, $r = 2M$, corresponds to $T = \pm X$ which is perfectly regular in the metric (10.37). So, the space (10.34) can be smoothly continued to a regular space containing no singularities. Thus we can conclude that the Schwarzschild solution can be smoothly expanded past the Schwarzschild radius so that there are no singularities at $r = R_S$.

10.2   Radial free fall in Schwarzschild spacetime

We will consider a radially falling particle in a Schwarzschild spacetime. The perhaps easiest way to calculate the equations of motion is to use Lagrange's equations. The Lagrangian of the particle is

$$
\mathcal{L} = -\frac{1}{2}\left(1 - \frac{R_S}{r}\right)c^2\dot{t}^2 + \frac{1}{2}\frac{\dot{r}^2}{\left(1 - \frac{R_S}{r}\right)}, \qquad (10.38)
$$

where a dot means derivative with respect to the proper time τ. The time coordinate is cyclic so its canonical momentum is a constant:

$$
p_t \equiv \frac{\partial\mathcal{L}}{\partial t} = -\left(1 - \frac{R_S}{r}\right)c^2\dot{t}. \qquad (10.39)
$$

Inserting this into the 4-velocity identity, $u^\mu u_\mu = -c^2$, gives an expression for $\dot{r}$:

$$
\dot{r}^2 - \frac{p_t^2}{c^2} = -\left(1 - \frac{R_S}{r}\right)c^2. \qquad (10.40)
$$

The value of p_t can be given in terms of the initial condition $r(0) = r_0$, $\dot{r}(0) = 0$. Using this initial condition we get the equation

$$
\dot{r} = c\left(\frac{R_S}{r_0}\right)^{\frac{1}{2}}\sqrt{\frac{r_0 - r}{r}}, \qquad (10.41)
$$

which can be integrated to give

$$
\tau = \frac{r_0}{c}\left(\frac{r_0}{R_S}\right)^{\frac{1}{2}}\left[\arccos\sqrt{\frac{r}{r_0}} + \sqrt{\frac{r}{r_0}}\sqrt{1 - \frac{r}{r_0}}\right]. \qquad (10.42)
$$

Here, τ is the proper time that a particle spends falling from rest at r_0 to r. The particle reaches the singularity $r = 0$ in a finite proper time given by

$$
\tau(r = 0) = \frac{\pi r_0}{2c}\sqrt{\frac{r_0}{R_S}} \qquad (10.43)
$$

Describing the same motion in terms of the coordinate time t we end up with the equation

$$t = \frac{1}{c} \left(\frac{r_0 - R_S}{R_S} \right)^{\frac{1}{2}} \int_{r_0}^{r} \frac{x^{\frac{3}{2}} dx}{(x - R_S)\sqrt{r_0 - x}}. \tag{10.44}$$

As we approach $r = R_S$, the integral on the right hand side diverges. Thus for an observer at infinity a particle falling towards the origin will only reach the Schwarzschild radius after an infinite amount of time has elapsed. The observer at infinity will never see it pass the Schwarzschild radius. An observer comoving with the particle, on the other hand, will not find anything particular happening at the Schwarzschild radius. It will pass the Schwarzschild radius and reach the singularity $r = 0$ in a finite proper time.

This is further evidence that the Schwarzschild radius is just a coordinate singularity – not a singularity in the spacetime itself. However, it is obvious that for an observer at infinity there is something fundamental about the Schwarzschild radius. Even though mathematically speaking the spacetime is perfectly regular at R_S, the radius R_S has deep consequences for the physics. We will see in the next section that at the Schwarzschild radius the observer at infinity observes a *horizon*. Nothing can escape this horizon, not even light. Once a photon has passed inside the horizon, it cannot get out. For this reason, the Schwarzschild metric describes a *black hole*. The radius of the black hole is the Schwarzschild radius. The inside of the black hole cannot, according to general relativity, communicate with the outside. Particles and light can get in, but there is nothing that can escape.

10.3   The light-cone in a Schwarzschild spacetime

We will now explore more of the significance of the horizon (which we will from now on call the surface given by $r = R_S$) and we will do so by studying the light-cone in the Schwarzschild spacetime. Light serves as an upper bound (except for so-called tachyons) for how fast particles can travel. It also serves as a measure of how fast information can travel. To get information about the life and times for some inhabitants of a planet outside the Solar system, say, the fastest way that we can get such information is by means of light signals. The light-cone tells us what region of spacetime we can get information from. If our world-line is outside the future light-cone of some event, then we can *never* get information about that event.

Consider radially moving light in a Schwarzschild spacetime. Radially moving means that the angular velocity is zero, so we will drop the angular part of the Lagrangian. Light has no proper time, so we will use the coordinate time as a time parameter. The four-velocity identity for light, $u^\mu u_\mu = 0$, yields

$$- \left(1 - \frac{2M}{r} \right) dt^2 + \frac{dr^2}{1 - \frac{2M}{r}} = 0. \tag{10.45}$$

Rearranging we get

$$\frac{rdr}{r - 2M} = \pm dt, \tag{10.46}$$

which can be integrated to yield

$$r \mp t + 2M \ln |r - 2M| = C, \tag{10.47}$$

where C is an integration constant. The inward moving photons have the positive sign, while the outward photons have the negative sign. If we introduce a time coordinate defined by

$$\tilde{t} = t + 2M \ln |r - 2M|, \tag{10.48}$$

the inward going photons have

$$\frac{dr}{d\tilde{t}} = -1. \tag{10.49}$$

The outward going photons, on the other hand, have

$$\frac{dr}{d\tilde{t}} = \frac{r - 2M}{r + 2M}. \tag{10.50}$$

The inward going photons have constant coordinate velocity, but the "outward going" photons are actually going inward for $r < 2M$. Thus with this time coordinate the light-cone inside the horizon will point inwards towards $r = 0$! (see Fig. 10.1) Light inside the horizon cannot escape the black hole. If light cannot escape, nothing can, according to general relativity. This is indeed the metric of a black hole. Also note that at $r = 0$ the light-cone collapses.

For an observer at infinity, who measures time in the parameter t the outward and inward going photons have

$$\frac{dr}{dt} = \pm \left(1 - \frac{2M}{r}\right). \tag{10.51}$$

Thus light is decelerated in the gravitational field, as the photons descend into a gravitational field their speed is decelerated. At the horizon, the light-cone collapses which indicates the strong significance the horizon has for an observer at infinity. One could also believe that the Special Theory of Relativity is violated since the observer at infinity sees light moving at a speed less than c. However, one must keep in mind that the Special Theory of Relativity is only valid *locally*.

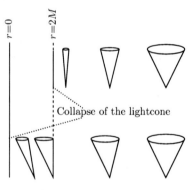

Figure 10.1: Illustration of light-cones in the two coordinate systems. The top one is in the Schwarzschild time coordinate t while the lower is in the coordinate $\tilde{t}$.

Example 10.1 (Time delay of radar echo)   Examples
Let us consider an experiment where we send light towards Mercury, say. The speed
of light in Schwarzschild coordinates is

$$\tilde{c} = \left(1 - \frac{2M}{r}\right).$$

(10.52)

The time used for light to travel from the Earth to Mercury and back is then given by
the integral (see figure 10.2)

$$
\begin{aligned}
t &= 2 \int_{-\lambda_E}^{\lambda_M} \frac{d\lambda}{1 - \frac{2M}{r}} \approx 2 \int_{-\lambda_E}^{\lambda_M} \left(1 + \frac{2M}{r}\right) d\lambda = 2 \int_{-\lambda_E}^{\lambda_M} \left(1 + \frac{2M}{\sqrt{b^2 + \lambda^2}}\right) d\lambda \\
&= 2 \left[\lambda_E + \lambda_M + 2M \ln\left(\frac{\sqrt{\lambda_M^2 + b^2} + \lambda_M}{\sqrt{\lambda_E^2 + b^2} - \lambda_E}\right)\right].
\end{aligned}
$$

(10.53)

The deceleration is the greatest when the Earth and Mercury are on the opposite sides
of the Sun. The impact parameter b is then very small compared to λ_E and λ_M. Thus
we can approximate $\lambda_M \approx r_M$, and $\lambda_E \approx r_E$ which yield to lowest order in $\frac{b}{\lambda}$:

$$t \approx 2 \left[r_E + r_M + 2M \ln\left(\frac{4 r_E r_M}{b^2}\right)\right].$$

(10.54)

The following data are given for the various parameters:
$2M$ = Sun's Schwarzschild radius ≈ 2 km
r_E = radius of Earth's orbit $\approx 15 \cdot 10^{10}$ m
r_M = radius of Mercury's orbit $\approx 5.8 \cdot 10^{10}$ m
b = Sun's radius $\approx 7 \cdot 10^8$ m.
Thus, theoretically, we get a time delay of

$$\Delta t = 2[t - (r_E + r_M)] = 2.2 \cdot 10^{-4} \text{ s}.$$

(10.55)

Shapiro *et al.* [SAI+71] managed to measure the time delay due to this effect by letting
radar signal bounce off Mercury's surface. Later, by using a transponder on the surface
of Mars, the theoretical prediction was confirmed within ±0.1% accuracy [RT02]. We
have not taken into account the curvature in the neighbourhood of the Sun in the sense
that we have assumed a straight path for the light. Atmospheric disruption, amongst
other things, of a light-signal must also be taken into account if such a delay would be
measured.

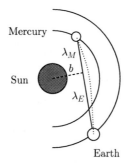

Figure 10.2: Path of a light ray between the Earth and Mercury. The true path is
indicated by the dashed line.

Using data from NASA's Cassini spacecraft an experiment by Italian scientists [BLT03] has confirmed the relativistic correction for the time delay of radar echo with a precision that is 50 times greater than the previous measurements. We can write

$$\Delta t = 2(1 + \gamma) M \ln \left(\frac{4 r_E r_C}{b^2} \right), \tag{10.56}$$

where r_C is Cassini's distance from the Sun, and $\gamma - 1$ measures deviation from the general relativistic prediction. The results of the measurement was

$$\gamma - 1 = (2.1 \pm 2.3) \cdot 10^{-5}.$$

Example 10.2 (The Hafele-Keating experiment)
Another measured effect is the difference in time shown on stationary and moving atomic clocks. By having one clock on an airplane circumnavigating the Earth in the western direction and one circumnavigating in the eastern direction, the time shown by these clocks was compared to a clock on the ground. Even though the time difference is minute, atomic clocks are accurate enough to measure this tiny time difference.

The proper time interval measured by a moving clock with a three-velocity $v^i = \frac{dx^i}{dt}$ in a coordinate system with metric $g_{\mu\nu}$ is given by

$$d\tau = \left(-\frac{1}{c^2} g_{\mu\nu} dx^\mu dx^\nu \right)^{\frac{1}{2}} = \left(-g_{00} - 2 g_{i0} \frac{v^i}{c} - \frac{v^2}{c^2} \right)^{\frac{1}{2}} dt, \tag{10.57}$$

where $v^2 = g_{ij} v^i v^j$. In the Schwarzschild metric this becomes

$$d\tau = \left(1 - \frac{2M}{r} - \frac{v^2}{c^2} \right)^{\frac{1}{2}} dt. \tag{10.58}$$

If we consider an idealized situation where a plane flies at a constant altitude h and with constant speed u along the equator, then if R and Ω are the Earth's radius and angular velocity respectively, the expression becomes to second order

$$\Delta \tau = \left(1 - \frac{Gm}{Rc^2} - \frac{R^2 \Omega^2}{2c^2} + \frac{gh}{c^2} - \frac{2R\Omega u + u^2}{2c^2} \right) \Delta t. \tag{10.59}$$

Here g is the acceleration of gravity at Earth's surface, and $u > 0$ if the plane is eastbound and $u < 0$ if it is westbound. A clock left on the ground on the airport has $h = u = 0$

$$\Delta \tau_0 = \left(1 - \frac{Gm}{Rc^2} - \frac{R^2 \Omega^2}{2c^2} \right) \Delta t. \tag{10.60}$$

Thus to the lowest order we get a relative time difference of the atomic clocks

$$\kappa = \frac{\Delta \tau - \Delta \tau_0}{\Delta \tau_0} = \frac{gh}{c^2} - \frac{2R\Omega u + u^2}{2c^2}. \tag{10.61}$$

If the planes have a travel time $\Delta \tau_0 = 1.2 \cdot 10^5$ s , then theoretically the eastbound plane will measure $\kappa_E = -1.0 \cdot 10^{-12}$ s while the westbound will measure $\kappa_W = 2.1 \cdot 10^{-12}$ s. The time difference for the two planes are approximately -120 ns and 250 ns respectively. These values were confirmed within 20% accuracy experimentally. Thus despite that these numbers are small and that they are far beyond the human detectability in everyday life, it can be observed with the aid of atomic clocks.

10.4   Particle trajectories in Schwarzschild spacetime

Einstein and his contemporaries were not in the possession of modern atomic clocks or even jet-planes when the General Theory of Relativity was in its infancy. But they where aware of something else; a part amounting to 43″ per century of the perihelion precession of Mercury that could not be explained by classical mechanics. Einstein soon realized that the General Theory of Relativity could explain this perihelion precession of the Mercurian orbit. We will in this section investigate particle trajectories in the Schwarzschild spacetime and see how general relativity explains this perihelion precession.

For a test particle outside a static spherically symmetric body we can use the Lagrangian

$$
\begin{aligned}
\mathcal{L} &= \frac{1}{2} g_{\mu\nu} u^\mu u^\nu \\
&= \frac{1}{2}\left[-\left(1 - \frac{2M}{r}\right)\dot{t}^2 + \frac{\dot{r}^2}{1 - \frac{2M}{r}} + r^2\dot{\theta}^2 + r^2\sin^2\theta\,\dot{\phi}^2 \right]
\end{aligned} \tag{10.62}
$$

In addition to the equations of motion derivable from this Lagrangian, we have the four-velocity identity $g_{\mu\nu}u^\mu u^\nu = -1$.

Both t and ϕ are cyclic coordinates, so their canonical momenta, p_t and p_ϕ respectively, are constants:

$$
p_t = \frac{\partial \mathcal{L}}{\partial \dot{t}} = -\left(1 - \frac{2M}{r}\right)\dot{t} \tag{10.63}
$$

$$
p_\phi = \frac{\partial \mathcal{L}}{\partial \dot{\phi}} = r^2\sin^2\theta\,\dot{\phi}. \tag{10.64}
$$

These constants of motion can be interpreted in the following way: p_ϕ is the angular momentum of the orbit of the particle and $-p_t$ is the energy of the particle as measured by an observer at infinity. These are also constants of motion in the Newtonian theory. Another constant of motion in the Newtonian theory is the z-component of the angular momentum, which is also a constant here. This is not difficult to see since we have a spherically symmetric Lagrangian, but let us still check this out by explicit calculation. The equation of motion for θ is

$$
\begin{aligned}
0 &= \frac{d}{d\tau}\left(\frac{\partial \mathcal{L}}{\partial \dot{\theta}}\right) - \frac{\partial \mathcal{L}}{\partial \theta} \\
&= \frac{d}{d\tau}\left(r^2\dot{\theta}\right) - r^2\sin\theta\cos\theta\,\dot{\phi}^2 \\
&= \frac{d}{d\tau}\left(r^2\dot{\theta}\right) + \frac{p_\phi^2\cos\theta}{r^2\sin^3\theta}.
\end{aligned} \tag{10.65}
$$

Multiplying by $r^2\dot{\theta}$, we end up with a total derivative,

$$
0 = \frac{d}{d\tau}\left(r^2\dot{\theta}\right)^2 + \frac{d}{d\tau}\left(\frac{p_\phi}{\sin\theta}\right)^2. \tag{10.66}
$$

The spherical symmetry allows us to impose the boundary condition $\theta(\tau_0) = \frac{\pi}{2}$ and $\dot{\theta}(\tau_0) = 0$ at some time τ_0. This is no loss of generality because there are no preferred direction in a spherically symmetric spacetime, the North and South can be anywhere. Integration then yields

$$
(r^2\dot{\theta})^2 = -p_\phi^2\cot^2\theta. \tag{10.67}
$$

The left hand side is never negative, while the right hand side is never positive. Hence, they both have to be zero. This implies $\theta = \frac{\pi}{2}$ and $\dot{\theta} = 0$ always. The orbit is therefore *planar*.

We therefore assume that the orbit is in the equatorial plane. The four-velocity identity then yields

$$-\frac{p_t^2}{1 - \frac{2M}{r}} + \frac{\dot{r}^2}{1 - \frac{2M}{r}} + \frac{p_\phi^2}{r^2} = -1, \tag{10.68}$$

which after a rearranging gives

$$\frac{1}{2}\dot{r}^2 + V(r) = E, \tag{10.69}$$

where

$$V(r) = -\frac{M}{r} + \frac{p_\phi^2}{2r^2} - \frac{Mp_\phi^2}{r^3}, \tag{10.70}$$

$$E = \frac{1}{2}(p_t^2 - 1). \tag{10.71}$$

In the Newtonian case the "potential" $V(r)$ is equal to the Newtonian potential

$$V_N(r) = -\frac{M}{r} + \frac{p_\phi^2}{2r^2}. \tag{10.72}$$

The term $-\frac{Mp_\phi^2}{r^3}$ is thus a relativistic effect which has some interesting consequences for the particle motion. First of all, it is this term that causes the famous perihelion precession of the Mercurian orbit. Secondly, for small enough r this term will dominate and, since it has a negative sign, a particle with angular momentum can still plunge into the singularity $r = 0$. This is not the case for Newtonian mechanics. The Newtonian potential has an infinitely high centrifugal barrier given by the angular momentum term (see Fig. 10.3).

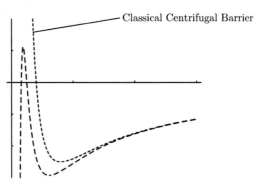

Classical Centrifugal Barrier

Figure 10.3: The graphs of the two potentials $V(r)$ and $V_N(r)$. Notice how the Newtonian potential has a centrifugal barrier for small r.

Circular motion can only exist where $\frac{\partial V}{\partial r} = 0$. Solving this equation for r gives two possibilities:

$$r_\pm = \frac{p_\phi}{2M}\left(1 \pm \sqrt{1 - 12\frac{M^2}{p_\phi^2}}\right). \tag{10.73}$$

The innermost radius, r_-, is unstable and a generic perturbation of this circular orbit will make it either plunge into the singularity or move outwards far away from its original circular orbit. The outermost radius, r_+, on the contrary, is stable. If $p_\phi^2 < 12M^2$ there exist no possibilities for a circular orbit, and all particles having $p_\phi^2 < 12M^2$ will plunge unconditionally into the singularity.

Let us instead write the radius as a function of ϕ: $r = r(\phi)$. This means that we make a transition from the equation of motion of the particle to an equation for the path it follows. Then

$$\dot{r} = \frac{dr}{d\phi}\dot{\phi} = \frac{p_\phi}{r^2}\frac{dr}{d\phi}. \tag{10.74}$$

It is also useful to introduce a new variable u by

$$u = \frac{1}{r}. \tag{10.75}$$

The four-velocity identity (or the energy equation) is then

$$\left(\frac{du}{d\phi}\right)^2 + (1 - 2Mu)\left(u^2 + p_\phi^{-2}\right) = p_t^2 p_\phi^{-2}. \tag{10.76}$$

If we differentiate this equation once, we get the simple form

$$\frac{d^2u}{d\phi^2} + u = \frac{M}{p_\phi^2} + 3Mu^2. \tag{10.77}$$

The last of the terms on the right hand side is the relativistic correction. This gives the deviation from a pure elliptic motion which follows from the laws of Kepler. The Newtonian potential has the peculiar feature that bound particles will have a closed orbit[1]. Any slight deviation from this potential will cause the orbit not to close and we will have a precession of the orbit.

The perihelion precession of Mercury

Let us solve the classical equation first, and then consider a small relativistic correction. The classical equation is

$$\frac{d^2u_0}{d\phi^2} + u_0 = \frac{M}{p_\phi^2}, \tag{10.78}$$

which has the solution

$$u_0 = \frac{M}{p_\phi^2}(1 + e\cos\phi). \tag{10.79}$$

Here, e is called the *eccentricity* of the orbit. For $0 \le e < 1$ the orbit is an ellipse, for $e = 1$ it is a parabola and for $e > 1$ it is a hyperbola. We are interested in the elliptic case, therefore we will assume $0 \le e < 1$. We can also write p_ϕ^2/M as

$$\frac{p_\phi^2}{M} = a(1 - e^2), \tag{10.80}$$

---

[1]This is because the r^{-1} and r^2 potentials have an accidental symmetry in their mechanics. All spherically symmetric systems have an $SO(3)$ symmetry group, but for these specific potentials there is an $SO(4)$ symmetry.

where a is the semi-major axis of the orbit.

Let us therefore make the ansatz

$$u = p^{-1}(1 + e\cos\omega\phi) \tag{10.81}$$

and assume that e is small and ω is close to 1. Inserting this into equation (10.77) we get

$$p^{-1}(1 + e(1 - \omega^2)\cos\omega\phi) \approx \frac{M}{p_\phi^2} + 3Mp^{-2}(1 + 2e\cos\omega\phi + e^2\cos^2\omega\phi) \tag{10.82}$$

To lowest order in e we have

$$p = \frac{p_\phi^2}{2M}\left(1 + \sqrt{1 - 12\frac{M^2}{p_\phi^2}}\right) \approx \frac{p_\phi^2}{M}$$

$$\frac{3M}{p} \approx (1 - \omega) = \delta\omega. \tag{10.83}$$

The precession angular velocity is given by $\omega_p = \frac{2\pi\delta\omega}{T}$ where T is the classical orbital period. From the 3rd law of Kepler, $4\pi^2 a^3 = MT^2$, we get finally the precession angular velocity (with c and G inserted):

$$\omega_p = \frac{2\pi\delta\omega}{T} = \frac{3(Gm)^{\frac{3}{2}}}{c^2(1 - e^2)a^{\frac{5}{2}}}. \tag{10.84}$$

Here we also have written the angular momentum p_ϕ in terms of a, m and e. This is the correct expression in terms of e as well, even though we assumed in our calculations that e was small.

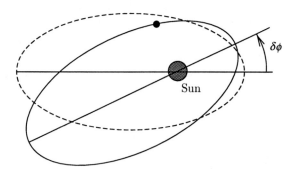

Figure 10.4: The precession of the Mercurian orbit.

For the planet Mercury this formula predicts a precession of 43 arc seconds per century (see Fig. 10.4). Even though this precession seems minute it caused problems for astronomers and physicists at that time. Of Mercury's total precession of approximately 500 arc seconds per century, Newtonian perturbation analysis explained most of this precession as due to the other planets, but about 40 arc seconds were unaccounted for. It was therefore a major breakthrough for the General Theory of Relativity that it predicted a precession of 43 arc seconds per century. It was another discovery however that would make the headlines in the newspapers of the world in 1919.

Deflection of light

We will now see how light is deflected in a gravitational field. Like ordinary matter, light is also under the influence of gravity. Since gravity curves space itself, it is no surprise that photons travelling in space have their trajectories curved when they move close to a massive body.

The orbit equation for light can be derived similarly as for a particle. The only difference is that the four-velocity identity is zero: $u^\mu u_\mu = 0$. The orbit equation is for light

$$\frac{d^2 u}{d\phi^2} + u = 3Mu^2. \tag{10.85}$$

To lowest order we can solve the equation

$$\frac{d^2 u_0}{d\phi^2} + u_0 = 0. \tag{10.86}$$

It has the solution

$$u_0 = \frac{1}{b} \cos \phi, \tag{10.87}$$

where b is *the impact parameter*. The integration constant is chosen so that $\phi = 0$ closest to the gravitating body. Since the configuration is symmetric about this point, we assume the perturbation is symmetric with respect to $\phi = 0$. We thus use the trial function

$$u = \frac{1}{b}(\cos \phi + B + A \sin^2 \phi) \tag{10.88}$$

to calculate the deflection angle to lowest order. Inserting this into equation (10.85), we get to lowest order

$$\frac{1}{b}\left(B + 2A - 3A \sin^2 \phi\right) = \frac{3M}{b^2}(1 - \sin^2 \phi). \tag{10.89}$$

Thus

$$B = A = \frac{M}{b}. \tag{10.90}$$

The solution is therefore

$$u = \frac{1}{b}\left[\cos \phi + \frac{M}{b}\left(1 + \sin^2 \phi\right)\right]. \tag{10.91}$$

The photon flies out towards radial infinity, i.e. at $u = 0$. The deflection angle $\delta\phi$ can therefore be determined from the equation (see Fig. 10.5)

$$u\left(\frac{\pi}{2} + \frac{\delta\phi}{2}\right) = 0. \tag{10.92}$$

Expanding the function u around $\frac{\pi}{2}$ we obtain

$$\boxed{\delta\phi = \frac{4M}{b}.} \tag{10.93}$$

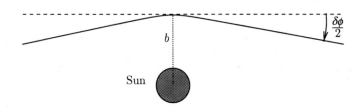

Figure 10.5: The Sun's gravitational field causes the light to deflect in the solar neighbourhood.

For light that just barely misses the Sun's surface the deflection angle turns out to be $\delta\phi = 1.75''$. During a solar eclipse in 1919, two British expeditions observed stars in the solar neighbourhood in the sky. The observers found that the position of the stars was slightly shifted compared to their star charts. This shift agreed with what the Theory of General Relativity had predicted. This observation of light deflecting in the Sun's gravitational field was seen upon as the final breakthrough of the Theory of General Relativity. The theory was not just a mathematical curiosity, it was a Theory that explained fundamental properties of Nature.

10.5   Analytical extension of the Schwarzschild spacetime

The geometry of the Schwarzschild spacetime is quite intriguing and has some nice properties which we shall explore in this section. Even though the Schwarzschild spacetime comes from a very simple ansatz, its geometry can be quite complex. We will explore some of the techniques often used in general relativity to find "exotic" spacetimes, mostly because the techniques themselves are highly general and are applicable to various problems related to geometry and physics. We will first see how the spatial hypersurfaces "look like".

Embedding of a space-like hypersurface of the Schwarzschild spacetime

Let us consider the three-dimensional spatial hypersurface given by $t = 0$ of the Schwarzschild spacetime. The metric for this hypersurface is

$$d\Sigma^2 = \frac{dr^2}{1 - \frac{2M}{r}} + r^2(d\theta^2 + \sin^2\theta d\phi^2). \qquad (10.94)$$

We will embed it in a four-dimensional Euclidean space $\mathbb{E}^4$. Since the metric is spherically symmetric we use cylindrical coordinates in four dimensions. The flat metric of the ambient space can be written

$$ds^2 = dz^2 + dr^2 + r^2(d\theta^2 + \sin^2\theta d\phi^2). \qquad (10.95)$$

We will try to find a hypersurface in $\mathbb{E}^4$ which is rotationally symmetric around the z-axis and has an induced metric equal to the metric (10.94). Since it is rotationally symmetric it should be possible, at least locally, to find a parameterization where the surface is given by $z(r)$. Then, we have

$$dz = \frac{dz}{dr}dr. \tag{10.96}$$

Thus the induced metric on the hypersurface is

$$d\Sigma^2 = \left(1 + \left(\frac{dz}{dr}\right)^2\right)dr^2 + r^2(d\theta^2 + \sin^2\theta d\phi^2). \tag{10.97}$$

For this to coincide with the metric (10.94) we require

$$\frac{dz}{dr} = \pm\sqrt{g_{rr} - 1}. \tag{10.98}$$

Integrating (choosing the positive sign) gives

$$z(r) = \int_{2M}^{r} dx\sqrt{\frac{x}{x - 2M} - 1} = \sqrt{8M(r - 2M)}. \tag{10.99}$$

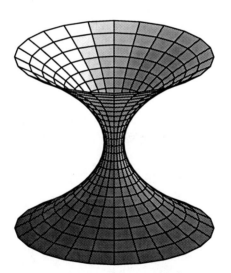

Figure 10.6: The embedding of a space-like hypersurface of the Schwarzschild spacetime. Depicted is Flamm's parabola which is two such hypersurfaces glued together along the horizon.

This is half of a parabola going in the r direction. The negative sign gives the other half of the parabola (see Fig. 10.6). The three-dimensional hypersurface can be either of these, they both yield the same induced metric. Note also that if we instead choose r to be a function of z we get simply

$$r(z) = \frac{1}{8M}z^2 + 2M \tag{10.100}$$

for both. This is *Flamm's parabola*. Thus we can analytically continue the spatial hypersurfaces to the whole parabola $r = \frac{1}{8M}z^2 + 2M$. Note that this surface is totally regular everywhere, there is nothing particular happening at $z = 0$ ($r = 2M$). This expansion of the Schwarzschild spacetime is called *the Einstein-Rosen bridge*. It describes two identical Schwarzschild spacetimes with a common horizon. Since the horizon acts as a one way membrane, the two exterior Schwarzschild solutions cannot communicate with each other. If anything would pass though the horizon it can only end in the singularity, not in the other "universe".

In a previous section we showed that the horizon was only a coordinate singularity, not a physical singularity. This fits well with the Einstein-Rosen bridge. However, we noted that the metric could also be expanded to the interior of the horizon. Both spacetimes in the Einstein-Rosen bridge are *exterior* solutions so there must be something more. To find the maximally extended Schwarzschild spacetime we must introduce a new set of coordinates which is well-behaved at the horizon.

Eddington-Finkelstein- and Kruskal-Szekeres-coordinates

We have already noted that infalling observers do not experience anything particular at the horizon. Let us therefore introduce a set of coordinates which is connected to infalling/outgoing photons.

The radially travelling photons are governed by the geodesic equation which reduces to (10.51):

$$\frac{dr}{dt} = \pm \left(1 - \frac{2M}{r}\right). \tag{10.101}$$

This equation can be integrated to yield

$$\pm t + r + 2M \ln\left|\frac{r}{2M} - 1\right| = C_\pm, \tag{10.102}$$

where $C_\pm$ are integration constants. For convenience, let us define

$$r^* = r + 2M \ln\left|\frac{r}{2M} - 1\right|, \tag{10.103}$$

so that

$$r^* \pm t = C_\pm. \tag{10.104}$$

The constant C_+ uniquely tells us when the photon was sent towards the horizon. We can therefore consider $v \equiv C_+$ as our new time coordinate. Then

$$dt = dv - dr^* = dv - \frac{dr}{1 - \frac{2M}{r}}, \tag{10.105}$$

which brings the Schwarzschild metric on the form

$$ds^2 = -\left(1 - \frac{2M}{r}\right)dv^2 + 2dvdr + r^2(d\theta^2 + \sin^2\theta d\phi^2). \tag{10.106}$$

We now have a non-singular description of particles falling inwards towards $r = 0$ from spatial infinity $r = \infty$. These coordinates are called *ingoing Eddington-Finkelstein-coordinates*.

Likewise, if we had chosen $u \equiv C_-$ as our new time coordinate we would have gotten the metric

$$ds^2 = -\left(1 - \frac{2M}{r}\right) du^2 - 2du\,dr + r^2(d\theta^2 + \sin^2\theta\,d\phi^2). \qquad (10.107)$$

These coordinates have a non-singular description of particles travelling *outwards*. Thus neither of the chosen time coordinates have a non-singular description for both outgoing *and* ingoing particles. Let us choose a combination

$$ t = \frac{1}{2}(v + u) \qquad (10.108)$$

$$ r^* = \frac{1}{2}(v - u), \qquad (10.109)$$

so that

$$ds^2 = -\left(1 - \frac{2M}{r}\right) du\,dv + r^2(d\theta^2 + \sin^2\theta\,d\phi^2). \qquad (10.110)$$

This does not quite take care of the problem at the horizon. However, introducing

$$ U = -e^{-\frac{u}{4M}} \qquad (10.111)$$
$$ V = e^{\frac{v}{4M}}, \qquad (10.112)$$

we get the result

$$ds^2 = -\frac{32M^3}{r} e^{-\frac{r}{2M}} dU\,dV + r^2(d\theta^2 + \sin^2\theta\,d\phi^2). \qquad (10.113)$$

These coordinates are called *Kruskal-Szekeres-coordinates* and are the maximally expanded Schwarzschild solution. It has no coordinate singularities except at $r = 0$ which corresponds to a physical singularity. These Kruskal-Szekeres-coordinates cover the whole spacetime and show explicitly that the horizon at $r = 2M$ is a mere coordinate singularity in the Schwarzschild coordinates.

In figure 10.7 we have illustrated the Kruskal-Szekeres diagram for the analytically extended Schwarzschild solution. The original metric covers the region I, while region II is the interior of the black hole. Region IV is the interior of a "white hole" while region III is just a copy of region I.

10.6   Charged and rotating black holes

The Schwarzschild solution for empty space is perhaps the simplest possible non-trivial solution to the Einstein equations. There are also similar solutions which describe black holes with a cosmological constant, black holes with an electric charge and with angular momentum. Let us investigate some of these solutions.

The Reissner-Nordström Black Hole

The Reissner-Nordström black hole is a spherically symmetric spacetime that has non-zero electric charge. If we start with an electromagnetic field one-form **A** given by

$$ \mathbf{A} = -\frac{q}{r} \mathbf{dt}, \qquad (10.114)$$

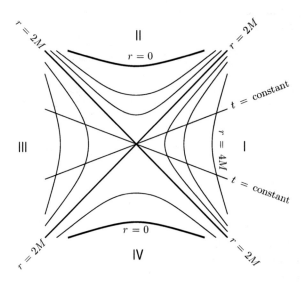

Figure 10.7: Kruskal-Szekeres diagram of the analytically extended Schwarzschild solution.

then the electromagnetic field-tensor becomes

$$\mathbf{F} = \mathbf{dA} = \frac{q}{r^2}\mathbf{dr} \wedge \mathbf{dt}. \tag{10.115}$$

The energy-momentum tensor is no longer zero; the spacetime is no longer a solution of Einstein's field equations for empty space.

Using eq. (8.49), the non-zero components of the energy-momentum tensor are

$$T_{\hat{t}\hat{t}} = T_{\hat{\theta}\hat{\theta}} = T_{\hat{\phi}\hat{\phi}} = \frac{q^2}{2r^4}, \quad T_{\hat{r}\hat{r}} = -\frac{q^2}{2r^4}. \tag{10.116}$$

From eqs. (10.17) and (10.18) we get by adding the $\hat{t}\hat{t}$- and $\hat{r}\hat{r}$-field equations, and integrating

$$\alpha(r) = -\beta(r). \tag{10.117}$$

Inserting this into the $\hat{t}\hat{t}$-equation, we get

$$\frac{1}{r^2}\left[r\left(1 - e^{-2\beta}\right)\right]' = \kappa\frac{q^2}{2r^4}. \tag{10.118}$$

This equation can be integrated to yield

$$e^{-2\beta} = 1 - \frac{2M}{r} + \frac{Q^2}{r^2}, \tag{10.119}$$

where we have defined $Q^2 \equiv \kappa q^2$. The general solution can thus be written

$$ds^2 = -\left(1 - \frac{2M}{r} + \frac{Q^2}{r^2}\right)dt^2 + \frac{dr^2}{1 - \frac{2M}{r} + \frac{Q^2}{r^2}} + r^2(d\theta^2 + \sin^2\theta d\phi^2). \tag{10.120}$$

This metric describes a black hole with an electric charge. The electromagnetic field tensor has a non-zero electric component. By inspecting the line-element (10.120) we see that this spacetime has *two* horizons at

$$r = M \pm \sqrt{M^2 - Q^2}. \tag{10.121}$$

These horizons merge into to one in the extremal limit $M = \pm Q$. For $M < |Q|$ there are no horizons, and the singularity at $r = 0$ becomes a so-called *naked singularity* because it has no surrounding horizons. This is however an unphysical situation[2] so we have the bound $M \geq |Q|$.

Also the coordinate singularity at the horizon for this metric can be removed by introducing Kruskal-Szekeres-coordinates. The horizons are only coordinate singularities, there are no physical singularities except at $r = 0$.

The axisymmetric and stationary line-element: The Ernst equation

A spacetime is called *stationary* if there exists a Killing vector $\boldsymbol{\xi}$ which is asymptotically time-like at spatial infinity. If, in addition, this Killing vector is orthogonal to some space-like three-surface then we say that the spacetime is *static*. The Schwarzschild and the Reissner-Nordström solutions are static, but in the following we will only assume the spacetime is stationary. We will consider axisymmetric spacetimes which possess an asymptotically time-like Killing vector $\boldsymbol{\xi}_t = \frac{\partial}{\partial t}$. This spacetime will also have a two-dimensional surface $^{(2)}\Sigma$ which is orthogonal to the Killing vectors $\boldsymbol{\xi}_t$ and $\boldsymbol{\xi}_\phi = \frac{\partial}{\partial \phi}$. Its metric can therefore be written

$$ds^2 = -V dt^2 + 2W dt d\phi + X d\phi^2 + e^{2\mu} \left[(dx^1)^2 + (dx^2)^2 \right], \tag{10.122}$$

where V, W, X and μ are functions of x^A, $A = 1, 2$ only. A coordinate transformation of the two-surface $^{(2)}\Sigma$ – which changes the coordinates x^A only – leaves the functions V, W and X invariant. Hence, they behave as scalars under such transformations. Note also that the metric (10.122) stays invariant under transformations

$$t \mapsto At + B\phi, \quad \phi \mapsto Ct + D\phi, \tag{10.123}$$

with A, B, C and D constants. The determinant of the metric of the two-dimensional surface spanned by (t, ϕ) is given by

$$-(VX + W^2) = -\rho^2, \tag{10.124}$$

where ρ is the coordinate distance from the axis. At the axis of symmetry, $W = X = 0$, so $\rho = 0$. Moreover, $\rho = 0$ on event horizons as can be shown. In addition, the Einstein's field equations for empty space imply that ρ satisfies the two-dimensional Laplace equation

$$^{(2)}\nabla_A {}^{(2)}\nabla^A \rho = 0, \tag{10.125}$$

on $^{(2)}\Sigma$. It can be shown that ρ can be taken as a variable on the two-dimensional space $^{(2)}\Sigma$, and, using the orthogonal direction z as the second variable, the metric on $^{(2)}\Sigma$ can be written

$$ds_2^2 = e^{2\mu(\rho, z)} \left(d\rho^2 + dz^2 \right). \tag{10.126}$$

---

[2]The statement that all physical singularities have to be surrounded by a horizon is referred to as *cosmic censorship*.

Introducing the metric functions h and γ by

$$W = hV, \quad X = V^{-1}\rho^2 - h^2 V, \quad e^{2\mu} = e^{2\gamma}V^{-1}, \tag{10.127}$$

the general axisymmetric metric (10.122) can be put onto the canonical form

$$ds^2 = -V\left(dt - hd\phi\right)^2 + V^{-1}\left[e^{2\gamma}(d\rho^2 + dz^2) + \rho^2 d\phi^2\right]. \tag{10.128}$$

Henceforth, we will only consider Einstein's equations for empty space, $R_{\mu\nu} = 0$. After a long algebraic manipulation, the form of the vacuum equations can be deduced. Those involving V and h are

$$V\bar{\nabla}^\alpha\bar{\nabla}_\alpha V = \left(\bar{\nabla}^\alpha V\right)\left(\bar{\nabla}_\alpha V\right) - \rho^{-2}V^4\left(\bar{\nabla}^\alpha h\right)\left(\bar{\nabla}_\alpha h\right) \tag{10.129}$$
$$\bar{\nabla}^\alpha\left(\rho^{-2}V^2\bar{\nabla}_\alpha h\right) = 0. \tag{10.130}$$

Here, the Greek indices and the covariant derivative $\bar{\nabla}^\alpha$ are with respect to the fictitious Euclidean metric

$$ds_3^2 = \rho^2 d\phi^2 + d\rho^2 + dz^2. \tag{10.131}$$

Eq. (10.130) can be written, using the metric (10.131),

$$\frac{\partial}{\partial x^\alpha}\left(\rho^{-1}V^2\frac{\partial h}{\partial x_\alpha}\right) = 0. \tag{10.132}$$

This implies that there exists a "potential" Φ' which is a function of ρ and z only, such that

$$\rho^{-1}V^2\frac{\partial h}{\partial x_\alpha} = \varepsilon^{\phi\alpha\beta}\frac{\partial\Phi'}{\partial x^\beta}, \tag{10.133}$$

where $\varepsilon^{\gamma\alpha\beta}$ is the totally antisymmetric tensor with $\varepsilon^{\phi\rho z} = 1$. Redefining $\Phi = -\Phi'$ we can write this as

$$V^{-2}\frac{\partial\Phi}{\partial x^\alpha} = -\rho^{-1}\varepsilon_{\alpha\phi\beta}\frac{\partial h}{\partial x_\beta}, \tag{10.134}$$

which implies that eq. (10.130) can be written

$$\bar{\nabla}^\alpha\left(V^{-2}\frac{\partial\Phi}{\partial x^\alpha}\right) = 0. \tag{10.135}$$

Further, this makes it possible to write eq. (10.129) as

$$\bar{\nabla}^\alpha\left[\frac{\bar{\nabla}_\alpha(V^2 + \Phi^2)}{V^2}\right] = 0. \tag{10.136}$$

Introducing the complex function ξ by

$$\frac{\xi - 1}{\xi + 1} = V + i\Phi, \tag{10.137}$$

eqs. (10.135) and (10.136) are encompassed in the single equation

$$\boxed{(\xi\xi^* - 1)\,\bar{\nabla}^\alpha\bar{\nabla}_\alpha\xi = 2\xi^*\left(\bar{\nabla}_\alpha\xi\right)\left(\bar{\nabla}^\alpha\xi\right),} \tag{10.138}$$

where * denotes complex conjugation. This equation is called *the Ernst equation*.

The Einstein equations for empty space are replaced in the axisymmetric and stationary case by the Ernst equation. By finding solutions to the Ernst equation we find the metric functions V and h (via Φ). The remaining metric function γ can be determined from the remaining field equations, which are equivalent to the equations

$$
\frac{\partial \gamma}{\partial \rho} = \frac{\rho}{(|\xi|^2 - 1)^2} \left(\frac{\partial \xi}{\partial \rho} \frac{\partial \xi^*}{\partial \rho} - \frac{\partial \xi}{\partial z} \frac{\partial \xi^*}{\partial z} \right)
$$

$$
\frac{\partial \gamma}{\partial z} = \frac{2\rho}{(|\xi|^2 - 1)^2} \mathrm{Re} \left(\frac{\partial \xi}{\partial \rho} \frac{\partial \xi^*}{\partial z} \right). \tag{10.139}
$$

An important class of solutions to the Ernst equation is when ξ is of the form

$$
\xi = e^{i\alpha} \coth \psi, \tag{10.140}
$$

where α is a constant, and ψ is a real function of ρ and z only. The function ψ obeys the linear differential equation

$$
\bar{\nabla}^\alpha \bar{\nabla}_\alpha \psi = 0. \tag{10.141}
$$

The Ernst equation, eq. (10.138), on the other hand is not linear, and thus for any two solutions ξ_1 and ξ_2, the coefficients α_1 and α_2 need to be constrained if the linear combination $\xi = \alpha_1 \xi_1 + \alpha_2 \xi_2$ is to be a solution.

The Kerr metric

The Kerr metric is due to Roy Kerr who in 1963 found an axisymmetric and stationary solution to Einstein's field equations for empty space [Ker63]. A couple of years later it was generalized by Newman *et al.* [NCC⁺65], but we will only consider the Kerr solution here. We will derive this solution using the Ernst equation to illustrate how one can generate solutions using this equation.

It is useful to introduce spheroidal coordinates x, y which are related to cylindrical ones ρ, z by

$$
\rho = k\sqrt{(x^2 - 1)(1 - y^2)}
$$

$$
z = kxy, \tag{10.142}
$$

where $|y| < 1 < |x|$ and k is a constant scale factor. The two-dimensional flat metric becomes in these coordinates

$$
d\rho^2 + dz^2 = k^2(x^2 - y^2) \left(\frac{dx^2}{x^2 - 1} + \frac{dy^2}{1 - y^2} \right). \tag{10.143}
$$

The surfaces of constant x and y are families of spheroids and hyperboloids, respectively. Using these coordinates the Ernst equation, eq. (10.138), can be written

$$
(\xi\xi^* - 1) \left\{ \frac{\partial}{\partial x} \left[(x^2 - 1) \frac{\partial \xi}{\partial x} \right] + \frac{\partial}{\partial y} \left[(1 - y^2) \frac{\partial \xi}{\partial y} \right] \right\}
$$

$$
= 2\xi^* \left[(x^2 - 1) \left(\frac{\partial \xi}{\partial x} \right)^2 + (1 - y^2) \left(\frac{\partial \xi}{\partial y} \right)^2 \right]. \tag{10.144}
$$

Let us seek solutions of the form

$$\xi = px + qy, \tag{10.145}$$

where p, q are complex constants. Inserting this trial function into eq. (10.144) yields

$$px - qy = (p^*x + q^*y)(p^2 - q^2), \tag{10.146}$$

which is equivalent to

$$\begin{aligned} p &= p^*(p^2 - q^2) \\ q &= -q^*(p^2 - q^2). \end{aligned} \tag{10.147}$$

We can first note that the Ernst equation, eq. (10.138), is invariant under a change of phase: $\xi \mapsto e^{i\alpha}\xi$. Thus there is no loss of generality to assume that $p = P$ where P is real.

Eq. (10.147) now implies $q = \pm iQ$ where Q is real, and

$$P^2 + Q^2 = 1. \tag{10.148}$$

The sign ambiguity in q corresponds to choosing the complex conjugate of ξ. Choosing $q = -iQ$, eq. (10.145) yields

$$\xi = Px - iQy. \tag{10.149}$$

The functions V and Φ can now be found from eq. (10.137):

$$V = \frac{P^2x^2 + Q^2y^2 - 1}{(Px + 1)^2 + Q^2y^2} \tag{10.150}$$

$$\Phi = -\frac{2Qy}{(Px + 1)^2 + Q^2y^2}. \tag{10.151}$$

It remains to find the metric functions h and γ. Eq. (10.133) relates Φ, V and h:

$$\begin{aligned} \sqrt{x^2 - 1}\frac{\partial\Phi}{\partial x} &= \frac{V^2}{\rho}\sqrt{1 - y^2}\frac{\partial h}{\partial y} \\ -\sqrt{1 - y^2}\frac{\partial\Phi}{\partial y} &= \frac{V^2}{\rho}\sqrt{x^2 - 1}\frac{\partial h}{\partial x}. \end{aligned} \tag{10.152}$$

Inserting for V and Φ gives

$$\begin{aligned} \frac{\partial h}{\partial y} &= \frac{4k(x^2 - 1)PQy(Px + 1)}{(P^2x^2 + Q^2y^2 - 1)^2} \\ \frac{\partial h}{\partial x} &= \frac{2k(1 - y^2)Q\left[(Px + 1)^2 - Q^2y^2\right]}{(P^2x^2 + Q^2y^2 - 1)^2}, \end{aligned} \tag{10.153}$$

which, upon integration, yields

$$h = -\frac{2kQ}{P}\frac{(Px + 1)(1 - y^2)}{(P^2x^2 + Q^2y^2 - 1)}. \tag{10.154}$$

Here, the integration constant has been determined by requiring that h vanishes on the axis of symmetry ($y = \pm 1$).

It remains only to find γ. We define a new variable by the relation

$$e^{2\gamma'} = e^{2\gamma}(x^2 - y^2).$$ (10.155)

Eq. (10.139) can now – through a lengthy but straightforward calculation – be written

$$\frac{\partial \gamma'}{\partial y} = \frac{Q^2 y}{P^2 x^2 + Q^2 y^2 - 1}$$
$$\frac{\partial \gamma'}{\partial x} = \frac{Q^2 x}{P^2 x^2 + Q^2 y^2 - 1}.$$ (10.156)

Integration yields

$$e^{2\gamma'} = C(P^2 x^2 + Q^2 y^2 - 1),$$ (10.157)

where C is an integration constant which will be determined later. All the metric functions are now determined. Using eqs. (10.142) and (10.155), the line element (10.128) can be written

$$ds^2 = -V (dt - h d\phi)^2 + V^{-1} \left[k^2 e^{2\gamma'} \left(\frac{dx^2}{x^2 - 1} + \frac{dy^2}{1 - y^2} \right) + \rho^2 d\phi^2 \right].$$ (10.158)

Due to the constraint (10.148) the metric depends on 2 parameters only. Let these be a and M and make the following parameter change

$$P = \sqrt{1 - \frac{a^2}{M^2}}, \quad Q = \frac{a}{M}, \quad k = \sqrt{M^2 - a^2}.$$ (10.159)

Introducing Boyer-Linquist coordinates r and θ by

$$r = (M^2 - a^2)^{1/2} x + M$$
$$\theta = \arccos y,$$ (10.160)

the metric functions become

$$V = \frac{\Delta - a^2 \sin^2 \theta}{\Sigma}$$ (10.161)

$$h = -\frac{2Mar \sin^2 \theta}{\Delta - a^2 \sin^2 \theta}$$ (10.162)

$$e^{2\gamma'} = \frac{C}{M^2} (\Delta - a^2 \sin^2 \theta),$$ (10.163)

where we have defined

$$\Sigma = r^2 + a^2 \cos^2 \theta$$
$$\Delta = r^2 + a^2 - 2Mr.$$ (10.164)

Finally, choosing $C = M^2/(M^2 - a^2)$ the metric can be written

$$ds^2 = -\frac{\Delta - a^2 \sin^2 \theta}{\Sigma} dt^2 - \frac{4Mar \sin^2 \theta}{\Sigma} dt d\phi$$
$$+ \frac{\Sigma}{\Delta} dr^2 + \left[\frac{(r^2 + a^2)^2 - \Delta a^2 \sin^2 \theta}{\Sigma} \right] \sin^2 \theta d\phi^2 + \Sigma d\theta^2.$$ (10.165)

This metric is called the *Kerr metric*. The physical interpretations of M and a are found in problem 9.3. The Kerr metric describes the spacetime outside a rotating mass distribution with mass M and angular momentum $J = Ma$. When $a = 0$ this metric reduces to the ordinary Schwarzschild vacuum solution, eq. (10.24). It behaves properly everywhere except where $\Delta = 0$ or $\Sigma = 0$. The equation $\Delta = 0$ describes a horizon and is no real singularity. However, the set of points given by the equation

$$\Sigma = r^2 + a^2 \cos^2 \theta = 0 \tag{10.166}$$

can by evaluation of curvature invariants like the Kretchmann scalar, for $M \neq 0$ be shown to be physical singularities. It seems a bit strange that the only solution to this equation is for $r = 0$, $\theta = \frac{\pi}{2}$. However, despite its immediate appearance, this is a *ring singularity*. If we set $M = 0$ and make the coordinate transformation

$$\begin{aligned} z &= r \cos \theta \\ R &= \sqrt{r^2 + a^2} \sin \theta, \end{aligned} \tag{10.167}$$

we recover Minkowski space in cylindrical coordinates. Thus for $M = 0$ the singularity $r = 0$, $\theta = \frac{\pi}{2}$ is no physical singularity, but merely a coordinate singularity. Since this set is a ring, the claim that the singularity for $M \neq 0$ is a ring singularity is reasonable. This also tells us that we should not trust blindly on the apparent topology for a spacetime based on some choice of coordinates.

The exterior solution of $\Delta = 0$ is

$$r_+ = M + \sqrt{M^2 - a^2}, \tag{10.168}$$

which is the radius of the horizon. The area of the horizon is

$$A = \int \omega^{\hat{\theta}} \wedge \omega^{\hat{\phi}} = (r_+^2 + a^2) \int_0^\pi \sin \theta d\theta \int_0^{2\pi} d\phi = 4\pi(r_+^2 + a^2). \tag{10.169}$$

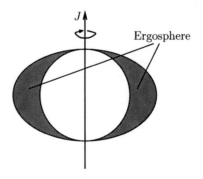

Figure 10.8: The ergosphere in the Kerr spacetime.

Stationary observers have a four-velocity proportional to the Killing vector $\xi_t = \frac{\partial}{\partial t}$ which in a Kerr spacetime has norm

$$\xi^\mu \xi_\mu = g_{tt} = -\frac{\Delta - a^2 \sin^2 \theta}{\Sigma}. \tag{10.170}$$

This becomes positive whenever

$$r^2 + a^2 \cos^2 \theta - 2Mr < 0. \tag{10.171}$$

If $a \neq 0$, part of this region is outside the horizon at r_+. This region of the Kerr spacetime is called the *ergosphere* (see Fig. 10.8). Thus if an observer is to remain stationary in this region, he has to travel faster than the speed of light. This is of course impossible. Thus in the region given by

$$r_+ < r < r_S, \tag{10.172}$$

where $r_S = M + \sqrt{M^2 - a^2 \cos^2 \theta}$, all particles and observers have to be dragged along around the black hole, they simply cannot remain stationary even though they are outside the black hole. The rotation drags the space surrounding it along with it. The surface $r = r_S$ is thus called the stationary limit.

This *inertial dragging* can be seen if we consider a freely falling observer in a Kerr spacetime. We use the Lagrangian

$$\mathcal{L} = \frac{1}{2}\left(\frac{ds}{d\tau}\right)^2, \tag{10.173}$$

and the metric (10.165). Since ϕ is a cyclic coordinate, its canonical momentum is a constant of motion

$$p_\phi = \frac{\partial \mathcal{L}}{\partial \dot{\phi}} = g_{t\phi}\dot{t} + g_{\phi\phi}\dot{\phi} = g_{\phi\phi}\dot{t}(\Omega - \omega), \tag{10.174}$$

where

$$\Omega = \frac{d\phi}{dt} \tag{10.175}$$

$$\omega = -\frac{g_{t\phi}}{g_{\phi\phi}} = \frac{a(r^2 + a^2 - \Delta)}{(r^2 + a^2)^2 - \Delta a^2 \sin^2 \theta}. \tag{10.176}$$

As $r \longrightarrow \infty$, $\omega \longrightarrow 0$. Thus if $p_\phi = 0$ at infinity, the infalling observer will experience an angular velocity given by

$$\Omega = \frac{d\phi}{dt} = \omega. \tag{10.177}$$

The Kerr spacetime in these coordinates is stationary, so an observer at infinity observes that the infalling particle obtains an angular velocity. Since the infalling observer carries a local inertial frame, local inertial frames are dragged around the source of the Kerr spacetime in the same direction as the source rotates. Furthermore, because $p_\phi = 0$, we say that the infalling observer is a *zero angular momentum observer* (ZAMO). In spite of this, the observer experiences an inertial dragging effect from the rotating body.

If we consider a satellite in a polar orbit around the Earth, the orbit of the satellite will precess due to the rotation of Earth. The Earth's diurnal rotation causes the space surrounding the Earth to be "dragged along" with it. The orbit of the satellite will therefore experience an inertial dragging of its orbit, and the orbit will precess in the same direction as the Earth's rotation. This effect is called the *Lense-Thirring effect*.

Example

Example 10.3 (The Lense-Thirring effect)
(see also section 9.4) In the weak-field approximation the angular velocity (10.176) can be approximated by

$$\omega \approx \frac{2Ma}{r^3}. \tag{10.178}$$

Considering a satellite in orbit around the Earth, we get

$$\omega \approx \frac{2GJ_E}{r^3} = 0.2\left(\frac{R_E}{r}\right)^3, \tag{10.179}$$

where J_E and R_E is the Earth's angular momentum and radius respectively. For the LAGEOS and LAGEOS II satellites the precession is about $1/20$ arc seconds per year. During a period of 4 years this rotation of the orbital plane has been measured with 20% accuracy [CPC$^+$98, CCV97, Ciu02]. This confirms that the space outside the Earth can be considered a Kerr spacetime.

The Penrose process

We shall here see how the rotational energy can be extracted from a rotating black hole [Pen69].

The energy of a free particle as measured by an observer in the asymptotic Minkowski spacetime far from the black hole is $E = -p_t$ where p_t is the covariant momentum conjugate to the time coordinate. Since the metric is stationary, t is a cyclic coordinate, and hence E is a constant of motion. As decomposed in an orthonormal ZAMO-field, $\mathbf{e}_t$ has a $\hat{t}$-component and a $\hat{\phi}$-component. Hence,

$$E = p^{\hat{t}}\omega^{\hat{t}}(\mathbf{e}_t) - p^{\hat{\phi}}\omega^{\hat{\phi}}(\mathbf{e}_t). \tag{10.180}$$

Since $\mathbf{p}$ is time-like, $p^{\hat{t}} > p^{\hat{\phi}}$. If $\mathbf{e}_t$ is time-like, then $\omega^{\hat{t}}(\mathbf{e}_t) > \omega^{\hat{\phi}}(\mathbf{e}_t)$ and thus $E > 0$. If $\mathbf{e}_t$ is space-like, then $\omega^{\hat{t}}(\mathbf{e}_t) < \omega^{\hat{\phi}}(\mathbf{e}_t)$ which permits $E \gtreqless 0$.

Outside the stationary limit, $g_{tt} = \mathbf{e}_t \cdot \mathbf{e}_t < 0$. In this region $\mathbf{e}_t$ is time-like. Since $\mathbf{p}$ is time-like, E is positive here. However, for $r_+ < r < r_S$ (in the ergosphere), $g_{tt} > 0$, and $\mathbf{e}_t$ is space-like. In the ergosphere there exist paths of particles with negative energy, i.e. the gravitational binding energy of the particle can be larger than the sum of its mass-energy and kinetic energy.

In order to find the paths of the particles with negative energy we decompose their four-velocity in an orthonormal ZAMO basis,

$$\mathbf{e}_{\hat{t}} = e^{-\nu}\left(\mathbf{e}_t + \omega\mathbf{e}_\phi\right), \quad \mathbf{e}_{\hat{r}} = e^{-\mu}\mathbf{e}_r,$$
$$\mathbf{e}_{\hat{\theta}} = e^{-\lambda}\mathbf{e}_\theta, \quad \mathbf{e}_{\hat{\phi}} = e^{-\psi}\mathbf{e}_\phi, \tag{10.181}$$

where

$$e^{2\nu} = g_{tt} + \omega^2 e^{2\psi}, \quad e^{2\mu} = g_{rr}, \quad e^{2\lambda} = g_{\theta\theta},$$
$$e^{2\psi} = g_{\phi\phi}, \quad \omega = -\frac{g_{t\phi}}{g_{\phi\phi}}. \tag{10.182}$$

Thus $\mathbf{e}_t = e^{\nu}\mathbf{e}_{\hat{t}} - \omega e^{\psi}\mathbf{e}_{\hat{\phi}}$. The four-velocity of the particle is

$$\mathbf{u} = u^{\hat{\mu}}\mathbf{e}_{\hat{\mu}} = \hat{\gamma}(1, \hat{\mathbf{v}}) = \hat{\gamma}\left(\mathbf{e}_{\hat{t}} + v^{\hat{\phi}}\mathbf{e}_{\hat{\phi}}\right), \tag{10.183}$$

where $\hat{\gamma} = (1 - \hat{v}^2)^{-1/2}$ is the usual relativistic factor used by an observer at rest in the orthonormal basis field. This gives for the four-momentum of the particle,

$$\mathbf{p} = m\mathbf{u} = \hat{\gamma}m \left(\mathbf{e}_{\hat{t}} + v^{\hat{\phi}} \mathbf{e}_{\hat{\phi}} \right). \tag{10.184}$$

The energy of the particle is

$$E = -\mathbf{p} \cdot \mathbf{e}_t = \hat{\gamma}m \left(e^\nu + v^{\hat{\phi}} \omega e^\psi \right). \tag{10.185}$$

Hence, the energy of the particle is negative if

$$v^{\hat{\phi}} < -\frac{1}{\omega} e^{\nu - \psi} = -\frac{\rho^2 \Delta^{1/2}}{2Mar \sin \theta}. \tag{10.186}$$

Such solutions are permitted in the ergosphere.

The following process is possible. A rocket ship moves into the ergosphere and fires a particle that enters a path with negative energy. Hence, the rocket ship emits a negative energy and thereby increases its energy. The rocket ship then moves away from the black hole with greater energy than when it entered the ergosphere. In this way it has extracted energy away from the Kerr black hole. The particle with negative energy is absorbed by the black hole. It has $v^{\hat{\phi}} < 0$, meaning that it rotates around the black hole in the opposite sense of the black hole. Absorbing this particle the rotational energy of the black hole decreases. Thus, the Penrose process is a mechanism for extracting rotational energy from a rotating black hole.

Before leaving this topic we shall discuss the question "Can particles really leave the ergosphere?" and try to understand how this can happen [Sch85].

Let us consider a photon moving in the equatorial plane of a Kerr black hole. The equation $\mathbf{p} \cdot \mathbf{p} = 0$ applied to this photon gives

$$-e^{-2\nu} E^2 + e^{-2\nu} 2\omega p_\phi E + \left(e^{-2\psi} - \omega^2 e^{-2\nu} \right) p_\phi^2 + e^{-2\mu} p_r^2 = 0. \tag{10.187}$$

Since $p_r = e^{2\mu} \dot{r}$ we obtain

$$\dot{r}^2 = e^{-2(\mu+\nu)} \left[E^2 - 2\omega p_\phi E + \left(\omega^2 - e^{2\nu - 2\psi} \right) p_\phi^2 \right], \tag{10.188}$$

which may be factorized as

$$\dot{r}^2 = e^{-2(\mu+\nu)} (E - V_+)(E - V_-), \tag{10.189}$$

where

$$V_\pm = \omega p_\phi \pm e^{\nu - \psi} |p_\phi| = \frac{2Ma p_\phi \pm r\Delta^{1/2} |p_\phi|}{r^3 + a^2 r + 2Ma^2}. \tag{10.190}$$

If there exist photon paths with a minimum for r in the ergosphere, the photons will be able to move outwards in the ergosphere. Then there will exist photon paths connecting an emitter in the ergosphere with an observer outside it. This requires $E \leq V_+$ or $E \leq V_-$.

Also, the energy of a photon as measured by an arbitrary observer must be positive. Consider a ZAMO-observer with four-velocity $\mathbf{U} = U^0 (\mathbf{e}_t + \omega \mathbf{e}_\phi)$. As measured by this observer the energy of the photon is

$$\hat{E} = -\mathbf{p} \cdot \mathbf{U} = U^0 (E - \omega p_\phi). \tag{10.191}$$

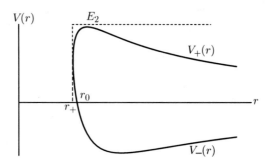

Figure 10.9: Effective potentials of a photon moving in the equatorial plane of a Kerr black hole. The photon moves in the same direction as the black hole rotates.

Hence, the constant energy measured by a far away observer must fulfill $E \leq \omega p_\phi$, which also requires $E \leq V_+$.

Consider a photon with angular velocity in the same direction as the black hole rotates. Then $a p_\phi > 0$ and $V_\pm(r)$ has the form shown in Fig. 10.9. At equator $\theta = \pi/2$ and the surface of infinite redshift is at $r_0 = 2M$. All paths with $E = E_2 > V_+$ have $E > 0$.

In the case of a photon with angular velocity in the opposite direction $a p_\phi < 0$ and $V_\pm(r)$ has the form shown in Fig. 10.10. In this case there exist

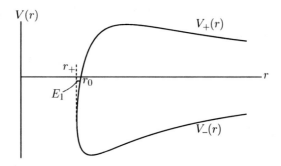

Figure 10.10: Effective potentials of a photon moving in the equatorial plane of a Kerr black hole. The photon moves in the opposite direction as the black hole rotates.

paths in the ergosphere with $E > V_+$ and $E < 0$. From Fig. 10.10 it is seen that this photon cannot move out of the ergosphere.

We thus have an electromagnetic version of the Penrose process. Two rays of electromagnetic radiation are emitted from a position in the ergosphere both with $E > V_+$, one with $E = E_1 < 0$, and the other with $E = E_2 > 0$. The first will be absorbed by the black hole and reduce its rotational energy, and the other will extract energy from the black hole.

It may be noted that an observer at the stationary limit, which is a surface of infinite redshift, will measure an infinitely large frequency for the radiation.

10.7   Black Hole thermodynamics

We have seen how black holes in the General Theory of Relativity act as a one-way membrane for particles and light. Matter can only go into a black hole, it cannot get out. A black hole is black, it does not radiate anything according to general relativity. The physicists were therefore quite surprised when Stephen Hawking discovered that black holes do actually radiate. This is due to quantum effects. When quantum mechanics is applied in areas where the gravity is strong like in the neighbourhood of a black hole horizon, it implies that an observer will see that black body radiation, usually of extremely low temperature, is being emitted from the black hole.

The first sign of a such a property was when Bekenstein [Bek73] conjectured that black holes have an entropy proportional to the black hole's surface area. Later Hawking took this idea seriously, discovered that black holes radiates and gave an exact relation between the entropy of the black hole and its surface area [Haw75]. The four laws of black hole thermodynamics were completed just a few years later by Gibbons and Hawking [GH77]. We will in this section review their results, but first we need to introduce the concept *surface gravity*.

Surface Gravity

The surface gravity is an expression of the acceleration of gravity at the horizon of a black hole. It can be defined and calculated in terms of the Killing vector which is orthogonal to the horizon, or alternatively, in terms of the four-acceleration and four-velocity of a free particle. Both procedures will be demonstrated. First we shall calculate the surface gravity of a Schwarzschild black hole by the Killing vector method, then that of a Kerr metric using the four-acceleration of a free particle.

The surface gravity of a Schwarzschild black hole

The horizon of a black hole is a *null surface*. This can most easily be seen by considering the black hole in Kruskal coordinates. That a surface is a null surface means that any vector normal to the surface is a null vector. Let us consider the Killing vector that generates time translations, $\boldsymbol{\xi} = \xi^\mu \mathbf{e}_\mu$. In the Schwarzschild spacetime this vector is simply $\boldsymbol{\xi} = \mathbf{e}_t$. This Killing vector is normal to the horizon so that[3] $\xi_\mu \xi^\mu = 0$. More specifically, $\xi_\mu \xi^\mu$ is constant on the horizon, thus the gradient $\nabla^\alpha (\xi_\mu \xi^\mu)$ is also normal to the horizon. Hence, there exists a function κ called *the surface gravity* such that

$$\nabla^\alpha (\xi_\mu \xi^\mu) = -2\kappa \xi^\alpha. \tag{10.192}$$

This equation can be rewritten

$$\xi^\nu \nabla_\mu \xi_\nu = -\xi^\nu \nabla_\nu \xi_\mu = -\kappa \xi_\mu. \tag{10.193}$$

A vector ξ_μ is hypersurface orthogonal if[4]

$$\xi_{[\mu} \nabla_\nu \xi_{\rho]} = 0, \tag{10.194}$$

---

[3]Note that in the Kerr spacetime this Killing vector's obvious generalization – also given by a pure time translation – is not orthogonal to the horizon. In the Kerr case we have to use another Killing vector which is a linear combination of $\frac{\partial}{\partial t}$ and $\frac{\partial}{\partial \phi}$.

[4]This follows from Frobenius' Theorem, see problem 7.12.

evaluated on the hypersurface. For a Killing vector, $\nabla_\mu \xi_\nu = -\nabla_\nu \xi_\mu$, so

$$\xi_\rho \nabla_\mu \xi_\nu = -2\xi_{[\mu} \nabla_{\nu]} \xi_\rho \qquad (10.195)$$

evaluated on the horizon. Contracting with $\nabla^\mu \xi^\nu$, we obtain

$$\xi_\rho (\nabla^\mu \xi^\nu)(\nabla_\mu \xi_\nu) = -2(\nabla^\mu \xi^\nu)(\xi_{[\mu} \nabla_{\nu]} \xi_\rho) = -2\kappa \xi^\mu \nabla_\mu \xi_\rho = -2\kappa^2 \xi_\rho, \quad (10.196)$$

where we have used eq. (10.193) successively. The surface gravity can thus be found from

$$\kappa^2 = -\frac{1}{2}(\nabla_\mu \xi_\nu)(\nabla^\mu \xi^\nu) \qquad (10.197)$$

evaluated at the horizon. Since the Schwarzschild metric is diagonal we have

$$\xi^\mu = \delta^\mu_t, \quad \xi_\mu = \delta_{\mu t} g_{tt}. \qquad (10.198)$$

The covariant derivative, $\nabla_\mu \xi_\nu$, is given in terms of the Christoffel symbols

$$\nabla_\mu \xi_\nu = \xi_{\nu,\mu} - \Gamma^\alpha_{\nu\mu} \xi_\alpha. \qquad (10.199)$$

The only nonzero $\xi_{\nu,\mu}$ is $\xi_{t,r} = g_{tt,r}$ since the metric components are dependent on r only. From Killing's equation, eq. (6.297),

$$\nabla_\mu \xi_\nu = -\nabla_\nu \xi_\mu, \qquad (10.200)$$

and – since in a coordinate basis the connection coefficients are symmetric in the lower indices – the only nonzero $\nabla_\mu \xi_\nu$ can be $\nabla_r \xi_t$ and $\nabla_t \xi_r$. Thus

$$\nabla_r \xi_t = \xi_{t,r} - \Gamma^t_{tr} \xi_t = -(\xi_{r,t} - \Gamma^t_{tr} \xi_t) = -\nabla_t \xi_r. \qquad (10.201)$$

Since $\xi_{r,t} = 0$ and

$$\Gamma^t_{tr} = +\frac{1}{2} g^{tt} g_{tt,r} \qquad (10.202)$$

we get

$$\kappa = \sqrt{-\frac{1}{2}(\nabla_\mu \xi_\nu)(\nabla^\mu \xi^\nu)} = \sqrt{-\frac{1}{4} g^{rr} g^{tt} (g_{tt,r})^2}. \qquad (10.203)$$

Evaluating this at $r = 2M$ the surface gravity of a Schwarzschild black hole is

$$\boxed{\kappa = \frac{1}{4M}.} \qquad (10.204)$$

The surface gravity of a Kerr black hole

Consider a particle with four-velocity $\mathbf{u} = u^t(\mathbf{e}_t + \omega \mathbf{e}_\phi)$. The components of its four-acceleration are $a^\mu = u^\mu_{;\nu} u^\nu$. The surface gravity is defined by

$$\kappa \equiv \lim_{r \to r_+} \frac{a}{u^t}, \quad a = (a_\mu a^\mu)^{\frac{1}{2}}. \qquad (10.205)$$

Here, r_+ is the radial coordinate of the horizon. The reason why we divide by $u^t = dt/d\tau$ is that the acceleration scalar is velocity change per unit time as measured by a clock moving along with the particle. Due to gravitational time dilatation this clock stands still at the horizon. Hence, a diverges here.

We now consider a particle moving along a path with constant r and θ, and with a constant angular velocity $\Omega = u^\phi/u^t$. The four-velocity of the particle has components

$$u^\alpha = \left[-\left(g_{tt} + 2g_{t\phi}\Omega + g_{\phi\phi}\Omega^2\right)\right]^{\frac{1}{2}} (1, \Omega). \tag{10.206}$$

Furthermore, the components of the four-acceleration are

$$a^\alpha = \left(u^\alpha{}_{,\nu} + \Gamma^\alpha{}_{\mu\nu}u^\mu\right) u^\nu. \tag{10.207}$$

Since $u^r = u^\theta = 0$, $u^t{}_{,t}$ and $\Omega = $ constant we have $u^\alpha{}_{,\nu}u^\nu = 0$. Hence, the components are

$$
\begin{aligned}
a^\alpha &= \left(\Gamma^\alpha{}_{tt} + 2\Gamma^\alpha{}_{t\phi}\Omega + \Gamma^\alpha{}_{\phi\phi}\Omega^2\right) \left(u^t\right)^2 \\
&= -\frac{1}{2}g^{\alpha\mu}\left(g_{tt,\nu} + 2g_{t\phi,\nu}\Omega + g_{\phi\phi,\nu}\Omega^2\right) \\
&= -\frac{1}{2}\left(u^t\right)^2 \left(\frac{1}{u^t}\right)^2_{,\mu} g^{\mu\alpha} = -\left(\ln u^t\right)_{,\mu} g^{\mu\alpha}. \tag{10.208}
\end{aligned}
$$

The acceleration scalar is thus

$$a = \left\{g^{rr}\left[\left(\ln u^t\right)_{,r}\right]^2 + g^{\theta\theta}\left[\left(\ln u^t\right)_{,\theta}\right]^2\right\}^{\frac{1}{2}}. \tag{10.209}$$

We now specialise to a zero-angular-momentum particle; i.e., the angular velocity is $\Omega = -g_{t\phi}/g_{\phi\phi}$. Using the expressions for the components of the metric tensor and differentiating gives

$$\left(\ln u^t\right)_{,\theta} = \frac{Mra^2(r^2 + a^2)\sin 2\theta}{\rho^2\left[(r^2 + a^2)^2 - \Delta^2\sin^2\theta\right]}, \tag{10.210}$$

$$
\begin{aligned}
\left(\ln u^t\right)_{,r} &= \frac{\rho^2(r - M)(r^2 + a^2)^2}{\Delta\rho^2\left[(r^2 + a^2)^2 - \Delta^2\sin^2\theta\right]} \\
&\quad - \frac{\left[r\Delta a^2\sin^2\theta + 2r\rho^2(r^2 + a^2) - r(r^2 + a^2)^2\right]}{\rho^2\left[(r^2 + a^2)^2 - \Delta^2\sin^2\theta\right]}. \tag{10.211}
\end{aligned}
$$

At the horizon, $\Delta = 0$, and the acceleration scalar is

$$a_+ = \frac{r_+ - M}{\rho_+\Delta^{\frac{1}{2}}}. \tag{10.212}$$

Moreover, the time component of the four-velocity at the horizon is

$$u^t_+ = \frac{r_+^2 + a^2}{\rho_+\Delta^{\frac{1}{2}}}. \tag{10.213}$$

Hence, the surface gravity of a Kerr black hole can be written

$$\boxed{\kappa = \frac{r_+ - M}{2Mr_+} = \frac{\sqrt{M^2 - a^2}}{2M\left(M + \sqrt{M^2 - a^2}\right)}.} \tag{10.214}$$

The Four Laws of Black Hole Thermodynamics

The expression (10.214) shows that the surface gravity has no angular dependence. The zeroth law of black hole thermodynamics follows immediately.

- **0th law:** κ *is constant over the horizon of a black hole.*

The first law of black hole thermodynamics is an expression of the energy conservation formulated in a similar way as the first law in ordinary thermodynamics. From eqs. (10.168) and (10.169) follow that the area of the horizon is

$$A = 8\pi \left(M^2 - \sqrt{M^4 - J^2} \right). \tag{10.215}$$

Due to a variation dM of its mass and dJ of its spin, the horizon area of a Kerr black hole changes by

$$dA = 8\pi \left(2 \frac{M\sqrt{M^4 - J^2} + M^3}{\sqrt{M^4 - J^2}} dM - \frac{J}{\sqrt{M^4 - J^2}} dJ \right). \tag{10.216}$$

Inserting the expression (10.214) of the surface gravity, and the angular velocity at the horizon given by

$$\Omega \equiv \omega(r_+) = \frac{a}{r_+^2 + a^2} = \frac{J\kappa}{\sqrt{M^4 - J^2}}, \tag{10.217}$$

eq. (10.216) takes the form

$$dA = \frac{8\pi}{\kappa} \left(dM - \Omega dJ \right). \tag{10.218}$$

This may be written

- **1st law:**

$$\boxed{dM = \frac{\kappa}{8\pi} dA + \Omega dJ,} \tag{10.219}$$

or in S.I. units

$$d(mc^2) = \frac{\kappa c^2}{8\pi G} dA + \Omega dJ. \tag{10.220}$$

This is the 1st law of black hole thermodynamics. ΩdJ is the work performed upon a black hole when its spin changes by dJ. Comparing with the first law of ordinary thermodynamics,

$$dU = TdS + dW, \tag{10.221}$$

Bekenstein [Bek74] tentatively suggested that one can associate a temperature T and entropy S with a black hole, such that $T \propto \kappa$ and $S \propto A$.

We shall go on and deduce the black hole analogue of the second law of black hole thermodynamics. We consider a free particle moving into a Kerr black hole. Since the Kerr metric is independent of the angular coordinate ϕ, the momentum p_ϕ of the particle is a constant of motion. This is utilized by writing

$$\mathbf{p} \cdot \mathbf{p} = g^{\mu\nu} p_\mu p_\nu = -m^2, \tag{10.222}$$

where m is the mass of the particle. Writing out this equation we get

$$-e^{-2\nu}E^2 + e^{-2\nu}2\omega p_\phi E + \left(e^{-2\psi} - \omega^2 e^{-2\nu}\right)p_\phi^2$$
$$+e^{-2\mu}p_r^2 + e^{-2\lambda}p_\theta^2 \;=\; -m^2, \quad (10.223)$$

where $E = -p_t$ is the constant energy of the particle. The solution of this equation corresponding to $E = +m$ for a particle at rest in the asymptotic far-away region is

$$E = \omega p_\phi + e^\nu \left(e^{-2\phi}p_\phi^2 + e^{-2\mu}p_r^2 + e^{-2\lambda}p_\theta^2 + m^2\right)^{\frac{1}{2}}. \quad (10.224)$$

Absorbing the particle the mass of the black hole changes by $\delta M = E$ and its spin by $\delta J = p_\phi$. Both δM and δJ may be either positive or negative. Since the particle has to pass through the horizon at $r = r_+$, we can calculate its energy by putting $r = r_+$ in eq. (10.224). On the horizon we also have $e^\nu = 0$. Hence, only terms in the square root that diverge at the horizon will contribute. The only such term is $e^{-2\mu}p_r^2 = m(\rho^2/\Delta)\dot{r}^2$ since $\Delta = 0$ at the horizon. Thus,

$$\delta M = \omega(r_+)\delta J + m\left(\frac{\rho^2}{\Sigma}|\dot{r}|\right)_{r_+}, \quad (10.225)$$

where $\omega(r_+) = a/(r_+^2 + a^2)$. This gives

$$\delta M = \frac{a\delta J + r_+^2}{r_+^2 + a^2} + \frac{a^2\cos^2\theta}{r_+^2 + a^2}m|\dot{r}|_{r_+}. \quad (10.226)$$

The change of mass is smallest if $\dot{r} = 0$ at the horizon. In this case the process is called *reversible*. Hence, for a reversible process,

$$M dM = \frac{J dJ}{r_+^2 + a^2}. \quad (10.227)$$

Using $r_+^2 + a^2 = 2M\left(M + \sqrt{M^2 - a^2}\right)$, the above equation takes the form

$$M dM = \frac{J dJ}{2M\left(M + \sqrt{M^2 - a^2}\right)}. \quad (10.228)$$

Integrating and rearranging gives

$$M = \frac{M_I}{\sqrt{1 - \frac{a^2}{4M_I^2}}}, \quad (10.229)$$

where M_I is a constant of integration. From eq. (10.229) it is seen that M_I is the mass of a black hole with $a = 0$; i.e. a non-rotating black hole. M_I is called the *irreducible mass* of a Kerr black hole since it is the mass that remains when all the rotational energy of the black hole is extracted by means of the Penrose process.

Inverting eq. (10.229) gives

$$M_I^2 = \frac{1}{2}\left(M^2 + \sqrt{M^4 - J^2}\right). \quad (10.230)$$

Hence,

$$M_I \delta M_I = \frac{(r_+^2 + a^2) M \delta M - J \delta J}{4\sqrt{M^4 - J^2}}. \tag{10.231}$$

For reversible processes dM is given by eq. (10.228) which implies $dM_I = 0$. For irreversible processes $\delta M_I > 0$. The irreducible mass of a black hole cannot decrease by any non-quantum mechanical process. Eq. (10.230) may be written

$$M_I^2 = \frac{1}{4}\left(r_+^2 + a^2\right) = \frac{A}{16\pi}. \tag{10.232}$$

Then we can state *the second law of black hole thermodynamics*:

- **2nd law:** *No classical process can make the horizon area of a black hole decrease.*

The third law of ordinary thermodynamics states that no system in thermodynamic equilibrium can have negative temperature. The corresponding law of black hole thermodynamics is an expression of the existence of a *cosmic censorship: No naked singularity with $J > M^2$ may exist.*

This follows from the expressions of the surface gravity and the horizon. A black hole with $T = 0$ has $\kappa = 0$; hence, $r_+ = M$. This corresponds to an extreme Kerr black hole with $J = M^2$. If $J^2 > M$ then κ and T would be negative, and the horizon would vanish. This is not possible according to the third law of black hole thermodynamics.

Hawking radiation from a black hole

The tentative formulation of black hole thermodynamics by J. Bekenstein got physical contents through a discovery by S.W. Hawking [Haw75]. Applying quantum field theory to the curved spacetime of a black hole he found that the black hole emits electromagnetic radiation with a temperature

$$\boxed{T = \frac{\hbar \kappa}{2\pi k_B c}}, \tag{10.233}$$

where $\hbar$ is the reduced Planck constant and k_B is the Boltzmann constant. In the case of a Schwarzschild black hole this expression reduces to

$$T = \frac{\hbar c^3}{8\pi G k_B m}. \tag{10.234}$$

Inserting the values for the constants gives $T = 10^{-7} \, (m_{\text{Sun}}/m) \, \text{K}$, where m_{Sun} is the mass of the Sun. The formula shows that the temperature of a black hole with mass like that of the Sun is extremely low. However, the temperature increases with decreasing mass. Hence, a black hole has negative heat capacity; giving away mass by radiation increases its temperature.

The energy loss when it radiates is given by the *Stefan-Boltzmann law*,

$$-\frac{\dot{E}}{A} = \sigma T^4, \tag{10.235}$$

where σ is Stefan-Boltzmann's constant. Integration of this equation is left to problem 10.4. It yields the following mass as a function of time:

$$m(t) = \left(m_0^3 - 3Kt\right)^{\frac{1}{3}}, \quad K = \frac{\hbar c^4}{15360\pi G^2}, \tag{10.236}$$

where $m_0 \equiv m(0)$. At a point of time, $t_1 = m_0^3/3K$ the black hole vanishes in a great flash. Hawking speculated whether we might observe such flashes from mini-black holes created shortly after the big bang and exploding now. Putting $t_1 = t_0 = 10^{18}$s, the age of the universe, we find $m_0 = 10^{12}$kg (see problem 10.4). For such black holes, we can write

$$m(t) = \left(1 - \frac{t}{t_0}\right)^{\frac{1}{3}} m_0. \tag{10.237}$$

Let Δt be the time interval from an arbitrary point of time t to the hole has exploded at t_0. Then $t = t_0 - \Delta t$, which gives

$$m(t) = \left(\frac{\Delta t}{t_0}\right)^{\frac{1}{3}} m_0. \tag{10.238}$$

Inserting $\Delta t = 1$s gives $m = 10^6$kg. During the last second the black hole radiates energy amounting to $mc^2 = 10^{23}$J. Hence, the average effect during the last second is 10^{23}W.

From eqs. (10.220), (10.221) and (10.233) the black hole possesses an entropy, S, given by

$$\boxed{S_{BH} = \frac{1}{4}\frac{k_B c^3}{G\hbar}A.} \tag{10.239}$$

This black hole entropy comes in addition to the ordinary entropy of the matter. This entropy completes the picture of the black hole as a thermodynamically interacting system.

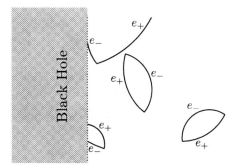

Figure 10.11: Hawking radiation: particle-anti-particle pair production in the neighbourhood of a black hole.

The radiation from a black hole has the name *Hawking Radiation* and is due to random processes in the quantum fields near the horizon (see Fig. 10.11). A striking thing about the radiation and the property of the black hole itself is that the black hole is completely determined by three parameters: the mass M, the charge Q and the angular momentum J. Thus the black hole is emitting equal amounts of matter and antimatter! If a star in our universe which almost entirely consists of matter (and not antimatter) is collapsing into a black hole huge amounts of information is lost in this process.

10.8   The Tolman-Oppenheimer-Volkoff equation

Until now we have only considered solutions for empty space (except the Reissner-Nordström black hole). An equally interesting task is to study solutions of the Einstein field equations in, for example, the interior of stars. Since the interior of stars is a highly complex system we have to do quite a few simplifications. In spite of these simplifications, some of the results obtained are quite fascinating and interesting; for example, they provide an upper limit on the mass of a star for it to avoid collapse to a black hole.

We consider the Einstein field equations inside a static, spherically symmetric distribution of perfect fluid. The line-element can be written

$$ds^2 = -e^{2\alpha}dt^2 + e^{2\beta}dr^2 + r^2(d\theta^2 + \sin^2\theta d\phi^2), \tag{10.240}$$

where $\alpha = \alpha(r)$ and $\beta = \beta(r)$. This is the same form as for the exterior Schwarzschild solution. In an orthonormal frame Einstein's field equations are

$$E_{\hat{\mu}\hat{\nu}} = 8\pi G T_{\hat{\mu}\hat{\nu}}. \tag{10.241}$$

The left hand side of these equations has already been calculated, while the right-hand side is diagonal

$$T_{\hat{\mu}\hat{\nu}} = \text{diag}(\rho, p, p, p), \tag{10.242}$$

where $\rho = \rho(r)$ and $p = p(r)$. The $\hat{t}\hat{t}$-component is

$$\frac{1}{r^2}\frac{d}{dr}\left[r\left(1 - e^{-2\beta}\right)\right] = 8\pi G\rho. \tag{10.243}$$

Introducing the mass inside a spherical shell of coordinate radius r by

$$m(r) = \int_0^r 4\pi\rho(r)r^2 dr, \tag{10.244}$$

the solution can be written as

$$e^{-2\beta} = (g_{rr})^{-1} = 1 - \frac{2Gm(r)}{r}. \tag{10.245}$$

Comparing this with the vacuum case, we see that this is of a similar form, except that the mass in this case is r-dependent.

From the $\hat{r}\hat{r}$-equations we get

$$\frac{2}{r}\alpha' e^{-2\beta} - \frac{1}{r^2}\left(1 - e^{-2\beta}\right) = 8\pi p. \tag{10.246}$$

Inserting the solution for β and rearranging, we end up with the equation for α:

$$\frac{d\alpha}{dr} = G\frac{m(r) + 4\pi r^3 p(r)}{r(r - 2Gm(r))}. \tag{10.247}$$

To relate the $p(r)$ and the $\rho(r)$ we can use that the energy-momentum tensor has to be divergence free, i.e. $T^{\hat{\mu}}{}_{\hat{\nu};\hat{\mu}} = 0$. If we let $\hat{\nu} = \hat{r}$, then

$$T^{\hat{\mu}}{}_{\hat{r};\hat{\mu}} = T^{\hat{\mu}}{}_{\hat{r},\hat{\mu}} - \Gamma^{\hat{\rho}}{}_{\hat{r}\hat{\mu}}T^{\hat{\mu}}{}_{\hat{\rho}} + \Gamma^{\hat{\rho}}{}_{\hat{\mu}\hat{\rho}}T^{\hat{\mu}}{}_{\hat{r}} = 0. \tag{10.248}$$

The first term of this equation is simply

$$T^{\hat{\mu}}_{\ \hat{r},\hat{\mu}} = T^{\hat{r}}_{\ \hat{r},\hat{r}} = e^{-\beta}\frac{dp}{dr}. \tag{10.249}$$

Using the connection forms eq. (10.11) we can write

$$
\begin{aligned}
-\Gamma^{\hat{\rho}}_{\ \hat{r}\hat{\mu}}T^{\hat{\mu}}_{\ \hat{\rho}} + \Gamma^{\hat{\rho}}_{\ \hat{\mu}\hat{\rho}}T^{\hat{\mu}}_{\ \hat{r}} &= -\Gamma^{\hat{\rho}}_{\ \hat{r}\hat{\rho}}T^{\hat{\rho}}_{\ \hat{\rho}} + \Gamma^{\hat{\rho}}_{\ \hat{r}\hat{\rho}}T^{\hat{r}}_{\ \hat{r}} \\
&= -\Omega^{\hat{\rho}}_{\ \hat{r}}(e_{\hat{\rho}})T^{\hat{\rho}}_{\ \hat{\rho}} + \Omega^{\hat{\rho}}_{\ \hat{r}}(e_{\hat{\rho}})T^{\hat{r}}_{\ \hat{r}} \\
&= e^{-\beta}(p+\rho)\frac{d\alpha}{dr}. \tag{10.250}
\end{aligned}
$$

So,

$$\frac{dp}{dr} + (p+\rho)\frac{d\alpha}{dr} = 0. \tag{10.251}$$

Inserting the equation for $\frac{d\alpha}{dr}$ we get the *Tolman-Oppenheimer-Volkoff (TOV) equation*:

$$\boxed{\frac{dp}{dr} = -G(p+\rho)\frac{m(r)+4\pi r^3 p(r)}{r(r-2Gm(r))}.} \tag{10.252}$$

In the Newtonian limit $(p \ll \rho,\ Gm(r) \ll r)$ the TOV equation reduces to the equation of hydrostatic equilibrium

$$\frac{dp}{dr} \approx -G\frac{\rho m(r)}{r^2}. \tag{10.253}$$

In order to see more clearly how the relativistic corrections appear in the TOV-equation we may write it in the form

$$\frac{dp}{dr} = -G\frac{\rho m(r)}{r^2}\left(1+\frac{p}{\rho}\right)\left(1+3\frac{p}{\bar{\rho}}\right)\left(1-\frac{r_S}{r}\right)^{-1}, \tag{10.254}$$

where $\bar{\rho} = (3/r^3)\int_0^r \rho(r)r^2 dr$, and $r_S = 2Gm(r)$ is the Schwarzschild radius of the mass inside r.

Note that the relativistic correction factors are all greater than one. This means that the relativistic gravity is stronger than Newtonian gravity at any r.

10.9   The interior Schwarzschild solution

Let us now consider an incompressible star with radius R, i.e. we consider a density distribution

$$\rho(r) = \rho = \text{constant} \tag{10.255}$$

for $r \leq R$. The mass function then becomes

$$m(r) = \frac{4}{3}\pi\rho r^3. \tag{10.256}$$

In the Newtonian limit the equation for the pressure yields

$$p_N(r) = \frac{2}{3}\pi G\rho^2(R^2 - r^2), \tag{10.257}$$

where the boundary condition $p(R) = 0$ has been imposed. Thus Newton's theory puts no upper bound on the mass of a star.

In the relativistic case, we must use the TOV equation. This equation can also be integrated exactly (which was first done by Schwarzschild in 1916) to yield the result

$$p(r) = \rho \left[\frac{\sqrt{1 - \frac{R_S}{R}} - \sqrt{1 - \frac{R_S r^2}{R^3}}}{\sqrt{1 - \frac{R_S r^2}{R^3}} - 3\sqrt{1 - \frac{R_S}{R}}} \right], \tag{10.258}$$

where $R_S = 2GM$ is the Schwarzschild radius of the star. This expression is valid for $r \leq R$. The central pressure is given by

$$p_c = p(0) = \rho \left[\frac{\sqrt{1 - \frac{R_S}{R}} - 1}{1 - 3\sqrt{1 - \frac{R_S}{R}}} \right]. \tag{10.259}$$

Note that p_c becomes negative when

$$1 - 3\sqrt{1 - \frac{R_S}{R}} < 0. \tag{10.260}$$

This means that the central region of the star collapses. The star will therefore collapse under its own gravity. Hence, according to the general theory of relativity, the requirement of hydrostatic equilibrium puts the bound

$$R > \frac{9}{8} R_S \tag{10.261}$$

for a star. This leads to the following restriction on the mass of a star with constant density and radius R,

$$m < \frac{8}{9} \frac{R}{R_{S,\text{Sun}}} m_{\text{Sun}}, \tag{10.262}$$

where $R_{S,\text{Sun}} = 3\text{km}$ and m_{Sun} are the Schwarzschild radius and mass of the Sun. For a neutron star with $R = 10\text{km}$, say, this gives $m < 3m_{\text{Sun}}$.

Eq. (10.251) can formally be solved in the general case

$$\alpha = -\int \frac{dp}{p + \rho}. \tag{10.263}$$

By scaling the time variable, the integration constant can be changed to an arbitrary value. In the case of an incompressible fluid, $\rho = $ constant, the integral leads to

$$e^\alpha = \frac{C}{\rho + p}. \tag{10.264}$$

Inserting the expression for the pressure and choosing the integration constant such that it matches with the exterior solution we get

$$e^\alpha = \frac{3}{2}\sqrt{1 - \frac{R_S}{R}} - \frac{1}{2}\sqrt{1 - \frac{R_S}{R^3} r^2}. \tag{10.265}$$

Hence, the line-element for the interior Schwarzschild solution is

$$ds^2 = -\left(\frac{3}{2}\sqrt{1 - \frac{R_S}{R}} - \frac{1}{2}\sqrt{1 - \frac{R_S}{R^3}r^2}\right)^2 dt^2 + \frac{dr^2}{1 - \frac{R_S}{R^3}r^2}$$
$$+ r^2(d\theta^2 + \sin^2\theta d\phi^2).\tag{10.266}$$

We shall now deduce how the Newtonian expression for the gravitational self-energy of a solid sphere with constant mass density is generalised relativistically. Let us first consider the Newtonian expression. Adding the self-energy of spherical shells surrounding a solid sphere we obtain

$$U_N = -G\int_0^R 4\pi r\rho(r)m(r)dr = -4\pi G\rho\int_0^R r\frac{4\pi}{3}\rho r^3 dr^2 = -G\frac{16\pi^2\rho^2}{15}R^5$$
$$= -\frac{3}{5}G\frac{M^2}{R}.\tag{10.267}$$

Relativistically, $M = \int_0^R 4\pi r^2\rho(r)dr$ is the mass of a star acting upon an asymptotically far-away observer. This mass or energy is not invariant; it includes the gravitational self-energy of the star. The energy M corresponds to $-p_t$ for a particle, which includes the potential energy due to its position in a gravitational field.

The invariant energy of the star is

$$E = \int_0^R T_{\hat{t}\hat{t}}\omega^{\hat{\phi}} \wedge \omega^{\hat{\theta}} \wedge \omega^{\hat{r}} = 4\pi\int_0^R r^2\rho(r)e^{\beta(r)}dr.\tag{10.268}$$

This energy does not include the self-energy, U. Since $M = E + U$ and $U < 0$ it follows that $M < E$. the energy E corresponds to $E_{\text{local}} = -\mathbf{p}\cdot\mathbf{u}$ for a particle with four-momentum $\mathbf{p}$ measured locally (at the position of the observer) by an observer with four-velocity $\mathbf{u}$. E_{local} does not include the potential energy of the particle.

For an incompressible star we get

$$M = \int_0^R 4\pi r^2\rho dr = \frac{4\pi}{3}\rho R^3,\tag{10.269}$$

$$E = \int_0^R \frac{4\pi r^2\rho dr}{\sqrt{1 - \frac{R_S}{r^3}}} = \frac{3}{2}\frac{R}{R_S}\left(\sqrt{\frac{R}{R_S}}\arcsin\sqrt{\frac{R_S}{R}} - \sqrt{1 - \frac{R_S}{R}}\right)M.\tag{10.270}$$

Making a series expansion of E to third order in R_S/R we get for $R \gg R_S$

$$E \approx M + \frac{3}{5}G\frac{M^2}{R},\tag{10.271}$$

which represents the Newtonian result. In the limiting case $R = R_S$ we get $E = (3\pi/4)M$ and a self-energy $U = -(3\pi/4 - 1)M$.

10.10   Relativistic gravitation versus Newtonian gravitation

It is often said that in the theory of relativity gravity is described as curved space-time. This is, however, not the whole truth. In this section we shall highlight the different conceptual contents of Newtonian gravity and the general theory of relativity.

Newton's theory is based upon the concepts of absolute space and time. Space obeys the rules of Euclidean geometry independently of the properties of matter and energy filling the space. Time has an absolute character, proceeding at a rate which does not dependent upon motion and matter. Furthermore, the concept inertial frame is essential. Inertial frames are defined as those frames in which Newton's 1st law is valid. Such frames have a global character according to Newton's theory. Different inertial frames are connected by the Galilean transformation, and the Newtonian dynamics is Galilei invariant. However, acceleration and rotation of a reference frame introduces fictitious forces such as the centrifugal force and the Coriolis force in in a rotating reference frame. Therefore Newtonian dynamics is not Galilei invariant. Consequently acceleration and rotation has an absolute character in Newtonian theory.

According to Newton's theory of gravity there exists a gravitational force acting between masses. Since mass is positive this force is always attractive. As shown in chapter 1 (eqs. (1.32) and (1.33)) the dynamics of Newton's theory of gravitation may be summarized by the statement: Mass generates a gravitational field according to Poisson's equation, and the gravitational field generates acceleration according to Newton's 2nd law. The acceleration of gravity experienced on the surface of the Earth is a result of the gravitational force due to the Earth. Furthermore inhomogeneities in the gravitational field produce tidal forces as given by eq. (1.38).

The concepts of space and time are unified in a four-dimensional space-time according to the general theory of relativity. Hence, space and time do not have an independent existence. Furthermore both space-time and space can be curved, and their properties depend upon the matter and energy-contents in the universe. In Einstein's theory the concept inertial frame may still be defined as a frame in which Newton's 1st law is valid. A body at rest, and not acted upon by any force, should remain at rest. Gravity is treated very differently in Newton's and Einstein's theories. According to the general theory of relativity there is no gravitational force. The principle of equivalence says that the physical effects of permanent gravitational fields and those due to the acceleration or rotation of the reference frame are equivalent.

Mathematically the acceleration of gravity comes from certain Christoffel symbols, as seen by applying the geodesic equation to a particle instantaneously at rest. They can be transformed away by introducing a co-moving coordinate system in a local non-rotating and freely falling reference frame. Hence in such a frame the acceleration of gravity vanishes. A free particle in such a frame obeys Newton's 1st law. Therefore, in the theory of relativity inertial reference frames are defined as local non-rotating freely falling frames. Hence, according to the general theory of relativity, the acceleration of gravity comes from staying in a non-inertial frame of reference.

The relativistic equation which generalizes the Newtonian equation (1.38) for tidal forces is the equation (7.106) for geodesic deviation. Comparison of these equations shows that the Newtonian tidal force is perceived as space-time curvature in the theory of relativity. This is mathematically represented by a tensor and cannot be transformed away.

Relativistic gravity has two important properties that vanish in the Newtonian limit. The first one is the phenomenon of inertial dragging, also called the Lense-Thirring effect, which was described in sections 9.4 and 10.6. This has a Machian character, and its cosmological significance will be taken up in section 12.11. It means that the average motion of the cosmic masses may determine the motion of the swinging plane of the Foucault pendulum at, say, the South Pole, such that it does not rotate relative to the stars (neglecting local effects due to the rotation of the Earth). Hence the state of "non-rotation" of inertial frames may be determined mainly by the cosmic masses. The question whether this effect makes it possible to extend the principle of relativity to encompass accelerated and rotating motion, is still not settled [GV99].

The second fundamental relativistic property of gravitation is that it depends not only upon masses, but also upon pressure or strain as we saw in section 10.8 and as is further explored in problem 10.16. Pressure increases the attractive gravity, and strain gives a repulsive contribution. Hence in the theory of relativity there is a possibility of repulsive gravity. In chapter 12 we will see that this is of fundamental significance for the large scale dynamics of the universe.

Problems

10.1. *The Schwarzschild metric in Isotropic coordinates*
We will in this problem find the Schwarzschild solution expressed in isotropic coordinates. The metric in isotropic coordinates has the form

$$ds^2 = -e^{2A}dt^2 + e^{2B}(dR^2 + R^2(d\theta^2 + \sin^2\theta d\phi^2)), \qquad (10.272)$$

where $A = A(R)$ and $B = B(R)$.

(a) Show that the transformation

$$r = \left(1 + \frac{M}{2R}\right)^2 R \qquad (10.273)$$

transforms the Schwarzschild metric from Schwarzschild coordinates to the isotropic form.

(b) Calculate the Schwarzschild metric in isotropic coordinates by solving the Einstein field equations.

10.2. *Embedding of the interior Schwarzschild metric*
Make an embedding of the three-dimensional spatial section $t = 0$ of the internal Schwarzschild solution. Join the resulting surface to the corresponding embedding of the external Schwarzschild solution.

10.3. *The Schwarzschild-de Sitter metric*
In this problem we will solve the Einstein equations with a cosmological constant. The Einstein equations with a cosmological constant Λ can be written

$$R_{\mu\nu} - \frac{1}{2}Rg_{\mu\nu} + \Lambda g_{\mu\nu} = \kappa T_{\mu\nu} \qquad (10.274)$$

(a) Use Schwarzschild coordinates and solve the Einstein vacuum equations with a cosmological constant.

(b) Show that in this case there are two horizons. Set the mass parameter equal to zero, and show that the spatial sections $dt = 0$ can be considered as a 3-sphere, S^3.

10.4. *The life time of a black hole*
Consider a black hole in a zero-temperature heat-bath. Including c, $\hbar$, k_B and G we have $M = Gmc^{-2}$, the black hole temperature is

$$T = \frac{\hbar}{2\pi k_B c}\kappa \tag{10.275}$$

and the Stefan-Boltsmann constant is

$$\sigma = \frac{\pi^2 k_B^4}{60\hbar^3 c^2} \tag{10.276}$$

(a) Assume that a black hole has a mass m_0 at $t = 0$. Find $m(t)$.

(b) If we consider a time span of approximately $\tau = 10^{10}$ years, what mass would the black hole have at $t = 0$ to have $m(\tau) = 0$?

(c) Show that a black hole of mass m cannot disintegrate into two smaller black holes of mass m_1 and m_2 where $m = m_1 + m_2$.

10.5. *A spaceship falling into a black hole*

(a) In this problem we will consider a spaceship (A) falling radially into a Schwarzschild black hole with mass $M = 5M_{\text{Sun}}$ (set $c = 1$). What is the Schwarzschild radius of the black hole? Find the equations of motion of the spaceship in Schwarzschild coordinates r and t, using the proper time τ as time parameter. At the time $t = \tau = 0$ the spaceship is located at $r = 10^{10}M$. The total energy is equal to its rest energy. Solve the equations of motion with these initial conditions. When (in terms of τ) does the spaceship reach the Schwarzschild radius? And the singularity?

(b) Show that the spaceship, from the Schwarzschild radius to the singularity, uses maximally $\Delta\tau = \pi GM$ no matter how it is maneuvered. How should the spaceship be maneuvered to maximalise this time?

(c) The spaceship (A) has radio contact with a stationary space-station (B) at $r_B = 1$ light years. The radio-signals are sent with intervals ΔT and with frequency ω from both A and B. The receivers at A and B receive signals with frequency ω_A and ω_B, respectively. Find ω_A and ω_B as a function of the position of the spaceship. Investigate whether something particular is happening as the spaceship passes the Schwarzschild radius. Discuss what these results tell us about how the events in the spaceship is described in the space station, and vice versa.

10.6. *The GPS Navigation System*
GPS uses a network of 24 satellites orbiting the Earth about 20 000 km above the ground with an orbital speed of 20 000 km/hour. On board each satellite is an atomic clock that ticks with an accuracy of 1 nanosecond. A GPS receiver determines its current position by comparing the time signals it receives from several of the satellites and triangulating on the known positions of each. The receiver can determine your position to within 10 meters in only a few seconds. To achieve this level of precision, the clocks on the GPS satellites must

be known to an accuracy of 30 nanoseconds. Determine, by the following calculations, whether it is necessary to take account of the kinematical- and the
gravitational time dilation in the construction of this system.

(a) How many nanoseconds does the clock on a satellite lag behind in 24
hours due to the velocity of the satellite?

(b) How many nanoseconds does it lie ahead due to the height of the satellite?

(c) Use eq. (10.58) to calculate the error which is made in 24 hours by neglecting relativistic effects upon the rate of the satellite clocks.

10.7. *Physical interpretation of the Kerr metric*
In this problem we shall use the linearised solution of the spacetime outside a
rotating shell derived in problem 9.3.

(a) Show that the Kerr metric (10.165) is reduced to the metric (9.146) in the
limit $r > R$ and $r \gg M$ and identify thereby the constant a with the
angular momentum per unit mass of the rotating shell. (Hint: Expand
the Kerr metric to first order in J/Mr, introduce isotropic coordinates
($r \to \rho$ see problem 10.1), and expand the result to first order in M/ρ).

(b) Find the angular velocity

$$\omega_L = -\frac{g_{0\phi}}{g_{\phi\phi}} \tag{10.277}$$

that local reference frames are rotating with, with respect to reference
frames at infinity.

10.8. *A gravitomagnetic clock effect*
This problem is concerned with the difference of proper time shown by two
clocks moving freely in opposite directions in the equatorial plane of the Kerr
spacetime outside a rotating body. The clocks move along a path with $r =$
constant and $\theta = \pi/2$.

(a) Show that in this case the radial geodesic equation reduces to

$$\Gamma^r{}_{tt}dt^2 + 2\Gamma^r{}_{\phi t}d\phi dt + \Gamma^r{}_{\phi\phi}d\phi^2 = 0.$$

(b) Calculate the Christoffel symbols and show that the equation takes the
form

$$\left(\frac{dt}{d\phi}\right)^2 - 2a\frac{dt}{d\phi} + a^2 - \frac{r^3}{M} = 0,$$

where M is the mass of the rotating body and a its angular momentum
per unit mass, $a = J/M$.

(c) Use the solution of the geodesic equation and the four-velocity identity
to show that the proper time interval $d\tau$ shown on a clock moving an
angle $d\phi$ is

$$d\tau = \pm\sqrt{1 - \frac{3M}{r}} \pm 2a\omega_0 d\phi,$$

where $\omega_0 = (M/r^3)^{1/2}$ is the angular velocity of a clock moving in the
Schwarzschild spacetime in accordance with Kepler's 3rd law. The plus
and minus sign apply to direct and retrograde motion, respectively.

(d) Show that to first order in a the proper time difference for one closed orbit ($\phi \to \phi + 2\pi$) in the direct and the retrograde direction is $\tau_+ - \tau_- \approx 4\pi a = 4\pi J/M$, or in S.I. units, $\tau_+ - \tau_- \approx 4\pi a = 4\pi J/mc^2$.

Estimate this time difference for clocks in satellites moving in the equatorial plane of the Earth. (The mass of the Earth is $m = 6 \cdot 10^{26}$kg and its angular momentum $J = 10^{34}$kg m^{2}s^{-1}.)

10.9. *The photon sphere radius of a Reissner-Nordström black hole*
Show that there exists a sphere of radius

$$r_{PS} = \frac{3M}{2}\left(1 + \sqrt{1 - \frac{8Q^2}{9M^2}}\right) \qquad (10.278)$$

in the Reissner-Nordström black hole spacetime where photons will have circular orbits around the black hole.

10.10. *Curvature of 3-space and 2-surfaces of the internal and the external Schwarzschild spacetimes*

(a) The 3-space of the internal Schwarzschild solution has a geometry given by the line-element

$$d\ell_I^2 = \frac{dr^2}{1 - \frac{R_S}{R^2}r^2} + r^2(d\theta^2 + \sin^2\theta d\phi^2)$$

where $R_S = 2M$ is the Schwarzschild radius of the mass distribution and R its radius. The corresponding line-element for the external Schwarzschild solution is

$$d\ell_E^2 = \frac{dr^2}{1 - \frac{R_S}{r}} + r^2(d\theta^2 + \sin^2\theta d\phi^2)$$

Find the spatial curvature $k = k(r) = \frac{1}{6}R$ of the 3-spaces, where R is the Ricci scalar.

(b) We shall now consider the equatorial surfaces $\theta = \pi/2$. The line-elements of these surfaces are, for the internal solution

$$d\sigma_I^2 = \frac{dr^2}{1 - \frac{R_S}{R^2}r^2} + r^2 d\phi^2,$$

and for the external solution

$$d\sigma_E^2 = \frac{dr^2}{1 - \frac{R_S}{r}} + r^2 d\phi^2.$$

For these line-elements the Gaussian curvatures of the surfaces they describe are given by

$$K = -\frac{1}{2g}g'_{\phi\phi} + \frac{g_{\phi\phi}}{4g^2}g'_{rr}g'_{\phi\phi} + \frac{g_{rr}}{4g^2}\left(g'_{\phi\phi}\right)^2,$$

where $g = g_{rr}g_{\phi\phi}$ and differentiation is with respect to r. Show that the Gaussian curvature of the equatorial surfaces are for
The internal solution: $K = R_S/R^3$. What sort of surface is this?
The external solution: $K = -(1/2)(R_S/r^3)$.

(c) The equatorial surfaces shall now be compared to the embedding surfaces. The Gaussian curvature of a surface of revolution given by $z = z(r)$, is

$$K = \frac{z'z''}{r(1 + z'^2)^2}.$$

Calculate the Gaussian curvatures of the embedding surfaces of the internal Schwarzschild solution, as given in problem 10.2, and of the external solution, as given in eq. (10.99). Compare the results with those of the previous point.

10.11. *Proper radial distance in the external Schwarzschild space*
Show that the proper radial distance from a coordinate position r to the horizon R_S in the external Schwarzschild space is

$$\ell_r = \sqrt{r}\sqrt{r - R_S} + R_S \ln\left(\sqrt{\frac{r}{R_S}} - \sqrt{\frac{r}{R_S} - 1}\right).$$

Find the limit of this expression for $R_S \ll r$.

10.12. *Gravitational redshift in the Schwarzschild spacetime*
Define z, describing the redshift of light, by

$$z = \frac{\Delta\lambda}{\lambda_e}, \tag{10.279}$$

where $\Delta\lambda$ is the change in the photons wavelength and λ_e the wavelength of the photon when emitted.

Show that the gravitational redshift of light emitted at r_E and received at r_R in the Schwarzschild spacetime outside a star of mass M is

$$z = \left(\frac{r_R - R_S}{r_E - R_S}\right)^{\frac{1}{2}} - 1,$$

where $R_S = 2M$ is the Schwarzschild radius of the star. What is the gravitational redshift of light emitted from the surface of a neutron star as observed by a faraway observer? A neutron star has typically a mass of 1.2 solar masses and a radius of about 20km.

10.13. *The Reissner-Nordström repulsion*
Consider a radially infalling neutral particle in the Reissner-Nordström spacetime with $M > |Q|$. Show that when the particle comes inside the radius $r = Q^2/M$ it will feel a repulsion away from $r = 0$ (i.e. that $d^2r/d\tau^2 < 0$ for τ the proper time of the particle). Is this inside or outside the outer horizon r_+? Show further that the particle can never reach the singularity at $r = 0$.

10.14. *Light-like geodesics in the Reissner-Nordström spacetime*
We will in this problem consider radial photon paths in the Reissner-Nordström spacetime. The horizons of this spacetime are at $r_\pm = M \pm \sqrt{M^2 - Q^2}$, and we will assume that $M > |Q|$.

(a) Show that the radial light rays obey the differential equation

$$\frac{dr}{dt} = \pm\left(1 - \frac{2M}{r} + \frac{Q^2}{r^2}\right).$$

(b) It is convenient to introduce two *null coordinates* u and v by

$$
\begin{aligned}
u &= t - r^* \\
v &= t + r^*,
\end{aligned}
$$ (10.280)

where

$$r^* = \int \frac{dr}{1 - \frac{2M}{r} + \frac{Q^2}{r^2}}.$$

Show that u is a constant of motion for outgoing photons, while v is a constant of motion for ingoing photons. Show further that

$$r^* = r + \frac{r_+^2}{r_+ - r_-} \ln\left[\frac{1}{2M}|r - r_+|\right] + \frac{r_-^2}{r_+ - r_-} \ln\left[\frac{1}{2M}|r - r_-|\right]. \quad (10.281)$$

(c) Draw the light-cones in the tr-plane for the three regions $r < r_-,\, r_- < r < r_+$ and $r_+ < r$.

10.15. *The Jebsen-Birkhoff theorem*

We will in this problem consider a spherically symmetric metric describing the spacetime external to some region. We will first assume that the metric is time dependent, but will show that, under some assumptions, that this cannot be possible.

A spherically symmetric metric outside a source can always be put onto the canonical form

$$ds^2 = -e^{2\alpha(r,t)}dt^2 + e^{2\beta(r,t)}dr^2 + r^2(d\theta^2 + \sin^2\theta d\phi^2). \quad (10.282)$$

Assume also that the spacetime is asymptotically flat; i.e.

$$\lim_{r\to\infty} \alpha(r,t) = \lim_{r\to\infty} \beta(r,t) = 0.$$

(a) Outside some r_0 we have $T_{\mu\nu} = 0$. Denote the derivative $\frac{\partial}{\partial r}$ with a prime and $\frac{\partial}{\partial t}$ with a dot. Show that Einstein's field equations in vacuum (for $r > r_0$) can be written as

$$e^{-2\beta}\left(\frac{2\alpha'}{r} + \frac{1}{r^2}\right) - \frac{1}{r^2} = 0 \quad (10.283)$$

$$e^{-2\beta}\left(\frac{1}{r^2} - \frac{2\beta'}{r}\right) - \frac{1}{r^2} = 0 \quad (10.284)$$

$$2e^{-2\beta}\frac{\dot\beta}{r} = 0 \quad (10.285)$$

$$e^{-2\beta}\left(\alpha'' + \alpha'^2 + \frac{\alpha' - \beta'}{r} - \alpha'\beta'\right)$$

$$-e^{-2\alpha}\left(\ddot\beta + \dot\alpha^2 - \dot\alpha\dot\beta\right) = 0. \quad (10.286)$$

(b) Show that for $r > r_0$ we have $\beta(r,t) = \beta(r)$. Show also that $\alpha' = -\beta'$, and by integrating, $\alpha(r,t) = -\beta(r)$ for $r > r_0$. Explain that the metric must have the static form (10.4).

This is what is called the Jebsen-Birkhoff theorem [Jeb21, Bir23]: *If a spacetime contains a region which is spherically symmetric, asymptotically flat, and empty ($T_{\mu\nu} = 0$) for $r > r_0$, then the metric in this region is time independent and hence independent of the dynamical properties of its source.*

10.16. *Gravitational mass*

(a) Use the line-element (10.4) and show that the surface gravity of a Schwarz-schild black hole can be written

$$\kappa = -e^{\alpha-\beta}\alpha'. \qquad (10.287)$$

(b) Show, using Einstein's field equations, that

$$4\pi r^2 e^{\alpha+\beta}\left(T^0_0 - T^1_1 - T^2_2 - T^3_3\right) = \left(r^2 e^{\alpha-\beta}\alpha'\right)'. \qquad (10.288)$$

Hence, deduce that the surface gravity can be written

$$\kappa = -\frac{4\pi}{r^2}\int_0^r \left(T^0_0 - T^1_1 - T^2_2 - T^3_3\right) e^{\alpha+\beta} r^2 dr. \qquad (10.289)$$

(c) Define the gravitational mass M_G inside a radius r of a spherical mass distribution by

$$\kappa = -\frac{M_G}{r^2}, \qquad (10.290)$$

and deduce that

$$M_G = 4\pi \int_0^r \left(T^0_0 - T^1_1 - T^2_2 - T^3_3\right) e^{\alpha+\beta} r^2 dr. \qquad (10.291)$$

This is the *Tolman-Whittaker* expression for the gravitational mass of a system.

What is the condition for repulsive gravitation?

10.17. *The river model for black holes*

In this problem you are going to picture space as flowing like a river into a Schwarzschild black hole [HL04]. "Space" is then represented by a continuum of local inertial frames falling freely from zero velocity at infinity.

(a) Show that the Schwarzschild metric, eq. (10.24), may be written in the Gullstrand-Painlevé form [Gul22, Pai21],

$$ds^2 = -d\tau^2 + (dr + \beta d\tau)^2 + r^2(d\theta^2 + \sin^2\theta d\phi^2), \quad \beta = \sqrt{\frac{r_S}{r}},$$

by introducing a new coordinate time

$$\tau = t + 2r_S \left(\frac{1}{\beta} - \operatorname{artanh}\beta\right).$$

(b) Show that $\beta(r) = -dr/d\tau$ is the velocity of an inertial frame falling freely from rest at infinity, i.e., the river velocity. What happens at the horizon of the black hole? Show that $d\tau$ is the proper time interval as measured by a clock comoving with the inertial frames that define the river model of the space.

Part IV

COSMOLOGY

11
Homogeneous and Isotropic Universe Models

One of the most successful and useful applications of Einstein's General Theory of Relativity is within the field of cosmology. Newton's theory of gravitation, involves attraction between celestial bodies. However, very little is said of the evolution of the universe itself. The universe was believed to be static, and its evolution was beyond any physical theory. But after the year 1917, things were different. Within two years after the birth of the General Theory of Relativity, Einstein realized that this theory actually could say something about the universe and constructed a static universe model as a solution of the relativistic field equations. The era of modern cosmology had begun, which would revolutionize our view of the universe.

11.1 The cosmological principles

Since medieval times, the universe was seen upon as something fixed, with the Earth itself at the centre. The Earth was a very special place in this geocentric universe; everything – the Moon, the Sun, the planets and even the stars – moved in perfect circles around the Earth. However, beginning with Copernicus, this view upon the universe was going to be drastically altered. Copernicus placed our Sun in the centre, not our Earth. As the observational techniques developed and improved, the centre of the universe was shifted further away, and today we believe that *there is no centre of the universe*. Even as late as 1920, cosmologists and astrophysicists thought that our Milky Way was the only galaxy in the universe. Now we know that our Milky Way is only one of billions of galaxies in the universe. The Milky Way is not a special galaxy, it is rather a typical one.

When we observe galaxies, there are a couple of things to note. Looking in different directions of the sky, the galaxies are evenly distributed at large scales. Large scales in this context, are not galactic scales, nor scales large as galactic clusters, but scales of the order of a billion light years. At this scale, the galaxies have an *isotropic* distribution; they are distributed evenly in the different directions in the sky. The galaxies are also evenly distributed in space, they are *homogeneously* distributed in the universe at large scales. These two apparent facts are referred to as the two cosmological principles:

- There is no special point in the universe, the galaxies are evenly distributed in space at large scales. The universe is said to be *homogeneous* at large scales.

- There is no special spatial direction in the universe, the galaxies are evenly distributed in different angular directions at large scales. The universe is said to be *isotropic*.

We know that these two principles are not true at small scales, there are some inhomogeneities at small scales. There are galaxies, there are Solar systems and planets. However, at the largest scales, the universe is said to be homogeneous and isotropic. This principle provides us with the simplest cosmological models, the homogeneous and isotropic universe models. They give us the simplest models of the evolution of the universe.

This was early realized by several physicists, most notably by Einstein himself. Einstein applied his equations to cosmology, and realized to his astonishment, that in general the field equations yield a dynamical universe. To Einstein, this could not be correct, so he inserted a term, now called the cosmological constant term, into the equations. The equations now yielded a static and fixed universe, more in agreement with Einstein's beliefs. However, later it was observationally verified that the universe was actually expanding, the universe was indeed dynamical. This was shown by Edwin Hubble in 1929, and Einstein had to withdraw his cosmological constant. Later, Einstein called the inclusion of the cosmological constant "the biggest blunder of his life". By including the cosmological constant, he produced what he thought was correct, but in this process failed to be the first to realize that the universe was expanding. We will see in the next chapter, that his "blunder" was not really as big a blunder as he thought; newer observational facts, have shown that a cosmological constant most probably is present and can be interpreted as representing Lorentz-invariant vacuum energy with constant density. In this chapter, on the other hand, we will assume that the universe is homogeneous and isotropic, and that the cosmological constant is absent.

11.2 Friedmann-Robertson-Walker models

Based on the assumption of spatial homogeneity and isotropy, the equations of motion of the universe will be deduced. This will be performed by applying the structural equations of Cartan to calculate the components of the Einstein tensor.

The assumption of spatial homogeneity and isotropy, implies that we can foliate our spacetime with spatial sections. Each of the spatial sections is labelled with a parameter t, which can be identified as "cosmic time". The assumption of isotropy allows us to assume that the time direction, denoted by the time-like vector e_t, is orthogonal to the spatial sections. Hence, if we foliate our spacetime as $\mathbb{R} \times \Sigma_t$ where $\mathbb{R}$ is the time direction and Σ_t are the spatial hypersurfaces, then e_t can be chosen to be orthogonal to Σ_t. If this had not been the case, the projection of the time-vector onto Σ_t would yield a preferred direction in space which would have violated the assumption of isotropy.

We can therefore assume that the line-element has the form

$$ds^2 = -dt^2 + a(t)^2 \left(d\chi^2 + r(\chi)^2(d\theta^2 + \sin^2\theta d\phi^2)\right), \tag{11.1}$$

where χ is the radial coordinate. Here, the function $a(t)$ is called the *expansion factor* or the *scale factor* since the proper distance in the radial direction is $dl_\chi = a(t)d\chi$. It is dimensionless and is normalised so that it has the value 1 at the present time, i.e. $a_0 = a(t_0) = 1$. This means that $a(t)$ is the ratio of the distance to a far away object in the universe (a galactic cluster) at an arbitrary point of time t and its present distance. We also use polar coordinates comoving with free reference particles without any motion except that due to the expansion of the universe. The function $r(\chi)$ has dimension length and will be determined by requiring that the model is isotropic in the three spatial directions, while the function $a(t)$ will be determined by Einstein's field equations. In principle, we can use any other time coordinate, this special choice where the metric is given by eq. (11.1), is called the *universal time gauge* or *cosmic time*. This is the proper time of the reference particles. There are other time coordinates that are more useful in other connections; some of them will be mentioned later.

The physical significance of our coordinates is the following. We first choose a set of reference particles defining a cosmic reference frame. You may think of these particles as galaxies without peculiar motions (see problem 11.3). Then χ, θ, ϕ are comoving coordinates in this reference frame, and t is the proper time shown by clocks carried by the galaxies. The coordinate χ of an object is its present distance from an observer at $\chi = 0$.

Let us introduce an orthonormal frame given by

$$\begin{aligned}
\omega^{\hat{t}} &= \mathbf{dt} \\
\omega^{\hat{\chi}} &= a\mathbf{d\chi} \\
\omega^{\hat{\theta}} &= ar\mathbf{d\theta} \\
\omega^{\hat{\phi}} &= ar\sin\theta\,\mathbf{d\phi}.
\end{aligned} \tag{11.2}$$

By exterior differentiation we get

$$\begin{aligned}
\mathbf{d}\omega^{\hat{t}} &= 0 \\
\mathbf{d}\omega^{\hat{\chi}} &= \frac{\dot{a}}{a}\omega^{\hat{t}} \wedge \omega^{\hat{\chi}} \\
\mathbf{d}\omega^{\hat{\theta}} &= \frac{\dot{a}}{a}\omega^{\hat{t}} \wedge \omega^{\hat{\theta}} + \frac{r'}{ra}\omega^{\hat{\chi}} \wedge \omega^{\hat{\theta}} \\
\mathbf{d}\omega^{\hat{\phi}} &= \frac{\dot{a}}{a}\omega^{\hat{t}} \wedge \omega^{\hat{\phi}} + \frac{r'}{ra}\omega^{\hat{\chi}} \wedge \omega^{\hat{\phi}} + \frac{1}{ar}\cot\theta\,\omega^{\hat{\theta}} \wedge \omega^{\hat{\phi}} \tag{11.3}
\end{aligned}$$

where overdot means derivative with respect to t, and prime means derivative with respect to χ. According to Cartan's first structural equation, eq. (6.182), the non-zero connection forms are

$$\Omega^{\hat{t}}_{\ \hat{\chi}} = \quad \Omega^{\hat{\chi}}_{\ \hat{t}} \quad = \frac{\dot{a}}{a}\omega^{\hat{\chi}}$$

$$\Omega^{\hat{\theta}}_{\ \hat{t}} = \quad \Omega^{\hat{t}}_{\ \hat{\theta}} \quad = \frac{\dot{a}}{a}\omega^{\hat{\theta}}$$

$$\Omega^{\hat{\phi}}_{\ \hat{t}} = \quad \Omega^{\hat{t}}_{\ \hat{\phi}} \quad = \frac{\dot{a}}{a}\omega^{\hat{\phi}}$$

$$\Omega^{\hat{\theta}}_{\ \hat{\chi}} = \quad -\Omega^{\hat{\chi}}_{\ \hat{\theta}} \quad = \frac{r'}{ra}\omega^{\hat{\theta}}$$

$$\Omega^{\hat{\phi}}_{\ \hat{\chi}} = \quad -\Omega^{\hat{\chi}}_{\ \hat{\phi}} \quad = \frac{r'}{ra}\omega^{\hat{\phi}}$$

$$\Omega^{\hat{\phi}}_{\ \hat{\theta}} = \quad -\Omega^{\hat{\theta}}_{\ \hat{\phi}} \quad = \frac{1}{ra}\cot\theta\,\omega^{\hat{\phi}}. \tag{11.4}$$

Using Cartan's second structural equation, eq. (7.47), the curvature forms are

$$\mathbf{R}^{\hat{t}}_{\ \hat{i}} = \frac{\ddot{a}}{a}\omega^{\hat{t}} \wedge \omega^{\hat{i}}$$

$$\mathbf{R}^{\hat{\theta}}_{\ \hat{\chi}} = \left(\frac{\dot{a}^2}{a^2} - \frac{r''}{ra^2}\right)\omega^{\hat{\theta}} \wedge \omega^{\hat{\chi}}$$

$$\mathbf{R}^{\hat{\phi}}_{\ \hat{\chi}} = \left(\frac{\dot{a}^2}{a^2} - \frac{r''}{ra^2}\right)\omega^{\hat{\phi}} \wedge \omega^{\hat{\chi}}$$

$$\mathbf{R}^{\hat{\theta}}_{\ \hat{\phi}} = \left(\frac{\dot{a}^2}{a^2} + \frac{1}{r^2a^2} - \frac{(r')^2}{r^2a^2}\right)\omega^{\hat{\theta}} \wedge \omega^{\hat{\phi}} \tag{11.5}$$

where $\hat{i}$ runs over the spatial coordinates.

From the assumption of isotropy, the three spatial directions should be equal for the orthonormal frame. Hence, the components of the curvature matrix should be equal for the three directions. This means that we must have

$$-rr'' = 1 - (r')^2 \quad \Leftrightarrow \quad \frac{r'r''}{(r')^2 - 1} = \frac{r'}{r}. \tag{11.6}$$

Integrating, one finds

$$\frac{dr}{d\chi} = \left(1 - \frac{\kappa}{R_0^2}r^2\right)^{\frac{1}{2}} = (1 - kr^2)^{\frac{1}{2}}, \quad k \equiv \frac{\kappa}{R_0^2}, \tag{11.7}$$

where κ is a dimensionless integration constant whose sign characterizes the solution. Here R_0 is a constant with dimension length, which represents the present value of the curvature radius of the 3-space $t = $ constant.

Integrating once more leads to

$$r(\chi) = R_0 S_k(\chi/R_0) \tag{11.8}$$

where the function $S_k(y)$ is defined by

$$S_k(y) = \begin{cases} \sin y, & k > 0 \\ y, & k = 0 \\ \sinh y, & k < 0 \end{cases} \tag{11.9}$$

Hence, the line-element (11.1) takes the form

$$ds^2 = -dt^2 + a(t)^2 \left(d\chi^2 + R_0^2 S_k^2(\chi/R_0)(d\theta^2 + \sin^2\theta\, d\phi^2)\right). \tag{11.10}$$

Using eq. (11.7) the line-element may be expressed in terms of the radial coordinate r as

$$ds^2 = -dt^2 + a(t)^2 \left(\frac{dr^2}{1 - kr^2} + r^2(d\theta^2 + \sin^2\theta d\phi^2) \right), \qquad (11.11)$$

where $k > 0$, $k = 0$ or $k < 0$. This is the Robertson-Walker form of the line-element. All of the homogeneous and isotropic universe models may be represented by this line-element. For $k > 0$ the spatial hypersurfaces have constant positive curvature and are usually called *closed* models. For $k = 0$ the spatial hypersurfaces are Euclidean and are called *flat* models. Lastly, for $k < 0$ the spatial hypersurfaces have constant negative curvature and are called *open* models. It should be noted that the so-called flat models will in general have curved spacetime.

The shape of the spatial space is not completely determined by the assumption of homogeneity and isotropy. There are basically three options. The function $a(t)$ has to be determined by Einstein's field equations. Let us derive the curvature part of Einstein's field equations, that is the left side of eq. (9.42).

From equations (11.5) and (11.9) we have the non-zero components of the Riemann tensor (no summation!)

$$R^{\hat{j}}_{\hat{i}\hat{j}\hat{i}} = \frac{\dot{a}^2 + k}{a^2} \quad (\hat{i} \neq \hat{j})$$

$$R^{\hat{t}}_{\hat{i}\hat{t}\hat{i}} = \frac{\ddot{a}}{a}. \qquad (11.12)$$

Hence, the non-vanishing components of the Einstein tensor are

$$E_{\hat{t}\hat{t}} = 3\frac{\dot{a}^2 + k}{a^2}$$

$$E_{\hat{i}\hat{i}} = -2\frac{\ddot{a}}{a} - \frac{\dot{a}^2 + k}{a^2}. \qquad (11.13)$$

These will be useful later on.

11.3 Dynamics of Homogeneous and Isotropic cosmologies

We have assumed that our universe is homogeneous and isotropic. To solve the Einstein field equations under this assumption we need an energy-momentum tensor which is also homogeneous and isotropic. A general form of the energy-momentum tensor which is compatible with homogeneity and isotropy, is

$$T_{\mu\nu} = (\rho + p)u_\mu u_\nu + pg_{\mu\nu}. \qquad (11.14)$$

Here, ρ is the proper energy (or mass) density of the fluid and p its pressure ($p > 0$) or strain ($p < 0$). Homogeneity implies that the pressure and density should be position independent on the spatial hypersurfaces. Hence, they can only be time dependent. If the vector u_μ, which is the four-velocity of the fluid, has a spatial component then the fluid has a special direction compared to the hypersurfaces Σ_t. This would violate our assumption of spatial isotropy. Thus the vector u_μ has only a time component; the fluid flow is orthogonal to the hypersurfaces.

The energy-momentum tensor is therefore diagonal in the coordinate system given by (11.11), and hence,

$$T_{\hat{\mu}\hat{\nu}} = \text{diag}(\rho, p, p, p).$$ (11.15)

The Einstein field equations with $\Lambda = 0$ now turn into

$$3\frac{\dot{a}^2 + k}{a^2} = 8\pi G\rho$$ (11.16)

$$-2\frac{\ddot{a}}{a} - \frac{\dot{a}^2 + k}{a^2} = 8\pi Gp.$$ (11.17)

These equations are called the *Friedmann equations*.

Inserting eq. (11.16) into eq. (11.17) yields

$$\boxed{\frac{\ddot{a}}{a} = -\frac{4\pi G}{3}(\rho + 3p).}$$ (11.18)

The effective gravitational energy is given by $\rho + 3p$; the pressure also contributes to gravitation. Note that $p < -\rho/3$ implies repulsive gravitation.

We also need an equation relating the energy, the pressure and the scale factor. The energy-momentum tensor has to be divergence-free which signals the conservation of energy. This follows automatically from the equations (11.16) and (11.17) from which we find the following relation

$$\dot{\rho} + 3\frac{\dot{a}}{a}(\rho + p) = 0,$$ (11.19)

which may be written

$$\frac{d}{dt}(\rho a^3) + p\frac{d}{dt}a^3 = 0.$$ (11.20)

Let $V = a^3$ be the volume of a region expanding together with the cosmic fluid; a so-called comoving volume. The energy in a comoving volume is $U = \rho a^3$. In terms of U and V, eq. (11.20) takes the form

$$\boxed{dU + pdV = 0.}$$ (11.21)

The first law of thermodynamics states for a fluid in equilibrium

$$TdS = dU + pdV,$$ (11.22)

where T is the temperature and S the entropy. A process which has $dS = 0$ is called an *adiabatic* process. Equation (11.21) shows that the isotropic and homogeneous universe models with a perfect fluid expand adiabatically. This is not surprising since homogeneity and isotropy imply no temperature gradients and hence no heat flow.

If we further assume that the perfect fluid obeys the barotropic equation of state

$$p = w\rho,$$ (11.23)

then equation (11.20) turns into

$$\frac{d}{dt}(\rho a^3) + w\rho\frac{d}{dt}a^3 = 0.$$ (11.24)

This equation admits the solution

$$\rho a^{3(w+1)} = \rho_0,$$
(11.25)

where ρ_0 is the present value of the density.

LIVE has $w = -1$ and in this case the density of matter is constant as a function of the volume. For dust we have $w = 0$ while for radiation we have $w = \frac{1}{3}$. Hence,

$$
\begin{aligned}
\rho_v &= \rho_{v0} & \text{for LIVE,} \\
\rho_m &= \rho_{m0}a^{-3} & \text{for dust,} \\
\rho_\gamma &= \rho_{\gamma 0}a^{-4} & \text{for radiation.}
\end{aligned}
$$
(11.26)

As the scale factor of the universe increases, the density of radiation will decrease faster than for dust. The Friedmann equation (11.16) for a universe dominated by a perfect fluid with equation of state (11.23) may be written

$$\left(\frac{\dot{a}}{a}\right)^2 = \frac{8\pi G}{3}\frac{\rho_0}{a^{3(w+1)}} - \frac{k}{a^2}.$$
(11.27)

This equation shows that the ultimate fate of the universe is determined by the spatial curvature if $w > -1/3$. Then a flat and a negatively universe will expand forever, while a positively curved universe will stop expanding and will recollapse to a Big Crunch. If $w < -1/3$ the expansion will proceed for all time independent of the curvature. The limiting case, $w = -1/3$, represents a universe dominated by a fluid with vanishing gravitational mass density. The expansion velocity $\dot{a}$ is constant in such a universe, just as it is in an empty universe.

11.4 Cosmological redshift and the Hubble law

The reason that the density of radiation decreases faster than that of dust is that each photon will also be redshifted during the expansion. As the universe expands the light-waves will be stretched towards the red part of the spectrum. This result will now be deduced.

Consider a galaxy far away from an observer who is located at $r = 0$. If the radial coordinate distance to the galaxy is χ, the proper distance is $d_P = a(t)\chi$. The velocity of the galaxy relative to the observer will be

$$v = \frac{d\,d_P}{dt} = \frac{\dot{a}}{a}a\chi = Hd_P$$
(11.28)

where $H = \frac{\dot{a}}{a}$ is called the *Hubble parameter*. Its present value is called the *Hubble constant* and is usually written

$$H_0 = hH_1$$
(11.29)

where $H_1 = 100\text{km s}^{-1}\text{Mpc}^{-1} \approx 30\text{km/s}$ per l.y. Recent measurements have indicated that $h \approx 0.7$.

Hubble's law states that the velocity of a galaxy is proportional to its distance

$$\boxed{v = Hd_P.}$$
(11.30)

This result was observationally obtained in 1929 and was taken as an evidence for an expanding universe. Until then many physicists had believed that the universe was static (including Einstein). However, after the observational evidence for a dynamical universe was put forward, they had to admit that this was not the case. The universe is dynamical and is in a state of expansion!

The Hubble parameter has the dimension of inverse time. The inverse of the Hubble parameter is called the *Hubble age* of the universe, $t_H \equiv 1/H$. It is the age of a universe expanding with constant velocity. Thus a universe with decelerated expansion has an age less than its Hubble age. The *Hubble sphere* is defined as a spherical region within a distance beyond which the recession velocity exceeds the speed of light, $d_{PHS} \equiv ct_H = c/H$. Inserting the measured value of H_0, the present Hubble age is $14 \cdot 10^9$ years.

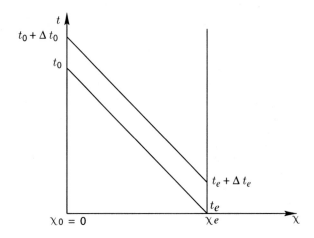

Figure 11.1: Cosmological redshift of light.

When light is travelling in an expanding universe, the light will be redshifted. Light moves along null geodesics. Hence, the world lines of light travelling towards us have $ds^2 = 0$, $d\theta = d\phi = 0$ and $dt = -a(t)d\chi$. Let Δt_e be the period of light at the emitter event and Δt_0 at the observation event (see Fig. 11.1). Consider two light signals emitted at t_e and $t_e + \Delta t_e$, respectively. Then

$$-\int_{\chi_e}^{0} d\chi = \chi_e = \int_{t_e}^{t_0} \frac{dt}{a(t)} = \int_{t_e+\Delta t_e}^{t_0+\Delta t_0} \frac{dt}{a(t)} \tag{11.31}$$

or

$$\int_{t_e+\Delta t_e}^{t_0+\Delta t_0} \frac{dt}{a(t)} - \int_{t_e}^{t_0} \frac{dt}{a(t)} = 0. \tag{11.32}$$

Hence,

$$\int_{t_0}^{t_0+\Delta t_0} \frac{dt}{a(t)} - \int_{t_e}^{t_e+\Delta t_e} \frac{dt}{a(t)} = 0. \tag{11.33}$$

Under the integration from t_e to $t_e + \Delta t_e$ and from t_0 to $t_0 + \Delta t_0$ the expansion factor can be considered constant with values $a(t_e)$ and $a(t_0)$ respectively. This gives

$$\frac{\Delta t_e}{a(t_e)} = \frac{\Delta t_0}{a(t_0)}. \tag{11.34}$$

Since the wavelength of light is $\lambda = c\Delta t$ we have

$$\frac{\lambda_0}{\lambda_e} = \frac{a(t_0)}{a(t_e)}. \tag{11.35}$$

This equation shows that the light waves are stretched by the expansion of the space. If λ is the corresponding wavelength of the light signal, the *redshift factor*, z, is defined by

$$\boxed{z \equiv \frac{\lambda_0 - \lambda_e}{\lambda_e} = \frac{a(t_0)}{a(t_e)} - 1} \tag{11.36}$$

If we make a Taylor expansion of $a(t_e)$ to second order

$$a(t_e) \approx a_0 + \dot{a}_0(t_e - t_0) + \frac{1}{2}\ddot{a}_0(t_e - t_0)^2, \tag{11.37}$$

and introduce the *deceleration parameter*,

$$q = -\frac{a\ddot{a}}{\dot{a}^2}, \tag{11.38}$$

the expansion factor can be written

$$a(t_e) \approx a_0 \left[1 + H_0(t_e - t_0) - \frac{1}{2}q_0 H_0^2 (t_e - t_0)^2 \right]. \tag{11.39}$$

Using eq. (11.36) this yields a power series for the redshift as a function of the time of flight $t_0 - t_e$,

$$z = H_0(t_0 - t_e) + \left(1 + \frac{q_0}{2} \right) H_0^2(t_0 - t_e)^2 + \cdots \tag{11.40}$$

Inverting this we obtain a formula for the time of flight in terms of the redshift

$$H_0(t_0 - t_e) = z - \left(1 + \frac{q_0}{2} \right) z^2 + \cdots \tag{11.41}$$

To the same order the comoving coordinate of the emitter is

$$\chi_e = \int_{t_e}^{t_0} \frac{dt}{a(t)} \approx \frac{t_0 - t_e}{a_0} \left[1 + \frac{1}{2}H_0(t_0 - t_e) \right]. \tag{11.42}$$

The proper distance of the emitter at the present time is $d_P = a_0 \chi_e$. From eqs. (11.40) and (11.41) we can relate the proper distance to the redshift for $z \ll 1$,

$$d_P = \frac{z}{H_0} \left[1 - \frac{z}{2}(1 + q_0) \right]. \tag{11.43}$$

This relationship is purely kinematical. We have not used Einstein's field equations. Hence, it is generally valid independently of the matter and energy content of the universe.

For small values of z we have approximately $z = H_0 d_P$. Interpreting the redshift as a Doppler effect, $z = v$, Hubble's law is recovered. However, in general relativity the cosmic redshift should be interpreted as an expansion effect. The quantity $1 + z$ is the ratio of distances at the time of arrival and the time of emission of a light signal. If, for example, $z = 1$, the cosmic distances have doubled during the time of travel of the light from the object to the observer. By measuring the distance and the redshifts to very distant objects, one can determine the deceleration parameter q_0. The deceleration parameter is positive if the expansion of the universe is decelerating. Recent measurements indicate that $q_0 < 0$! The universe seems to be in a state of accelerated expansion! This will be taken up in the next chapter.

Eq. (11.26) implies that the density of radiation decreases faster than the density of matter in an expanding universe. From the temperature $T = 2.726$K of the cosmic microwave background radiation one finds that its present density is $\rho_{\gamma 0} = 4.8 \cdot 10^{-31} \text{kg/m}^3$. The present density of matter is $\rho_{m0} = 6.0 \cdot 10^{-27} \text{kg/m}^3$. Radiation emitted at the point of time of equal density has a redshift

$$z_{eq} \approx \frac{a_0}{a_{eq}} = \frac{\rho_{m0}}{\rho_{\gamma 0}} = 1.25 \cdot 10^4. \tag{11.44}$$

For a flat universe $k = 0$ the Friedmann equation (11.16) reduces to

$$H^2 = \frac{8\pi G}{3} \rho_c , \tag{11.45}$$

where the density in the flat universe has been denoted by ρ_c and is called the *critical density*. Its present value is given in terms of the Hubble parameter by

$$\rho_c = \rho_1 h^2, \tag{11.46}$$

where $\rho_1 = 3H_1^2/8\pi G = 4 \cdot 10^{-26} \text{kg/m}^3$. With $h = 0.7$ the critical density is $\rho_c = 2 \cdot 10^{-26} \text{kg/m}^3$. If the mass density is larger than ρ_c the universe has positive spatial curvature, $k > 0$, and is thus closed. If it is less the curvature is negative, $k < 0$, and the universe is open.

The density relative to the critical density is denoted by Ω and is called the *density parameter* or the *relative density*; i.e.

$$\Omega = \frac{\rho}{\rho_c}. \tag{11.47}$$

Defining a spatial curvature parameter

$$\Omega_k = -\frac{k}{H^2 a^2} \tag{11.48}$$

the Friedmann equation (11.16) takes the form

$$\boxed{\Omega + \Omega_k = 1,} \tag{11.49}$$

where Ω is the total relative density of energy and matter. Since for an open model $\Omega_k < 0$, a flat model $\Omega_k = 0$ and for a closed model $\Omega_k > 0$ we have

$$\Omega \begin{cases} > 1, & \text{for } k > 0, \\ = 1, & \text{for } k = 0, \\ < 1, & \text{for } k < 0. \end{cases} \tag{11.50}$$

Hence, in principle we can measure the matter content of the universe and determine its geometry. From eqs. (11.48), (11.49), the last of eqs. (11.7), and defining the Hubble distance $\ell_H = ct_H$, where $t_H = 1/H_0$ is the Hubble age of the universe, we find that the present curvature radius of the cosmic 3-space is given by

$$R_0 = \ell_H \sqrt{\frac{\kappa}{\Omega_0 - 1}}. \qquad (11.51)$$

Recent measurements of the temperature variations in the background radiation and the apparent luminosity and redshift of supernovae of type Ia [UKE03] show that $\Omega_0 = 1.04\pm^{0.02}_{0.04}$. Hence, $R_0 > 4\ell_H = 56 \cdot 10^9$l.y.. This means that the 3-space of the universe is close to Euclidean on large scales.

Fig. 11.2 shows Ω_m and Ω_k as a function of $\ln a$ in an open model filled with dust (matter). We see that the universe at small scales is matter-dominated, while at late times and large scales it is curvature-dominated.

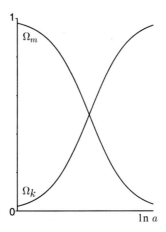

Figure 11.2: Ω_m and Ω_k as a function of $\ln a$ in the matter-dominated open model $(k = -1)$.

Introducing the relative density of radiation, Ω_γ, matter, Ω_m, and vacuum energy, Ω_v, and using eq. (11.26), the Friedmann equation (11.16) can be written

$$\left(\Omega_{v0}a^4 + \Omega_{k0}a^2 + \Omega_{m0}a + \Omega_{\gamma0}\right)^{-\frac{1}{2}} a \, da = H_0 dt, \qquad (11.52)$$

where the index 0 denotes the present value. This equation may also be expressed as a redshift-time relationship

$$\int_0^z \frac{(1+z)dz}{[\Omega_{v0} + \Omega_{k0}(1+z)^2 + \Omega_{m0}(1+z)^3 + \Omega_{\gamma0}(1+z)^4]^{\frac{1}{2}}} = H_0(t_0 - t_e) \quad (11.53)$$

where t_0 is the present age of the universe and t_e the emission time of radiation observed with redshift z. The parameters Ω_{v0}, Ω_{k0}, Ω_{m0} and $\Omega_{\gamma0}$ are not independent due to the constraint (11.49). The present age of the universe is given in terms of Hubble parameter, the present values of the relative densities, and the curvature parameter by

$$t_0 = \frac{1}{H_0} \int_0^1 \frac{a\,da}{\left(\Omega_{v0}a^4 + \Omega_{k0}a^2 + \Omega_{m0}a + \Omega_{\gamma 0}\right)^{\frac{1}{2}}}. \tag{11.54}$$

It may also be noted that the Friedmann equation (11.18) may be expressed as a relation between the deceleration parameter and the relative densities,

$$q = \frac{\Omega_m}{2} + \Omega_\gamma - \Omega_v. \tag{11.55}$$

11.5   Radiation dominated universe models

Let us now solve Einstein's field equations for a radiation fluid. Even though the radiation is not the dominant fluid at the present epoch, the radiation was dominant for a redshift $z > 1.25 \cdot 10^4$.

We start with the Friedmann equation, eq. (11.16)

$$\frac{\dot{a}^2}{a^2} = \frac{8\pi G}{3}\rho - \frac{k}{a^2}. \tag{11.56}$$

From eq. (11.26) we have that $\rho_\gamma = \rho_{\gamma 0}a^{-4}$ for radiation. Inserting this into the Friedmann equation we get

$$\dot{a}^2 = Ca^{-2} - k \tag{11.57}$$

where $C = \frac{8\pi G}{3}\rho_{\gamma 0}$. One can either integrate this equation directly or we can take the trace of the Friedmann equations (11.16) and (11.17). Since $p = \frac{1}{3}\rho$ for radiation, we get

$$a\ddot{a} + \dot{a}^2 + k = \frac{d}{dt}(a\dot{a} + kt) = 0. \tag{11.58}$$

Integrating twice, we obtain

$$a^2 + kt^2 = 2Bt + B' \tag{11.59}$$

where B and B' are integration constants. By choosing $a(0) = 0$ we can set $B' = 0$. Hence, the solution for a radiation dominated universe is

$$a(t) = \sqrt{2Bt - kt^2} \tag{11.60}$$

Inserting this into eq. (11.57) we can relate B and C:

$$C = B^2. \tag{11.61}$$

From eq. (11.56) follows $H_0 = \sqrt{k/(\Omega_{\gamma 0} - 1)}$ $(k \neq 0)$, where H_0 and $\Omega_{\gamma 0}$ are the present values of the Hubble parameter and the relative density of the radiation. Eqs. (11.57) and (11.61) give $H_0^2 = B^2 - k$. Hence,

$$B = \left(H_0^2 + k\right)^{\frac{1}{2}} = \left(k\frac{\Omega_{\gamma 0}}{\Omega_{\gamma 0} - 1}\right)^{\frac{1}{2}}. \tag{11.62}$$

Note that the expansion velocity $\dot{a}$ is

$$\dot{a} = \frac{B - kt}{\sqrt{t(2B - kt)}} \tag{11.63}$$

which diverges as $t \longrightarrow 0$. Hence, the expansion of the universe is infinite as we approach the initial $t = 0$. Even though a particle in our space cannot exceed the speed of light relative to any observer, the expansion of the universe can be of arbitrary velocity. This initial point where $t = 0$ is called the *Big Bang*.

Note also that for a closed universe $(k > 0)$, the universe has a turning point at $t = B$ where the universe stops expanding and begins to contract. The universe will end its days in a *Big Crunch* at $t = 2B$.

As the universe becomes bigger during the expanding phase, one expects that the radiation would cool to lower temperatures. From quantum statistical mechanics of massless particles we know that (Stephan-Boltzmann's law)

$$\rho \propto T^4. \tag{11.64}$$

Close to $t = 0$ all of the models (11.60) behave similarly to the flat case

$$a(t) = \sqrt{2Bt}. \tag{11.65}$$

The radiation density will decrease as

$$\rho_\gamma \propto t^{-2}. \tag{11.66}$$

This means that during the radiation era, the temperature will fall as

$$T \propto \rho_\gamma^{\frac{1}{4}} \propto \frac{1}{\sqrt{t}}. \tag{11.67}$$

This relation may be written $t = (T_1/T)^2 t_1$ where $T_1 = 10^{10}$K and $t_1 = 1$s. The highest energies accessible to terrestrial experiments correspond to a temperature of about 10^{15}K, which was attained when the universe was about 10^{-10}s old.

If a universe at $t = 1$s had a temperature of $T = 10^{10}$K, say, then the temperature would have dropped by a factor of 10 to $T = 10^9$K at $t = 100$s. The initial universe was a hot universe[1] dominated by radiation. However, today the universe is more dominated by matter (dust) than radiation.

The transition from a radiation to a dust dominated model, is believed to have happened around $t =$44 000 years (see below). Since this time, the dynamics of the universe has been driven by matter and vacuum energy. As the temperature of the radiation cooled the radiation reached a point where it did not have enough energy to keep the atoms ionised. At around $t = 400$ 000 years matter and radiation *decoupled*. During the period before this time the radiation was thermalized and in thermal equilibrium with the matter. But at this point, the free electrons could bind to a nucleus and form a neutral atom. Hence, the first atoms in the universe were created about 400 000 years after the Big Bang. The photons moved freely after this time; there were no free electrons to Compton-scatter them. Effectively, the universe became transparent! This time in the history of the universe is called the *recombination*. Since this point in time the photons have travelled more or less freely

---

[1]At least in the Hot Big Bang model. It is this model we will consider in the present book.

in space. These photons are what make out the *cosmic microwave background radiation* (CMB). Today the cosmic microwave background radiation has a temperature of about 2.7K but the radiation was emitted approximately 400 000 years after the Big Bang at a temperature of $T = 3000$K. Hence, this radiation is the relics of the universe, when it was only 400 000 years old. Thus by studying the CMB we can learn much about the state of our universe in its childhood.

Examples   **Example 11.1 (The temperature in the radiation dominated epoch)**
Cosmologists believe that the universe was radiation dominated in the period between $t_0 = 10^{-33}$s and $t_1 = 10^{11}$s. Assuming that the temperature at t_1 was $T_1 = 10^3$K we can estimate the temperature at the start of the radiation era. Since

$$T \propto t^{-\frac{1}{2}} \tag{11.68}$$

during a radiation dominated epoch, we have

$$\frac{T_1}{T_0} = \left(\frac{t_1}{t_0}\right)^{-\frac{1}{2}}. \tag{11.69}$$

Hence the temperature at t_0 was

$$T_0 = T_1 \left(\frac{t_1}{t_0}\right)^{\frac{1}{2}} \approx 10^{25}\,\text{K}. \tag{11.70}$$

At these temperatures, all atoms will be completely ionised; there will only be a soup of protons, neutrons and electrons. At some time during this radiation dominated period, one believes that the temperature was sufficiently low to allow for the lightest atoms to form. This is what cosmologists call the period of *nucleosynthesis*. During nucleosynthesis the lightest elements like Hydrogen, Helium, Beryllium and Lithium, formed. This process requires a temperature of about $T_n = 10^9$K which corresponds to a time

$$t_n = t_1 \frac{T_1^2}{T_n^2} \approx 1\text{s} \tag{11.71}$$

after the Big Bang.

Example 11.2 (The redshift of the cosmic microwave background)
The temperature in the cosmic microwave background has decreased from $T_e = 3000$K to $T_0 = 2.7$K since its emission. The frequency of a photon gas is directly related to its temperature:

$$k_B T = \hbar \nu. \tag{11.72}$$

The redshift is found via

$$z = \frac{\nu_e}{\nu_0} - 1 = \frac{T_e}{T_0} - 1 \approx 10^3. \tag{11.73}$$

The microwave background has been redshifted approximately by a factor thousand since its emission!

11.6 Matter dominated universe models

We will now turn our attention to universe models dominated by pressure-free matter with density ρ_m. The Friedmann equation is

$$\dot{a}^2 + k = \frac{8\pi G}{3}\rho a^2. \tag{11.74}$$

Multiplying by a and using eq. (11.26), gives

$$a\dot{a}^2 + ka = \frac{8\pi G}{3}\rho_m a^3 = C = H_0^2 \Omega_{m0} \tag{11.75}$$

where H_0 and Ω_{m0} are the present values of the Hubble parameter and the relative matter density.

Let us introduce a dimensionless *conformal time coordinate*, η, by

$$\frac{1}{t_0}\frac{dt}{d\eta} = a(\eta). \tag{11.76}$$

So

$$\dot{a} = \frac{d\eta}{dt}\left(\frac{da}{d\eta}\right) = \frac{1}{a}\left(\frac{da}{d\eta}\right)\frac{1}{t_0}. \tag{11.77}$$

Equation (11.75) gives

$$\frac{1}{a}\cdot\frac{1}{t_0^2}\left(\frac{da}{d\eta}\right)^2 = C - ka \tag{11.78}$$

which can be rewritten as

$$\frac{1}{a}\frac{da}{d\eta} = \frac{1}{t_0}\sqrt{\frac{C}{a}}\sqrt{1 - \frac{ka}{C}}. \tag{11.79}$$

By making the substitution $a = Cx^2$ we can readily integrate this equation. Using that $a(0) = 0$ we obtain

$$k < 0: \quad \begin{cases} a(\eta) = \frac{\Omega_{m0}}{2(1-\Omega_{m0})}(\cosh\eta - 1) \\ t(\eta) = \frac{\Omega_{m0}}{2H_0(1-\Omega_{m0})^{3/2}}(\sinh\eta - \eta) \end{cases} \tag{11.80}$$

$$k = 0: \quad a(t) = \left(\frac{t}{t_0}\right)^{\frac{2}{3}} \tag{11.81}$$

$$k > 0: \quad \begin{cases} a(\eta) = \frac{\Omega_{m0}}{2(\Omega_{m0}-1)}(1 - \cos\eta) \\ t(\eta) = \frac{\Omega_{m0}}{2H_0(\Omega_{m0}-1)^{3/2}}(\eta - \sin\eta). \end{cases} \tag{11.82}$$

Note from eq. (11.75) that $H_0^2(\Omega_{m0} - 1) = k = \kappa/R_0^2$. Inserting eq. (11.80) or (11.82) into eq. (11.79) gives $R_0 = t_0$.

For $k > 0$, the solution is that of a cycloid. The universe expands, reaches a maximum size, and recollapses. The big crunch happens at a point of time $t_C = \pi\Omega_{m0}/(\Omega_{m0} - 1)^{3/2}(1/H_0)$. The flat model ($k = 0$) is called the *Einstein-de Sitter model*. It is ever-expanding but its expansion velocity reaches zero in the far future, $\dot{a} \to 0$ as $t \to \infty$. The open model is also ever-expanding and for large values of t the scale factor grows as $a(t) = t$. Hence, the flat matter dominated model is just on the borderline between ever-expanding

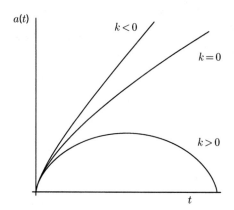

Figure 11.3: The cosmological scale factor for the open ($k < 0$), flat ($k = 0$) and closed ($k > 0$) models.

and recollapsing. The scale factor as a function of time is depicted in Fig. 11.3 for the three cases.

We shall calculate the point of time, t_{eq}, for the transition from a radiation dominated period to a matter dominated period [Ryd03]. Let us consider a flat universe and neglect vacuum energy. Then eq. (11.52) reduces to

$$a\dot{a} = H_0 \left(\Omega_{m0}a + \Omega_{\gamma0} \right)^{\frac{1}{2}}. \tag{11.83}$$

Integration with $a(0) = 0$ gives

$$H_0 t = \frac{4}{3} \frac{\Omega_{\gamma0}^{3/2}}{\Omega_{m0}^2} + \frac{2}{3} \frac{(\Omega_{m0}a - 2\Omega_{\gamma0})\sqrt{\Omega_{m0}a + \Omega_{\gamma0}}}{\Omega_{m0}^2}. \tag{11.84}$$

From eq. (11.26) follows that the scale factor at equal density of matter and radiation is $a_{eq} = \Omega_{\gamma0}/\Omega_{m0}$. This gives

$$t_{eq} = \frac{2}{3} \left(2 - \sqrt{2} \right) \frac{\Omega_{\gamma0}^{3/2}}{\Omega_{m0}^2} t_H, \tag{11.85}$$

where $t_H = 1/H_0$ is the Hubble age of the universe. Inserting the measured values $\Omega_{\gamma0} = 8.4 \cdot 10^{-5}$, $\Omega_{m0} = 0.3$ and $t_H = 14 \cdot 10^9$yr gives $t_{eq} = 47 \cdot 10^3$yr. The value $\Omega_{m0} = 0.3$ is, however, inconsistent with the assumption that the universe model is flat. One should insert $\Omega_{m0} \approx 1$ which gives $t_{eq} = 34 \cdot 10^3$yr. The corresponding result if one assumes an Einstein-de Sitter universe before the point of time $t = t_{eq}$ is $t_{eq} = (2/3)(\Omega_{\gamma0}/\Omega_{m0})^{3/2}t_H = 44 \cdot 10^3$yr. The result obtained for our universe with numerical integration of

$$H t_{eq} = \int_0^{\Omega_{\gamma0}/\Omega_{m0}} \frac{a\,da}{\sqrt{\Omega_{v0}a^4 + \Omega_{m0}a + \Omega_{\gamma0}}},$$

with $\Omega_{v0} = 0.7$ and $\Omega_{m0} = 0.3$ is $t_{eq} = 47 \cdot 10^3$yr. Note that the vacuum energy does not affect the result significantly due to the small value of $\Omega_{v0}a^4$ in the radiation dominated period before $t = t_{eq}$.

Example 11.3 (Age-redshift relation in the Einstein-de Sitter universe)
In the Einstein-de Sitter model (dust dominated with $k = 0$) we can now find a useful relation between the age of the universe and the redshift.
From eq. (11.81) we find

$$H = \frac{2}{3t}. \tag{11.86}$$

Let t_0 be the present time with the corresponding Hubble factor H_0. The Hubble-time $t_H = H_0^{-1}$ is the age of the universe if the expansion rate has been constant. We see that

$$t_0 = \frac{2}{3}t_H. \tag{11.87}$$

Inserting $t_H = 14 \cdot 10^9 \text{yr}$ we find that the age of the Einstein-de Sitter universe with the present rate of expansion is $t_0 = 9.3 \cdot 10^9 \text{yr}$.
The redshift is given by

$$1 + z = \frac{a_0}{a} = \left(\frac{t_0}{t}\right)^{\frac{2}{3}}. \tag{11.88}$$

The time difference between emission and receiving the photons is called the *lookback time*. From eq. (11.88) follows that it is given by

$$\Delta t = t_0 - t = t_0 \left[1 - \frac{1}{(1 + z)^{3/2}}\right]. \tag{11.89}$$

This can be written as

$$\Delta t = t_0 - t = \frac{2}{3}t_H \left[1 - \frac{1}{(1 + z)^{3/2}}\right]. \tag{11.90}$$

The age of the universe can be found by taking the limit for infinite redshift

$$\Delta t = \lim_{z \to \infty} \frac{2}{3}t_H \left[1 - \frac{1}{(1 + z)^{3/2}}\right] = \frac{2}{3}t_H \tag{11.91}$$

in accordance with eq. (11.87).

11.7   The gravitational lens effect

We have seen how a mass can deflect light towards it. Similarly, a concentration of masses, like a galaxy, will deflect light rays and may cause some interesting effects, such as making several images, and changing the intensity of the images of, for example, quasars lying behind the galaxy. It is called the *gravitational lens effect* [MHL89, MFS89].

Interestingly, this effect – as shown by Sjur Refsdal in 1964 [Ref64a, Ref64b] – can be used to determine the Hubble parameter. We will here show the idea behind Refsdal's derivation [GR92].

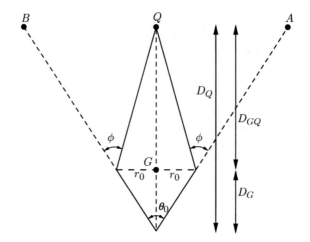

Figure 11.4: The Einstein ring. Here, Q is the observed object, and G is the gravitational lens.

Quasar masses determined from gravitational lens pictures

We will first consider the symmetrical case where the observed object is situated directly behind the gravitational lens, as shown in Fig. 11.4.

Let Q be the observed object, and G the gravitational lens. Usually, Q is a far-away quasar and G is a more nearby galaxy. The distance from the observer to Q and G are D_Q and D_G, respectively, and the distance between Q and G is D_{GQ}. Two light-rays are depicted; one on each side of the lens. The shortest distance between the rays and G is r_0, and the angle between the light-rays at the observer is θ_0. It is the measured angle between the objects. The deflection angle of the light-rays is

$$\phi = \frac{4Gm}{c^2 r_0} \tag{11.92}$$

where m is the mass of the lens, which is considered as a point mass. In this case the picture of the lens is a circle, called the *Einstein ring*, with angular radius θ_0 which is called the *Einstein radius*. In the case of the Sun this angle is less than 1.75 arc seconds. Also in the case of a galaxy, this deflection angle is small. Hence, using radians, we can assume that $\theta_0 \ll 1$. Inspecting Fig. 11.4, we get

$$\theta_0 = \frac{2r_0}{D_G} \quad \Rightarrow \quad r_0 = \frac{D_G \theta_0}{2}, \tag{11.93}$$

and by expressing the distance BQ in two different ways, we get

$$D_{GQ}\phi = D_Q \frac{\theta_0}{2} \quad \Rightarrow \quad \phi = \frac{D_Q \theta_0}{2 D_{GQ}}. \tag{11.94}$$

Inserting this into eq. (11.92), yields

$$\theta_0 = 4 \left(\frac{D_{GQ}}{D_G D_Q} \cdot \frac{Gm}{c^2} \right)^{\frac{1}{2}}. \tag{11.95}$$

Due to the expansion of the universe the received light will be redshifted by a factor z. To lowest order the redshift is given by the Hubble law, and hence,

$$cz_Q = H_0 D_Q, \qquad cz_G = H_0 D_G. \tag{11.96}$$

Inserting this into eq. (11.95) yields

$$\theta_0 = 4 \left(\frac{z_Q - z_G}{z_G z_Q} \cdot \frac{G H_0 m}{c^3} \right)^{\frac{1}{2}}. \tag{11.97}$$

For a massive galaxy with a mass $m = 10^{12} m_S$, where m_S is the mass of the Sun, at redshift $z_G = 0.5$ and an object, say a quasar, at redshift $z_Q = 2.0$ in a universe with a Hubble parameter $H_0 = 15$km/s per million light years, the Einstein radius is $\theta_0 \approx 1.8 \, (m/m_S)^{1/2} \cdot 10^{-6}$ arc seconds. In the case of so-called *microlensing* in which stars in the disk of the Milky Way act as lenses for stars close to the centre of the Milky Way, the angular scale defined by the Einstein radius is $\theta_0 \approx 0.5 \, (m/m_S)^{1/2}$ arc seconds.

Solving eq. (11.97) with respect to m, gives the mass of the object in terms of observable quantities

$$m = \frac{c^3}{16 G H_0} \frac{z_G z_Q}{z_Q - z_G} \theta_0^2. \tag{11.98}$$

Microlensing

When a star moves in front of another star it may act as a gravitational lens and magnify the star behind. In the case that the lensing star passes the line of sight of the far away star the intensity of the star will change with time in a characteristic way. We shall now deduce the shape of the light-curve.

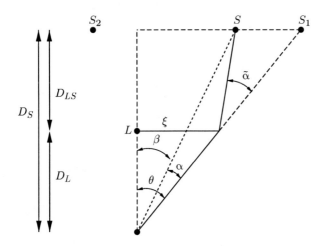

Figure 11.5: Microlensing. Here, L is the lens and S is the observed star.

Consider the situation shown in Fig. 11.5, where angles are so small that we can use the approximations $\tan x \approx x \approx \sin x$ and $\cos x \approx 1$. Here, L is the star which acts as a lens, and S the observed star. From the figure it is seen that

$$\beta = \theta - \alpha. \tag{11.99}$$

The deflection angle is

$$\tilde{\alpha} = \frac{4Gm}{c^2}\frac{1}{\xi}. \tag{11.100}$$

The angle α is

$$\alpha = \frac{D_{LS}}{D_S}\tilde{\alpha}. \tag{11.101}$$

Furthermore, $\xi = D_L \tan\theta \approx D_L\theta$. Hence, we get

$$\beta = \theta - \frac{D_{LS}}{D_L D_S}\frac{4Gm}{c^2\theta}. \tag{11.102}$$

Inserting the Einstein radius θ_0 from eq. (11.95), leads to the lens equation

$$\beta = \theta - \frac{\theta_0^2}{\theta}. \tag{11.103}$$

Solving this for θ one finds that a gravitational point lens produces two images of a background source (except in the case that the lens in positioned on the line of sight of the background source, when an Einstein ring appears). The positions of the images are given by the two solutions

$$\theta_{1,2} = \frac{1}{2}\left(\beta \pm \sqrt{\beta^2 + 4\theta_0}\right). \tag{11.104}$$

The *magnification* of an image is defined as the ratio between the solid angles of the images and the source. Hence the magnification is given by

$$\mu \equiv \frac{d\Omega_{Si}}{d\Omega_S} = \frac{\sin\theta d\theta}{\sin\beta d\beta} \approx \frac{\theta d\theta}{\beta d\beta}. \tag{11.105}$$

Using eq. (11.103) we get

$$\mu_{1,2} = \left(1 - \frac{\theta_0^4}{\theta_{1,2}^4}\right)^{-1}. \tag{11.106}$$

The sum of the absolute values of the two image magnifications is the measurable total magnification μ. Using that $\theta_1\theta_2 = \theta_0^2$, we find

$$\mu = \mu_1 - \mu_2 = \frac{\theta_2^4 - \theta_1^4}{2\theta_0 - (\theta_1^4 + \theta_2^4)}. \tag{11.107}$$

Inserting the solutions (11.104), and introducing a parameter $u \equiv \beta/\theta_0$, leads to

$$\boxed{\mu = \frac{u^2 + 2}{u\sqrt{u^2 + 4}}.} \tag{11.108}$$

(It may be noted that the difference between the two image magnifications is unity, $\mu_1 + \mu_2 = 1$.)

Magnification of stars caused by gravitational microlensing by Massive Astronomical Compact Halo Objects (MACHOs) has been used in the search for dark matter in the universe.

The Hubble parameter determined from the gravitational lens effect

By considering the non-symmetrical situation, Refsdal showed that – by measuring the time difference in the two lightpaths – it is possible to determine the Hubble parameter. This is a direct way of measuring the Hubble parameter and thus avoids the problem of finding "standard candles". However, an accurate determination of the value of H requires a good model for the gravitational lens and the galactic cluster of which a galaxy is usually a member.

Consider the non-symmetrical situation shown in Fig. 11.6 where the gravitational lens is not quite on the line-of-sight to the observed object Q.

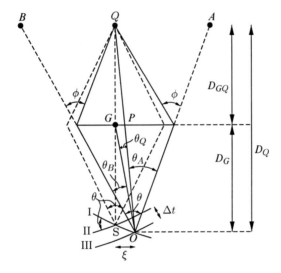

Figure 11.6: Typical gravitational lens situation.

We will assume that the gravitational lens has a mass distribution $M(r) \propto r$ in our derivation.

The main idea of the calculation is the following. The redshift of the quasar and the galactic lens can be measured with great accuracy. Since the redshift is proportional to the distance from us, one determines in this way the ratio between these distances. Furthermore, the angles θ_A and θ_B are observed. Thus a correct picture of the geometry can be drawn. The only missing piece of information to specify the figure completely, is the correct scaling factor. What are the actual distances? It suffices to know one of the distances to know all the others. This is just what a measurement of Δt – the difference in travel time for light-rays travelling on opposite side of the lens – provides, as we shall see.

Consider the wave-fronts in Fig. 11.6. The travel-time from Q to the point of symmetry S is the same for all light-signals reaching S. Since the wave-fronts I and II intersect at S, all points on these wavefronts must correspond to the same travel-time, $t_I = t_{II}$. Inspecting Fig. 11.6 yields

$$\Delta t = t_{III} - t_I = t_{III} - t_{II} = \theta \xi. \tag{11.109}$$

Here, θ is the angle between the two wave-fronts intersecting at S. Since the deflection angle is independent of the impact parameter for the gravitational model under consideration, the angle between the pictures is equal to θ. Furthermore, ξ is the distance between the observer and S. Thus it is possible to determine ξ from measuring Δt and θ, and hence, the scale of the figure can be determined.

The distance GP can be expressed in two ways, giving

$$\theta_G D_G = \frac{D_{GQ}}{D_Q}\xi. \tag{11.110}$$

The angle θ_G is not directly measurable, but θ_A and θ_B are. From Fig. 11.6 it is seen that $AQ = BQ$. Thus $\theta_A - \theta_Q = \theta_B + \theta_Q$ giving

$$\theta_Q = \frac{1}{2}\left(\theta_A - \theta_B\right). \tag{11.111}$$

Inserting eqs. (11.110) and (11.111), together with $D_{GQ} = D_Q - D_G$ and $\theta = \theta_A + \theta_B$ (see Fig. 11.6) into eq. (11.109) gives

$$\Delta t = \frac{D_Q D_G}{D_{GQ}}\theta\theta_G = \frac{1}{2}\frac{D_Q D_G}{D_Q - D_G}\left(\theta_A^2 - \theta_B^2\right). \tag{11.112}$$

Expressing the distances D_Q and D_G by the corresponding redshifts z_Q and z_G using Hubble's law, gives

$$\Delta t = \frac{1}{2H_0}\frac{z_Q z_G}{z_Q - z_G}\left(\theta_A^2 - \theta_B^2\right), \tag{11.113}$$

and hence,

$$\boxed{H_0 = \frac{1}{2\Delta t}\frac{z_Q z_G}{z_Q - z_G}\left(\theta_A^2 - \theta_B^2\right).} \tag{11.114}$$

This equation may be called *Refsdal's equation* and says how we can determine the Hubble parameter using the gravitational lens effect.

Refsdal's equation is derived using the following assumptions.

1. The gravitational lens has a mass proportial to the radius.

2. Possible modifications of the Hubble law depending on the universe model, have been neglected.

3. The lensing galaxy is not a member of a cluster of galaxies.

A more careful derivation of the Hubble parameter where these assumptions are not needed, gives the expression

$$H_0 = K_L K_U \left(\frac{v}{v_0}\right)^2 \frac{z_Q z_G}{z_Q - z_G}\left(\theta_A^2 - \theta_B^2\right)\frac{1}{\Delta t}. \tag{11.115}$$

Here, K_L is a numerical factor representing the mass distribution of the lens; K_U is a factor representing the geometric properties of the universe; v/v_0 is a factor representing the effect of the lensing galaxy upon the travel-time of the light-rays. More specifically, v is the radial velocity dispersion of the stars in the lensing galaxy, and v_0 is the radial velocity dispersion of an imaginary galaxy which is so massive that it can produce the lensing alone.

For example, if the mass distribution of the lensing galaxy is $M(r) \propto r^n$, we have $K_L = 1 - n/2$. The factors K_U and v/v_0 are usually slightly less than 1. For more details, consult for example [GR92].

The first gravitational lens images were detected in 1979. This was a double image of the quasar Q0957+561. The two images of this quasar have a redshift of about $z_Q = 1.41$ while the lens itself is at a redshift $z_G = 0.36$. The angular separations from the centre of the lens are $\theta_A = 5.24$ arc seconds and $\theta_B = 0.9$ arc seconds respectively. By studying the variation of the light intensity, one has established that $\Delta t = 1.4$ years. Using a model for the lens one finds $M(r) \propto r^{4/3}$ and thus $K_L = 1/3$. The factor K_U representing the geometry of space is probably close to unity, so $K_U = 1$. Furthermore, the velocities v and v_0 have been estimated to have the values $v = 360$km s^{-1} and $v_0 = 390$km s^{-1}. Inserting these values into eq. (11.115) gives the value of the Hubble parameter $H_0 = 57$km s^{-1}Mpc^{-1}.

11.8   Redshift-luminosity relation

Consider radiation emitted from a coordinate distance r_e at a point of time t_e and received at $r = 0$ at a point of time t_0. The radiation is detected by a telescope with proper area A. The light rays that just graze the mirror form a cone at the light source with solid angle $A/a^2(t_0)r_e^2 = A/r_e^2$ where we have used the normalization $a(t_0) = 1$. The fraction of the isotropically emitted radiation that reaches the mirror is the ratio of this solid angle to 4π, or $A/4\pi r_e^2$. Light with frequency ν_e is redshifted to frequency $\nu_e/(1+z)$, and light emitted during a time interval dt_e is received during a time interval $dt_0 = dt_e(1 + z)$ since the observer is moving away from the emitter. Thus the power P received by the mirror is equal to the power emitted by the source, i.e., its *absolute luminosity* L times the factor $(1 + z)^{-2}$ times the fraction $A/4\pi r_e^2$,

$$P = L(1+z)^{-2}\frac{A}{4\pi r_e^2}. \tag{11.116}$$

The *apparent luminosity*, l, is defined as the received power per unit area,

$$l \equiv \frac{P}{A} = \frac{L}{(1+z)^2 4\pi r_e^2}. \tag{11.117}$$

The *luminosity distance*, d_L, of a light source is defined as

$$d_L \equiv \left(\frac{L}{4\pi l}\right)^{\frac{1}{2}}. \tag{11.118}$$

Hence,

$$d_L = (1+z)r_e. \tag{11.119}$$

We must now evaluate the function $r_e(z)$. From the redshift formula $1 + z = a_0/a = 1/a$ follows

$$dz = -\frac{\dot{a}}{a^2}dt = -\frac{H}{a}dt. \tag{11.120}$$

For light moving towards the observer

$$dt = -\frac{adr}{\sqrt{1 - kr^2}} \tag{11.121}$$

which gives

$$\frac{dr}{\sqrt{1-kr^2}} = \frac{dz}{H(z)}.$$ (11.122)

From Friedmann's equation (11.16) and eq. (11.25) follow

$$H^2(z) = H_0^2 E^2(z)$$ (11.123)

where

$$E^2(z) \equiv \Omega_{k0}(1+z)^2 + \sum_i \Omega_{i0}(1+z)^{3(1+w_i)}$$ (11.124)

and Ω_{i0} is the present relative density at t_0 of a cosmic fluid with equation of state $p_i = w_i \rho_i$, where $w_i = $ constant. In particular, for $z = 0$ this gives

$$\Omega_0 + \Omega_{k0} = 1, \qquad \Omega_0 \equiv \sum_i \Omega_{i0}.$$ (11.125)

Inserting eq. (11.123) into eq. (11.122), using $-k = \Omega_{k0}H_0^2$, and integrating leads to

$$\int_0^{r_e(z)} \frac{dr}{\sqrt{1-kr^2}} = \int_0^{r_e(z)} \frac{dr}{\sqrt{1+H_0^2\Omega_{k0}r^2}} = \frac{1}{H_0}\int_0^z \frac{dy}{E(y)},$$ (11.126)

and thus, using the change of variables $R = H_0|\Omega_{k0}|^{\frac{1}{2}}r$, we get

$$H_0|\Omega_{k0}|^{\frac{1}{2}}r_e(z) = S_k[\chi(z)],$$ (11.127)

where

$$\chi(z) = \sqrt{|\Omega_{k0}|}\int_0^z \frac{dy}{E(y)},$$ (11.128)

and $S_k(\chi)$ is given in eq. (11.9). Inserting this into eq. (11.119) finally leads to the general redshift-luminosity relation

$$d_L = \frac{1+z}{H_0\sqrt{|\Omega_{k0}|}}S_k\left[\sqrt{|\Omega_{k0}|}\int_0^z \frac{dy}{E(y)}\right],$$ (11.129)

where the function $E(z)$ is defined in eq. (11.124). In the case of a flat universe the relation reduces to

$$d_L = \frac{1+z}{H_0}\int_0^z \frac{dy}{E(y)}.$$ (11.130)

To second order in z this gives

$$d_L \approx \frac{z}{H_0}\left[1 + \frac{z}{2}(1-q)\right].$$ (11.131)

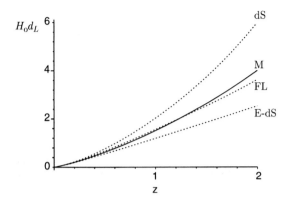

Figure 11.7: Luminosity distance as a function of cosmic redshift for four universe models; a de Sitter model (dS) with $\Omega_{v0} = 1$, $\Omega_{m0} = 0$ and a Friedmann-Lemaître model (FL) with $\Omega_{v0} = 0.7$ and $\Omega_{m0} = 0.3$ having $\ddot{a} > 0$; the Milne model (M) with $\Omega_{v0} = \Omega_{m0} = 0$ having $\ddot{a} = 0$; the Einstein-de Sitter model (E-dS) with $\Omega_{v0} = 0$ and $\Omega_{m0} = 1$ having $\ddot{a} < 0$.

In Fig. 11.7 the luminosity distance is plotted as a function of redshift for some universe models.

Two other concepts of cosmic distance should be mentioned. *Proper distance*, d_P, is the distance at time t from source to observer. It is given by

$$d_P(t) = a(t) \int_0^{r_e} \frac{dr}{\sqrt{1 - kr^2}} = \frac{a(t)}{H_0|\Omega_{k0}|^{\frac{1}{2}}} [S^{-1}]_k (H_0|\Omega_{k0}|^{\frac{1}{2}} r_e)$$

$$= \frac{\chi(z)}{H_0|\Omega_{k0}|^{\frac{1}{2}}(1 + z)}, \tag{11.132}$$

where

$$[S^{-1}]_k(y) = \begin{cases} \arcsin y, & k > 0 \\ y, & k = 0 \ . \\ \text{arsinh}\, y, & k < 0 \end{cases} \tag{11.133}$$

In a flat universe the proper distance can be expressed very simply in terms of conformal time. Then the proper distance of a source at the time that the signal is received, is $d_P(t_0) = r_e = \eta_0 - \eta_e$. Note that the coordinate distance of an object is its present distance in a flat universe.

Finally, the *angular-diameter distance*, d_A is based on the fact that the length l subtends a smaller angle θ the further away it is. This distance is given by

$$d_A = \frac{d_L}{(1 + z)^2}. \tag{11.134}$$

It is the angular-diameter distance which is used to derive the power spectrum of the cosmic microwave background radiation. In a flat universe it is simply

$$d_A = \frac{\eta_0 - \eta_e}{1 + z}. \tag{11.135}$$

Example

Example 11.4 (Redshift-luminosity relations for some universe models)
We shall first consider an empty universe model (the so-called Milne universe model) which has $\Omega_k = 1$ and $\Omega_i = 0$. Then eq. (11.129) takes the form

$$d_L = \frac{(1+z)}{H_0} \sinh \chi, \qquad \chi = \int_0^z \frac{dy}{(1+y)}. \qquad (11.136)$$

Hence,

$$H_0 d_L = z + \frac{1}{2}z^2. \qquad (11.137)$$

Consider now a flat universe dominated by a perfect fluid with equation of state $p = w\rho$ where w is constant. Then eq. (11.129) takes the form

$$d_L = \frac{(1+z)}{H_0} \int_0^z (1+y)^{-\frac{3}{2}(1+w)} dy, \qquad (11.138)$$

which leads to

$$H_0 d_L = \frac{2}{1+3w} \left[1 + z - (1+z)^{\frac{1}{2}(1-3w)} \right]. \qquad (11.139)$$

Measuring the luminosity distance and red-shift of far-away standard radiation sources allows one to determine the difference between the cosmic density of matter, Ω_{m0}, and vacuum energy, Ω_{v0}.

The best candidates for standard sources are supernovae of type Ia. They are thought to be white dwarf stars that accumulate gas from a companion star. When the mass of the white dwarf exceeds the Chandrasekhar mass of $1.4 M_{Sun}$, where $4 M_{Sun}$ is the mass of the Sun, the white dwarf becomes unstable against the force of gravity, and it collapses. This leads to a supernova explosion with an absolute luminosity that is determined mainly by the mass of the star. Since this is around $1.4 M_{Sun}$ for all such stars their luminosity does not vary very much. The peak luminosity reaches about $10^{10} L_{Sun}$ where L_{Sun} is the absolute luminosity of the Sun. Hence, they are visible at cosmological distances.

The luminosity of an object is usually expressed by a quantity called magnitude. The *apparent magnitude* is defined in terms of the apparent luminosity by

$$m = -2.5 \log_{10} l + C, \qquad (11.140)$$

where C is a constant. The *absolute magnitude* M corresponds to the luminosity of some source if it were at a distance $10\text{pc} = 32.6\text{l.y.}$ from the observer. The difference $m - M$ is called the *distance modulus* and is given by

$$m - M = 5 \log_{10} d_L - 5 \qquad (11.141)$$

where the luminosity distance, d_L, is expressed in parsecs, pc.

Using eq. (11.131) with $q_0 = (1/2)\Omega_{m0} - \Omega_{v0}$ if we neglect radiation, we obtain second order in z for the distance modulus [Wei72]

$$m - M = 25 - 5 \log_{10} H_0 + 5 \log_{10} z + 1.086 \left(1 + \Omega_{v0} - \frac{\Omega_{m0}}{2} \right) z. \qquad (11.142)$$

In 1998 two groups of astronomers, the Supernova Cosmology Project [*Pet. al.* 99] and the High-*z* supernova search team [*Set. al.*98, *Ret. al.*98] published results of determining the distance modulus and red-shift of around fifty supernovae of type Ia with red-shifts in the range $z \sim 0.4 - 0.8$. The results indicated that we live in a flat universe with $\Omega_{m0} \approx 0.28$ and $\Omega_{v0} \approx 0.72$. Although there are some uncertainties, the results of the measurements strongly excluded that the cosmic density of the vacuum energy is zero. On the contrary, the measurements indicate that we live in a vacuum dominated universe.

11.9   Cosmological horizons

Light moves along null geodesics. Hence for light moving from an object towards an observer at $r = 0$, $ds^2 = d\theta = d\phi = 0$ so that

$$\frac{dr}{(1 - kr^2)^{\frac{1}{2}}} = -\frac{dt}{a(t)}. \tag{11.143}$$

The *particle horizon* is a surface dividing space into a region that can be seen at a time t by an observer at the origin, and one that cannot yet be seen. The coordinate radius of this surface is given by

$$\int_0^{r_{PH}} \frac{dr}{(1 - kr^2)^{\frac{1}{2}}} = \begin{cases} \int_0^t \frac{dt}{a(t)} \\ \int_{-\infty}^t \frac{dt}{a(t)} \end{cases} \tag{11.144}$$

where the lower limit in the integral on the right hand side depends upon the region of definition of t. The particle horizon represents the extension of causally connected regions in the universe, or how far an observer can see.

The *event horizon* of an observer also divides the space into two regions: one from which light can reach the observer at a finite time, and one from which this is not possible. Light emitted at the event horizon takes an infinite time to reach the observer. The coordinate radius of the event horizon is given by

$$\int_0^{r_{EH}} \frac{dr}{(1 - kr^2)^{\frac{1}{2}}} = \int_t^{\infty} \frac{dt}{a(t)}. \tag{11.145}$$

The proper distance to the horizons are

$$\ell_{PH} = a(t) \int_0^{r_{PH}} \frac{dr}{(1 - kr^2)^{\frac{1}{2}}} = \begin{cases} a(t) \int_0^t \frac{dt}{a(t)} \\ a(t) \int_{-\infty}^t \frac{dt}{a(t)} \end{cases} \tag{11.146}$$

and

$$\ell_{EH} = a(t) \int_0^{r_{EH}} \frac{dr}{(1 - kr^2)^{\frac{1}{2}}} = a(t) \int_t^{\infty} \frac{dt}{a(t)}. \tag{11.147}$$

From eq. (11.36) follows $dz = -(H/a)dt$. Hence,

$$\int \frac{dt}{a(t)} = -\int \frac{dz}{H(z)}. \tag{11.148}$$

Applying eqs. (11.123) and (11.124) we can calculate the proper distances to the horizons by solving the integral $\int \frac{dz}{E(z)}$. The radius of the observable universe is equal to the proper distance to the particle horizon.

Summarizing, one may say that the event horizon defines a region from which we will never receive information in the future about events occuring now, while the particle horizon defines a region from which information has not yet been received about events that occurred in the past.

Example   **Example 11.5 (Particle horizon for some universe models)**
Consider a flat universe model dominated by a perfect fluid with equation of state $p = w\rho$ where $w \geq -1$ is a constant. Using eq. (11.25) the Friedmann equation (11.16) takes the form

$$a^{\frac{1}{2}(1+3w)} \dot{a} = H_0. \tag{11.149}$$

Integration with normalization $a(t_0) = 1$ gives

$$a(t) = \left(\frac{t}{t_0}\right)^{\frac{2}{3(1+w)}}, \tag{11.150}$$

where the age of the universe is

$$t_0 = \frac{2}{3(1+w)H_0}. \tag{11.151}$$

For such a universe model the proper distance to the particle horizon is

$$l_{PH} = \frac{3(1+w)}{1+3w}t. \tag{11.152}$$

Eqs. (11.150) and (11.152) show that the scale factor $a(t)$ increases faster than the particle radius if $w < -1/3$. Note that this is just the condition that the expansion of the universe model accelerates. Hence, in a universe model dominated by quintessence energy with $w < -1/3$ one would observe fewer quasars in the far future.

11.10   Big Bang in an infinite Universe

We shall discuss the compatibility of the concept of a homogeneous open universe model having infinite spatial extension with the concept of a point like big bang event in spacetime. Our treatment is based upon an analysis by W. Rindler [Rin00].

Consider the questions: Does the point-like character of the big bang event contradict the homogeneity of the Friedmann model? Did big bang happen everywhere?

In order to answer these questions it is sufficient to consider a simple universe model. In 1933 E.A. Milne presented a homogeneous, isotropic, expanding universe model as a solution of Einstein's field equations for empty space. The line-element of the model is

$$ds^2 = -dt^2 + \left(\frac{t}{t_0}\right)^2 \left(\frac{dr^2}{1+\frac{r^2}{R_0^2}} + r^2 d\Omega^2\right). \tag{11.153}$$

This is called the *Milne model*. The expansion factor is $a(t) = t/t_0$, and the 3-space has negative curvature; it is an open universe model.

Introducing the coordinates

$$R = \frac{rt}{t_0}, \quad T = t\left(1 + \frac{r^2}{t_0^2}\right)^{\frac{1}{2}}, \tag{11.154}$$

the line-element takes the form

$$ds^2 = -dT^2 + dR^2 + R^2 d\Omega^2, \tag{11.155}$$

which represents Minkowski spacetime in spherical coordinates.

Let us find the shape of the coordinate curves $t = $ constant in an (T, R)-Minkowski diagram. From the transformation (11.154) follows

$$R = \frac{r}{\sqrt{1 + \frac{r^2}{t_0^2}}} \frac{T}{t_0}, \quad T^2 - R^2 = t^2. \tag{11.156}$$

Hence, the curves $r = $ constant are straight lines in the diagram. They are the world lines of the reference particles in an expanding reference frame in Minkowski spacetime with r as a comoving radial coordinate. The simultaneity space of this reference frame, $t = $ constant, are hyperbolae. In the (r, t)-coordinate system a part of the Minkowski spacetime is represented as an expanding universe model with 3-space defined by constant cosmic time $t = $ constant.

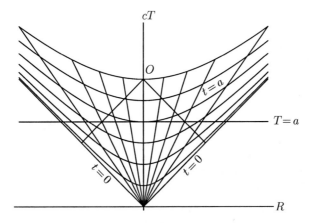

Figure 11.8: The (R, T)-system is the comoving stationary inertial system of an arbitrarily chosen test particle in the Milne universe. The figure shows the world lines and the simultaneity curves of the expanding reference frame in the Minkowski diagram referring to the stationary system.

Figure 11.8 shows the world lines and simultaneity curves of the (r, t)-system in a Minkowski diagram which refers to the static, inertial (T, R)-system. Also indicated on the figure is the intersection of the backwards lightcone of an observer at O with the plane of the paper.

W. Rindler [Rin00] pointed out that instead of regarding the Milne universe as empty, it may be useful to imagine it filled with test particles that do not gravitate. Each particle carries with it a clock showing the time t since the big bang event of the universe model. The particles have constant velocities, but particles with all velocities between 0 and c were "emitted" at the big bang event.

Since each test particle move with a constant velocity the distance between two particles is equal to their relative velocity times the cosmic time. Hence each particle moves away from an arbitrary chosen "origin-particle" with a velocity proportional to the distance to the "origin-particle". This is Hubble's law which describes the expansion of Milne's universe model.

In Fig. 11.8 the vertical axis, $R = 0$, is the world line of the chosen origin particle, which we will call the observer. All clocks in the comoving inertial frame of the observer show the proper time, T, of the observer. Hence, each horizontal line $T = a$ in Fig. 11.8 represent the simultaneity space of the observer at different points of time. Milne called this the *private space* of the observer.

The 3-space of the Milne universe, however, is defined at constant *cosmic time*. This space, at different cosmic time, is represented by the hyperbolae $t = a$ in Fig. 11.8. Milne called this the *public space*. Each hyperbola has infinite length. This means that the (negatively curved) 3-space of the Milne universe has infinite extension. Note that this is the case for every point of cosmic time $t > 0$.

The lines representing the backwards light-cone of the observer at the event O comes from the lines $t = 0$. Hence in the Milne universe model an observer can in principle observe the big bang event. Since the direction of the radial coordinate is arbitrary in an isotropic universe model, the big bang would be observed in every direction, and because the observer is arbitrary the big bang event may be observed from every position. In this sense big bang happened everywhere.

However with reference to the comoving frame of the observer big bang was a point event. Furthermore, the 3-space at constant T of that part of Minkowski spacetime that makes up the Milne universe has finite extension, $\ell = cT$, which decreases to zero at the big bang event.

The main point of this discussion is valid for open universe models in general, namely: The simultaneity of a chosen particle in an expanding universe model is different from the simultaneity of all the reference particles of the model. Hence, the corresponding 3-spaces are different. The big bang was a point event. The 3-space of a universe model with a big bang has finite extension as defined by the simultaneity of a fixed observer. However, as measured by constant cosmic time space may have infinite extension.

Problems

11.1. *Physical significance of the Robertson-Walker coordinate system*
Show that the reference particles with fixed spatial coordinates move along geodesic world lines, and hence are free particles.

11.2. *The volume of a closed Robertson-Walker universe*
Show that the volume of the region contained inside a radius $r = a\chi = a$ arcsin r is

$$V = 2\pi a^3 \left(\chi - \frac{1}{2} \sin 2\chi \right).$$

Find the maximal volume. Find also an approximate expression for V when $\chi \ll R$.

11.3. *The past light-cone in expanding universe models*

(a) Show that the radial standard coordinate of the past light-cone is

$$\chi_{lc}(t_e) = c \int_{t_e}^{t_0} \frac{dt}{a(t)}. \qquad (11.157)$$

(b) The proper distance at a point of time t to a particle at a radial coordinate χ is $d = a(t)\chi$. Differentiation gives $\dot{d} = \dot{a}\chi + a\dot{\chi}$ which may be written

$$v_{tot} = v_{rec} + v_{pec}, \qquad (11.158)$$

where v_{tot} is the total radial velocity of the particle; v_{rec} is its *recession velocity*; and v_{pec} is its *peculiar velocity*.

Show that the recession velocity of a light source with redshift z is given by

$$v_{rec}(z) = c\frac{E(z)}{1+z} \int_{0}^{z} \frac{dy}{E(y)}. \qquad (11.159)$$

Can this velocity be greater than the speed of light? What is the total velocity of a photon emitted towards $\chi = 0$? Is it possible to observe a galaxy with recession velocity greater than the speed of light?

(c) Make a plot of the past light-cone; i.e. of t_e as a function of the proper distance

$$d_{lc} = a(t_e)\chi_{lc}(t_e),$$

for a flat, matter dominated universe model. Explain the shape of the light-cone using that its slope is equal to the total velocity of a photon emitted towards an observer at the origin.

(d) Introduce conformal time and calculate the coordinate distance of the past light-cone as a function of conformal time for the flat, matter dominated universe model. Make a plot of the past light-cone in these variables.

11.4. *Lookback time*
The lookback time of an object is the time required for light to travel from an emitting object to the receiver. Hence, it is $t_L \equiv t_0 - t_e$, where t_0 is the point of time that the object is observed and t_e is the point of time the light was emitted.

(a) Show that the lookback time is given by

$$t_L = \frac{1}{H_0} \int_{0}^{z} \frac{dy}{(1+y)E(y)}, \qquad (11.160)$$

where z is the redshift of the object.

(b) Show that $t_L = t_0 \left[1 - (1+z)^{-3/2}\right]$, where $t_0 = 2/(3H_0)$, in a flat, matter-dominated universe.

(c) Show that the lookback time in the Milne universe model with $a(t) = (t/t_0)$, $k < 0$, is

$$t_L = \frac{1}{H_0} \frac{z}{1+z}.$$

(d) Make a plot with t_L as a function of z for the last two universe models.

11.5. *The FRW-models with a w-law perfect fluid*
In this problem we will investigate FRW models with a perfect fluid. We will assume that the perfect fluid obeys the equation of state

$$p = w\rho \qquad (11.161)$$

where $-1 \leq w \leq 1$.

(a) Write down the Friedmann equations for a FRW model with a w-law perfect fluid. Express them in terms of the scale factor a only.

(b) Assume that $a(0) = 0$. Show that when $-1/3 < w \leq 1$, the closed model will recollapse. Explain why this does not happen in the flat and open models.

(c) Solve the Friedmann equation for a general $w \neq -1$ in the flat case. What is the Hubble parameter and the deceleration parameter? Write also down the time evolution for the matter density.

(d) Find the particle horizon distance in terms of H_0, w and z.

(e) Specialize the above to the dust and radiation dominated universe models.

11.6. *Radial motion in a Friedmann universe* [GE07]

(a) Calculate the Christoffel symbols $\Gamma^\chi_{t\chi}$ and $\Gamma^t_{\chi\chi}$ from the line-element (11.10).

(b) Consider the radial motion of a free particle in the space-time described by this line-element. Show that the geodesic equations then reduce to

$$\frac{d^2\chi}{d\tau^2} = -2\frac{\dot{a}}{a}\frac{\dot{\chi}}{\dot{t}^2}, \qquad \frac{d^2t}{d\tau^2} = -\frac{a\dot{a}}{c^2}\frac{\dot{\chi}^2}{\dot{t}^2}, \qquad (11.162)$$

where the dot denotes differentiation with respect to coordinate time, and τ is the proper time of the particle.

(c) Express $\dot{\chi}$ and $\ddot{\chi}$ in terms of derivatives with respect to τ and show that the equation of motion of the particle takes the form

$$\ddot{\chi} = \frac{a\dot{a}}{c^2}\dot{\chi}^3 - 2\frac{\dot{a}}{a}\dot{\chi}. \qquad (11.163)$$

Introduce the dimensionless variable $y = c^2/\dot{\chi}^2$ and show that the equation can be written

$$\dot{y} - (\ln a^4)\dot{}y = -(a^2)\dot{}. \qquad (11.164)$$

(d) Find the general solution of this equation and show that the radial coordinate of the particle is

$$\chi = \pm \int \frac{cdt}{a\sqrt{1 + Aa^2}}, \qquad (11.165)$$

where A is a positive constant.

(e) Using the result $a = (t/t_0)^{2/3(1+w)}$ from Problem 11.5c, show that in a flat universe with a perfect fluid with equation of state $p = w\rho$, $w = $ constant, the result in (d) reduces to

$$\chi = \chi_0 \pm \frac{3}{2}(1+w)ct_0 \int_1^a \frac{x^{\frac{1}{2}(3w-1)}}{\sqrt{1+Ax^2}}dx, \qquad (11.166)$$

where the initial condition is $\chi = \chi_0$ and $t = t_0$.

(f) We shall now assume that the particle is initially at rest relative to an observer at $\chi = 0$ and is then let free. The proper distance of the particle from the observer is $\ell = a\chi$. Show that the initial coordinate velocity of the particle is

$$\dot{\chi}_0 = -H_0\chi_0,$$

where H_0 is the present value of the Hubble parameter.

(g) Differentiate $\ell = a\chi$ twice and use the equation of motion of the particle together with $\dot{\chi}_0 = -H_0\chi_0$ to show that the initial acceleration of the particle after it has been let free is

$$\ddot{\ell}_0 - \left(\ddot{a}_0 - \frac{H_0^4}{c^2}\chi_0^2\right)\chi_0.$$

Find what happens to a particle which is let free at $\chi_0 > \frac{c\sqrt{|\ddot{a}_0|}}{H_0^2}$ and try to explain its behaviour.

11.7. Age-density relations

(a) Show that the age of a radiation dominated universe model is given by

$$t_0 = \frac{1}{H_0} \cdot \frac{1}{1 + \sqrt{\Omega_{\gamma 0}}}, \qquad (11.167)$$

for all values of k.

(b) Show that the age of a matter dominated universe model with $k > 0$ may be expressed by

$$t_0 = \frac{\Omega_{m0}}{2H_0(\Omega_{m0}-1)^{\frac{3}{2}}} \left[\arccos\left(\frac{2}{\Omega_{m0}} - 1\right) - \frac{2}{\Omega_{m0}}(\Omega_{m0}-1)^{\frac{1}{2}}\right] \quad (11.168)$$

and of a matter dominated universe model with $k < 0$

$$t_0 = \frac{\Omega_{m0}}{2H_0(1-\Omega_{m0})^{\frac{3}{2}}} \left[\frac{2}{\Omega_{m0}}(1-\Omega_{m0})^{\frac{1}{2}} - \text{arcosh}\left(\frac{2}{\Omega_{m0}} - 1\right)\right] \quad (11.169)$$

(c) Show that the lifetime of the closed universe is

$$T = \frac{\pi}{H_0} \frac{\Omega_{m0}}{(\Omega_{m0}-1)^{\frac{3}{2}}}, \qquad (11.170)$$

and that the scale factor at maximum expansion is

$$a_{\max} = \frac{\Omega_{m0}}{\Omega_{m0}-1}. \qquad (11.171)$$

11.8. *Redshift-luminosity relation for matter dominated universe*
Show that the luminosity distance of an object with redshift z in a matter dominated universe with relative density Ω_0 and Hubble parameter H_0 is

$$d_L = \frac{2c}{H_0 \Omega_0^2}\left[\Omega_0 z + (\Omega_0 - 2)(\sqrt{1 + \Omega_0 z} - 1)\right]. \tag{11.172}$$

For the Einstein-de Sitter universe, with $\Omega_0 = 1$, this relation reduces to

$$d_L = \frac{2c}{H_0}(1 + z - \sqrt{1 + z}). \tag{11.173}$$

Plot this distance in light years as a function of z for a universe with Hubble
parameter $H_0 = 20$km/s per light years.

11.9. *Newtonian approximation with vacuum energy*
Show that Einstein's linearised field equations for a static spacetime containing dust with density ρ and vacuum energy with density ρ_Λ takes the form of
a modified Poisson equation

$$\nabla^2 \phi = 4\pi G(\rho - 2\rho_\Lambda). \tag{11.174}$$

11.10. *Universe with multi-component fluid*
Consider a FRW universe model with perfect fluids

$$\Omega_i, \quad \rho_i = w_i p_i. \tag{11.175}$$

Show that the deceleration parameter q can be written as

$$q = \frac{1}{2E^2}\sum_i \Omega_{i0}(1 + z)^{3(1+w_i)}(1 + 3w_i) \tag{11.176}$$

where E is defined in eq. (11.124). What is q for $z = 0$?

Consider a universe with cold dark matter (dust) and vacuum energy. Find
the redshift z_1 at which the universe went from cosmic retardation to acceleration. Express z_1 in terms of the Ω_{i0}'s.

11.11. *Gravitational collapse*
In this problem we shall find a solution to Einstein's field equations describing a spherical symmetric gravitational collapse. The solution shall describe
the spacetime both exterior and interior to the star. From the Jebsen-Birkhoff
theorem [Jeb21], stated in problem 10.15 on page 262, the exterior metric
is the Schwarzschild metric. But to connect the exterior and interior solutions,
the metrics must be expressed in the same coordinate system. We will assume that the interior solution has the same form as a Friedmann solution.
The Friedmann solutions are expressed in comoving coordinates, thus freely
falling particles have constant spatial coordinates.

Let (ρ, τ) be the infalling coordinates. τ is the proper time to a freely falling
particle starting at infinity with zero velocity. These coordinates are connected
to the Schwarzschild coordinates via the requirements

$$\begin{aligned}
\rho &= r, \quad \text{for} \quad \tau = 0 \\
\tau &= t, \quad \text{for} \quad r = 0.
\end{aligned}$$

$$\tag{11.177}$$

(a) Show that the transformation between the infalling coordinates and the Schwarzschild coordinates is given by

$$\tau = \frac{2}{3}(2M)^{-\frac{1}{2}}\left(\rho^{\frac{3}{2}} - r^{\frac{3}{2}}\right),$$

$$t = \tau - 4M\left(\frac{r}{2M}\right)^{\frac{1}{2}} + 2M\ln\left[\frac{\left(\frac{r}{2M}\right)^{\frac{1}{2}} + 1}{\left(\frac{r}{2M}\right)^{\frac{1}{2}} - 1}\right],$$

where M is the Schwarzschild mass of the star. Show that the Schwarzschild metric in these coordinates takes the form

$$ds^2 = -d\tau^2 + \left[1 - \frac{3}{2}(2M)^{\frac{1}{2}}\tau\rho^{-\frac{3}{2}}\right]^{-\frac{2}{3}}d\rho^2$$

$$+ \left[1 - \frac{3}{2}(2M)^{\frac{1}{2}}\tau\rho^{-\frac{3}{2}}\right]^{\frac{4}{3}}\rho^2\left(d\theta^2 + \sin^2\theta d\phi^2\right). \quad (11.178)$$

Show that the metric is not singular at the Schwarzschild radius. Where is it singular?

(b) Assume the star has a position dependent energy-density $\varrho(\tau)$, and that the pressure is zero. Assume further that the interior spacetime can be described with a Friedmann solution with Euclidean geometry ($k = 0$). Find the solution when the radius of the star is R_0 at $\tau = 0$.

11.12. Cosmic redshift
We shall in this problem study the cosmic redshift in an expanding FRW universe and show that this redshift, for small distances between emitter and receiver, can be split into a gravitational and a kinematic part.

(a) Show that the assumption that the distance between emitter and receiver is small, can be expressed as

$$H_0(t_0 - t_e) \ll 1.$$

Here, the lower index of 0 and e mean evaluated at the receiver and emitter, respectively.

In the following, include only terms to 2nd order in $H_0(t_0 - t_e)$.

(b) Light is emitted at wavelength λ_e and received at λ_0. Show that the redshift, z, can be written as

$$z = H_0(t_0 - t_e) + \left(1 + \frac{q_0}{2}\right)H_0^2(t_0 - t_e)^2, \quad (11.179)$$

where q is the deceleration parameter.

(c) Use the relativistic formula for the Doppler shift,

$$z_D = \sqrt{\frac{1 + v_e}{1 - v_e}} - 1, \quad (11.180)$$

where v_e (in units where $c = 1$) is the velocity of the source when it emitted the light, together with the formula for the expansion velocity, $v_e = \dot{a}_e\chi_e$, to show that to second order in v_e the formula for the Doppler shift takes the form

$$z_D = \dot{a}_e\chi_e + \frac{1}{2}\dot{a}_e^2\chi_e^2. \quad (11.181)$$

(d) Use a Taylor expansion of $\dot{a}$ to first order in $H_0(t_0 - t_e)$ and the expression (11.42) for χ_e in terms of $H_0(t_0 - t_e)$ to show that

$$z_D = H_0(t_0 - t_e) + (1 + q_0)H_0^2(t_0 - t_e)^2, \qquad (11.182)$$

to second order in $H_0(t_0 - t_e)$.

(e) The Newtonian expression for the gravitational frequency shift is

$$z_G = -\phi_e, \qquad (11.183)$$

where ϕ_e is the gravitational potential at the emitter, and the potential has been defined so that $\phi_0 = 0$ at the observer. Assuming that the universe is dominated by dust with density ρ_0, show that

$$\phi_e = \frac{2\pi G}{3}\rho_0\chi_e^2. \qquad (11.184)$$

(f) Show that the Friedmann equations give

$$\frac{4\pi G}{3}\rho_0 = q_0 H_0^2, \qquad (11.185)$$

for a flat, dust dominated universe. Using this and that $\chi_e = t_0 - t_e$ to lowest order, show that the gravitational frequency shift is

$$z_G = -\frac{1}{2}q_0 H_0^2(t_0 - t_e)^2. \qquad (11.186)$$

Is this a redshift or a blueshift? Explain!

(g) Show from the previous results that for small values of z the cosmic redshift can be separated into a Doppler effect and a gravitational frequency shift, as follows:

$$z = z_D + z_G. \qquad (11.187)$$

11.13. *Universe models with constant deceleration parameter*

(a) Show that the universe with constant deceleration parameter q has expansion factor

$$a = \left(\frac{t}{t_0}\right)^{\frac{1}{1+q}}, \quad q \neq -1, \text{ and } a \propto e^{Ht}, \quad q = -1.$$

(b) Find the Hubble length $\ell_H = H^{-1}$ and the radius of the particle horizon as functions of time for these models.

11.14. *Relative densities as functions of the expansion factor*
Show that the relative densities of matter and vacuum energy as functions of a are

$$\Omega_v = \frac{\Omega_{v0}a^3}{\Omega_{v0}a^3 + (1 - \Omega_{v0} - \Omega_{m0})a + \Omega_{m0}}$$

$$\Omega_m = \frac{\Omega_{m0}}{\Omega_{v0}a^3 + (1 - \Omega_{v0} - \Omega_{m0})a + \Omega_{m0}} \qquad (11.188)$$

What can you conclude from these expressions concerning the universe at early and late times?

11.15. *FRW universe with radiation and matter*
Show that the expansion factor and the cosmic time as functions of conformal time of a universe with radiation and matter are

$$k > 0: \quad \begin{cases} a = a_0 \left[\alpha(1 - \cos\eta) + \beta\sin\eta\right] \\ t = a_0 \left[\alpha(\eta - \sin\eta) + \beta(1 - \cos\eta)\right] \end{cases} \tag{11.189}$$

$$k = 0: \quad \begin{cases} a = a_0 \left[\tfrac{1}{2}\alpha\eta^2 + \beta\eta\right] \\ t = a_0 \left[\tfrac{1}{6}\alpha\eta^3 + \tfrac{1}{2}\beta\eta^2\right] \end{cases} \tag{11.190}$$

$$k < 0: \quad \begin{cases} a = a_0 \left[\alpha(\cosh\eta - 1) + \beta\sinh\eta\right] \\ t = a_0 \left[\alpha(\sinh\eta - \eta) + \beta(\cosh\eta - 1)\right] \end{cases} \tag{11.191}$$

where $\alpha = a_0^2 H_0^2 \Omega_{m0}/2$ and $\beta = (a_0^2 H_0^2 \Omega_{\gamma 0})^{1/2}$, and $\Omega_{\gamma 0}$ and Ω_{m0} are the present relative densities of radiation and matter, H_0 is the present value of the Hubble parameter.

12

Universe Models with Vacuum Energy

Soon after Einstein had introduced the cosmological constant he withdrew it and called it "the biggest blunder" of his life. However, there has been developments in the last decades that have given new life to the cosmological constant. Firstly, the idea of *inflation* gave cosmology a whole new view upon the first split second of our universe. A key ingredient in the inflationary model is the behaviour of models that have a cosmological constant-like behaviour. Secondly, recent observations may indicate that we live in an accelerated universe. The inclusion of a cosmological constant can give rise to such behaviour as we will show in this chapter. We will first start with the static solution that Einstein found and was the reason that Einstein introduced the cosmological constant in the first place.

12.1 Einstein's static universe

The Einstein field equations with a cosmological constant are (see eq. (9.42))

$$R_{\mu\nu} - \frac{1}{2}g_{\mu\nu}R + \Lambda g_{\mu\nu} = 8\pi G T_{\mu\nu}. \tag{12.1}$$

We assume that the space-time is homogeneous and isotropic as in the previous chapter. The line-element has the form

$$ds^2 = -dt^2 + a(t)^2 \left(\frac{dr^2}{1 - kr^2} + r^2(d\theta^2 + \sin^2\theta d\phi^2) \right). \tag{12.2}$$

The components of the Einstein tensor were calculated in the previous chapter for this metric, eq. (11.13). Using this the field equations are

$$3\frac{\dot{a}^2 + k}{a^2} = 8\pi G\rho + \Lambda \tag{12.3}$$

$$-2\frac{\ddot{a}}{a} - \frac{\dot{a}^2 + k}{a^2} = 8\pi G p - \Lambda. \tag{12.4}$$

Note that eq. (11.19) is still valid with a non-vanishing Λ. Einstein also assumed that we lived in a static matter dominated universe where $p = 0$. This immediately leads to

$$\Lambda = 4\pi\rho = \frac{k}{a^2}. \tag{12.5}$$

Thus the only possibility is that the universe is closed, $k > 0$, and that $a^2 = \Lambda^{-1}$. Einstein's static solution is therefore given by

$$ds^2 = -dt^2 + \frac{1}{\Lambda}\left(\frac{dr^2}{1 - \frac{r^2}{R_0^2}} + r^2(d\theta^2 + \sin^2\theta d\phi^2)\right). \tag{12.6}$$

The metric is often written in Schwarzschild coordinates. Rescaling the radial coordinate by defining $R = ra$, we get for the Einstein's static universe

$$\boxed{ds^2 = -dt^2 + \frac{dR^2}{1 - \Lambda R^2} + R^2(d\theta^2 + \sin^2\theta d\phi^2).} \tag{12.7}$$

The later observations made by Edwin Hubble that the universe was expanding, dethroned this model as the model for our universe. In addition to this, physicists noticed that this model is unstable. The configuration between matter and the cosmological constant in this model is highly fine-tuned. Einstein's field equations show that for any small perturbation away from this configuration the universe tends to enlarge this perturbation even more. If the matter has a slightly higher density, then the universe will recollapse! Hence, the static universe is unstable and is therefore unphysical.

12.2   de Sitter's solution

We will now solve the vacuum ($T_{\mu\nu} = 0$) field equations for a homogeneous and isotropic model with a positive cosmological constant. The solutions we find will be the simplest inflationary solutions and are highly interesting for various reasons.

   The Einstein field equations are

$$3\frac{\dot{a}^2 + k}{a^2} - \Lambda = 0$$

$$-2\frac{\ddot{a}}{a} - \frac{\dot{a}^2 + k}{a^2} + \Lambda = 0. \tag{12.8}$$

The first of the above equations can be written

$$\dot{a}^2 - \omega^2 a^2 = -k \tag{12.9}$$

where $\omega^2 = \frac{\Lambda}{3}$. This equation has the following solution (see Fig. 12.1)

$$a(t) = \begin{cases} \frac{\sqrt{k}}{\omega}\cosh\omega t, & k > 0 \\ e^{\omega t}, & k = 0 \\ \frac{\sqrt{|k|}}{\omega}\sinh\omega t, & k < 0. \end{cases} \tag{12.10}$$

Note that in the closed case, $k > 0$, the universe obtains an expansion minimum at $t = 0$ which is not zero. The universe contracts for $t < 0$, reaches

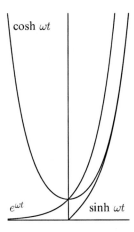

Figure 12.1: The scale factor as a function of time for de Sitter's solutions.

a minimum size at $t = 0$ and thereafter expands for all time. The universe "bounces" and is not reaching a singularity. The cases $k = 0$ and $k < 0$ are more subtle. For $k = 0$ the universe is not really reaching zero size until $t \longrightarrow -\infty$. Hence, the universe has no singularity for any finite t. The case $k < 0$ has what appears to be a zero-size singularity at $t = 0$, but as we shall see later, this can be seen upon as a coordinate singularity.

Let us consider the flat case $k = 0$. When we calculate the Hubble parameter, we get

$$H = \frac{\dot{a}}{a} = \omega. \tag{12.11}$$

Hence, in this case the Hubble parameter is a constant, and the metric can be written

$$ds^2 = -dt^2 + e^{2Ht}\left(dx^2 + dy^2 + dz^2\right). \tag{12.12}$$

Typical for the de Sitter models is that they possess horizons. The flat model, for instance, has an event horizon at a distance $1/H$ (see Fig. 12.2). Also particle horizons are present for the de Sitter solutions, as we will see in the next example.

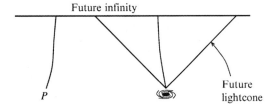

Figure 12.2: A cosmological horizon: Galaxies farther away than $1/H$ in the flat de Sitter universe model is hidden from our view.

Example

Example 12.1 (The particle horizon of the de Sitter universe)
We shall first consider the particle horizon for the de Sitter universe models with $k > 0$. The expansion factor is $a(t) = (1/\omega)\cosh\omega t$ where $\omega = (\Lambda/3)^{1/2}$. The coordinate radius of the particle horizon as a function of time is given in eq. (11.144), which in this case takes the form

$$\int_0^{r_{PH}} \frac{dr}{R_0\sqrt{1 - \frac{r^2}{R_0^2}}} = \int_{-\infty}^{t} \frac{\omega dt}{\cosh\omega t} \tag{12.13}$$

which gives

$$\arcsin\frac{r_{PH}}{R_0} = \arcsin(\tanh\omega t) - \frac{\pi}{2}$$

where we have used that $\tanh(-\infty) = -1$ and $\arcsin(-1) = 3\pi/2$. This gives

$$\frac{r_{PH}}{R_0} = \sin u \cos v - \cos u \sin v$$

where $u = \arcsin(\tanh\omega t)$, $v = 3\pi/2$. Hence $\sin u = \tanh\omega t$, $\cos u = 1/\cosh\omega t$ which leads to

$$\frac{r_{PH}}{R_0} = \frac{1}{\cosh\omega t}. \tag{12.14}$$

The proper distance to the particle horizon is

$$d_{PH} = a(t)\arcsin\frac{r_{PH}}{R_0} = \frac{1}{\omega}\cosh\omega t \cdot \arcsin\left(\frac{1}{\cosh\omega t}\right). \tag{12.15}$$

These equations tell that during the contracting period, $t < 0$, this universe model has a particle horizon with a proper distance which increases from 1 at $\omega t = -\infty$ to $\pi/2$ at $t = 0$, while the coordinate distance of the horizon increases from zero at $t = -\infty$ to 1 at $t = 0$. At the moment $t = 0$ the whole 3-space from $r = 0$ to $r = 1$ is inside the particle horizon. An observer at $r = 0$ is from now on able to see the whole of this space. For positive values of t there is no particle horizon.

The integrals on the right hand side of eq. (11.144) diverge for the vacuum dominated models with $k = 0$ and $k < 0$. They have no particle horizon.

An event horizon in the de Sitter models can be seen when we use Schwarzschild coordinates. As we will show in the next section, the de Sitter space in Schwarzschild coordinates is

$$ds^2 = -\left(1 - H^2 r^2\right)dt^2 + \frac{dr^2}{1 - H^2 r^2} + r^2\left(d\theta^2 + \sin^2\theta d\phi^2\right). \tag{12.16}$$

Clearly, at $r = 1/H$ the metric is singular. This singularity is much like the coordinate singularity at the horizon in the Schwarzschild spacetime except that we are now *inside* the horizon. We can send signals out through the horizon at $r = 1/H$, but there is no way we can receive information from outside the horizon.

The similarity between the horizon in the Schwarzschild spacetime and the horizon in de Sitter's solution is striking. Both of them can be assigned a temperature and an entropy. Their nature are quite different – one is a black hole horizon while the other is a cosmological horizon – but still they have many of the same thermodynamical properties.

12.3   The de Sitter hyperboloid: The many guises of the de Sitter spacetime

Consider the hyperboloid given by

$$-T^2 + X^2 + Y^2 + Z^2 + W^2 = R_0^2, \tag{12.17}$$

embedded in flat 5-dimensional Minkowski space

$$ds_5^2 = -dT^2 + dX^2 + dY^2 + dZ^2 + dW^2. \tag{12.18}$$

The radius R_0 is related to Λ via $R_0^2 = 3/\Lambda$. First of all we note that the de Sitter hyperboloid is invariant under Lorentz transformations in 5-dimensional Minkowski space with respect to the origin of the Minkowski space. This can be seen as follows. The Lorentz transformations L are linear transformations

$$\bar{\mathbf{X}} = \mathsf{L}\mathbf{X}, \tag{12.19}$$

that leaves the metric invariant, i.e.

$$\mathsf{L}^t \eta \mathsf{L} = \eta. \tag{12.20}$$

We note that the hyperboloid eq. (12.17) can be written

$$\mathbf{X}^t \eta \mathbf{X} = R_0^2. \tag{12.21}$$

From eq. (12.19) we have

$$\mathbf{X}^t \eta \mathbf{X} = \mathbf{X}^t \mathsf{L}^t \eta \mathsf{L} \mathbf{X} = (\mathsf{L}\mathbf{X})^t \eta (\mathsf{L}\mathbf{X}) = \bar{\mathbf{X}}^t \eta \bar{\mathbf{X}}. \tag{12.22}$$

Both the metric of the ambient space and the hyperboloid are invariant under such transformations. Thus the Lorentz transformations have to be *isometries* for the hyperboloid. Lorentz transformations in 5-dimensional Minkowski spacetime form a 10 dimensional space; hence, this hyperboloid is a maximally symmetric space.

First we choose a set of global coordinates. Let $d\Omega_3^2$ be the metric on the unit 3-sphere and define R as

$$R^2 = X^2 + Y^2 + Z^2 + W^2. \tag{12.23}$$

The de Sitter hyperboloid then becomes

$$-T^2 + R^2 = R_0^2, \tag{12.24}$$

which can be parameterized by

$$T = R_0 \sinh t \tag{12.25}$$
$$R = R_0 \cosh t. \tag{12.26}$$

Inserting this into the 5-dimensional Minkowski metric, we get the induced metric

$$ds^2 = R_0^2 \left(-dt^2 + \cosh^2 t \, d\Omega_3^2 \right). \tag{12.27}$$

By a rescaling of t, this is the same metric as the de Sitter's solution eq. (12.10) with $k > 0$ and $R_0 = 1/\omega$. The other solutions for $k = 0$ and $k < 0$ can also be found, but these cover only part of the de Sitter hyperbola (see Fig. 12.3).

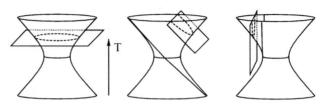

Figure 12.3: The de Sitter hyperboloid: The different de Sitter solutions are different sections of this hyperboloid. From left to right: closed, flat and open spatial sections.

The flat solution $k = 0$ are the sections given by

$$
\begin{aligned}
T - X &= R_0 e^t \\
T + X &= R_0(r^2 e^t - e^{-t}) \\
Y &= R_0 r e^t \cos\phi\cos\theta \\
Z &= R_0 r e^t \cos\phi\sin\theta \\
W &= R_0 r e^t \sin\phi.
\end{aligned}
\tag{12.28}
$$

The hyperbolic $k < 0$ sections are

$$
\begin{aligned}
T &= R_0\sqrt{1 + r^2}\sinh t \\
X &= R_0\cosh t \\
Y &= R_0 r \sinh t \cos\phi\cos\theta \\
Z &= R_0 r \sinh t \cos\phi\sin\theta \\
W &= R_0 r \sinh t \sin\phi.
\end{aligned}
\tag{12.29}
$$

These different sections of de Sitter space are similar to the conical sections from classical geometry. We can also parameterize the hyperboloid as

$$
\begin{aligned}
T &= R_0\sqrt{1 - r^2}\sinh t \\
X &= R_0 r \\
Y &= R_0\sqrt{1 - r^2}\cosh t \cos\phi\cos\theta \\
Z &= R_0\sqrt{1 - r^2}\cosh t \cos\phi\sin\theta \\
W &= R_0\sqrt{1 - r^2}\cosh t \sin\phi.
\end{aligned}
\tag{12.30}
$$

Inserting this into the metric (12.18) we see that this is the static de Sitter space-time, eq. (12.16). Hence, we have shown that all of de Sitter's solutions can be seen upon as different foliations of the same space! In particular, this shows that the singularity at $t = 0$ for the $k < 0$ de Sitter solution, eq. (12.10) can be viewed upon as a coordinate singularity.

A thorough discussion of the de Sitter spacetime as described in various coordinate systems is found in [EG95].

12.4   The horizon problem and the flatness problem

The cosmological constant laid almost dead for several decades. Not many physicists or astronomers believed that the cosmological constant had any-thing to do with the real world. Measurements of the evolution of the uni-verse and the matter content in it, showed that the cosmological constant was very close to zero. Therefore it was assumed that $\Lambda = 0$.

However, there were a couple of observations that puzzled the physicists for a long time. Gravity has a tendency to clump matter together and form inhomogeneities. This process causes galaxies to form and stars and planets to form. It is an irreversible process that has been going on since the beginning of time. Hence, if gravity steadily is clumping matter together and forms inhomogeneities, then the universe must have been in an extreme state of homogeneity initially. This seems very unlikely because one expects that the universe was formed in a rather arbitrary state. A homogeneous and isotropic state is quite special; an inhomogeneous state is by far a more general state than a homogeneous one.

Also the *horizon problem* disturbed the cosmologists. The cosmic microwave background was seen to be very homogeneous and isotropic. Actually it is the most perfect blackbody known to man. The isotropy of the radiation indicates that the radiation had thermal contact once in the past, before it was emitted. But there was no universe model which seemed to explain this; the radiation of the cosmic microwave background coming from one direction could not have been in thermal contact with the radiation in different directions (see Fig. 12.4). This is what is called the *horizon problem* because the particle horizon to each photon in the last scattering surface[1] only covers a small patch of the sky.

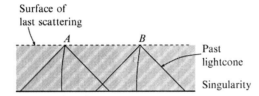

Figure 12.4: The Horizon problem: The horizon of the photons in the last scattering surface covers only a very small patch of the sky.

In order to find a simple quantitative expression of the horizon problem we shall consider the Einstein-de Sitter universe model. For this model the proper distance to the particle horizon is

$$\ell_{PH} = t^{\frac{2}{3}} \int_0^t x^{-\frac{2}{3}} dx = 3t. \tag{12.31}$$

The volume inside the horizon is therefore $V_{PH} \propto t^3$. Hence, the "horizon volume" at the time of decoupling is

$$(V_{PH})_d = \left(\frac{t_d}{t_0}\right)^3 V_0, \tag{12.32}$$

where V_0 is the present magnitude of the horizon volume i.e of the presently observable part of the universe. Events inside the horizon volume are causally connected. A volume with size $(V_{PH})_d$ may be in thermal equilibrium at the time of decoupling.

---

[1]The last scattering surface is the three dimensional spatial hypersurface for which the universe became transparent. At the last scattering surface the photons decoupled from the matter in the universe and became more or less free photons.

Let $(V_0)_d$ be the magnitude at the time of the decoupling, of V_0. Then

$$(V_0)_d = \frac{a^3(t_d)}{a^3(t_0)}V_0 = \left(\frac{t_d}{t_0}\right)^2 V_0. \tag{12.33}$$

From eqs. (12.32) and (12.33) follow

$$\frac{(V_{PH})_d}{(V_0)_d} = \frac{t_d}{t_0}. \tag{12.34}$$

Inserting the time of decoupling $t_d = 3 \cdot 10^5$ years and the present age of the universe $t_0 = 15 \cdot 10^9$ years we get $(V_{PH})_d/(V_0)_d = 2 \cdot 10^{-5}$. This shows that at the time of decoupling the volume of a causally connected region was only a $2 \cdot 10^{-5}$ part of the region representing our observable part of the universe. This is the quantitative expression of the horizon problem.

We may also deduce a quantitative expression of the *flatness problem*. From eqs. (11.48) and (11.49) follow that the time evolution of the total relative density is given by

$$\Omega - 1 = \frac{k}{\dot{a}^2}. \tag{12.35}$$

From eq. (11.63) follows that in the case of a radiation dominated universe with near critical density

$$\frac{\Omega - 1}{\Omega_0 - 1} = \frac{t}{t_0}. \tag{12.36}$$

The order of magnitude of $\Omega_0 - 1$ is not larger than one, maybe less. When we are going to stipulate initial values for the universe it is natural to consider the Planck time, $t_P = 10^{-43}$s. It follows that the magnitude of $\Omega - 1$ was less than 10^{-60} at the Planck time [2]. Such an extreme adjustment could not be explained within the old standard Big Bang model of the universe. However the problems were solved in a natural way within the frame of the inflationary universe models. We shall give a brief summary of the physical ideas behind these universe models and start by considering gauge theories and spontaneous breaking of symmetries.

12.5   Inflation

The idea of "gauge invariance" was first proposed by Herman Weyl in 1919. The only known elementary particles were the electron and the proton, and the only known fundamental forces were gravitation and electromagnetism. These were described by the general theory of relativity and Maxwell's theory of electromagnetism, respectively. A fundamental principle in the theory of relativity is that the laws of physics should be formulated in a coordinate independent way. As we have seen this requires that partial derivatives are replaced by covariant derivatives, which requires the introduction of connection coefficients, $\Gamma^\alpha_{\mu\nu}$, that also appear in the equations of motion of particles in a gravitational field.

---

[2]This argument is potentially flawed. In fact, allowing for anisotropies, the value $\Omega = 1$ is unstable both in the future and in the past. Notwithstanding this, why its value is so close to unity still has to be explained.

Weyl wanted to formulate a unified theory of gravitation and electromagnetism. His idea was that since the motion of particles in a gravitational field is determined by connections introduced in the covariant derivative, there should also exist a connection determining the motion of charged particles in electromagnetic fields. He suggested that the laws of nature should be formulated in a scale invariant, or gauge invariant, way and introduced a gauge-covariant derivative, $D_\mu = \partial_\mu - iqA_\mu$, in order to formulate the laws in a gauge invariant way. Here, A_μ are the covariant components of the electromagnetic vector potential.

It was later pointed out that Weyl's idea of scale invariance was in conflict with quantum mechanics as the Compton-wavelength of a particle, $\lambda = h/mc$, defines a position independent scale for the particle. However, with the development of quantum mechanics Weyl, Fock, and London could in 1927 and 1928 give the mathematical formalism of Weyl's theory a new meaning. The new idea was to require that the laws of Nature should be represented by equations independent of the phase of the wave-function of the particle. The gauge-covariant derivative was replaced by a "quantum operator", $-i\hbar\nabla_\mu - eA_\mu$. In this way one was able to formulate a theory which contained the principle of gauge invariance, interpreted as a phase invariance.

The next main idea in the conceptual evolution towards the inflationary universe models was the introduction of the Higgs mechanism in order to explain the masses of the gauge bosons mediating the weak interaction. The main idea is that bosons that are originally massless, obtain an effective mass by interacting with vacuum. In this theory the energy of the vacuum is represented by certain fields, ϕ, called Higgs fields. Different values of the Higgs fields cause different energies of the vacuum.

This can happen when the Higgs field has a temperature dependent potential. Let us, as an illustration, consider a real scalar field with the Lagrange density

$$\mathcal{L} = \frac{1}{2}\frac{\partial\phi}{\partial x^\mu}\frac{\partial\phi}{\partial x_\mu} - V(\phi), \quad V(\phi) = -\frac{1}{2}\mu^2\phi^2 + \frac{1}{4}\lambda\phi^4 \tag{12.37}$$

The potential $V(\phi)$ is shown in Fig. 12.5 for two different temperatures.

The sign of μ^2 and thereby of the form of the potential depends upon the temperature. If it is above a critical temperature T_C the potential has the form of Fig. 12.5a. Then there is a stable minimum at $\phi = 0$. If it is less the form is that of Fig. 12.5b and there are stable minima at $\phi = \pm\phi_0 = \pm|\mu|/\sqrt{\lambda}$ and an unstable maximum at $\phi = 0$. The true vacuum state of the system is a stable minimum for the potential. For $T > T_C$ the minimum is in the symmetrical state $\phi = 0$. But for $T < T_C$ the state $\phi = 0$ is unstable and is therefore called a "false vacuum". The system will then pass over to one of the stable minima at $\pm\phi_0$.

Regardless of the sign of μ^2 the potential is invariant under the symmetry transformation $\phi \leftrightarrow -\phi$. But when the system is in one of the minima at $\pm\phi_0$ it is no longer invariant under a change of sign of the field $\delta \equiv \phi \mp \phi_0$. Such a symmetry, which is not present in the actual state of the vacuum, is said to be spontaneously broken. From Fig. 12.5 is seen that the energy in a false vacuum is greater than in a true vacuum.

The false vacuum at $\phi = 0$ in Fig. 12.5 is classically unstable. There may also be a false vacuum at a local minimum of the potential, as shown in Fig. 12.6. Such a state is classically stable. But it may nevertheless be quantum mechanically

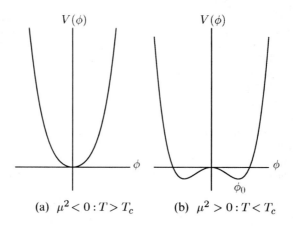

Figure 12.5: The shape of the potential $V(\phi) = -\frac{1}{2}\mu^2\phi^2 + \frac{1}{4}\lambda\phi^4$. Depicted are two different potentials corresponding to the different signs of μ^2.

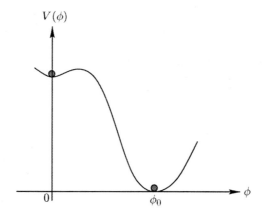

Figure 12.6: The effective potential to a system with a false vacuum at $\phi = 0$ and true vacuum at $\phi = \phi_0$.

unstable because of a finite probability of quantum tunnelling through the potential barrier.

The main idea behind the inflationary universe models was to take into consideration the consequences of the gauge theories of the fundamental interactions when constructing relativistic universe models. According to the Friedmann models the temperature was extremely high early in the history of the universe. The Higgs field of the Grand Unified Theories of the electromagnetic, weak and strong interactions has a critical temperature T_C corresponding to the energy $k_B T_C = 10^{14}$GeV. This was the temperature of the universe 10^{-35}s after the hypothetical Big Bang singularity. At this time there existed a false vacuum with mass density of the order $\rho_v = 10^{76}$kg/m^3 and radiation energy with about the same density. While the radiation energy density decreased as a^{-4} the vacuum energy density remained constant until a transition to true vacuum with a much smaller energy density. According to calculations based on suitably chosen potentials of the Higgs field this happened about 10^{-33}s after $t = 0$.

According to the original inflationary universe models the inflationary era lasted from 10^{-35}s to 10^{-33}s. However, there now exist several inflationary scenarios. In some of them the universe entered a vacuum dominated era already at the Planck time as a result of a quantum fluctuations.

We shall consider a simple inflationary scenario which illustrates the main properties of most models. In this connection it is usual to employ units so that the velocity of light and the reduced Planck constant is put equal to one and Newton's constant of gravity is related to the Planck mass by $G = m_{Pl}^{-2}$. Inserting eq. (8.84) into eq. (11.19) we obtain the equation of motion of the scalar field

$$\ddot{\phi} + 3H\dot{\phi} = -V'(\phi).\tag{12.38}$$

Assuming that the universe is flat and dominated by the scalar field, the energy density and pressure are given by

$$\rho = \frac{1}{2}\dot{\phi}^2 + V, \quad p = \frac{1}{2}\dot{\phi}^2 - V.\tag{12.39}$$

Thus eqs. (11.16) and (11.18), respectively, take the form

$$H^2 = \frac{8\pi}{3m_{Pl}^2}\left(\frac{1}{2}\dot{\phi}^2 + V\right),\tag{12.40}$$

and

$$\frac{\ddot{a}}{a} = -\frac{8\pi}{3m_{Pl}^2}\left(\dot{\phi}^2 - V\right).\tag{12.41}$$

Accelerated expansion, i.e. inflation, happens whenever the potential dominates, $V > \dot{\phi}^2$. This is realized in the case of the *slow-roll approximation*, when the term $\ddot{\phi}$ is neglected in eq. (12.38) and the term $(1/2)\dot{\phi}^2$ in eq. (12.40). Hence these equations reduce to

$$3H\dot{\phi} = -V',\tag{12.42}$$

and

$$H^2 = \frac{8\pi}{3m_{Pl}^2}V.\tag{12.43}$$

Differentiating eq. (12.42) leads to

$$\ddot{\phi} = -\frac{V''\dot{\phi}}{3H} + \frac{\dot{H}\dot{\phi}}{H}.\tag{12.44}$$

Using eqs. (12.42) and (12.43) this can be written

$$\ddot{\phi} = (-\eta + \varepsilon)H\dot{\phi},\tag{12.45}$$

where the slow-roll parameters η and ε are defined by

$$\eta = \frac{m_{Pl}^2}{8\pi}\frac{V''}{V}, \quad \varepsilon = \frac{m_{Pl}^2}{16\pi}\left(\frac{V'}{V}\right)^2.\tag{12.46}$$

Hence the slow-roll condition $|\eta| \ll 1$, $\varepsilon \ll 1$ secures that $\ddot{\phi} \ll H\dot{\phi}$ so that $\ddot{\phi}$ can be neglected in eq. (12.38). Furthermore, from the dominating terms, i.e., from eqs. (12.42) and (12.43) follow

$$\dot{\phi}^2 = \frac{V'^2}{9H^2} = \frac{m_{Pl}^2 V'^2}{24\pi V} = \frac{2}{3}\varepsilon V \ll V,, \tag{12.47}$$

which shows that the term $(1/2)\dot{\phi}^2$ of eq. (12.40) and $\dot{\phi}^2$ in eq. (12.41) can be neglected in the slow-roll approximation. The slow-roll parameter ε also tells how fast the Hubble parameter changes. This can be seen as follows. Differentiating eq. (12.43) gives

$$\frac{\dot{H}}{H} = \frac{4\pi}{3m_{Pl}^2} \frac{V'}{H^2}\dot{\phi} = \frac{1}{2}\frac{V'}{V}\dot{\phi}. \tag{12.48}$$

Dividing by H, substituting for $\dot{\phi}$ from eq. (12.42) and then for H^2 from (12.43) leads to

$$\frac{\dot{H}}{H^2} = -\frac{3m_{Pl}^2}{16\pi}\left(\frac{V'}{V}\right)^2. \tag{12.49}$$

Hence,

$$\dot{H} = -\varepsilon H^2. \tag{12.50}$$

This shows that the Hubble parameter changes very slowly during a period when the scalar field rolls slowly.

The number of e-foldings during inflation is

$$N = \ln\left(\frac{a_f}{a_i}\right) \tag{12.51}$$

where a_i and a_f are the initial and final values of the expansion factor. In the slow-roll approximation the potential is approximately constant in time, and hence, according to eq. (12.43), the Hubble parameter is also constant, and there is exponential expansion (see eq. (12.12)). Then the number of e-foldings is given by

$$N = \int_{t_i}^{t_f} H dt. \tag{12.52}$$

From eq. (12.42), $dt = -(3H/V')d\phi$. Combining this with eq. (12.43) leads to

$$N = -\frac{8\pi}{m_{Pl}^2}\int_{\phi_i}^{\phi_f} \frac{V}{V'}d\phi. \tag{12.53}$$

If $V = \lambda\phi^\nu$ we get

$$N = \frac{4\pi}{m_{Pl}^2\nu}\left(\phi_i^2 - \phi_f^2\right). \tag{12.54}$$

Example 12.2 (Polynomial inflation)   Example
A simple inflationary model arises when one chooses the polynomial potential of a
massive non-interacting field

$$V = \frac{1}{2}m^2\phi^2. \tag{12.55}$$

In this case the slow roll parameters are

$$\eta = \varepsilon = \frac{m_{Pl}^2}{4\pi\phi^2}. \tag{12.56}$$

Hence inflation can happen if $|\phi| > m_{Pl}/\sqrt{4\pi}$. The solutions of the slow-roll equations
are now

$$\phi(t) = \phi_i - \frac{mm_{Pl}}{2\sqrt{3\pi}}t,$$

$$a(t) = a_i \exp\left[\sqrt{\frac{4\pi}{3}}\frac{m}{m_{Pl}}\left(\phi_i t - \frac{mm_{Pl}}{4\sqrt{3\pi}}t^2\right)\right]. \tag{12.57}$$

The number of e-foldings is

$$N = 2\pi\frac{\phi_i^2}{m_{Pl}^2} - \frac{1}{2}. \tag{12.58}$$

During the vacuum dominated inflationary era the dominating gravitational mass density was negative, $\rho_G = \rho_v + 3p_v = -2\rho_v < 0$. Hence the
dynamical evolution of the universe during this era was dominated by the
repulsive gravitation of the vacuum energy. The observed expansion of the universe is a remnant of the accelerated expansion during this incredibly short
era.

In the approximation that the density of radiation energy is neglected the
expansion factor evolved according to eq. (12.10). If the spatial curvature was
positive or vanishing there was no initial singularity in any finite past time.
The Big Bang was then just the explosive inflationary era that lasted for about
10^{-33}s.

Let us now see how the existence of an inflationary era solves the flatness-
and horizon problems. The flatness problem was that the present average relative density of the cosmic energy and matter is so close to unity in spite of
the fact that the total density evolves away from the critical density according
to the pre inflationary cosmological models. The inflationary models give another result. All three expressions (12.10) for the expansion factor approach
the exponential form for $t \gg \sqrt{3/8\pi G\rho_v}$, i.e. $t > 1.5 \cdot 10^{-35}$s when we use
the GUT value above for the density of the vacuum energy. Inserting this into
eq. (12.35) we get

$$\Omega - 1 = \frac{k}{\omega^2}e^{-2\omega t}. \tag{12.59}$$

Eq. (12.59) shows that during the inflationary era the total density approaches
exponentially to the critical density. The quotient between the values of $\Omega-1$ at
the end and the beginning of the inflationary era (assuming minimal duration
from 10^{-35}s to 10^{-33}s) is

$$\frac{\Omega_2 - 1}{\Omega_1 - 1} = e^{-2\omega(t_2-t_1)} = 10^{-56}. \tag{12.60}$$

Within a large range of initial conditions this implies that the total density is extremely close to the critical density at the end of the inflationary era, and that it is close to the critical density also at the present time. The physical picture of this mechanism is illustrated in Fig. 12.7.

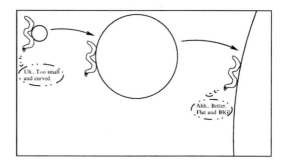

Figure 12.7: The resolution of the flatness problem in the inflationary scenario.

Let us now consider how the inflationary universe models solve the horizon problem. In order to obtain the order of magnitude of the horizon radius at the initial point of time $t_1 = 10^{-35}$s of the original inflationary universe models, we may assume that the universe is flat. Then we get

$$\ell_{PH}(t_1) = 2t_1 = 6 \cdot 10^{-27}\text{m}. \tag{12.61}$$

We shall compare this with the radius d_1 at t_1 of the region which is inside the horizon today. The present radius of this region is $\ell_{PH}(t_0) = 15 \cdot 10^9$ly $= 1.4 \cdot 10^{26}$m. Since the density is close to the critical density after the inflationary era we may put $k = 0$. Then $a \propto t^{1/2}$ in the radiation dominated era from $t_2 = 10^{-33}$s to $t_3 = 10^{11}$s and $a \propto t^{2/3}$ in the matter dominated era from t_3 to the present time $t_0 = 10^{17}$s. Hence we obtain

$$d_1 = e^{\omega(t_1 - t_2)} \left(\frac{t_2}{t_3}\right)^{\frac{1}{2}} \left(\frac{t_3}{t_0}\right)^{\frac{2}{3}} r_{PH} = 1.4 \cdot 10^{-28}\text{m}. \tag{12.62}$$

We see that the horizon radius at the point of time t_1 of the beginning of the inflationary era was greater than the radius at t_1 of the region inside the present horizon. Hence thermal equilibrium could have been established in our observable part of the universe already before the inflationary era started. The reason is that the vast expansion during the inflationary era implies that our observable region of the universe was much smaller at points of time before inflation than it would have been at the same points of time according to universe models without inflation. One more question that the pre-inflationary universe models could not answer was where did the density fluctuations that developed into stars and galaxies come from? One had to postulate a suitable initial fluctuation spectrum. According to the inflationary universe models however, there appeared density fluctuations early in the inflationary era due to quantum mechanical fluctuations, and these were greatly expanded during the inflationary era. Inflationary cosmology predicts a scale-invariant spectrum for the fluctuations. This corresponds very well with the Harrison-Zel'dovich spectrum [Har70, Zel70] of the observed distribution of matter in the universe.

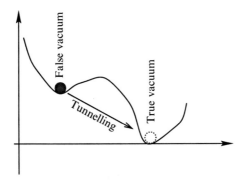

Figure 12.8: Guth's Inflation was driven by a false vacuum.

Even though the mechanism that causes inflation is probably more sophisticated than the initial proposal[3], inflation is a wonderful idea which solves many questions. By this date, there has not been a single idea that has challenged inflation when it comes to explaining many of the features of our universe; its flatness, its homogeneity, the homogeneity in the CMB, etc.

12.6   The Friedmann-Lemaître model

At the turn of the millennium observations of supernovae of type Ia indicated that the universe is currently in a state of accelerated expansion. This means that $\ddot{a} > 0$. From eq. (11.18) then follows that

$$\rho + 3p < 0. \tag{12.63}$$

All known forms of matter has $\rho > 0$, thus to obey the above inequality the pressure must be negative! The simplest model that can describe such a behaviour is a model with a cosmological constant. A cosmological constant represents LIVE with constant density. The equation of state for vacuum energy is

$$p = -\rho, \tag{12.64}$$

and hence,

$$\rho + 3p = -2\rho < 0, \tag{12.65}$$

which means that LIVE produces repulsive gravity. Models with matter and a cosmological constant are termed *Friedmann-Lemaître universe models*.

The Friedmann equations with cosmological constant for homogeneous and isotropic universe models with pressure-free matter are

$$\begin{aligned} \frac{\ddot{a}}{a} &= \frac{\Lambda}{3} - \frac{4\pi G}{3}\rho, \\ H^2 &= \frac{\dot{a}^2}{a^2} = \frac{\Lambda}{3} - \frac{k}{a^2} + \frac{8\pi G}{3}\rho. \end{aligned} \tag{12.66}$$

---

[3]The first inflationary model was put forward by Alan Guth in 1981 [Gut81] (see Fig. 12.8).

The critical mass density ρ_c is defined by

$$\rho_c = \frac{3H^2}{8\pi G}.$$
(12.67)

Defining the parameters Ω_m, Ω_k and Ω_Λ by

$$\Omega_m = \frac{\rho}{\rho_c}$$

$$\Omega_k = -\frac{k}{a^2 H^2}$$

$$\Omega_\Lambda = \frac{\Lambda}{3H^2},$$
(12.68)

we may write the Friedmann equation, eq. (12.66), as

$$\boxed{1 = \Omega_\Lambda + \Omega_k + \Omega_m.}$$
(12.69)

This is a very useful form of the Friedmann equation. It tells us that we have to add up the contribution from both the cosmological constant and the matter term to find out whether the universe is closed $(k > 0)$, flat $(k = 0)$ or open $(k < 0)$. Present day observations indicate that $\Omega_{k0} \approx 0$ and $\Omega_{m0} \approx 0.3$. Hence, this indicates that $\Omega_{\Lambda 0} \approx 0.7$. This fits well with the observation that the universe is currently accelerating. The present epoch can have a significant contribution from a cosmological constant! Einstein's "blunder" has resurrected and shown its place in the physical world.

Figure 12.9 shows different possibilities for the cosmological expansion as a function of matter density and vacuum energy. The line $1 = \Omega_{\Lambda 0} + \Omega_{m0}$

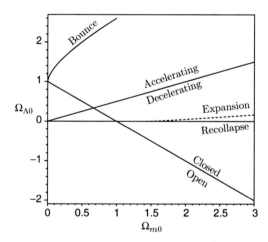

Figure 12.9: Different expansion histories depending upon the densities of matter and vacuum energy.

represents a flat universe separating open from closed universes. The line $\Omega_{\Lambda 0} = \Omega_{m0}/2$ corresponds to uniform expansion with vanishing deceleration parameter separating decelerating from accelerating universes (see eq. (11.55)). The dashed, nearly horizontal curve corresponds to critical universes separating eternal expansion from recollapse in the future. The upper dotted curve corresponds to $t_0 H_0 = \infty$ where t_0 is the present age of the universe and H_0 the present value of the Hubble parameter. If the density of the vacuum energy is above this curve there will be no Big Bang. An initially collapsing universe of this type will bounce and then expand with increasing velocity.

A universe expanding forever is called *hyperbolic*. In a hyperbolic ΛCDM universe the vacuum energy will eventually dominate and give accelerated expansion. A universe that eventually collapses is called *elliptic*. The border-line between elliptic and hyperbolic universes represents *critical* universes; i.e. universe models with vanishing expansion velocity as $t \to \infty$. The border-line between hyperbolic and bouncing universes represents *loitering* universes; i.e. universes that are nearly static for a period before the vacuum energy becomes dominant and the expansion becomes accelerating.

The mathematical expressions of the curves representing bounce and criti-cal universes, respectively, are found by considering the conditions for a bounce and an expansion that instantaneously stops, i.e. for a vanishing Hubble parameter. Neglecting radiation the Friedmann equation (11.16) can be written

$$H = H_0 a^{-\frac{3}{2}} \left(\Omega_{\Lambda 0} a^3 + \Omega_{k0} a + \Omega_{m0} \right)^{\frac{1}{2}}. \tag{12.70}$$

The condition that the Hubble parameter vanishes gives the cubic equation

$$\Omega_{\Lambda 0} a^3 + \Omega_{k0} a + \Omega_{m0} = 0. \tag{12.71}$$

If $\Omega_{\Lambda 0} < 0$ the universe recollapses. If $\Omega_{\Lambda 0} > 0$ and $\Omega_{m0} < 1$ the universe expands to infinity independently of the curvature. If $\Omega_{m0} > 1$ recollapse is only avoided if $\Omega_{\Lambda 0}$ exceeds a critical value[4]

$$\Omega_{\Lambda 0} = 4\Omega_{m0} \sin^3 \left[\frac{1}{3} \arcsin \left(1 - \Omega_{m0}^{-1} \right) \right]. \tag{12.72}$$

This is the equation of the dashed curve in Fig. 12.9. For $\Omega_{\Lambda 0}$ larger than

$$\Omega_{\Lambda 0} = 4\Omega_{m0} \left[f\left(\frac{1}{3} f^{-1} \left(\Omega_{m0}^{-1} - 1 \right) \right) \right]^3,$$

$$f(x) = \begin{cases} \cosh x, & \Omega_{m0} < \frac{1}{2} \\ \cos x, & \Omega_{m0} > \frac{1}{2} \end{cases} \tag{12.73}$$

a universe which initially collapses bounces, and there is no Big Bang. Eq. (12.73) is the equation of the slightly bent curve in the upper part of Fig. 12.9.

Let us now consider the flat ($k = 0$) model. The matter density can be written as $\frac{8\pi G}{3}\rho = Ka^{-3}$, where K is a constant. Introducing $v = a^{\frac{3}{2}}$ we can write the Friedmann equation, eq. (12.66) with $k = 0$ as

$$\dot{v}^2 = \frac{9}{4} \left(\frac{\Lambda}{3} v^2 + K \right). \tag{12.74}$$

This equation can be solved to yield

$$v = \left(\frac{3K}{\Lambda} \right)^{\frac{1}{2}} \sinh\left(t/t_\Lambda \right), \quad t_\Lambda \equiv \frac{2}{\sqrt{3\Lambda}}. \tag{12.75}$$

---

[4]To obtain this expression we have utilized the identities

$$\sin 3x = 3 \sin x - 4 \sin^3 x, \quad \cos 3x = 4 \cos^3 x - 3 \cos x.$$

Let us normalize the scale factor $a(t)$ such that $a(t_0) = 1$. This implies $v(t_0) = 1$ and $K = \frac{8\pi G}{3}\rho_0$. Hence, using eqs. (12.68) and (12.69) with $\Omega_k = 0$, we can write

$$A \equiv \frac{3K}{\Lambda} = \frac{\Omega_{m0}}{\Omega_{\Lambda 0}} = \frac{1 - \Omega_{\Lambda 0}}{\Omega_{\Lambda 0}}. \tag{12.76}$$

The scale factor is now given by

$$\boxed{a(t) = A^{\frac{1}{3}} \sinh^{\frac{2}{3}}(t/t_\Lambda).} \tag{12.77}$$

The age t_0 of the universe is found by the requirement $a(t_0) = 1$ (or equivalently $v(t_0) = 1$). By use of the identity $\operatorname{artanh} x = \operatorname{arsinh}(x/\sqrt{1-x^2})$ we get the expression

$$t_0 = t_\Lambda \operatorname{artanh}\sqrt{\Omega_{\Lambda 0}}. \tag{12.78}$$

Inserting the values $t_0 = 13.7 \cdot 10^9$ years and $\Omega_{\Lambda 0} = 0.7$ found from the WMAP measurements of temperature fluctuations in the cosmic microwave background radiation, and from the determination of the luminosity-redshift relationship of supernovae of type Ia, we get $A = 0.43$, $t_\Lambda = 11.2 \cdot 10^9$ years and $\Lambda = 1.1 \cdot 10^{-20}$ years^{-2}. With these values the expansion factor is $a(t) = 0.75 \sinh^{2/3}(1.2t/t_0)$. This function is plotted in Fig. 12.10.

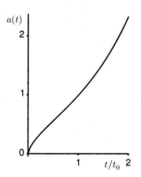

Figure 12.10: The expansion factor as function of cosmic time in units of the age of the universe.

The Hubble parameter as a function of time is

$$H = \left(\frac{\Lambda}{3}\right)^{\frac{1}{2}} \coth(t/t_\Lambda). \tag{12.79}$$

In Fig. 12.11 the function $Ht_0 = 0.8 \coth(1.2t/t_0)$ is plotted.

The Hubble parameter decreases all the time and approaches a constant value

$$H_\infty = \sqrt{\frac{\Lambda}{3}}, \tag{12.80}$$

in the infinite future. The present value of the Hubble parameter is

$$H_0 = \sqrt{\frac{\Lambda}{3\Omega_{\Lambda 0}}}. \tag{12.81}$$

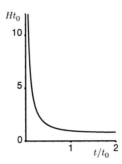

Figure 12.11: The Hubble parameter of a flat universe model with dust and LIVE as a function of cosmic time.

The corresponding Hubble age is $t_{H0} = \sqrt{\frac{3\Omega_{\Lambda 0}}{\Lambda}}$. Inserting numerical values gives $H_0 = 69\,\mathrm{km\,s^{-1}Mpc^{-1}}$ and $t_{H0} = 14.1 \cdot 10^9$ years. This value of the Hubble constant is in good agreement with the results of the measurements by the Key Project Group of the Hubble Space Telescope. In this universe model the age of the universe is nearly as large as the Hubble age, while in the Einstein-de Sitter model the corresponding age is $t_{EdS0} = \frac{2}{3}t_{H0} = 9.4 \cdot 10^9$ years (see Example 11.3). The reason for this difference is that in the Einstein-de Sitter model the expansion is decelerated all the time, while in the Friedmann-Lemaître model the repulsive gravitation due to the vacuum energy have made the expansion accelerate lately. Hence, for a given value of the Hubble constant the previous velocity was larger in the Einstein-de Sitter model than in the Friedmann-Lemaître model.

The age of the universe increases with increasing density of vacuum energy. In the limit that the density of the vacuum approaches the critical density, there is no dark matter, and the universe approaches the de Sitter model with exponential expansion and no Big Bang. This model behaves in the same way as the Steady State cosmological model and is infinitely old.

A dimensionless quantity representing the rate of change of the cosmic expansion velocity is the deceleration parameter, which is defined in eq. (11.38). For the present universe model the deceleration parameter as a function of time is

$$q = \frac{1}{2}\left[1 - 3\tanh^2\left(t/t_\Lambda\right)\right], \tag{12.82}$$

which is shown graphically in Fig. 12.12.

The inflection point of time t_1 when the deceleration turned into acceleration is given by $q(t_1) = 0$. This leads to

$$t_1 = t_\Lambda \operatorname{artanh}\left(\frac{1}{\sqrt{3}}\right), \tag{12.83}$$

or expressed in terms of the age of the universe

$$t_1 = \frac{\operatorname{artanh}\left(\frac{1}{\sqrt{3}}\right)}{\operatorname{artanh}\left(\sqrt{\Omega_{\Lambda 0}}\right)}\,t_0. \tag{12.84}$$

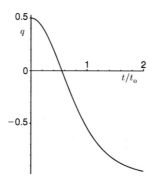

Figure 12.12: The deceleration parameter as a function of cosmic time.

The Hubble parameter at this time is $H(t_1) = \sqrt{\Lambda}$. The corresponding cosmic redshift is

$$z(t_1) = \frac{a_0}{a(t_1)} - 1 = \left(\frac{2\Omega_{\Lambda 0}}{1 - \Omega_{\Lambda 0}}\right)^{\frac{1}{3}} - 1. \tag{12.85}$$

Example

Example 12.3 (Transition from deceleration to acceleration for our universe)
Let us use the observational data to find out when our universe turned from an deceleration to an acceleration state. Inserting $\Omega_{\Lambda 0} = 0.7$ gives $t_1 = 0.54t_0$ and a redshift of $z(t_1) = 0.67$.

The results of analysing the observations of supernova SN 1997 at $z = 1.7$, corresponding to an emission time $t_e = 0.30t_0 = 4.1 \cdot 10^9$ years, have provided evidence that the universe was decelerated at that time [*Ret. al.*01]. M. Turner and A.G. Reiss[TR02] have argued that the other supernovae data favour a transition from deceleration to acceleration for a redshift around $z = 0.5$.

Note that the expansion velocity given by Hubble's law, $v = H\ell$, always decreases as seen in Fig. 12.11. This is the velocity away from the Earth of the cosmic fluid at a fixed physical distance ℓ from Earth. The quantity $\dot{a}$ on the other hand, is the velocity of a fixed fluid particle comoving with the expansion with the universe. If such a particle accelerates, the expansion of the universe is said to accelerate. While $\dot{H}$ tells how fast the expansion velocity changes at a fixed distance from the Earth, the quantity $\ddot{a}$ represents the acceleration of a free particle comoving with the expanding universe. The connection between these two quantities is $\ddot{a} = a\left(\dot{H} + H^2\right)$. Note from eqs. (12.3) and (12.4) that

$$\dot{H} = -4\pi G(\rho + p), \tag{12.86}$$

for a flat universe. Hence, in order that $\dot{H} > 0$ the universe must be dominated by a fluid with $p < -\rho$, which has been called *phantom energy* (see problem 12.9). In order that $\ddot{a} > 0$ it is sufficient that $p < -\rho/3$, as is seen from eq. (11.18).

It may be noted that the critical density is given by

$$\rho_{cr} = \rho_\Lambda \tanh^{-2}(t/t_\Lambda), \quad \rho_\Lambda \equiv \frac{\Lambda}{8\pi G}, \tag{12.87}$$

showing that the critical density decreases with time.

Using eqs. (12.68) and (12.79) the relative density of the vacuum energy is found to be

$$\Omega_\Lambda = \tanh^2 (t/t_\Lambda).$$

(12.88)

Hence, the relative density of the matter is

$$\Omega_m = 1 - \Omega_\Lambda = \cosh^{-2} (t/t_\Lambda),$$

(12.89)

for the flat Friedmann-Lemaître universe model. These densities are depicted in Fig. 12.13.

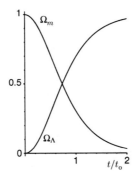

Figure 12.13: The relative densities in the Friedmann-Lemaître model as a function of cosmic time.

Note from eqs. (12.82) and (12.88) that in the case of a flat Friedmann-Lemaître universe model, the deceleration parameter may be expressed in terms of the relative density of vacuum only, $q = \frac{1}{2} (1 - 3\Omega_\Lambda)$. Hence, if observations show that the universe is accelerating, then this alone means that $\Omega_{\Lambda 0} > \frac{1}{3}$, presupposing a homogeneous universe model.

12.7   Universe models with quintessence energy

As noted in section 12.5 one of the predictions of the inflationary universe models is that the average total density of the cosmic matter and energy is close to the critical density. This prediction has been confirmed by the recent WMAP-measurements of the temperature fluctuations in the cosmic microwave background radiation. However, measurements of the large scale distribution and motions of the matter in the universe indicate that the average cosmic density of the gravitating mass is only 30% of the critical mass density. Hence, 70% is missing. Furthermore, measurements of redshifts and distances to supernovae of type Ia have shown that the universe is in a state of accelerated expansion. Thus at recent times the universe must be dominated by a sort of energy causing repulsive gravitation.

We shall assume that this energy can be described as a perfect fluid with equation of state $p = w\rho$. From eq. (11.18) follows that the dark energy must have $w < -1/3$, hence it must be in a state of tension.

One candidate for the missing energy is LIVE for which $w = -1$, corresponding to a cosmological constant, Λ. The resulting universe model, called the ΛCDM-model, consists of a mixture of vacuum energy and cold, dark mater.

Two difficulties arise from this scenario. The first is the *fine-tuning problem*: Why is the missing energy density today so small compared to typical particle physics scales? The missing energy density is of order 10^{-47}GeV^4 which is 14 orders of magnitude smaller than the electroweak scale. The second difficulty is the *cosmic coincidence problem*: Since the missing energy density and the matter energy density decrease at different rates as the universe expands, their ratio must be specified incredibly accurately in the very early universe in order for the two densities to nearly coincide today.

There have been several attempts to resolve the fine-tuning problem. A relatively recent attempt comes from string theory (and the so-called *"landscape"*) [Bou05]. String theory contains a dense discretuum of metastable vacua and the vacuum energy can vary discretely among widely separated regions. Structures, such as galaxies, will only form if the vacuum energy fulfills the bounds

$$-10^{-123} < \rho_\Lambda < 10^{-121}.$$

Since galaxies are presumably a requirement for the existence of observers, we can only exist in a region that fulfills this bound. Such an argument, which essentially says that "if the universe was different we would not have been here to observe it", is often called an *anthropic* argument [BT86].

We will focus on an alternative candidate for the dark energy called *quintessence*. This is a perfect fluid model with $-1 < w < -1/3$. The value of w may vary with the cosmic time. This energy comes from a scalar field with a potential that is introduced in an ad hoc way to suit observations. It has a phenomenological character, and may be regarded as a first step in order to find some properties that dark energy may have.

There exists a class of quintessence models of dark energy called tracker models [ZWS99]. They are constructed to solve the coincidence and fine-tuning problems. Let us consider universe models containing dark energy and one other type of energy or matter, say radiation or cold dark matter. The tracker models have the property that the energy density of the tracker field approaches that of the other component from a wide variety of initial conditions. A special type of tracker field, called k-essence [APMS01], behaves as follows. The k-essence energy density catches up and overtakes the matter-density, typically several billions of years after matter-domination, driving the universe into a period of accelerated expansion. In this scenario, we observe cosmic acceleration today because the time for human evolution and the time for k-essence to overtake the matter density are both several billions of years after matter-radiation equality.

A full description of the tracker models requires both analytical and numerical calculations. In this section we shall only consider a rather simple class of quintessence models following [ZP01] in order to illustrate some properties of such models at late times with constant ratio between energy density and matter density.

The total density and pressure of the cosmic fluid are

$$\rho = \rho_s + \rho_m, \qquad p = p_s, \qquad (12.90)$$

where the indices s and m refer to a scalar field component and a cold matter component, respectively. The energy density and pressure of the scalar field are

$$\rho_s = \frac{1}{2}\dot{\phi}^2 + V(\phi), \quad p_s = \frac{1}{2}\dot{\phi}^2 - V(\phi). \tag{12.91}$$

We do not assume that the dark energy and the matter evolve independently, but allow on interaction between them, described by a source (loss) term δ in the energy conservation equations,

$$\dot{\rho}_m + 3H\rho_m = \delta, \tag{12.92}$$

and

$$\dot{\rho}_s + 3H(\rho_s + p_s) = -\delta. \tag{12.93}$$

The last equation may be written

$$\dot{\phi}\left(\ddot{\phi} + 3H\dot{\phi} + \frac{dV}{d\phi}\right) = -\delta, \tag{12.94}$$

which generalizes eq. (12.38) to allow for energy transfer between the dark energy and the cosmic matter.

We shall now study a model with constant ratio $r \equiv \rho_m/\rho_s$. Then eqs. (12.92) and (12.93) lead to

$$\dot{\rho}_s + 3H\rho_s + \frac{3}{1+r}Hp_s = 0. \tag{12.95}$$

Assuming $p_s = w_s p_s$ with $w_s = $ constant, this equation takes the form

$$\frac{\dot{\rho}_s}{\rho_s} + n\frac{\dot{a}}{a} = 0, \quad n = 3\left(1 + \frac{w_s}{1+r}\right). \tag{12.96}$$

Integration with $a_0 = 1$ gives

$$\rho_s = \rho_{s0}a^{-n}. \tag{12.97}$$

Inserting this into eq. (12.93) gives for the interaction term

$$\delta = -\frac{3rw_s}{1+r}H\rho_s. \tag{12.98}$$

For the present universe model the Friedmann equation (11.16) takes the form

$$\dot{a}^2 = \frac{8\pi G}{3}\rho_0 a^{2-n}, \quad \rho_0 = \rho_{m0} + \rho_{s0}. \tag{12.99}$$

Integration with $a(t_0) = 1$ yields

$$a(t) = \left(\frac{t}{t_0}\right)^{\frac{2}{n}}. \tag{12.100}$$

The deceleration parameter, defined in eq. (11.38), is

$$q = \frac{1}{2}\left(1 + \frac{3w_s}{1+r}\right). \tag{12.101}$$

Observations of supernovae of type Ia and of temperature fluctuations in the cosmic background radiation indicate $r \approx 1/3$. Using this value the condition for accelerated expansion, $q < 0$, becomes $w_s < -4/9$.

From eqs. (12.97) and (12.100) follow

$$\rho_s = \rho_{s0} \left(\frac{t_0}{t} \right)^2. \tag{12.102}$$

During this late era the densities of the dark energy and the cold matter both decrease as t^{-2}. The potential as a function of the time is found from eqs. (12.91) and (12.102),

$$V = \frac{1}{2}(1 - w_s)\rho_{s0} \left(\frac{t_0}{t} \right)^2. \tag{12.103}$$

Differentiation gives

$$\dot{V} = \dot{\phi} \frac{dV}{d\phi} = -\frac{2}{t}V. \tag{12.104}$$

From eqs. (12.91) and (12.102) we also have

$$\dot{\phi} = \sqrt{(1 + w_s)\rho_{s0}} \frac{t_0}{t}. \tag{12.105}$$

Hence,

$$\frac{dV}{V} = -\lambda d\phi, \qquad \lambda^2 = \frac{4}{(1 + w_s)\rho_{s0}t_0}, \tag{12.106}$$

giving

$$V(\phi) = V_0 e^{-\lambda\phi}. \tag{12.107}$$

This shows that a scenario with a late-time constant ratio between the densities of the dark energy and matter can be realized by a quintessence energy with a simple exponential potential.

12.8   Dark energy explored by means of supernova observations and the statefinder diagnostic

Observations of supernovae of type Ia have shown that the universe is dominated by dark energy causing accelerated expansion. V. Sahni and coworkers [ASSS03] have recently introduced a pair of of parameters $\{r, s\}$ called *statefinders*, which are useful to distinguish different types of dark energy.

The Friedmann-Robertson-Walker models of the universe have earlier been characterized by the Hubble parameter and the deceleration parameter, depending upon the first and second derivatives of the scale factor.

If the satellite SNAP works according to the planes, we shall eventually have accurate determinations of the luminosity distance and redshift of more than 5000 supernovae of type Ia. These data will permit a very precise determination of $a(z)$. Then it will be important to include also the third derivative of the scale factor in our characterization of different universe models.

The statefinders were introduced to characterize primarily flat universe models with cold dark matter (dust) and dark energy. They were defined as

$$r \equiv \frac{\dddot{a}}{aH^3}, \tag{12.108}$$

$$s \equiv \frac{r-1}{3\left(q - \frac{1}{2}\right)}. \tag{12.109}$$

For the present universe models the Friedmann equation takes the form

$$H^2 = \frac{8\pi g}{3}\left(\rho_m + \rho_x\right). \tag{12.110}$$

If the equation of state of the dark energy has the form $p_x = w\rho_x$ with $w = $ constant, the energy conservation equation implies

$$\rho_m = \rho_{m0}a^{-3}, \qquad \rho_x = \rho_{x0}a^{-3(1+w)}. \tag{12.111}$$

Introducing the cosmic redshift by $1 + z = a^{-1}$ we obtain

$$H(y) = H_0\left[\Omega_{m0}y^3 + \Omega_{x0}y^{3(1+w)}\right]^{\frac{1}{2}}, \quad y \equiv 1 + z. \tag{12.112}$$

Using $\dot{H} = -H'Ha$, where $H' \equiv dH/dy$, the deceleration parameter is given by

$$q(y) = \frac{H'}{H}y - 1. \tag{12.113}$$

For flat universe models

$$\rho_x = \rho_{cr} - \rho_m = \frac{3H^2}{8\pi G}(1 - \Omega_m) = \frac{3}{8\pi G}\left(H^2 - \Omega_{m0}H_0^2 y^3\right). \tag{12.114}$$

From Friedmann's acceleration equation (11.18) follows

$$p_x = \frac{H^2}{4\pi G}\left(q - \frac{1}{2}\right) = \frac{3}{8\pi G}\left[\frac{1}{3}(H^2)'y - H^2\right]. \tag{12.115}$$

Hence,

$$w(y) = \frac{\frac{1}{3}(H^2)'y - H^2}{H^2 - H_0^2\Omega_{m0}y^3}. \tag{12.116}$$

Calculating r, and using $a' = -a^2$, we obtain

$$r(y) = 1 - 2\frac{H'}{H}y + \left(\frac{(H')^2}{H^2} + \frac{H''}{H}\right)y^2. \tag{12.117}$$

The state finder $s(y)$ is found by inserting the expressions (12.113) and (12.117) into eq. (12.109).

If the luminosity distance d_L is found as a function of y for standard light sources, the Hubble parameter may be calculated by solving eq. (11.130) with respect to H, giving

$$H(y) = \left[\left(\frac{d_L}{y}\right)'\right]^{-1}. \tag{12.118}$$

We shall now calculate the localization of universe models with different types of dark energies in the statefinder diagrams. Let us first consider dark energy obeying an equation of state of the form $p = w\rho$. The formalism of Sahni and coworkers will here be generalized to permit curved universe models. In this case the definition of s is generalized to

$$s = \frac{r - \Omega}{3\left(q - \frac{\Omega}{2}\right)}. \tag{12.119}$$

The deceleration parameter may be expressed as

$$q = \frac{1}{2}\left[\Omega_m + (1 + 3w)\Omega_x\right]. \tag{12.120}$$

hence, if Ω_x, Ω and q are determined by measurements the equation of state factor w may be found from

$$w = \frac{2q - \Omega}{3\Omega_x}. \tag{12.121}$$

Differentiation of eq. (11.38), together with eq. (12.108) leads to

$$r = 2q^2 + q - \frac{\dot{q}}{H}. \tag{12.122}$$

From eq. (12.120) we have

$$\dot{q} = \frac{1}{2}\dot{\Omega}_m + \frac{1}{2}(1 + 3w)\dot{\Omega}_x + \frac{3}{2}w\Omega_x. \tag{12.123}$$

Furthermore,

$$\dot{\Omega} = \frac{\dot{\rho}}{\rho_{cr}} - \frac{\rho}{\rho_{cr}^2}\dot{\rho}_{cr}, \tag{12.124}$$

with

$$\dot{\rho}_{cr} = \frac{3H\dot{H}}{4\pi G}, \tag{12.125}$$

and

$$\dot{H} = -H^2(1 + q). \tag{12.126}$$

Hence,

$$\dot{\rho}_{cr} = -2(1 + q)H\rho_{cr}, \tag{12.127}$$

which leads to

$$\dot{\Omega} = \frac{\dot{\rho}}{\rho_{cr}} + 2(1 + q)H\Omega. \tag{12.128}$$

For cold matter $\dot{\rho}_m = -3H\rho_m$ giving

$$\dot{\Omega}_m = (2q - 1)H\Omega_m, \tag{12.129}$$

and for dark energy $\dot{\rho}_x = -3(1 + w)H\rho_x$ giving

$$\dot{\Omega}_x = (2q - 1 - 3w)H\Omega_x. \tag{12.130}$$

Inserting eqs. (12.129) and (12.130) into eq. (12.123) and the resulting expression into eq. (12.122) finally leads to

$$r = \Omega_m + \left[1 + \frac{9}{2}w(1+w)\right]\Omega_x - \frac{3}{2}\frac{\dot{w}}{H}\Omega_x. \qquad (12.131)$$

Inserting the expression (12.131) into eq. (12.119) gives

$$s = 1 + w - \frac{1}{3}\frac{\dot{w}}{wH}. \qquad (12.132)$$

It may be noted that a universe with cold dark matter and LIVE has $r = \Omega$, $s = 0$. For a flat universe $\Omega_m + \Omega_x = 1$ and the expression reduces to

$$r = 1 + \frac{9}{2}w(1+w)\Omega_x - \frac{3}{2}\frac{\dot{w}}{H}\Omega_x. \qquad (12.133)$$

If the dark energy is due to a scalar field the equation of state factor w is given by

$$w = \frac{\dot{\phi}^2 - 2V(\phi)}{\dot{\phi}^2 + 2V(\phi)}. \qquad (12.134)$$

Differentiation gives

$$\dot{w}\rho_x = \frac{2\dot{\phi}(2\dot{\phi}V - \dot{\phi}\dot{V})}{\dot{\phi}^2 + 2V(\phi)}. \qquad (12.135)$$

Using the equation of motion for the scalar field

$$\ddot{\phi} = -3H\dot{\phi} - V', \qquad (12.136)$$

and that $\dot{V} = V'\dot{\phi}$ in eq. (12.135) and inserting the result into eq. (12.131) we obtain

$$r = \Omega + 12\pi G\frac{\dot{\phi}^2}{H^2} + 8\pi G\frac{\dot{V}}{H^3}. \qquad (12.137)$$

Furthermore,

$$q - \frac{\Omega}{2} = \frac{3}{2}w\Omega_x = 4\pi G\frac{p_x}{H^2} = \frac{4\pi G}{H^2}\left(\frac{1}{2}\dot{\phi}^2 - V\right). \qquad (12.138)$$

Hence, the statefinder s is

$$s = \frac{2\left(\dot{\phi}^2 + \frac{2}{3}\frac{\dot{V}}{H}\right)}{\dot{\phi}^2 - 2V(\phi)}. \qquad (12.139)$$

We shall now find expressions for r and s that are valid even if the dark energy does not fulfill an equation of state of the form $p = w\rho$. The expression for the deceleration parameter may be written as

$$q = \frac{1}{2}\left(1 + 3\frac{p_x}{\rho_x}\right)\Omega. \qquad (12.140)$$

Using this in eq. (12.122) we find

$$r = \left(1 - \frac{3}{2}\frac{\dot{p}_x}{H\rho_x}\right)\Omega, \tag{12.141}$$

$$s = -\frac{1}{3H}\frac{\dot{p}_x}{p_x}. \tag{12.142}$$

If the universe contains only dark energy with an equation of state $p = p(\rho)$, then

$$\dot{p} = \frac{\partial p}{\partial \rho}\dot{\rho} = -3H(\rho + p)\frac{\partial p}{\partial \rho}, \tag{12.143}$$

which leads to

$$r = \left[1 + \frac{9}{2}\left(1 + \frac{p}{\rho}\right)\frac{\partial p}{\partial \rho}\right]\Omega, \tag{12.144}$$

$$s = \left(1 + \frac{\rho}{p}\right)\frac{\partial p}{\partial \rho}. \tag{12.145}$$

If the universe contains cold matter and dark energy these expressions are generalized to

$$r = \left(1 + \frac{9}{2}\frac{\rho_x + p_x}{\rho_m + \rho_x}\frac{\partial p_x}{\partial \rho_x}\right)\Omega, \tag{12.146}$$

$$s = \left(1 + \frac{\rho_x}{p_x}\right)\frac{\partial p_x}{\partial \rho_x}. \tag{12.147}$$

Examples

Example 12.4 (Universe model with Chaplygin gas)
Let us consider a universe model containing only Chaplygin gas. Then

$$p = -\frac{A}{\rho}. \tag{12.148}$$

The energy conservation equation then takes the form

$$\dot{\rho} = -3\frac{\dot{a}}{a}\left(\rho - \frac{A}{\rho}\right). \tag{12.149}$$

Integration gives

$$\rho = \sqrt{A + \frac{B}{a^6}}, \tag{12.150}$$

where B is a constant of integration. Imposing the standard normalization of the scale factor, $a(t_0) = 1$, we obtain

$$A = \rho_\infty^2, \qquad B = \rho_0^2 - \rho_\infty^2, \tag{12.151}$$

where ρ_0 is the current density of the Chaplygin gas and ρ_∞ in its asymptotically far future. For the Chaplygin gas

$$\frac{\partial p}{\partial \rho} = \frac{A}{\rho^2} = -\frac{p}{\rho}, \tag{12.152}$$

giving

$$r = \left[1 - \frac{9}{2}s(1 + s)\right]\Omega, \tag{12.153}$$

$$s = -\left(1 + \frac{p}{\rho}\right). \tag{12.154}$$

In the case of a universe model with cold dark matter and Chaplygin gas we have

$$r = \left[1 - \frac{9}{2}\frac{s(1+s)}{1 + \frac{\rho_m}{\rho_x}}\right]\Omega, \tag{12.155}$$

$$s = -\left(1 + \frac{p_x}{\rho_x}\right). \tag{12.156}$$

Here

$$\frac{\rho_m}{\rho_x} = \frac{\rho_{m0}}{\sqrt{Aa^6 + B}} = \kappa\sqrt{-s}, \tag{12.157}$$

where

$$\kappa = \frac{\rho_{m0}}{\sqrt{B}} = \frac{\rho_{m0}}{\sqrt{\rho_0^2 - \rho_\infty^2}} = \left(\frac{\rho_m}{\rho_x}\right)_{a=0}. \tag{12.158}$$

This gives

$$r = \left[1 - \frac{9}{2}\frac{s(1+s)}{1 + \kappa\sqrt{-s}}\right]\Omega. \tag{12.159}$$

Example 12.5 (Third order luminosity redshift relation)
We shall find a series expansion of the luminosity distance to third order in the cosmic redshift. To this order eq. (11.129) gives

$$H_0 d_L \approx (1 + z)I\left(1 + \frac{1}{6}\Omega_{k0}I^2\right). \tag{12.160}$$

Making use of a series expansion of the Hubble parameter to 2nd order in z,

$$H(z) \approx H_0 + \left(\frac{dH}{dz}\right)_0 z + \frac{1}{2}\left(\frac{d^2H}{dz^2}\right)_0 z^2, \tag{12.161}$$

we have

$$I \approx \int_0^z \frac{dy}{1 + ay + by^2}, \quad a = \frac{1}{H_0}\left(\frac{dH}{dz}\right)_0, \quad b = \frac{1}{2H_0}\left(\frac{d^2H}{dz^2}\right)_0. \tag{12.162}$$

To third order in z this leads to

$$(1 + z)I \approx z + \left(1 - \frac{1}{2}a\right)z^2 - \left(\frac{1}{2}a + \frac{1}{3}b - \frac{1}{3}a^2\right)z^3. \tag{12.163}$$

Using that

$$\dot{H} = -(1 + q)H^2, \quad \ddot{H} = (r + 3q + 2)H^3, \tag{12.164}$$

and

$$\frac{d}{dz} = -\frac{1}{(1 + z)H}\frac{d}{dt}, \tag{12.165}$$

we obtain

$$a = 1 + q_0, \quad b = \frac{1}{2}(r_0 - q_0^2). \tag{12.166}$$

Inserting these expressions in eq. (12.163) and the resulting expression in eq. (12.160) finally gives the luminosity redshift relation

$$H_0 d_L \approx z\left[1 + \frac{1}{2}(1 - q_0)z - \frac{1}{6}\left(1 + r_0 - q_0 - 3q_0^2 - \Omega_{k0}\right)z^2\right]. \tag{12.167}$$

12.9   Cosmic density perturbations

We shall present the most simple aspects of the cosmological perturbation theory (see also [LL00, Ama03]). This will provide a background for describing acoustic oscillations in the plasma that existed during the first 400 000 years of our universe, and which produced temperature fluctuations in the cosmic microwave background.

The physical universe is described as a Friedmann-Robertson-Walker universe that is perturbed due to density perturbations in the cosmic fluid. Since we are concerned with fluctuations at early times when the universe was very close to flat, we choose to consider only flat universe models. The line-element of the unperturbed universe can then be written

$$ds^2 = a^2(\eta)\left(-d\eta^2 + \delta_{ij}dx^i dx^j\right),\qquad(12.168)$$

where η is conformal time which is related to the cosmic time t by

$$t = \int_0^\eta a(\eta')d\eta'.\qquad(12.169)$$

In the so-called Newtonian gauge, and assuming that there is no shear in the cosmic fluid, the line-element of the perturbed universe can be written

$$ds^2 = a^2(\eta)\left[-(1+2\Phi)d\eta^2 + (1-2\Phi)\delta_{ij}dx^i dx^j\right],\qquad(12.170)$$

where the perturbing function Φ satisfies $|\Phi| \ll 1$.

Calculating the components of the Einstein tensor from the line-element to 1st order in Φ, one finds the zeroth order components,

$$^{(0)}E^0_{\ 0} = -\frac{3}{a^2}\mathcal{H}^2,\quad ^{(0)}E^0_{\ i} = 0,\quad ^{(0)}E^i_{\ j} = -\frac{1}{a^2}(\mathcal{H}^2 + 2\dot{\mathcal{H}})\delta^i_{\ j},\qquad(12.171)$$

where

$$\mathcal{H} = \frac{1}{a}\frac{da}{d\eta} = \frac{1}{a}\frac{da}{dt}\frac{dt}{d\eta} = aH.\qquad(12.172)$$

Here, H is the usual Hubble parameter. The first order components are

$$\delta E^0_{\ 0} = -\frac{2}{a^2}\left[\nabla^2\Phi - 3\mathcal{H}(\dot{\Phi} + \mathcal{H}\Phi)\right],\qquad(12.173)$$

$$\delta E^0_{\ i} = -\frac{2}{a^2}\left(\dot{\Phi} + \mathcal{H}\Phi\right),\qquad(12.174)$$

$$\delta E^i_{\ j} = \frac{2}{a^2}\left[(\mathcal{H}^2 + 2\dot{\mathcal{H}})\Phi + \ddot{\Phi} + 3\mathcal{H}\dot{\Phi}\right]\delta^i_{\ j}.\qquad(12.175)$$

The energy-momentum tensor of the cosmic fluid can be split into zeroth and first order parts as follows

$$^{(0)}T^0_{\ 0} = -\rho_0,\qquad \delta T^0_{\ 0} = -\delta\rho_0,\qquad(12.176)$$
$$^{(0)}T^0_{\ i} = 0,\qquad \delta T^0_{\ i} = (\rho_0 + p_0)a\delta u_i,\qquad(12.177)$$
$$^{(0)}T^i_{\ j} = p_0\delta^i_{\ j},\qquad \delta T^i_{\ j} = \delta p\delta^i_{\ j}.\qquad(12.178)$$

We will assume that the equation of state is $p = w\rho$, and that there are entropy perturbations. Hence, the speed of sound on the fluid is given by $c_s^2 = \partial p/\partial\rho = w$, and the pressure perturbation is $\delta p = w\delta\rho$.

The zeroth order Einstein equations are

$$\mathcal{H}^2 = \frac{8\pi G}{3}a^2\rho_0, \tag{12.179}$$

$$\mathcal{H}^2 + 2\dot{\mathcal{H}} = -8\pi Ga^2 p_0 = -8\pi Ga^2 w\rho_0. \tag{12.180}$$

Next we consider the first order Einstein equations. The time-time component is

$$\nabla^2\Phi - 3\mathcal{H}(\dot{\Phi} + \mathcal{H}\Phi) = 4\pi Ga^2\delta\rho. \tag{12.181}$$

Taking the Newtonian limit of this equation by letting $a \to 1$ and $\mathcal{H} \to 0$, it reduces to $\nabla^2\Phi = 4\pi G\delta\rho$, which is just the Poisson equation (1.32) of Newtonian gravitational theory, where Φ is the gravitational potential due to mass-inhomogeneity $\delta\rho$.

It is often convenient to define the *density contrast*, δ, by

$$\delta \equiv \frac{\delta\rho}{\rho_0}. \tag{12.182}$$

Introducing this in eq. (12.181) and eliminating ρ_0 by means of eq. (12.179) we obtain

$$\nabla^2\Phi - 3\mathcal{H}(\dot{\Phi} + \mathcal{H}\Phi) = \frac{3}{2}\mathcal{H}^2\delta. \tag{12.183}$$

The first order time-space equations are

$$\left(\dot{\Phi} + \mathcal{H}\Phi\right)_{,i} = -\frac{3}{2}\mathcal{H}^2(1+w)a\delta u_i, \tag{12.184}$$

while the space-space equations are

$$(\mathcal{H}^2 + 2\dot{\mathcal{H}})\Phi + \ddot{\Phi} + 3\mathcal{H}\dot{\Phi} = \frac{3}{2}\mathcal{H}^2 w\delta. \tag{12.185}$$

We shall now find solutions to these equations, and start with the zeroth order equations. They are simply the Friedmann equations for flat universe models expressed in conformal time. Eqs. (12.179) and (12.180) can be combined to give

$$\dot{\mathcal{H}} = -\frac{1}{2}\mathcal{H}^2(1+3w). \tag{12.186}$$

Integration gives

$$\frac{1}{\mathcal{H}} = \frac{1}{2}(1+3w)\eta + C_0, \tag{12.187}$$

where C_0 is a constant of integration. Inserting this into eq. (12.172) and integrating leads to

$$a = C_1 \left[\frac{1}{2}(1+3w)\eta + C_0\right]^{\frac{2}{1+3w}}, \tag{12.188}$$

where C_1 is a constant. For a universe model where $w \neq -1$ we can impose the condition $a(0) = 0$ which implies $C_0 = 0$. Using the standard normalization $a(\eta_0) = 1$, the scale factor may be written

$$a = \left(\frac{\eta}{\eta_0}\right)^{\frac{2}{1+3w}}. \tag{12.189}$$

The "conformal Hubble parameter" is

$$\mathcal{H} = \frac{2}{1+3w}\frac{1}{\eta}, \quad w \neq -1. \tag{12.190}$$

Inserting these expressions into eq. (12.179) we find the unperturbed density as a function of the conformal time

$$\rho_0(\eta) = \frac{3\eta_0^{\frac{2}{1+3w}}}{2\pi G(1+3w)^2}\eta^{-\frac{6(1+w)}{1+3w}}. \tag{12.191}$$

Substituting expression (12.189) into eq. (12.169) and integrating, we find the cosmic time in terms of the conformal time,

$$t = t_0\left(\frac{\eta}{\eta_0}\right)^{\frac{3(1+w)}{1+3w}}, \quad t_0 = \frac{1+3w}{3(1+w)}\eta_0. \tag{12.192}$$

From eqs. (12.189) and (12.192) we obtain for the scale factor as a function of cosmic time

$$a(t) = \left(\frac{t}{t_0}\right)^{\frac{2}{3(1+w)}}, \tag{12.193}$$

where $a(t_0) = 1$.

The results are not valid for universe models where $w = -1$. Hence, they are not valid for a universe dominated by LIVE. In this case eq. (12.188) gives

$$a(\eta) = \frac{C_1}{C_0 - \eta}; \quad \mathcal{H} = \frac{1}{C_0 - \eta}. \tag{12.194}$$

From this and eq. (12.179) follows that $\rho_0 =$ constant. Defining $H_\Lambda = (8\pi G\rho_0/3)^{1/2}$ and choosing again $a(\eta_0) = 1$, we get

$$a(\eta) = \frac{1}{1 - H_\Lambda(\eta - \eta_0)}, \tag{12.195}$$

$$t(\eta) = t_0 - \frac{1}{H_\Lambda}\ln[1 - H_\Lambda(\eta - \eta_0)], \tag{12.196}$$

or

$$1 - H_\Lambda(\eta - \eta_0) = e^{-H_\Lambda(t-t_0)}, \tag{12.197}$$

giving

$$a(t) = e^{H_\Lambda(t-t_0)}. \tag{12.198}$$

Let us now consider the first order equations. First we consider a dust dominated model, i.e. a model in which $w = 0$. Then the scale factor and the Hubble parameter are

$$a(\eta) = a_0\eta^2, \quad \mathcal{H} = \frac{2}{\eta}. \tag{12.199}$$

In this case eq. (12.185) reduces to

$$\ddot{\Phi} + \frac{6}{\eta}\dot{\Phi} = 0. \tag{12.200}$$

The general solution of this equation is

$$\Phi(\mathbf{x}, \eta) = c_1(\mathbf{x}) + c_2(\mathbf{x})\eta^{-5}. \tag{12.201}$$

The density contrast is found by inserting this expression into eq. (12.183), which leads to

$$\delta(\mathbf{x}, \eta) = \frac{1}{6\eta^3}\nabla^2 c_2(\mathbf{x}) + \frac{3}{\eta^5}c_2(\mathbf{x}) + \frac{1}{6}\eta^2\nabla^2 c_1(\mathbf{x}) - 2c_1(\mathbf{x}). \tag{12.202}$$

This solution consists of two types of terms; those depending on $c_1(\mathbf{x})$ are growing with time, and those depending on $c_2(\mathbf{x})$ are decaying. We are interested in the growing solutions with $c_2(\mathbf{x}) = 0$,

$$\Phi(\mathbf{x}, \eta) = c_1(\mathbf{x}); \quad \delta(\mathbf{x}, \eta) = \frac{1}{6}\eta^2\nabla^2 c_1(\mathbf{x}) - 2c_1(\mathbf{x}). \tag{12.203}$$

Hence, the metric perturbation of this kind is constant in time although the density perturbation grows.

For dust the conformal time is related to the cosmic time as $\eta \sim t^{1/3}$. This means that the growing term of δ increases proportionally to $t^{2/3}$. The important conclusion is that there are growing density perturbation in a matter dominated FRW universe.

We shall now investigate if there are growing density perturbations in a radiation dominated universe model, too. For this purpose we can assume that the spatial variation of the perturbation has the form of a plane sinus wave,

$$\Phi(\mathbf{x}, \eta) = f(\eta)\sin(\mathbf{k}\cdot\mathbf{x}). \tag{12.204}$$

In the case of a radiation dominated universe, $w = 1/3$, and we get the following zeroth order parameters,

$$a = a_0\eta; \quad \mathcal{H} = \frac{1}{\eta}; \quad \dot{\mathcal{H}} = -\frac{1}{\eta^2}. \tag{12.205}$$

Inserting eqs. (12.204) and (12.205) into eqs. (12.183) and (12.185) we get

$$k^2\eta^2 f + 3\eta\dot{f} + 3f = -\frac{3}{2}\delta, \tag{12.206}$$

$$\eta^2\ddot{f} + 3\eta\dot{f} - f = \frac{1}{2}\delta. \tag{12.207}$$

These equations can be combined to give

$$\ddot{f} + \frac{4}{\eta}\dot{f} + \omega^2 f = 0, \quad \omega^2 \equiv \frac{k^2}{3}. \tag{12.208}$$

The general solution of this equation is

$$f(\eta) = c_1\frac{\omega\eta\cos\omega\eta - \sin\omega\eta}{\eta^3} + c_2\frac{\omega\eta\sin\omega\eta + \cos\omega\eta}{\eta^3}, \tag{12.209}$$

where c_1 and c_2 are integration constants. Inserting this expression into eq. (12.207) we find the time evolution of the amplitude $g(\eta)$ of the density contrast

$$\begin{aligned} g(\eta) = \frac{4}{\eta^3}\Big\{ &c_1\Big[(\omega^2\eta^2 - 1)\sin\omega\eta + \omega\eta\Big(1 - \frac{1}{2}\omega^2\eta^2\Big)\cos\omega\eta\Big] \\ &+ c_2\Big[(1 - \omega^2\eta^2)\cos\omega\eta + \omega\eta\Big(1 - \frac{1}{2}\omega^2\eta^2\Big)\sin\omega\eta\Big]\Big\}. \end{aligned} \tag{12.210}$$

The amplitude of the density contrast consists of terms that are proportional to $1, \eta^{-1}, \eta^{-2}$ and η^{-3} times a trigonometric function. This leads to the conclusion that in a radiation dominated universe model perturbations in the density of radiation do not grow with time.

12.10   Temperature fluctuations in the cosmic microwave background (CMB)

As noted in the last sections the photons of the CMB have moved freely since the universe was about 400 000 years old. The emitter events form a spherical shell around us with a radius around 13.7 billion light years and a thickness about 50 million light years.

In a flat matter dominated universe the horizon radius at a point of time t is $3ct$. Hence, the horizon radius at the time of decoupling is about one million years. As seen from our position this distance extends over an angle about $1°$ on the sky.

Measuring the temperature fluctuations in the CMB we obtain a map showing the physical conditions in the surface of last scattering. The observed temperature variations are due to density fluctuations in the shell of last scattering.

The original density fluctuations are thought to have their origin in quantum fluctuations that happened extremely early in the history of the universe, during the inflationary era.

Physical effects causing CMB-temperature fluctuations

One can describe the statistical properties of these fluctuations and the corresponding temperature fluctuations. In order to give a mathematical description of the CMB-temperature fluctuations, one utilizes that they are observed on a spherical surface. Hence, they can be written as a sum of spherical harmonic functions

$$\frac{\Delta T}{T}(\theta, \phi) = \sum_{\ell=0}^{\infty} \sum_{m=-\ell}^{\ell} a_{\ell m} Y_{\ell m}(\theta, \phi). \qquad (12.211)$$

One then introduces the expectation value of $|a_{\ell m}|^2$,

$$c_\ell \equiv \left\langle |a_{\ell m}|^2 \right\rangle. \qquad (12.212)$$

The power per logarithmic interval in ℓ is defined as

$$P^2(\ell) = \frac{\ell(\ell+1)}{2\pi} c_\ell. \qquad (12.213)$$

Here, ℓ is the multipole number which is related to the angular extension on the sky, so that $\theta = (\pi/\ell)\text{radians} = 180°/\ell$. The function $P(\ell)$ is called the *power spectrum*. It represents the average of the root mean square temperature difference ΔT in two directions separated by an angle $\theta = 180°/\ell$. Hence, $\Delta T = P(\ell)T_{CMB}$, where T_{CMB} is the average temperature of the CMB.

At scales larger than the horizon radius the fluctuations have not been modified by causal, dynamical processes since they were created. Hence, at

such scales the power spectrum shows the spectrum of the original fluctuations that were created quantum mechanically early in the inflationary era. It
has been shown that this part of the spectrum should be scale invariant. This is
one of the predictions of inflationary cosmology. Hence, the power spectrum
of the CMB-temperature fluctuations is expected to be flat for angles greater
than about $2°$; i.e. for values of ℓ less than about 100. Measurements by COBE
have confirmed this prediction and determined the magnitude of the power
in the flat part of the spectrum, $\Delta T = 27.9 \pm 2.5\mu K$ [Met. al.94, Bet. al.96].

One may distinguish between primary and secondary fluctuations. The
primary fluctuations are a result of processed happening before the universe became transparent for the CMB-radiation. The *secondary* fluctuations are
due to changes of the frequency (apart from that caused by the expansion of
the universe) while the photons move from the shell of last scattering to the
detector. We shall first consider the three most important effects causing the
primary fluctuations.

The Sachs-Wolfe effect

The Sachs-Wolfe effect is due to spatial variations of the gravitational potential in the shell of last scattering. This has two effects. (*i*) Photons from regions
with high density at last scattering loose energy as they move out of the gravitational field (moving "upwards"), and hence, they get a redshift. This gives
a temperature decrease

$$\left(\frac{\Delta T}{T}\right)_I = -\frac{\Delta\phi}{c^2}, \tag{12.214}$$

where $\Delta\phi$ is the difference of gravitational potential at the emitter position and
an observer position far away from the emitter. (*ii*) Due to gravitational time
dilatation the time proceeds at a slower rate far down in a gravitational field.
Looking toward a region with a deeper potential than at the surroundings,
we observe a region where time goes slower. So we seem to be looking at
a younger, and hence hotter region of the universe where there is an over-
density. The time dilatation is

$$\frac{\Delta t}{t} = -\frac{\Delta\phi}{c^2}. \tag{12.215}$$

The density ρ_γ of radiation is related to the expansion factor by $\rho_\gamma a^4 = \rho_{\gamma 0}$,
and according to Stefan-Boltzmann's law $\rho_\gamma \propto T^4$. Hence, $aT = \text{constant}$,
independent of the time dependence of the scale factor $a(t)$. We thus have

$$\left(\frac{\Delta T}{T}\right)_{II} = -\frac{\Delta a}{a}. \tag{12.216}$$

Assuming a flat universe dominated by a fluid with equation of state $p = w\rho$,
the time difference of the scale factor is $a \sim t^{2/3(1+w)}$. This leads to

$$\frac{\Delta a}{a} = \frac{2}{3(1+w)}\frac{\Delta t}{t} = -\frac{2}{3(1+w)}\frac{\Delta\phi}{c^2}. \tag{12.217}$$

Hence,

$$\left(\frac{\Delta T}{T}\right)_{II} = \frac{2}{3(1+w)}\frac{\Delta\phi}{c^2}. \tag{12.218}$$

The Sachs-Wolfe temperature fluctuations are

$$\boxed{\left(\frac{\Delta T}{T}\right)_{SW} = \left(\frac{\Delta T}{T}\right)_I + \left(\frac{\Delta T}{T}\right)_{II} = -\frac{1+3w}{3(1+w)}\frac{\Delta\phi}{c^2}.}$$
(12.219)

For a matter dominated universe, $w = 0$, which gives

$$\left(\frac{\Delta T}{T}\right)_{SW} = -\frac{1}{3}\frac{\Delta\phi}{c^2}.$$
(12.220)

The internal adiabatic effect

This effect is due to a coupling between the photon gas and the matter [Pea98]. The photon gas is compressed in regions with a large mass density. If the density fluctuations are adiabatic the density fluctuations of the photon gas and the matter are related by

$$\left(\frac{\Delta\rho}{\rho}\right)_\gamma = \frac{4}{3}\left(\frac{\Delta\rho}{\rho}\right)_m.$$
(12.221)

In a region with increased density of the photon gas there is higher temperature. On the other hand, the surface of last scattering is determined by the ionization potential of the hydrogen molecule, and thereby represents a surface of constant temperature. However, at a given point of time regions with larger density will have higher temperature. Hence, the surface of last scattering does not represent a set of simultaneous events. Since the temperature decreases with time, one observes later emitter events in direction of mass concentrations. The cosmic redshift is therefore less in these directions, and one observes a higher temperature,

$$\left(\frac{\Delta T}{T}\right)_A = -\frac{\Delta z}{1+z} = \frac{\Delta\rho}{\rho},$$
(12.222)

where the last equality assumes linear growth $\Delta\rho \sim (1+z)^{-1}$.

Doppler effect

This is the effect upon the observed temperature of the CMB of the peculiar velocity of that part of the surface of last scattering which is along the line of sight. The temperature change due to this effect is

$$\left(\frac{\Delta T}{T}\right)_D = \frac{\mathbf{v}\cdot\mathbf{n}}{c},$$
(12.223)

where $\mathbf{n}$ is a unit vector along the line of sight.

Acoustic oscillations in the early cosmic plasma

The Sachs-Wolfe, the adiabatic, and the Doppler effect tell us how the CMB-temperature result from fluctuations of the gravitational potential, the density and the velocity in the shell of last scattering. We shall now consider the physical mechanism behind these three types of fluctuations.

The most important mechanism is associated with the so-called *acoustic oscillations*. The photon and mass densities are assumed to be coupled adiabatically, so that $n_\gamma \sim n_m \sim T^3$. Hence, the temperature fluctuations and density fluctuations are related by

$$\frac{\Delta T}{T}(\mathbf{x}, \eta) = \frac{1}{3}\delta(\mathbf{x}, \eta). \tag{12.224}$$

In order to obtain a mathematical description of the fluctuations the fractional perturbations of the temperature are expanded in Fourier modes, with a corresponding Fourier expansion of the density contrast,

$$\delta(\mathbf{x}, \eta) = \frac{1}{(2\pi)^3} \int \delta_\mathbf{k}(\eta) e^{-i\mathbf{k}\cdot\mathbf{x}} d^3\mathbf{k}, \tag{12.225}$$

where $\mathbf{k}$ is a wave-number vector. One can then study each mode separately.

Ignoring, for the moment, the matter, one can deduce the equation of motion for the photon gas from the Euler equation and the equation of continuity. Denoting the fractional temperature fluctuations due to the Sachs-Wolfe effect by θ, i.e.

$$\theta \equiv \left(\frac{\Delta T}{T}\right)_{SW}, \tag{12.226}$$

the equation of motion of the density fluctuations in the photon gas can be written

$$\ddot{\theta} + c_s^2 k^2 \theta = 0, \tag{12.227}$$

where the adiabatic sound speed is defined by

$$c_s^2 \equiv \frac{\dot{p}_\gamma}{\dot{\rho}_\gamma}. \tag{12.228}$$

Since $p = \rho/3$ the sound speed is $c_s^2 = 1/3$ (corrections due to matter are considered in Example 12.6). The pressure waves propagate extremely fast. The equation of motion describes oscillations, i.e. sound waves in the photon gas. Hence, one calls the temperature fluctuations due to this effect for *acoustic peaks* in the power spectrum of the fluctuations.

The general solution of eq. (12.227) is

$$\theta(\eta) = \theta(0)\cos(ks) + \frac{\dot{\theta}(0)}{kc_s}\sin(ks), \tag{12.229}$$

where s is the sound horizon, defined by

$$s \equiv \int_0^\eta c_s d\eta. \tag{12.230}$$

All modes are frozen in at recombination, at η_{rc}, yielding temperature perturbations of different amplitude for different modes. For adiabatic oscillations with $\dot{\theta}(\eta_{rc}) = 0$,

$$\theta(\eta_{dc}) = \theta(0)\cos(ks_{dc}). \tag{12.231}$$

Modes with extrema of their oscillations at the surface of last scattering have $k_n s_{dc} = n\pi$. This introduces a fundamental scale related to the inverse sound horizon, $k_{dc} = \pi/s_{dc}$.

The fundamental physical scale is translated into a fundamental angular scale by a simple triangulation. The angle subtended by the proper value of the fundamental scale,

$$l_{Pdc} \approx a(t_{dc})\frac{2\pi}{k_{dc}} = \frac{2s_{dc}}{1+z} \approx \frac{2\eta_{dc}}{\sqrt{3}(1+z)}, \tag{12.232}$$

at the angular diameter distance of the surface of last scattering (see eq. (11.135))

$$l_{Adc} = \frac{\eta_0 - \eta_{dc}}{1+z} \approx \frac{\eta_0}{1+z}, \tag{12.233}$$

is

$$\theta_{dc} = \frac{2}{\sqrt{3}}\frac{\eta_{dc}}{\eta_0}. \tag{12.234}$$

The corresponding value of the spherical harmonic index is

$$\ell_{dc} = \frac{2\pi}{\theta_{dc}} \approx \pi\sqrt{3}\frac{\eta_0}{\eta_{dc}}. \tag{12.235}$$

In a matter dominated universe model, $\eta \propto a^{1/2}$, so

$$\ell_{dc} \approx \pi\sqrt{\frac{3}{a_{dc}}} \approx \pi\sqrt{3z_{dc}}. \tag{12.236}$$

Inserting $z_{dc} = 1100$ gives $\ell_{dc} = 180$. In this region of the power spectrum one expects a transition from a flat spectrum due to the original scale invariant fluctuations to a part of the spectrum containing acoustic peaks.

In order to obtain an accurate CMB spectrum one must perform some rather complex calculations. Computer based packages performing such calculations have been developed. One of the most used packages is CMB-FAST. However, some properties of the spectrum can be found analytically.

Let us first investigate how the positions of the peaks depend upon the curvature of space. Consider a space with constant positive curvature. Then the proper distance, d_P, is replaced by $R\sin(d_P/R)$ where R is the curvature radius of the space. Hence, the ratio of the angle subtended by a physical scale λ in the curved space and in flat space is

$$\frac{\theta_+}{\theta_0} = \frac{d_P}{R\sin\frac{d_P}{R}}. \tag{12.237}$$

Assuming that $d_P \ll R$ we obtain to 3rd order in d_P/R,

$$\frac{\theta_+}{\theta_0} \approx 1 + \frac{1}{6}\left(\frac{d_P}{R}\right)^2. \tag{12.238}$$

The curvature radius of space in a Friedmann universe model with relative density Ω_{tot} is (see eq. (11.51))

$$R = \left(Ha\sqrt{|\Omega_k|}\right)^{-1} = \left(Ha\sqrt{|\Omega_{tot} - 1|}\right)^{-1}. \tag{12.239}$$

At the present time

$$R_0 = \left(H_0\sqrt{|\Omega_{tot0} - 1|}\right)^{-1}. \tag{12.240}$$

The proper distance to the surface of last scattering is presently

$$d_P = \eta_0 - \eta_{rc} \approx \eta_0 \approx 3t_0, \tag{12.241}$$

where the last approximate equality is valid for a flat mass dominated universe. We consider universe models that are nearly flat. Using that $t_0 \approx (2/3)H_0^{-1}$ gives

$$\frac{\theta_+}{\theta_0} \approx \frac{1}{3}\left(1 + 2\Omega_{tot}\right). \tag{12.242}$$

One can show that the same expression is valid in a negatively curved universe. The expression shows that the same physical distance subtends a larger angle in a closed universe, $\Omega_{tot} > 1$, than in a flat universe, and a smaller angle in a negatively curved universe, $\Omega_{tot} < 1$.

We shall now consider the effects of baryons upon the CMB-spectrum. Baryons add inertia to the cosmic fluid. There are three effects of raising the baryon density: an amplitude increase, a zero-point shift towards higher compression, and a frequency decrease. The magnitude of these effects are given by the factor $1 + r$ where r is the density plus momentum ratio of baryons and photons,

$$r = \frac{\rho_b + p_b}{\rho_\gamma + p_\gamma} \approx \frac{3\rho_b}{4\rho_\gamma}, \tag{12.243}$$

which may be expressed as

$$r = \frac{3\Omega_{b0}}{4\Omega_{\gamma0}}a. \tag{12.244}$$

Inserting the measured values $\Omega_{b0} = 0.04$ and $\Omega_{\gamma0} = 5 \cdot 10^{-5}$, gives at the time of last scattering $r_{rc} \approx 0.5$. More accurate calculations give a somewhat smaller number. However, the CMB-temperature fluctuations are a good baryometer. The recent very accurate measurements by the WMAP-mission have given the result, $\Omega_b = 0.044 \pm 0.004$. This is in very good agreement with results of measurements of the cosmic abundances of the lightest element combined with the theory of their production in the cosmic nucleosynthesis during the first ten minutes of our universe.

Example 12.6 (The velocity of sound in the cosmic plasma)   Example
The *bulk modulus* of a fluid gives the relative change of volume dV/V of a fluid element due to a change of pressure, dp. It is defined as

$$\kappa = -V\frac{dp}{dV} = -V\frac{\dot{p}}{\dot{V}}. \tag{12.245}$$

The negative sign indicates that the volume decreases when the pressure increases. The mass of a fluid element is constant. Hence,

$$(\rho V)^{\cdot} = \dot{\rho}V + \rho\dot{V} = 0, \qquad \dot{V} = -\frac{V}{\rho}\dot{\rho}, \tag{12.246}$$

giving

$$\kappa = \rho \frac{\dot{p}}{\dot{\rho}}. \tag{12.247}$$

We shall now deduce the equation of pressure waves in a fluid. Let $s(z,t)$ be the displacement of fluid particles in the direction of motion of a plane pressure wave. The corresponding change of pressure at a wave front with area A is

$$\delta p = -\kappa \frac{\delta V}{V} = -\kappa \frac{A\left[s(z + \delta z) - s(z)\right]}{A \delta z} \xrightarrow[\delta z \to 0]{} -\kappa \frac{\partial s}{\partial z}. \tag{12.248}$$

The variation of the pressure in the z-direction is

$$\frac{\partial \delta p}{\partial z} = -\kappa \frac{\partial^2 s}{\partial z^2}. \tag{12.249}$$

The corresponding variation of the pressure force is

$$dF = -A d\delta p = -A \frac{\partial \delta p}{\partial z} dz = \kappa A \frac{\partial^2 s}{\partial z^2} dz. \tag{12.250}$$

Newton's 2nd law applied to the fluid element gives

$$dF = a dm = \frac{\partial^2 s}{\partial t^2} A \rho dz, \tag{12.251}$$

which leads to

$$\frac{\partial^2 s}{\partial t^2} = \frac{\kappa}{\rho} \frac{\partial^2 s}{\partial z^2}. \tag{12.252}$$

This is the equation of motion for the pressure waves.

In general, the wave equation has the form

$$\frac{\partial^2 s}{\partial t^2} = c_s^2 \frac{\partial^2 s}{\partial z^2}, \tag{12.253}$$

where c_s is the velocity of propagation of the waves. The pressure waves are often called *acoustic waves* or *sound waves*. From eqs. (12.252) and (12.253) we get

$$c_s^2 = \frac{\kappa}{\rho}, \tag{12.254}$$

which together with eq. (12.247) gives

$$\boxed{c_s^2 = \frac{\dot{p}}{\dot{\rho}}.} \tag{12.255}$$

We shall find the velocity of sound in a cosmic fluid consisting of cold matter, LIVE and radiation. The relative densities of the components of the cosmic fluid are $\Omega_m, \Omega_\Lambda,$ and Ω_γ. The total relative density is $\Omega = \Omega_m + \Omega_\Lambda + \Omega_\gamma$.

Using that the deceleration parameter is

$$q = \frac{1}{2}\left(1 + 3\frac{p}{\rho}\right)\Omega, \quad p = \sum p_i, \quad \rho = \sum \rho_i, \tag{12.256}$$

the statefinder r, using eq. (12.122), is

$$r = \left[1 + \frac{9}{2}\left(1 + \frac{p}{\rho}\right)\frac{\dot{p}}{\dot{\rho}}\right]\Omega. \tag{12.257}$$

Using eq. (12.255), this may be written

$$r = \left[1 + \frac{9}{2}\left(1 + \frac{p}{\rho}\right)c_s^2\right]\Omega. \tag{12.258}$$

From eqs. (12.256) and (12.258) we get

$$\boxed{c_s^2 = \frac{1}{3}\frac{r - \Omega}{q + \Omega}.}$$
(12.259)

In the case of a flat universe this reduces to

$$c_s^2 = \frac{1}{3}\frac{r - 1}{q + 1}.$$
(12.260)

We now proceed with the general case permitting curved universe models. If the components of the cosmic fluid have relative densities Ω_i and equations of state $p_i = w_i \rho_i$, the deceleration parameter and the statefinder may be expressed as

$$q = \frac{1}{2}\sum_i (1 + 3w_i)\Omega_i,$$
(12.261)

$$r = \sum_i \left[1 + \frac{9}{2}w_i(1 + w_i)\right]\Omega_i.$$
(12.262)

In a universe with cold matter, LIVE and radiation we get

$$q = \frac{1}{2}(\Omega - 3\Omega_\Lambda + \Omega_\gamma), \quad r = \Omega + 2\Omega_\gamma,$$
(12.263)

which gives

$$c_s^2 = \frac{4}{9}\frac{\Omega_\gamma}{\Omega_m + \frac{4}{3}\Omega_\gamma}.$$
(12.264)

This may be written

$$\frac{1}{c_s^2} = 3\left(1 + \frac{3}{4}\frac{\Omega_m}{\Omega_\gamma}\right).$$
(12.265)

Note that c_s is the total matter sound speed. The photon-baryon sound speed squared is given by the above expressions replacing Ω_m by the relative density of baryons, Ω_b. It is the photon-baryon sound speed which appears in the theory of the temperature fluctuations in the cosmic microwave background radiation [Väl99].

12.11 Mach's principle

This principle was introduced in section 1.9. The effect that inertial frames are dragged by rotating and accelerating mass distributions was mentioned. This effect has the potential of explaining the fixed direction of the swinging plane of a Fouceault pendulum relative to the stars as a gravitational effect due to the remote mass in the Universe. The phenomenon of perfect dragging inside a massive shell with radius equal to its Schwarchild radius is significant in this connection. We shall here consider rotational inertial dragging in the context of cosmological perturbation theory, following a work by C. Schmid [Sch04]. Until now we have only considered isotropic perturbations with a line element of the form (12.170). This must now be generalized in order to permit vorticity perturbations. The unperturbed universe models are here assumed to be flat. The line-element is written

$$ds^2 = -dt^2 + (ah_i)^2 (dx^i)^2 + 2(ah_i)^2\beta^i dx^i dt,$$
(12.266)

Here $a(t)$ is the scale factor, $h_\chi = 1$, $h_\theta = \chi$, $h_\phi = \chi \sin\theta$ and β^i are the components of the so-called shift vector. We consider vector perturbations in an asymptotic isotropic universe, i.e. $\beta^i \to 0$ for $r \to \infty$. The coordinate system is comoving with galaxies having no peculiar velocities in the unperturbed universe. The spatial coordinate basis vectors, $\mathbf{e}_i = \partial/\partial x^i$ point along geodesics from an observer point P to the galaxies. In the unperturbed universe the observers are at rest in the coordinate system. An orthonormal basis field with time-like basis vector $\mathbf{e}_{\hat{0}} = \mathbf{u}$, is comoving with the observers, where $\mathbf{u}$ is the four-velocity field of the observers.

The perturbation introduces a vorticity field with a 3-velocity having coordinate components β^i. This is the velocity field of the observers in the perturbed universe. The velocity of a freely falling particle in the cosmic fluid has covariant components $v_{\hat{i}}$ in the orthonormal basis field of the observers, and a gyroscope has corresponding spin components $S_{\hat{i}}$.

The laws of linearized gravity shall now be formulated in a similar way as the laws of electromagnetism in a $(3 + 1)$-formalism. The *gravitoelectric field strength* is defined by considering a freely falling particle instantaneously at rest,

$$\dot{v}_{\hat{i}} \equiv E_{\hat{i}}^g. \tag{12.267}$$

Hence $\mathbf{E}_g = \mathbf{g}$, where $\mathbf{g}$ is the acceleration of gravity. The *gravitomagnetic field strength* is defined by

$$\dot{S}_{\hat{i}} \equiv -\frac{1}{2}[\mathbf{B}_g \times \mathbf{S}]_{\hat{i}}. \tag{12.268}$$

This equation gives the angular velocity of precession of the gyroscope spin axis relative to the orthonormal frame of the observers,

$$\mathbf{\Omega}_{\text{gyro}} = -\frac{1}{2}\mathbf{B}_g. \tag{12.269}$$

In the linear field approximation the equations of motion for freely falling test particles and for the spin axes of the gyroscopes carried along by the observers are

$$\dot{v}^{\hat{a}} + \Gamma^{\hat{a}}_{\hat{b}\hat{c}}v^{\hat{b}}u^{\hat{c}} = 0, \quad \dot{S}^{\hat{i}} + \Gamma^{\hat{i}}_{\hat{j}\hat{0}}S^{\hat{j}} = 0, \tag{12.270}$$

where $\Gamma^{\hat{a}}_{\hat{b}\hat{c}}$ are the connection coefficients of the basis comoving with the observers. Hence

$$\Gamma_{\hat{i}\hat{0}\hat{0}} = -E_{\hat{i}}^g, \quad \Gamma_{\hat{i}\hat{j}\hat{0}} = -\frac{1}{2}B_{\hat{i}\hat{j}}^g, \tag{12.271}$$

where $B_{\hat{i}\hat{j}} \equiv \varepsilon_{\hat{i}\hat{j}\hat{k}}B_{\hat{k}}$ with summation over $\hat{k}$.

To first order in β^i the basis vectors of the observer's orthonormal basis may be expressed in terms of the coordinate vectors as

$$\mathbf{e}_{\hat{0}} = \frac{\partial}{\partial t}, \quad \mathbf{e}_{\hat{k}} = \frac{1}{ah_k}\left(\frac{\partial}{\partial x^k} + \beta_k \frac{\partial}{\partial t}\right), \quad \text{(no summation over } k\text{)} \tag{12.272}$$

From this one may calculate the connection coefficients (no summation over i

and j, summation over k)

$$\Gamma_{\hat{i}\hat{0}\hat{0}} = -E_i^g = \frac{1}{ah_i}\dot{\beta}_i, \tag{12.273}$$

$$\Gamma_{\hat{i}\hat{j}\hat{0}} = = -\varepsilon_{\hat{i}\hat{j}\hat{k}}B_{\hat{k}}^g = -\frac{1}{2a^2h_ih_j}\left(\frac{\partial\beta_i}{\partial x^j} - \frac{\partial\beta_j}{\partial x^i}\right), \tag{12.274}$$

$$\Gamma_{\hat{i}\hat{0}\hat{j}} = = -\varepsilon_{\hat{i}\hat{j}\hat{k}}B_{\hat{k}}^g + \delta_{\hat{i}\hat{j}}H, \tag{12.275}$$

$$\Gamma_{\hat{i}\hat{j}\hat{k}} = \frac{\delta_{\hat{i}\hat{j}}}{ah_j}\left(H\beta_j + \frac{\partial L_{\hat{i}}}{\partial x^j}\right) - \frac{\delta_{\hat{j}\hat{k}}}{ah_j}\left(H\beta_i + \frac{\partial L_{\hat{j}}}{\partial x^i}\right), \tag{12.276}$$

where $L_{\hat{i}} \equiv \ln h_i$ and $H \equiv \dot{a}/a$ is the Hubble parameter.

We now identify the shift vector β with the gravitomagnetic vector potential given by $\mathbf{B}_g = \nabla \times \mathbf{A}_g$, i.e. $\beta = \mathbf{A}_g$. Then eqs. (12.273) and (12.274) may be written

$$\mathbf{E}_g + \frac{1}{a}\frac{\partial}{\partial t}(a\mathbf{A}_g) = 0, \tag{12.277}$$

$$\nabla \times \mathbf{E}_g + \frac{1}{a^2}\frac{\partial}{\partial t}(a^2\mathbf{B}_g) = 0, \tag{12.278}$$

corresponding to the homogeneous Maxwell equations in electromagnetism.

Inserting the connection coefficients (12.273) – (12.276) into the first one of eqs. (12.270) and writing the component equations as a vector equation one obtains

$$\frac{1}{a}\frac{\partial}{\partial t}(a\mathbf{v}) = \mathbf{E}_g + \mathbf{v} \times \mathbf{B}_g + H\mathbf{v} \times (\beta \times \mathbf{v}). \tag{12.279}$$

The first term on the right hand side represents the acceleration of gravity, i.e. the acceleration of a freely falling particle instantaneously at rest. From eq. (12.269) is seen that the second term is the Coriolis acceleration. The third term is an acceleration of the test particle due to the expansion of the Universe. The expansion velocity of the particle increases as it moves farther away from the observer.

The connection forms have components given in eqs. (12.273) – (12.276). Cartan's second equation is used to calculate the curvature forms, from which the components of the Einstein tensor are found.

For the vorticity perturbations the relevant Einstein equation is $E_{\hat{0}\hat{i}} = \kappa T_{\hat{0}\hat{i}}$, which takes the form

$$\nabla \times \mathbf{B}_g - 4\dot{H}\mathbf{A}_g = -2\kappa\mathbf{J}, \tag{12.280}$$

where

$$J^{\hat{i}} \equiv T^{\hat{0}\hat{i}} = (\rho + p)v^{\hat{i}} \tag{12.281}$$

and we have used $\beta = \mathbf{A}_g$.

From $\dot{H} = -(\kappa/2)(\rho + p)$ we see that $\dot{H} \le 0$ for $p \ge -\rho$. Hence, we may define $\mu^2 \equiv -4\dot{H}$. Inserting the vector potential $\mathbf{A}_g$ into eq. (12.280), and using that $\nabla \cdot \mathbf{A}_g = 0$ so that $\nabla \times \mathbf{B}_g = \nabla \times (\nabla \times \mathbf{A}_g) = -\nabla^2\mathbf{A}_g$, the equation may be written

$$\left(-\nabla^2 + \mu^2\right)\mathbf{A}_g = -2\kappa\mathbf{J}. \tag{12.282}$$

The solution of this equation is the Yukawa potential for $\mathbf{A}_g$ in terms of the sources $\mathbf{J}$ at the same fixed time

$$\mathbf{A}_g(\mathbf{r}, t) = -\frac{\kappa}{2\pi} \int \mathbf{J}(\mathbf{r}', t) \frac{\exp(-\mu |\mathbf{r} - \mathbf{r}'|)}{|\mathbf{r} - \mathbf{r}'|} d^3 r'. \tag{12.283}$$

We shall now consider the dragging of gyroscope axes by a homogeneous rotation of the matter in the Universe out to distances significantly beyond the $\dot{H}$-radius . Then the second term in the parenthesis of eq. (12.282) dominates over the first one. Hence, in this case we may write

$$\mu^2 \mathbf{A}_g = -2\kappa \mathbf{J}, \tag{12.284}$$

which gives $\beta = -\mathbf{v}$. Using eq. (12.269) we then have

$$\mathbf{\Omega}_{\text{gyro}} = \frac{1}{2} \nabla \times \mathbf{v} = \mathbf{\Omega}_{\text{fluid}}. \tag{12.285}$$

This shows that there is perfect dragging in this case with a homogeneous rotation of the cosmic fluid beyond the $\dot{H}$-radius.

We now calculate $\mathbf{B}_g = \nabla \times \mathbf{A}_g$ from eq. (12.283) and set $r = 0$, which gives $\mathbf{B}(r = 0) = -(1/2)\mathbf{\Omega}_{\text{gyro}}$. Then we obtain $\mathbf{\Omega}_{\text{gyro}}$ in terms of the sources at a fixed time

$$\mathbf{\Omega}_{\text{gyro}} = \frac{\kappa}{2\pi}(\rho + p) \int_0^\infty \frac{(1 + \mu r)\exp(-\mu r)}{r^3} \mathbf{r} \times \mathbf{v} d^3 r. \tag{12.286}$$

This is an expression for Mach's principle telling that the motion of the axis of a gyroscope – or the plane of a Fouceault pendulum – here at $r = 0$ is determined by a specific average of the energy flow in the Universe. Using that $2\kappa(\rho + p) = -\mu^2$ this average may be written

$$\mathbf{\Omega}_{\text{gyro}} = \frac{\mu^2}{3} \int_0^\infty \mathbf{\Omega}_{\text{fluid}}(r) r(1 + \mu r)\exp(-\mu r) dr. \tag{12.287}$$

For the special case that $\mathbf{\Omega}_{\text{fluid}}(r)$ is independent of r the integral (12.287) gives $\mathbf{\Omega}_{\text{gyro}} = \mathbf{\Omega}_{\text{fluid}}$, i.e., perfect dragging.

12.12   The History of our Universe

We have gone through the most important concepts in the standard model of our universe. We will now give a short outline of the history of our universe, from the Big Bang to the present time and beyond.

The Planck era: $t < 10^{-43}\text{s}$, $T > 10^{32}\text{K}$.

The laws of physics, as we know then, may describe the universe backwards until a time Δt after the point of time of a theoretical and singular Big Bang event. The time Δt may be estimated heuristically from Heisenberg's uncertainty relation, $\Delta E \Delta t \leq \hbar$ where ΔE is the energy fluctuation during a time interval Δt. The energy fluctuation has an extension $\Delta x = c\Delta t$ so that $\Delta E = \hbar c/\Delta x$. If ΔE is equal to or larger than the gravitational self-energy of the fluctuation, $Gm^2/\Delta x$, then the fluctuations are so significant that the spacetime cannot be described without a quantum theory of gravity. No generally accepted theory of this type has been constructed. The *Planck mass* is defined by the limiting case, $Gm_{Pl}^2/\Delta x = \hbar c/\Delta x$, which yields

$$m_{Pl} = \sqrt{\frac{\hbar c}{G}} = 2.2 \cdot 10^{-8} \text{kg.} \tag{12.288}$$

The corresponding *Planck time* is

$$t_{Pl} = \frac{\hbar}{m_{Pl}c^2} = \sqrt{\frac{\hbar G}{c^3}} = 5.4 \cdot 10^{-44} \text{s.} \tag{12.289}$$

So close to the Big Bang singularity can we in principle describe the universe, but not closer. The corresponding *Planck length* is $\ell_{Pl} = ct_{Pl} \approx 1.6 \cdot 10^{-35}$m, the Planck temperature is $T_{Pl} = m_{Pl}c^2/k_B \approx 1.5 \cdot 10^{32}$K, where k_B is Boltzmann's constant, and the Planck energy density is $\rho_{Pl} = m_{Pl}c^2/\ell_{Pl}^3 \approx 10^{97}$kg/m^3.

The time before t_{Pl} is called the *Planck era*. At this time the Universe was filled with a plasma of relativistic elementary particles, including quarks, leptons, gauge bosons and possibly Higgs bosons. This spacetime cannot be described by means of the presently known laws of physics. However, one might guess that the universe existed in a state of fluctuating chaos during this era. Time was not a well defined quantity, and the curvature and even the topology of space fluctuated wildly.

A fluctuation may have happened so that a region of space became dominated by vacuum energy.

The inflationary era: $10^{-43} < t < 10^{-33}$s

Due to repulsive gravity this region got an exponentially accelerated expansion (see section 12.5) which lasted for 10^{-33}s. However, although recent observations of the temperature variations in the cosmic microwave background indicate that the universe has really passed through an inflationary era, the question when and how this started (and ended) can only be answered by educated guesses. We have no knowledge about this. Maybe it started due to a phase transition at a temperature 10^{27}K at the GUT point of time, $t_{GUT} \approx 10^{-35}$s, when the strong force separated from the electroweak force. Thinking about this possibility the period from 10^{-43} to 10^{-35}s is called the GUT era. During the GUT-era the quarks and leptons were indistinguishable since quarks and leptons exchanged X-bosons which changed their identities: quarks became leptons and vice versa.

Exponential expansion starts slowly. Hence, during the first part of the inflationary era the radiation within our observable part of the universe came into a state of thermal equilibrium. This explains the isotropy of the microwave background radiation. But still there were quantum fluctuations, and they are the seeds from which the galaxies evolved much later.

During the inflationary era the total density of the cosmic energy approached exponentially towards the critical density, corresponding to a flat universe. Hence, a prediction of the inflationary cosmological models is that the universe should still be extremely flat within the limits of observational accuracy. This seems now to be confirmed by, among others, the WMAP-observations of the cosmic microwave background temperature fluctuations.

The density of the vacuum energy remained constant during the cosmic expansion in the inflationary era. Hence, a vast amount of vacuum energy was produced. Still the energy of the inflating part of the universe may be considered to be constant. This may be understood by considering an expanding surface bounding a finite volume of space. Due to the negative pressure of the vacuum energy the thermodynamic work at the surface transports energy through the surface in the opposite direction of its motion. Hence, there was an energy flux from the region outside the surface to the region inside it. This accounts for the increase of vacuum energy inside the comoving surface during the inflationary era.

At about 10^{-33}s the vacuum energy field began to oscillate, and vacuum energy was transformed into radiation and elementary particles.

Baryongenesis

If equal amounts of matter and antimatter were created, the antimatter would rapidly annihilate the matter, and the end result would be a universe filled with radiation and no matter. Hence, in order to arrive at the present universe without antimatter and about 10^9 more photons than baryons there must have been created slightly more matter than antimatter. Since all the antibaryons annihilated together with an equal amount of baryons leaving an excess number of baryons, we can calculate the magnitude of the original asymmetry in terms of the present ratio of baryon and photon numbers [Ham].

Let the baryon number density be n_b, the number density of antibaryons be $\bar{n}_b$, and the number density of photons be n_γ. Present values are denoted by an index 0. Before the annihilation there were approximately equal numbers of baryons, antibaryons and photons, $n_b \approx \bar{n}_b \approx n_\gamma$. From baryon number conservation we have a preserved comoving number density,

$$(n_b - \bar{n}_b)a^3 = (n_{b0} - \bar{n}_{b0})a_0^3 = n_{b0}a_0^3. \qquad (12.290)$$

Similarly, for photons

$$n_\gamma a^3 = n_{\gamma 0}a_0^3. \qquad (12.291)$$

Hence,

$$\frac{n_b - \bar{n}_b}{n_b + \bar{n}_b} \approx \frac{n_{b0}}{2n_{\gamma 0}} \approx 10^{-9}, \qquad (12.292)$$

showing that the original baryon asymmetry was very small. For every billion antibaryons in the early universe there were one billion and one baryons. This was an asymmetry of the order one part in a billion. It is believed that this asymmetry were generated dynamically at some very early time in the history of the universe. However, one still does not know when and how the baryon asymmetry in the universe was produced. Two possibilities have been considered: that it happened either at the beginning or at the end of the electroweak era.

It was argued by Andrei Sakharov in 1967 that three conditions must be fulfilled in order to produce matter-antimatter asymmetry.

1. There must exist a C and CP violation of one of the fundamental interactions.

2. Non-conservation of baryon number must be possible.

3. There must have existed a state of thermodynamical non-equilibrium.

The basic statement of the rule that baryon number is conserved is that no physical process can change the net number of quarks. To understand why we need baryon non-conserving processes to generate a baryon asymmetry from an initial symmetric state, one may suppose that all physical processes obey the rule of baryon number conservation. Then the net baryon number zero of the initial state cannot be changed. Hence, the universe would always be baryon symmetric.

Imagine that nature allows for a baryon non-conserving process where a massive gauge particle with baryon number $B = 0$ decays into a proton with $B = 1$ and an electron with $B = 0$. The initial net baryon number is zero and the final $+1$. Suppose that a second process in which the particles are replaced by their antiparticles, occurs at the same rate as the first process. Then the change of baryon number produced by the two processes would cancel, and the universe would remain baryon symmetric.

We now turn to the question of whether the cancelling "anti-reaction" would take place. There are three fundamental symmetry transformations in classical physics: charge conjugation, C, parity transformation, P, and time reversal, T. The operation of charge conjugation reverses the signs of all the internal quantum numbers in a system leaving the mass, energy, momentum and spin unchanged. A neutrino, for example, carries a non-zero internal quantum number called the lepton number. Charge conjugation changes the sign of the lepton number which means that it changes a neutrino to its antineutrino without changing its spin.

The parity operation is essentially a mirror reflection. The effect of the parity operation on a right-handed neutrino is to turn it into a left-handed neutrino. Under time reversal the motion reverses while the internal properties remain unchanged. Hence, for a right-handed neutrino we find that time reversal gives us a right-handed neutrino travelling in the opposite direction.

Under a combined CP-transformation a right-handed neutrino becomes a left-handed antineutrino. Getting back to our hypothetical reaction capable of generating baryon asymmetry, it turns out that the "anti-reaction" is just the CP-transformed reaction of the original process we considered. Therefore, suppression of the "anti-reaction" requires CP-violation in this situation. Hence, CP-violation allows for a preference of matter over antimatter in some processes. CP-violation must therefore have been an essential ingredient in generating the baryon asymmetry.

The Grand Unified Theories unifies the electroweak force with the strong force between quarks. One expects the gauge bosons of the GUT-theory to mediate interactions mixing leptons and quarks, thereby allowing non-conservation of baryon number. CP-violation is a feature of the simplest GUT-theory. A problem, however, is that this CP-violation provides far too small a contribution to account for successful baryogenesis. Extensions of the GUT-theories have been constructed that provide sufficient CP-violation. The earliest attempts at constructing a model of baryogenesis therefore incorporated the GUT-theories. In these models the baryogenesis happened before the GUT phase transition at $t = 10^{-35}$s. At this early time the expansion of the universe was sufficiently fast to allow deviation from thermodynamical equilibrium. However, there exists a serious problem for the GUT baryogenesis. The difficulty arises from the subsequent inflation that lasts for 10^{-33}s. The inflation will dilute the generated net baryon density, so that the baryon density

Particle created	Energy	Temperature	Time
proton ⎫ neutron ⎭	1GeV	10^{13}K	10^{-6}s
muon	50MeV	$5 \cdot 10^{11}$K	$4 \cdot 10^{-4}$s
electron	0.5MeV	$5 \cdot 10^9$K	4s

Table 12.1: Particle creation in the early universe.

becomes too small to account for its presently observed value. Hence, the baryo-genesis must have happened at the end of the inflationary era if the reheating at this point of time was sufficiently strong. However, it can be shown that the temperature expected during reheating is not sufficiently high to reignite the GUT process.

Due to these difficulties with GUT-baryogenesis, one has focused on baryo-genesis at much lower energies[5]. In particular, one has studied the possibil-ity of baryogenesis at the electroweak symmetry breaking, which happened about 10^{-10}s after the big bang. The electroweak vacuum allows processes that violate baryon number conservation. At this time deviations from ther-modynamical equilibrium happened due to rapid changes of the properties of the vacuum. However, whether the CP-violation during these processes is sufficiently effective to account for successful baryogenesis is still an open question.

Cosmic time and temperature for annihilation of particle species

In order to create particle-antiparticle pair of particles, with mass m, from the photon energy in an hot mixture of plasma and radiation the tempera-ture must fulfill $k_B T > 2mc^2$. Inserting numerical values for c and k_B gives approximately $T > (m/1\text{MeV})10^{10}$K, where the mass is measured in MeV.

In a flat, radiation dominated universe model the cosmic time correspond-ing to annihilation at temperature T is

$$t = \frac{2.3}{\sqrt{g_{\text{eff}}}} \left(\frac{10^{10}\text{K}}{T}\right)^2 \text{s} = \frac{2.3}{\sqrt{g_{\text{eff}}}} \left(\frac{1\text{MeV}}{m}\right)^2 \text{s}, \qquad (12.293)$$

where g_{eff} is the effective degrees of freedom. The particle content around 1s after the big bang gives $g_{\text{eff}} = 5.4$ leading to $t \approx (10^{10}\text{K}/T)^2\text{s} = (1\text{MeV}/m)^2\text{s}$. We can use this relation also at other energy scales than 1MeV to estimate typical points of time for different processes. The result is shown in Table 12.1. The last protons and neutrons were created about 10^{-6}s after the big bang, and the final large scale electron-positron annihilation happened about 10s later.

The electro-weak era: $10^{-33} < t < 10^{-10}$s

During this period, which started at the end of the inflationary era, the elec-tromagnetic and weak force were unified into an electro-weak force. In this era the temperature was above 10^{15}K. This corresponds to energies which are much higher than the energies represented by the masses of the $W^\pm$ and Z^0 bosons that mediate the weak force. Hence in this era the masses of these

---

[5]It should be pointed out that baryogenesis is also possible at intermediate energies. For example, baryogenesis via leptogensis is also possible at intermediate scale energies [BPY05].

bosons can be neglected so that the weak interaction can be considered as being mediated by massless spin 1 particles, like the photons that mediate the electromagnetic force.

When the temperature dropped below 10^{15}K the bosons acquired mass by interacting with the vacuum via the Higgs mechanism. Then the weak force separated from the electromagnetic force and became a short range force. The universe was filled with hadrons, leptons, weak bosons and photons.

If one characterizes the universe by its dominating matter contents, and not by the type of fundamental interaction, the first part of the electroweak era is instead called the *quark era* (there is no general agreement on the use of this term, however). The dominating form of matter was now quark-antiquark-gluon plasma.

Usually this era is said to last until the thermal energy is no longer sufficiently large to produce quark-antiquark pairs, at 10^{-10}s. However, there has been some speculation that the quark era was replaced by a hadron dominated era at the time 10^{-23}s when the observable universe became larger than the size of a nucleon [Har81] . If that happened, the quark era existed at 10^{-33}s $< t < 10^{-26}$s and the hadron era 10^{-26}s $< t < 10^{-6}$s. However, there is still no established theory for baryogenesis at the time 10^{-26}s.

The hadron era: $10^{-10} < t < 10^{-6}$s

If the baryogenesis happened at the time of electro-weak symmetry breaking, this time marks the end of the quark era and the beginning of the hadron era. The dominating form of matter was now protons and neutrons with an equal number of pions. The last protons and neutrons were made 10^{-6}s after the big bang. Then the proton-antiproton and neutron-antineutron pairs annihilated and left their energy to photons and lighter particles that were produced in this process. The baryon asymmetry secured that the later universe had sufficient baryonic matter to evolve stars and eventually life.

The lepton era: $10^{-6} < t < 10$s

During this era the temperature decreased from about 10^{12}K to $6 \cdot 10^{9}$K. The thermal energy of the cosmic plasma was no longer large enough to create quark-antiquark pairs. The quarks were from now on confined in baryons and mesons. However, the dominating form of matter was electron-positron pairs. At the initial time of the lepton era the average density of the cosmic plasma was 10^{17}kg/m^3. The hadrons were buried in a dense lepton-gas. To each hadron there existed roughly one billion photons, electron-positron pairs and neutrino-antineutrino pairs. Everything was in thermal equilibrium and there were approximately equal numbers of photons, electrons and neutrinos, and initially also of muons. However, at about 10^{-3}s there was no longer sufficient energy to create muon-antimuon pairs.

Although the muons had decayed, the much lighter μ-neutrinos were still present and continued to interact with the electrons via the neutral-current weak interactions

$$e^+ + e^- \leftrightharpoons \nu_i + \bar{\nu}_i, \qquad i = e, \mu.$$

In order to judge the importance of this reaction at a certain cosmic time, we must take into consideration that all reactions in the universe will have a certain reaction rate, and, the inverse of this, a characteristic reaction time-scale.

If the reaction time is longer than the age of the universe at the epoch in question, then the reaction can be considered not to be occurring.

This certainly applies to the reaction above. The reaction rate can be expressed as the product of a velocity, v a number-density, n, and a reaction cross section, σ. From weak interaction physics we get $t_{weak} = 1/n\sigma c$ with $n = 2 \cdot 10^{-31}(T/10^{10})^3 \text{cm}^{-3}$, $\sigma = 10^{-44}(T/10^{10})^2 \text{cm}^2$, which leads to $t_{weak} = 160(T/10^{10})^5 \text{s}$. These neutral-current reactions occur typically at a temperature around $5 \cdot 10^{10}\text{K}$ corresponding to a time $4 \cdot 10^{-2}\text{s}$. After this time the muon-neutrinos effectively interact no further with the rest of the universe except gravitationally.

The electron neutrinos, ν_e, can continue to interact with the electrons and positrons through the charged-current weak interactions,

$$p^+ + e^- \leftrightharpoons n^0 + \nu_e, \qquad n^0 + e^+ + \leftrightharpoons p^+ + \bar{\nu}_e.$$

These reactions have slightly shorter reaction time-scale than the neutral-current weak interactions, around $(10^{10}\text{K})^5\text{s}$. Hence, these reactions proceed until the temperature has fallen below 10^{10}K., at about 1s after the big bang. After this the neutrinos do not interact with the rest of the universe. Hence, the universe became transparent for the neutrinos about 1s after the big bang.

These neutrinos now form a background neutrino gas. In order to calculate the present temperature of this gas, we must consider what happened to the electron-positron pairs a little later. About 3s after the big bang the temperature became lower than $6 \cdot 10^9\text{K}$. Then the photon-energy was not large enough to produce electron-positron pairs. Hence, the electron and positrons started to annihilate and produced photons. Most of the cosmic electromagnetic background radiation was produced at this time. The present temperature of this radiation has been measured with great accuracy, and is 2.728K.

The energy released by the electron-positron annihilation slowed down the rate at which the electromagnetic radiation cooled, but the decoupled neutrinos did not get any of this extra heat. Hence, the neutrino gas became colder than the electromagnetic radiation.

Let us find the relationship between the temperature of the photon-gas before and after the electron-positron annihilation. The photons and the electron-positron pairs are in thermodynamic equilibrium during the annihilation process. The gas expands adiabatically. The total entropy is therefore conserved. The entropy of a gas with density ρ, pressure p and temperature T in a comoving volume $V = a^3$ is

$$S = (\rho + p)\frac{V}{T}. \tag{12.294}$$

Radiation and ultra-relativistic gas of electrons and neutrinos have $p = (1/3)\rho$, so that

$$S = \frac{4}{3}\frac{a^3}{T}\rho. \tag{12.295}$$

The photons are bosons and obey the Bose-Einstein statistics which leads to the Planck spectrum. The energy density per unit frequency interval is

$$u(\omega) = \frac{\hbar\omega^3}{\pi^2 c^3}\frac{1}{e^{\frac{\hbar\omega}{k_B T}} - 1}. \tag{12.296}$$

The density of the photon gas is

$$\rho = \int_0^\infty u(\omega)d\omega = \frac{4\sigma}{c}T^4,$$
(12.297)

where $\sigma = \pi^2 k_B^4/60\hbar c^2$ is Stephan's constant. This is the *Stephan-Boltzmann radiation law*. Thus the entropy of the photon gas is

$$S_\gamma = \frac{16}{3}\frac{\sigma}{c}a^3 T^3.$$
(12.298)

The electrons and neutrinos are fermions obeying Fermi-Dirac statistics with +1 in the denominator of eq. (12.296) instead of −1. Calculating the energy density one finds in the relativistic limit, where the rest mass can be neglected,

$$\rho_e = \rho_\nu = \frac{7}{2}\frac{\sigma}{c}T^4.$$
(12.299)

Hence, the entropy of a relativistic electron and neutrino gas is

$$S_e = S_\nu = \frac{14}{3}\frac{\sigma}{c}a^3 T^3.$$
(12.300)

The total entropy of a gas consisting of photons and ultra-relativistic electrons and positrons is then

$$S_1 = S_\gamma + S_{e^+} + S_{e^-} = \frac{44}{3}\frac{\sigma}{c}a_1^3 T_1^3,$$
(12.301)

where a_1 and T_1 are the expansion factor and temperature at the start of the annihilation process. At the end of the annihilation process, where the expansion factor is a_2 and the temperature T_2, the energy is dominated by a photon gas with entropy S_γ, as given in (12.298). Since the total entropy has been conserved, it follows that

$$\frac{a_2 T_2}{a_1 T_1} = \left(\frac{11}{4}\right)^{\frac{1}{3}}.$$
(12.302)

This is the relationship between the temperature of the cosmic gas before and after the electron-positron annihilation.

At the same time the neutrino gas expanded freely. From eqs. (11.65) and (11.67) follow that aT = constant. Hence, the temperature, T_ν, of the neutrino gas at the time when the electron-positron annihilation had finished, was

$$T_\nu = \frac{a_1}{a_2}T_1.$$
(12.303)

Here we have used that the neutrino gas had the same temperature as the rest of the universe before the annihilation. Thus the ratio between the temperatures of the photon gas and the neutrino gas after the annihilation is

$$\frac{T_2}{T_\nu} = \frac{a_2 T_2}{a_1 T_1} = \left(\frac{11}{4}\right)^{\frac{1}{3}} \approx 1.4.$$
(12.304)

This ratio has not been changed during the later history of the universe.

Since the present temperature of the electromagnetic background is 2.728K we find that the present temperature of the neutrino gas is 1.95K. These neutrinos have been moving freely since they decoupled 1s after the big bang. This means that if we can observe the state of the cosmic neutrinos, we will be able to observe the state of the universe about 1s after the big bang. However, this low temperature neutrino gas is extremely difficult to observe.

The neutrinos are very numerous. About a million trillion cosmic neutrinos pass through each human body every second. In every cubic centimetre of the universe there are now 600 neutrinos. They are so numerous that even if only one type of neutrinos has mass, a very modest rest mass of around 90eV would suffice to make the universe flat. The average density of the neutrino gas would then equal the critical mass density. Recent measurements at the Super Kamiokande in Japan have shown that neutrinos have indeed a rest mass. However, the measurements indicate a neutrino mass much less than 90eV. Probably the neutrinos contribute to less than 0.5% of the critical density to the cosmic gas.

The number density of the photons per unit frequency interval is $n(\omega) = u(\omega)/\hbar c$, where $u(\omega)$ is given in eq. (12.296). Hence, the number density of photons of all frequencies is

$$n_\gamma = \int_0^\infty n(\omega)d\omega = 20.3T^3\text{cm}^{-3}\text{K}^{-3}.$$ (12.305)

Similarly, one finds, using the Fermi-Dirac distribution, that in the ultra-relativistic limit, the number density of the electrons and positrons is

$$n_{e^-} \approx n_{e^+} \approx \frac{3}{4}n_\gamma = 15.3T^3\text{cm}^{-3}\text{K}^{-3}.$$ (12.306)

The annihilation starts at a temperature $T = 6 \cdot 10^9$K, around 3s after the big bang. Then, according to the above formula, the number density of the electrons was $3 \cdot 10^{30}\text{cm}^{-3}$.

Let us now calculate the number density of the electrons after the annihilation. The annihilation finished at a temperature $T = 10^9$K, about 3 minutes after the big bang. From eq. (12.305) the number density of the photons at this point of time was $n_\gamma = 2 \cdot 10^{28}\text{cm}^{-3}$. After the electron-positron annihilation there were no cosmic production of photons. Hence, both photon number and baryon number in a comoving volume a^3 were constant during the expansion to the present time. From observation of baryon mass density and the energy density of the cosmic microwave background follow that the present ratio of baryon number and photon number is 10^{-9}. Hence, this ratio had the same value just after the annihilation. Furthermore, since the universe is electrically neutral, the electron number density is equal to the proton number density, which was one billionth of the photon number density. It follows that the electron number density just after the annihilation was $n_{e^-} = 2 \cdot 10^{19}\text{cm}^{-3}$. Comparing with the corresponding number $3 \cdot 10^{20}\text{cm}^{-3}$ before the annihilation, we see that only a very small part of the electrons that existed 1s after the big bang was left intact after the annihilation.

However, there were sufficiently many energetic electrons left to make the universe opaque for the electromagnetic radiation.

One more process of great significance happened in the lepton era: the neutron-proton ratio was 'frozen'. This ratio can be calculated as follows.

We consider the conditions a little earlier than 1s after the big bang when the temperature was a little higher than 10^{10}K. Then the baryons were non-relativistic. Their number density at thermodynamical equilibrium is given by the Boltzmann distribution. The neutron-proton ratio is therefore given by

$$r = \frac{n_n}{n_p} = e^{-\frac{\Delta m c^2}{k_B T}}, \tag{12.307}$$

where $\Delta m = m_n - m_p = 1.29$MeV.

Due to the temperature decrease when the universe expands the reaction time of reactions such as $p + e^- \leftrightarrows n + \nu_e$ increases. As long as the reaction times are shorter than the age of the universe these reactions maintain thermodynamic equilibrium. Eventually the reactions were so slow that thermodynamical equilibrium was lost, and the neutron-proton ratio was frozen. This happened at a temperature $T^* = 10^{10}$K, corresponding to a cosmic time about one second. After this time the neutron-proton ratio has been constant and equal to $r^* = r(T^*) = 0.21$.

Primordial cosmic nucleosynthesis: $1s < t < 12$min

As mentioned above, during the lepton era protons and neutrons were able to transform into each other through the following weak interactions,

$$p^+ + \bar{\nu}_e \leftrightarrows n^0 + e^+, \qquad p^+ + e^- \leftrightarrows n^0 + \nu_e.$$

The weak interaction time scale for these interactions exceeds the the expansion timescale when the temperature falls below 10^{10}K, about 1s after the big bang. Then the reactions effectively cease, and the neutron fraction is frozen in the value it had at this time.

Free neutrons, which are unstable to β-decay with a half life of a approximately 10.6 minutes, unless they are bound to protons in stable atomic nuclei, would eventually decay into protons. However, nuclear reactions occur which bind the neutrons and protons into stable nuclei before this β-decay of free neutrons had progressed very far. The first process of interest is

$$p^+ + n^0 \leftrightarrows {}^2\text{H} + \gamma,$$

i.e. a proton and a neutron form a deuterium nucleus with emission of electromagnetic radiation.

Now ^{2}H has a binding energy of only 2.2MeV and there were enough high energy photons present to photo-dissociate ^{2}H until the temperature dropped to around 10^9K, about 3min after the big bang. During this period the neutron fraction decreased due to the β-decay of free neutrons. After 3min and 46s the temperature was $0.9 \cdot 10^9$K. Now the photons were sufficiently soft such that the ^{2}H-nuclei could survive. Then several nuclear reactions, building heavier elements from protons and neutrons, took place. The hold off of the fusion processes in the first three minutes due to the photo-dissociation of ^{2}H is called the 'deuterium bottleneck'.

After the deuterium bottleneck the following chain reactions took place. Deuterium nuclei collided with protons and neutrons, forming Helium-3 (^{3}He) and tritium (^{3}H). Finally, the Helium-3 collided with a neutron, and the tritium could collide with a proton, in both cases forming a nucleus of ordinary helium (^{4}He), consisting of two protons and two neutrons.

Let us calculate the fraction by weight of helium,

$$f \equiv \frac{m_{He}}{m_{He} + m_H} = \frac{1}{1 + \frac{m_H}{m_{He}}}. \tag{12.308}$$

Since each helium nucleus contains 2 neutrons it is possible to create a number density equal to $n_n/2$ of helium nuclei. Each has mass approximately equal to $4m_p$. Hence,

$$\frac{m_H}{m_{He}} = \frac{(n_p - n_n)m_p}{\frac{n_n}{2} \cdot 4m_p} = \frac{n_p - n_n}{2n_n}, \tag{12.309}$$

which leads to

$$f = \frac{2n_n}{n_p + n_n} = \frac{2r}{1 + r}, \tag{12.310}$$

where $r = n_n/n_p$. After the time t^* the ratio r has been constant and equal to $r^* = 0.21$. Inserting this in the above equation gives for the helium-hydrogen mass ratio $f = 0.35$. This is only an approximate result. Taking into account the β-decay of free neutrons one obtains a mass ratio around $f = 0.25$. This is rather close to the observed ratio $f_{obs} = 0.24$.

It may be noted that this prediction is essentially independent of the total present density of all forms of matter and energy, because whatever the present value of the total relative density, Ω_{tot0}, the value of Ω_{tot} at such early cosmic time will be extremely close to 1. There is, however, a weak dependence on $\Omega_b h^2$. A higher value of $\Omega_b h^2$, which may be due to higher density of baryons and to faster expansion, means that the deuterium bottleneck is overcome earlier and hence there will be less free-neutron decay. Then the neutron fraction will be higher, which results in a slightly higher ^{4}He abundance.

Apart from minute amounts of ^{7}Li big bang nucleosynthesis stops at ^{4}He because there are no stable nuclei with mass numbers 5 and 8. Heavier elements are produced in stars.

The last scattering surface of the microwave background

When the nucleosynthesis had finished around twelve minutes after the big bang, nothing qualitatively new happened during the next 300 000 years. Then the temperature had decreased to 3000K, and a new process started. At this time the first neutral atoms were formed.

Looking back along the cosmic light paths one can calculate the optical depth to Thompson scattering. In order to carry out the integral one must first find how the ionization fraction of the matter depends upon the the cosmic redshift due to reionization happening between 300000 years and 400000 years after the big bang. The result of the calculation is that the optical depth to scattering is

$$\tau(z) = 0.37 \left(\frac{z}{1000}\right)^{14.25}. \tag{12.311}$$

For $\tau = 1$ there is a high probability that light is scattered. Inserting $\tau = 1$ in the above equation gives the redshift of the *last scattering surface*, $z_{LS} = 1055$. In reality the probability of scattering increases from near zero to near one

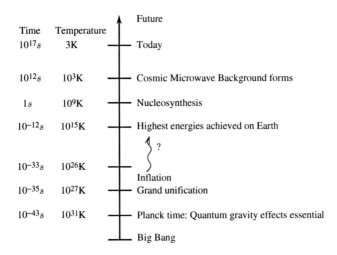

Figure 12.14: The history of our universe.

over a finite interval. The chance that a photon was last scattered at a redshift less than z is

$$P(< z) = e^{-\tau(z)}. \tag{12.312}$$

Hence, there is 25% chance that a photon is last scattered for $z < 980$ and 75% for $z < 1100$. So half of the photons in the microwave background were last scattered at $980 < z < 1100$. Observations of temperature fluctuations in the cosmic microwave background tell us about the physical properties in the universe during this redshift interval.

The future?

As we now have presented the past history of our universe, which is schematically illustrated in Fig. 12.14, we might wonder what can be said about the future? Observations seem to indicate that we already have entered a new era. The universe is currently accelerating. An effective cosmological constant is once again dominating the universe. However will this endure for ever? If it does, the universe will best be described by a de Sitter universe at late times. The universe will expand for all time, growing larger and larger at an exponential rate.

Or maybe, for some reason, the effective cosmological constant is turned off again. Maybe a curvature dominated epoch will follow? Or perhaps a dark matter domination? Or maybe the universe will recollapse in a Big Crunch? There are certainly many open questions still left in cosmology.

Problems

12.1. *Matter-vacuum transition in the Friedmann-Lemaître model*
Find the point of time of transition from matter domination to vacuum domi-

nation of the flat Friedmann-Lemaître universe model and the corresponding redshift.

12.2. Event horizons in de Sitter universe models
Show that the coordinate distances to the event horizons of the de Sitter universe models with $k > 0$, $k = 0$ and $k < 0$ are (assuming $\Lambda = 3$)

$$
\begin{aligned}
r_{EH} &= \frac{1}{\cosh t}, & k &> 0, & t &\geq 0 \\
r_{EH} &= e^{-t}, & k &= 0, & & \\
r_{EH} &= \frac{1}{\sinh t}, & k &< 0, & t &\geq 0
\end{aligned}
$$

respectively.

12.3. Light travel time
In this problem you are going to calculate the light travel time of light from an object with redshift z in a flat Friedmann-Lemaître model with age t_0 and a present relative density of LIVE, $\Omega_{\Lambda 0}$.

Show that the point of time of the emission event is

$$
t_e = t_0 \frac{\operatorname{arsinh}\sqrt{\frac{\Omega_{\Lambda 0}}{(1-\Omega_{\Lambda 0})(1+z)^3}}}{\operatorname{arsinh}\sqrt{\frac{\Omega_{\Lambda 0}}{1-\Omega_{\Lambda 0}}}}, \tag{12.313}
$$

and calculate the light travel time $t_0 - t_e$. Make a plot of $(t_0 - t_e)/t_0$ as a function of z.

12.4. Superluminal expansion
Show from Hubble's law that all all objects in a flat Friedmann-Lemaître model with redshifts $z > z_c$ are presently receding faster than the speed of light, where z_c is given by

$$
\int_0^{1+z_c} \frac{dy}{\sqrt{\Omega_{\Lambda 0} + \Omega_{m0} y^3}} = 1. \tag{12.314}
$$

Find z_c for a universe model with $\Omega_{\Lambda 0} = 0.7$ and $\Omega_{m0} = 0.3$.

12.5. Flat universe model with radiation and vacuum energy

(a) Find the expansion factor as a function of time for a flat universe with radiation and Lorentz-invariant vacuum energy represented by a cosmological constant Λ, and with present relative density of vacuum energy Ω_{v0}.

(b) Calculate the Hubble parameter, H, as a function of time, and show that the model approaches a de Sitter model in the far future. Find also the deceleration parameter, $q(t)$.

(c) When is the inflection point, t_1, for which the universe went from deceleration to acceleration? What is the corresponding redshift observed at the time t_0?

12.6. Creation of radiation and ultra-relativistic gas at the end of the inflationary era
Assume that the vacuum energy can be described by a decaying cosmological parameter $\Lambda(t)$. Show from energy conservation that if the density of radiation

and gas is negligible at the final period of the inflationary era compared to after it, then the density immediately after the inflationary era is

$$\rho = \frac{1}{8\pi Ga(t)^4} \int_{t_1}^{t_2} \dot{\Lambda} a(t)^4 dt \qquad (12.315)$$

where $t_2 - t_1$ is the duration of the period with $\dot{\Lambda} \neq 0$.

12.7. *Universe models with Lorentz invariant vacuum energy (LIVE).*
(see [Sil02]) We shall here consider universe models with LIVE and a perfect fluid with equation of state $p = w\rho$, $\rho_v = $ constant. The density of the LIVE is constant and related to a cosmological constant Λ by $\Lambda = 8\pi G\rho_v$.

(a) Show that the mass of the fluid M inside a comoving volume a^3 is

$$M = \rho a^{3(1+w)}.$$

(b) Introduce a rescaled time variable $\tau = \sqrt{\Lambda/3}t$, a rescaled expansion factor $y(\tau) = (\Lambda/8\pi GM)^{1/3(1+w)} a$, and the parameters $n = 1 + 3w$ and $\omega = (3/\Lambda)(\Lambda/8\pi GM)^{2/3(1+w)}$, and show that the Friedmann equation (12.3) takes the form

$$\dot{y}^2 = y^{-n} + y^2 - k\omega \qquad (12.316)$$

where the dot denotes derivative with respect to τ. We shall consider solutions with the initial condition $y(0) = 0$. The equation can be integrated in terms of elementary functions in the following four cases.

(c) *Flat universe:* $k = 0$.
Show that in this case the solution is:

$$y(\tau) = \left[\sinh\left(\frac{2+n}{2}\tau\right)\right]^{\frac{2}{2+n}}.$$

Find the Hubble parameter, and the deceleration parameter as a function of time for the models with $n = 1$ (dust) and $n = 2$ (radiation), and calculate the age of the models assuming that the present value of the Hubble parameter is $H_0 = 20$km/s per million light years.

(d) Show that the age of the flat universe models may be expressed as

$$t_0 = \frac{t_\Lambda}{1+w} \operatorname{artanh}\sqrt{\Omega_{\Lambda 0}}, \quad t_\Lambda = \frac{2}{\sqrt{3\Lambda}},$$

where $\Omega_{\Lambda 0}$ is the present value of the relative density of the LIVE.

(e) *Universe with radiation and LIVE:* $n = 2$ ($w = 1/3$)
Show that in this case the solution is:

$$y(\tau) = \sqrt{\sinh(2\tau) - k\omega \sinh^2 \tau}.$$

(f) *Universe with "string fluid" and LIVE:* $n = 0$, $(w = -1/3)$
Show that in this case the solution is:

$$y(\tau) = \sqrt{1 - k\omega} \sinh \tau.$$

(g) *Universe with "domain-wall fluid" and LIVE:* $n = -1$, $(w = -2/3)$
Show that in this case the solution is:

$$y(\tau) = \sqrt{-k\omega}\sinh\tau + \sinh^2(\tau/2).$$

These universe models only exist for $k = 0$, $k < 0$.

(h) Show that for all the universe models with LIVE and perfect fluid obeying the equation of state $p = w\rho$ the ratio of Ω_Λ and Ω_M is

$$\frac{\Omega_\Lambda}{\Omega_M} = y^{n+2}.$$

12.8. Cosmic strings

At the end of the inflationary era there was a phase transition from a false to a true vacuum with very high energy density for the false vacuum and low energy density for the true vacuum. Due to the topological properties of the vacuum field long stable strings of false vacuum may have been formed at this time. These objects are called cosmic strings.

In this problem you are going to find a solution of Einstein's field equations describing the gravitational field of a thin, static, straight string lying along the z-axis. The energy- momentum tensor of the string is

$$(T^\mu_\nu) = \lambda\delta(\rho)\text{diag}(1, 0, 0, 1),$$

where $\sigma = d\mu/dA = \lambda\delta(\rho)$ is the mass per unit volume of the string, μ its mass per unit length, and $\delta(\rho)$ is Dirac's delta function. Choosing coordinate time equal to the proper time measured with clocks at rest, the line element for the static, cylindrically symmetric space may be written

$$ds^2 = -dt^2 + d\rho^2 + B^2(\rho)d\phi^2 + dz^2.$$

(a) Show that Einstein's field equations reduce to the single equation

$$\frac{1}{B}\frac{d^2B}{d\rho^2} = -8\pi G\sigma.$$

(b) Find $B(\rho)$, determining a constant of integration by demanding Minkowski metric in the absence of a string, and showing that $B(0) = \mu/2\pi\lambda$. Introduce a new radial coordinate

$$\bar\rho = \rho + \frac{\mu}{2\pi\lambda(1 - 4G\mu)}$$

where G is Newton's constant of gravitation, and show that the line element takes the form

$$ds^2 = -dt^2 + d\bar\rho^2 + (1 - 4G\mu)^2\bar\rho^2 d\phi^2 + dz^2.$$

(c) Introduce a new angular coordinate

$$\bar\phi = (1 - 4G\mu)\phi.$$

What does the new form of the line element tell you about the spacetime outside the string?

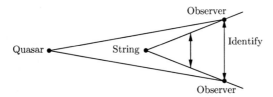

Figure 12.15: A cosmic string.

(d) The old usual angular coordinate varies in the range $0 \leq \phi < 2\pi$. Hence the new angular coordinate varies in the range $0 \leq \bar{\phi} < 2\pi(1 - 4G\mu)$. Thus there is an angular 'deficit angle'

$$\Delta\phi = 8\pi G\mu$$

which shows that a surface of constant t and z has the topology of a cone rather than that of a plane, as illustrated in Fig. 12.15.

An observer photographs a quasar at a distance d_Q. Assume that there is a cosmic string between the quasar and the observer at a distance d_S from the observer, orthogonal to the direction of sight of the quasar. Describe the picture qualitatively and quantitatively.

12.9. *Phantom Energy*
Consider a flat universe model dominated by quintessence energy with equation of state, $p = w\rho$, $w = $ constant. We shall consider 'phantom energy' with $w < -1$ (see also [CKW03]).

(a) Use the normalization $a(t_0) = 1$ and find the scale factor a as a function of cosmic time.

(b) Find the energy density as a function of time.

(c) Show that the scale factor and the density blows up to infinity at a time

$$t_r = t_0 - \frac{2}{3(1+w)H_0\sqrt{\Omega_{P0}}}, \qquad (12.317)$$

where Ω_{P0} is the relative density of the phantom energy at the present time. The cosmic catastrophe at the time t_r is called 'the Big Rip'. What is $t_r - t_0$ for $H_0 = 20$km/s per 10^6l.y., $\Omega_{P0} = 0.64$, and $w = -3/2$?

(d) A planet in an orbit of radius R around a star of mass M will become unbound roughly when $-(4\pi/3)(\rho + 3p)R^3 \approx M$, where ρ and p are the density and the pressure of the phantom energy.

Show that a gravitationally bound system of mass M and radius R will be stripped at a time t_s before the big rip, given by

$$t_s \approx -\frac{\sqrt{-2(1+3w)}}{6\pi(1+w)}T,$$

where T is the period of a circular orbit with radius R around the system. Find t_s for the Milky Way galaxy with $w = -3/2$.

12.10. *Velocity of light in the Milne universe*
We consider light emitted towards an observer at the spatial origin of a two dimensional Milne universe with line-element (14.44). The physical distance

from the origin to a point with coordinate x is $\ell = ax$ where $a = t$ is the expansion factor. The velocity of an object is $\dot{\ell} = \dot{a}x + a\dot{x}$. The first term is the velocity due to the expansion of the universe and the second term is the *peculiar velocity*.

(a) Show that the peculiar velocity of light emitted towards the origin is

$$\dot{x} = -\frac{1}{\ell}.$$

(b) Show that the physical distance from the observer of a pulse of light emitted from a position x_e at a point of time t_e towards the observer is

$$\ell = x_e t - t \ln(t/t_e).$$

Make a plot of ℓ as a function of time.
Explain the shape of the curve. How can it be that there is a maximal distance where the light instantaneously has unchanging distance from the observer?

12.11. *Universe model with dark energy and cold dark matter*
We shall here consider a universe model with dark energy having equation of state $p_s = w_s \rho_s$ and with dust with energy density ρ_m.

(a) Show that the deceleration parameter in this universe model is

$$q = \frac{1}{2}\left[\Omega_m + (1 + 3w_s)\Omega_s\right].$$

(b) Show that a transition from decelerated to accelerated expansion happens at a redshift

$$z_1 = \left[-(1 + 3w_s)\frac{\Omega_{s0}}{\Omega_{m0}}\right]^{-\frac{1}{3w_s}} - 1, \tag{12.318}$$

where Ω_{s0} and Ω_{m0} are the present values of the relative densities. Make a plot of $z_1(w_s)$ for $\Omega_{s0}/\Omega_{m0} = 3$.

(c) Show that the dark energy and the cold matter has equal density at a redshift

$$z_{\text{eq}} = \left(\frac{\Omega_{s0}}{\Omega_{m0}}\right)^{-\frac{1}{3w_s}} - 1. \tag{12.319}$$

Plot z_{eq} for $\Omega_{s0}/\Omega_{m0} = 3$ in the same diagram as $z_1(w_s)$.

12.12. *Luminosity-redshift relations*

(a) Show that the luminosity distance of an object with redshift z in a Milne universe is

$$d_{L,\text{Milne}} = \frac{z^2 + 2z}{2H_0}.$$

(b) Show that the luminosity-distance of an object with redshift z in a flat universe model with a single perfect fluid having equation of state $p = w\rho$ is

$$d_L = -\frac{2(1 + z)}{H_0(1 + 3w)}\left[(1 + z)^{-\frac{1}{2}(1+3w)} - 1\right].$$

12.13. *Cosmic time dilation*
The clocks showing the time T of eq. (14.46) are at rest in Minkowski space-time. Their rate of time is position independent.

(a) How does the rate of time t, as measured by co-moving clocks in the Milne universe, depend upon the position? Make a plot to visualize the effect. Discuss whether this effect contradicts the homogeneity of the Milne universe.

(b) Consider light waves emitted from a source at x to an observer at $x = 0$. No waves disappear. Use this, together with eq. (11.34), to obtain a simple expression for the cosmic time dilation in terms of the scale factor.

12.14. *Chaplygin gas*
The discovery that the expansion of the universe is accelerating has stimulated the search for new types of matters or fields that can behave like a cosmological constant, by combining positive energy density and negative pressure. A so-called Chaplygin gas has this property. It is defined as a perfect fluid having the equation of state $p = -A/\rho$, where A is a positive constant. We shall consider a flat Robertson-Walker universe model dominated by a Chaplygin gas [KMP01].

(a) Show from the Friedmann equation (11.16) and the energy conservation equation (11.19) that the density of the gas depends on the scale factor as follows: $\rho = (A + B/a^6)^{1/2}$, where B is an integration constant.

(b) Describe the behaviour of ρ for small and large values of a, and find the expansion factor as a function of time in these limits.

(c) Make a series expansion of ρ and p in a^{-6} including only the first two terms. Consider the Chaplygin gas to consist of two fluids corresponding to the two terms and find the equation of state, $p = p(\rho)$, of these fluids.

12.15. *The perihelion precession of Mercury and the cosmological constant*

(a) Generalize the Schwarzschild solution to the case where a non-zero cosmological constant, Λ, is present.

(b) Show that the equation for the orbit of a test particle in the spacetime found above is (see section 10.4)

$$\frac{d^2u}{d\phi^2} + u = \frac{M}{p_\phi^2} + 3Mu^2 + \frac{\Lambda}{3p_\phi^2 u^3}, \qquad (12.320)$$

where p_ϕ is the conjugated momentum of the angular variable ϕ.

(c) Find the perihelion precession of the orbit per orbit according to the above equation. Assume as in section 10.4 that the eccentricity is very small, $e \ll 1$.

(d) The results for the Mercurian orbit excluding the Λ-term agree with the the observational data to an accuracy of less that 1 arc second per century. What upper bound does this give on the value of Λ?

13

Anisotropic and Inhomogeneous Universe Models

In this chapter we will investigate anisotropic and inhomogeneous universe models. If we relax the cosmological principles a bit we can get new and interesting models of our universe. Actually, one of the main goals of cosmology today is to *explain* the isotropy and homogeneity the universe has and in order to explain a certain property of the universe one has to consider sufficiently general models that need not have this property. In this chapter we will investigate the Bianchi type I universe model and the inhomogeneous Lemaître-Tolman-Bondi (LTB) universe models. The Bianchi type I model is the simplest of the spatially homogeneous models which allows for anisotropy and the LTB-models are inhomogeneous universe models with spherically symmetric three-space.

13.1 The Bianchi type I universe model

The Bianchi type I universe model is the generalization of the flat Friedmann-Robertson-Walker model. It has a metric

$$ds^2 = -dt^2 + a(t)^2 dx^2 + b(t)^2 dy^2 + c(t)^2 dz^2. \tag{13.1}$$

In this case there are three functions, $a(t)$, $b(t)$ and $c(t)$, to be determined by the Einstein equations. All the scale factors in different directions are allowed to vary independently of each other.

The universe is still spatially homogeneous, because we can find three Killing vectors given by

$$\xi_1 = \frac{\partial}{\partial x}, \qquad \xi_2 = \frac{\partial}{\partial y}, \qquad \xi_3 = \frac{\partial}{\partial z}, \tag{13.2}$$

which form a basis for the spatial hypersurfaces $t = $ constant. These Killing vectors correspond to translation in the spatial directions. The Bianchi type I universe is *translation invariant*.

Let us find the curvature tensors for the metric (13.1). It is useful to introduce the following parametrisation:

$$
\begin{aligned}
a(t) &= e^{\alpha(t)+a_1(t)}, \\
b(t) &= e^{\alpha(t)+a_2(t)}, \\
c(t) &= e^{\alpha(t)+a_3(t)},
\end{aligned}
\tag{13.3}
$$

where

$$
\begin{aligned}
a_1 &= \beta_+ + \sqrt{3}\beta_-, \\
a_2 &= \beta_+ - \sqrt{3}\beta_-, \\
a_3 &= -2\beta_+.
\end{aligned}
\tag{13.4}
$$

In this way we separate the anisotropic expansion and the volume expansion. We see that $a_1 + a_2 + a_3 = 0$, so the comoving volume is given by

$$
abc = e^{3\alpha}.
\tag{13.5}
$$

We can define a Hubble factor in each of the three different directions

$$
H_1 = \frac{\dot{a}}{a}, \quad H_2 = \frac{\dot{b}}{b}, \quad H_3 = \frac{\dot{c}}{c},
\tag{13.6}
$$

and an average Hubble factor

$$
H = \frac{1}{3}(H_1 + H_2 + H_3) = \dot{\alpha}.
\tag{13.7}
$$

These will be useful later on.

The following will hold

$$
\begin{aligned}
\sum_i a_i &= 0, \\
\sum_i a_i^2 &= 6(\beta_+^2 + \beta_-^2), \\
\sum_j \sum_{i \neq j} a_i a_j &= -6(\beta_+^2 + \beta_-^2).
\end{aligned}
\tag{13.8}
$$

To avoid confusion, there is *no* summation over Latin indices unless explicit specified.

We introduce the orthonormal frame

$$
\begin{aligned}
\omega^{\hat{t}} &= \mathbf{dt}, \\
\omega^{\hat{i}} &= e^{\alpha} e^{a_i} \mathbf{dx}^i.
\end{aligned}
\tag{13.9}
$$

Thus,

$$
\begin{aligned}
\mathbf{d}\omega^{\hat{t}} &= 0, \\
\mathbf{d}\omega^{\hat{i}} &= (\dot{\alpha} + \dot{a}_i)\,\omega^{\hat{t}} \wedge \omega^{\hat{i}}.
\end{aligned}
\tag{13.10}
$$

Then by *Cartan's first structural equation,*

$$
\mathbf{d}\omega^{\hat{\mu}} = -\sum_{\hat{\nu}} \Omega^{\hat{\mu}}{}_{\hat{\nu}} \wedge \omega^{\hat{\nu}},
$$

the non-trivial connection forms are:

$$\Omega^{\hat{i}}_{\ \hat{t}} = \Omega^{\hat{t}}_{\ \hat{i}} = (\dot{\alpha} + \dot{a}_i)\,\omega^{\hat{i}}.$$

By *Cartan's second structural equation*,

$$\mathbf{R}^{\hat{\mu}}_{\ \hat{\nu}} = \mathbf{d}\Omega^{\hat{\mu}}_{\ \hat{\nu}} + \sum_{\hat{\lambda}} \Omega^{\hat{\mu}}_{\ \hat{\lambda}} \wedge \Omega^{\hat{\lambda}}_{\ \hat{\nu}},$$

the non-vanishing curvature forms are

$$\begin{aligned}
\mathbf{R}^{\hat{i}}_{\ \hat{t}} = \mathbf{R}^{\hat{t}}_{\ \hat{i}} &= \left[\ddot{\alpha} + \ddot{a}_i + (\dot{\alpha} + \dot{a}_i)^2\right]\boldsymbol{\omega}^{\hat{t}} \wedge \boldsymbol{\omega}^{\hat{i}}, \\
\mathbf{R}^{\hat{i}}_{\ \hat{j}} = -\mathbf{R}^{\hat{j}}_{\ \hat{i}} &= (\dot{\alpha} + \dot{a}_i)(\dot{\alpha} + \dot{a}_j)\boldsymbol{\omega}^{\hat{i}} \wedge \boldsymbol{\omega}^{\hat{j}}.
\end{aligned} \tag{13.11}$$

Hence, the non-vanishing components of the Riemann tensor are

$$\begin{aligned}
R^{\hat{i}}_{\ \hat{t}\hat{i}\hat{t}} &= -\left[\ddot{\alpha} + \ddot{a}_i + (\dot{\alpha} + \dot{a}_i)^2\right], \\
R^{\hat{i}}_{\ \hat{j}\hat{i}\hat{j}} &= (\dot{\alpha} + \dot{a}_i)(\dot{\alpha} + \dot{a}_j).
\end{aligned} \tag{13.12}$$
$$i \neq j$$

By contraction we find the Ricci tensor

$$\begin{aligned}
R_{\hat{t}\hat{t}} &= \sum_{\hat{i}} R^{\hat{i}}_{\ \hat{t}\hat{i}\hat{t}} = -3\left[\ddot{\alpha} + \dot{\alpha}^2 + 2(\dot{\beta}^2_+ + \dot{\beta}^2_-)\right], \\
R_{\hat{j}\hat{j}} &= \sum_{i \neq j} R^{\hat{i}}_{\ \hat{j}\hat{i}\hat{j}} + R^{\hat{t}}_{\ \hat{j}\hat{t}\hat{j}} = 3\dot{\alpha}^2 + 3\dot{\alpha}\dot{a}_j + \ddot{\alpha} + \ddot{a}_j.
\end{aligned} \tag{13.13}$$

Hence, the scalar curvature is

$$R = \sum_{\hat{\mu}} R^{\hat{\mu}}_{\ \hat{\mu}} = 6\left[\ddot{\alpha} + 2\dot{\alpha}^2 + (\dot{\beta}^2_+ + \dot{\beta}^2_-)\right]. \tag{13.14}$$

The Einstein tensor can now readily be calculated

$$\begin{aligned}
E_{\hat{t}\hat{t}} &= -3\left[-\dot{\alpha}^2 + \dot{\beta}^2_+ + \dot{\beta}^2_-\right], \\
E_{\hat{j}\hat{j}} &= -3\dot{\alpha}^2 - 2\ddot{\alpha} + 3\dot{\alpha}\dot{a}_j + \ddot{a}_j - 3\left(\dot{\beta}^2_+ + \dot{\beta}^2_-\right).
\end{aligned} \tag{13.15}$$

It is useful to define E_+ and E_- by

$$\begin{aligned}
E_+ &= \frac{1}{6}\left(E_{\hat{1}\hat{1}} + E_{\hat{2}\hat{2}} - 2E_{\hat{3}\hat{3}}\right), \\
E_- &= \frac{1}{2\sqrt{3}}\left(E_{\hat{1}\hat{1}} - E_{\hat{2}\hat{2}}\right).
\end{aligned} \tag{13.16}$$

Using eq. (13.4) we find

$$E_\pm = 3\dot{\alpha}\dot{\beta}_\pm + \ddot{\beta}_\pm. \tag{13.17}$$

For a Bianchi type I model we define the *shear scalar* as[1]

$$\sigma^2 = \frac{1}{2}\sum_i \dot{a}_i^2 = 3\left(\dot{\beta}^2_+ + \dot{\beta}^2_-\right). \tag{13.18}$$

The physical interpretation of the shear scalar is that is measures the degree of anisotropy of the spacetime. For an isotropic spacetime, we have $\sigma^2 = 0$, while for spacetimes that expands anisotropically the shear will be non-zero.

---

[1] We will define the shear tensor and the shear scalar more rigorously in a later chapter.

13.2 The Kasner solutions

As a first step we derive the vacuum solutions for the Bianchi type I model. These solutions are named the *Kasner solutions* due to E. Kasner who first found them in 1921 [Kas21].

In this case the energy-momentum tensor vanishes so according to Einstein's equations, the Einstein tensor has to vanish as well. Hence, from the requirement $E_{\mu\nu} = 0$, we get

$$E_{\hat{t}\hat{t}} = -3\left[-\dot{\alpha}^2 + \dot{\beta}_+^2 + \dot{\beta}_-^2\right] = 0,$$
$$E_\pm = 3\dot{\alpha}\dot{\beta}_\pm + \ddot{\beta}_\pm = 0. \tag{13.19}$$

Multiplying the latter equation with $e^{3\alpha}$, the equation can be written

$$\frac{d}{dt}\left(\dot{\beta}_\pm e^{3\alpha}\right) = 0, \tag{13.20}$$

which admits the first integral

$$\dot{\beta}_\pm e^{3\alpha} = p_\pm. \tag{13.21}$$

We define the *anisotropy parameter* by

$$A^2 = p_+^2 + p_-^2. \tag{13.22}$$

The equation $E_{\hat{t}\hat{t}} = 0$ can now be written

$$\dot{\alpha}^2 = A^2 e^{-6\alpha} \quad \Rightarrow \quad 3\dot{\alpha}e^{3\alpha} = 3A. \tag{13.23}$$

This equation yields the integral

$$e^{3\alpha} = 3At + C. \tag{13.24}$$

By a translation of time, $t \mapsto t - t_0$, we can set the integration constant C to zero. Note that the shear is now given by

$$\sigma^2 = \frac{1}{3}\cdot\frac{1}{t^2}. \tag{13.25}$$

We can now integrate the equations for $\beta_\pm$

$$\dot{\beta}_\pm = \frac{p_\pm}{e^{3\alpha}} = \frac{p_\pm}{3A}\cdot\frac{1}{t} \quad \Rightarrow \quad \beta_\pm = \frac{p_\pm}{3A}\ln t. \tag{13.26}$$

Since $A^2 = p_+^2 + p_-^2$, we can introduce an angular variable ϕ defined by

$$p_+ = A\cos\phi, \qquad p_+ = A\sin\phi. \tag{13.27}$$

The expressions for $\beta_\pm$ are now simply

$$\beta_+ = \frac{1}{3}\cos\phi\ln t,$$
$$\beta_- = \frac{1}{3}\sin\phi\ln t. \tag{13.28}$$

The anisotropy parameter is only present in the expression for α. By a rescaling of the metric (13.1), A can be set to whatever we like. We will therefore choose $3A = 1$ for simplicity. Using eq. (13.4) and trigonometric identities we have

$$
\begin{aligned}
a_1 &= \beta_+ + \sqrt{3}\beta_- = \frac{2}{3}\cos\left(\phi + \pi/3\right)\ln t, \\
a_2 &= \beta_+ - \sqrt{3}\beta_- = \frac{2}{3}\cos\left(\phi - \pi/3\right)\ln t, \\
a_3 &= -2\beta_+ = -\frac{2}{3}\cos(\phi)\ln t.
\end{aligned}
\tag{13.29}
$$

The *Kasner solutions* can now be written

$$
ds^2 = -dt^2 + t^{\frac{2}{3}}\left[t^{\frac{4}{3}\cos(\phi+\pi/3)}dx^2 + t^{\frac{4}{3}\cos(\phi-\pi/3)}dy^2 + t^{-\frac{4}{3}\cos(\phi)}dz^2\right]. \tag{13.30}
$$

Since this metric is parametrised with an angular variable, this set of solutions is sometimes called the *Kasner circle*.

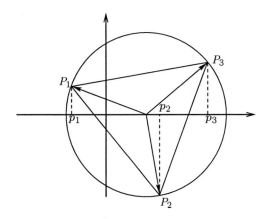

Figure 13.1: A geometrical representation of the Kasner solutions.

A useful representation of the Kasner solutions is illustrated in Fig. 13.1 and goes as follows. Draw a circle in the xy-plane centered at $(1/3, 0)$ with radius $2/3$. Draw an equilateral triangle inside this circle, with the vertices on the circle. Call the vertices P_1, P_2 and P_3. If the P_i's has the x-components p_i, then the metric

$$
ds^2 = -dt^2 + t^{2p_1}dx^2 + t^{2p_2}dy^2 + t^{2p_3}dz^2, \tag{13.31}
$$

with $\sum_i p_i = \sum_i p_i^2 = 1$ is one of the Kasner solutions. Since the orientation of the equilateral triangle can be given by any angle ϕ the whole of the Kasner circle is represented this way.

13.3   The energy-momentum conservation law in an anisotropic universe

We know that in our real universe there are matter. We should therefore include some non-trivial energy-momentum tensor.

Let us assume that the energy-momentum tensor is of the form

$$T_{\mu\nu} = (\rho + p)u_\mu u_\nu + p g_{\mu\nu} + \pi_{\mu\nu}. \tag{13.32}$$

The tensor $\pi_{\mu\nu}$ is called the anisotropic stress tensor and has the properties

$$\pi_{\mu\nu} = \pi_{\nu\mu}, \quad \pi^\mu_{\ \mu} = 0, \quad \pi_{\mu\alpha}u^\alpha = 0. \tag{13.33}$$

Using the energy-momentum conservation equation $T^\mu_{\ \nu;\mu} = 0$ we will derive an equation that must be fulfilled by the fluid.

Since the Einstein tensor is diagonal, the energy-momentum tensor must also be so in the orthonormal frame. Taking the covariant divergence, we get

$$\begin{aligned} T^{\hat\mu}_{\ \hat t;\hat\mu} &= = T^{\hat\mu}_{\ \hat t,\hat\mu} - \Gamma^{\hat\rho}_{\ \hat t\hat\mu}T^{\hat\mu}_{\ \hat\rho} + \Gamma^{\hat\rho}_{\ \hat\mu\hat\rho}T^{\hat\mu}_{\ \hat t} \\ &= -\dot\rho - \Omega^{\hat\rho}_{\ \hat t}(e_{\hat\mu})T^{\hat\mu}_{\ \hat\rho} + \Omega^{\hat\rho}_{\ \hat\mu}(e_{\hat\rho})T^{\hat\mu}_{\ \hat t}. \end{aligned} \tag{13.34}$$

The second term can be simplified as

$$\Omega^{\hat\rho}_{\ \hat t}(e_{\hat\mu})T^{\hat\mu}_{\ \hat\rho} = \Omega^{\hat i}_{\ \hat t}(e_{\hat j})T^{\hat j}_{\ \hat i} = 3\dot\alpha p + \sum_i \dot a_i \pi^{\hat i}_{\ \hat i}, \tag{13.35}$$

while the third term is

$$\Omega^{\hat\rho}_{\ \hat\mu}(e_{\hat\rho})T^{\hat\mu}_{\ \hat t} = -\Omega^{\hat i}_{\ \hat i}(e_{\hat i})\rho = -3\dot\alpha\rho. \tag{13.36}$$

Hence, the energy-momentum conservation equation turns into

$$\dot\rho + 3\dot\alpha(\rho + p) + \sum_i \dot a_i \pi^{\hat i}_{\ \hat i} = 0. \tag{13.37}$$

The last term on the left hand side can be interpreted as an energy-production term. Let us define

$$\frac{dE}{dt} = -\sum_i \dot a_i \pi^{\hat i}_{\ \hat i}. \tag{13.38}$$

The energy conservation law can now be written as

$$\dot\rho + 3\dot\alpha(\rho + p) = \dot E. \tag{13.39}$$

For a perfect fluid the left side can be interpreted as the change in entropy:

$$dS = dU + p\,dV \tag{13.40}$$

Inserting $U = V \cdot \rho$ where $V = e^{3\alpha}$ is the comoving volume, we get

$$dS = V(\dot\rho + 3\dot\alpha(\rho + p))dt, \tag{13.41}$$

which is the left side of eq. (13.39). For a perfect fluid the change in entropy is zero. If we consider viscous and dissipative fluids there is a change of entropy:

$$dS = \dot E V\,dt, \tag{13.42}$$

where $\dot E$ is given in eq. (13.38). According to the second law of thermodynamics, the entropy must increase for any physical process. Hence, for viscous fluids we have to assume that

$$\dot E \geq 0. \tag{13.43}$$

We would stress that this is only true for irreversible processes, and strictly speaking, only for processes close to equilibrium. There are a lot of fluids (for instance a cosmic magnetic field) that violate this inequality, so we have to utilize care when using this equation.

13.4   Models with a perfect fluid

We will first consider the simplest example, a universe with a w-law perfect fluid. In this case we have

$$p = w\rho, \qquad \pi_{\mu\nu} = 0. \tag{13.44}$$

The energy-momentum conservation equation turns now simply into

$$\dot{\rho} + 3(w+1)\dot{\alpha} = 0. \tag{13.45}$$

This equation can readily be solved to yield

$$\rho = K e^{-3(w+1)\alpha}, \tag{13.46}$$

where K is an integration constant. Since the three spatial pressures are equal, we have

$$T_+ \equiv \frac{1}{6}\left(T_{\hat{1}\hat{1}} + T_{\hat{2}\hat{2}} - 2T_{\hat{3}\hat{3}}\right) = 0,$$

$$T_- \equiv \frac{1}{2\sqrt{3}}\left(T_{\hat{1}\hat{1}} - T_{\hat{2}\hat{2}}\right) = 0, \tag{13.47}$$

and hence Einstein's field equations yield[2]

$$E_\pm = T_\pm = 0. \tag{13.48}$$

These equations are the same as in the vacuum case. Integration gives

$$\dot{\beta}_\pm e^{3\alpha} = p_\pm. \tag{13.49}$$

The E_{tt} equation is

$$-3\left[-\dot{\alpha}^2 + \dot{\beta}_+^2 + \dot{\beta}_-^2\right] = \rho, \tag{13.50}$$

which, by using eqs. (13.46) and (13.49), turns into

$$\dot{\alpha}^2 = \frac{K}{3}e^{-3(w+1)\alpha} + A^2 e^{-6\alpha}, \tag{13.51}$$

and can be solved in quadrature for a general w. Let us solve this equation for two particular cases, namely for $w = -1$ (vacuum dominated) and $w = 0$ (dust).

Vacuum dominated Bianchi type I model

The equation for α is now

$$\dot{\alpha}^2 = \frac{\Lambda}{3} + A^2 e^{-6\alpha}, \tag{13.52}$$

where the Lorentz invariant vacuum energy with a constant energy density has been represented by a cosmological constant. This equation can be integrated by means of the substitution $e^{3\alpha} = A\sqrt{3/\Lambda}x$. The solution is

$$e^{3\alpha} = A\sqrt{\frac{3}{\Lambda}}\sinh(\sqrt{3\Lambda}t). \tag{13.53}$$

---

[2] In this chapter we will set $8\pi G = 1$.

We have set the integration constant equal to zero because it only corresponds to a shift in the Big Bang time. The equations for $\beta_\pm$ turn into

$$\dot{\beta}_\pm = \frac{p_\pm}{A}\sqrt{\frac{\Lambda}{3}}\frac{1}{\sinh(\sqrt{3\Lambda}t)}. \tag{13.54}$$

This equation can be readily integrated. The result is

$$\beta_\pm = \frac{p_\pm}{3A}\ln\left[\tanh\left(\frac{\sqrt{3\Lambda}}{2}t\right)\right]. \tag{13.55}$$

Hence, the line-element can be written

$$ds^2 = -dt^2 + \left(\sinh(\sqrt{3\Lambda}t)\right)^{\frac{2}{3}} \tag{13.56}$$
$$\times\left[\Theta(t)^{\frac{4}{3}\cos(\phi+\pi/3)}dx^2 + \Theta(t)^{\frac{4}{3}\cos(\phi-\pi/3)}dy^2 + \Theta(t)^{-\frac{4}{3}\cos(\phi)}dz^2\right].$$

Here, $\Theta(t)$ is given by

$$\Theta(t) = \tanh\left(\frac{\sqrt{3\Lambda}}{2}t\right). \tag{13.57}$$

Furthermore, the shear scalar is given by

$$\sigma^2 = \frac{\Lambda}{\sinh^2(\sqrt{3\Lambda}t)}. \tag{13.58}$$

At late times,

$$e^{2\alpha} = \left(\sinh(\sqrt{3\Lambda}t)\right)^{\frac{2}{3}} \approx 2^{-2/3}\exp\left(2\sqrt{\frac{\Lambda}{3}}t\right)$$
$$\Theta(t) = \tanh\left(\frac{\sqrt{3\Lambda}}{2}t\right) \approx 1$$
$$\sigma^2 \approx 4\Lambda e^{-2\sqrt{3\Lambda}t}. \tag{13.59}$$

Thus at late times this line-element approaches that of the flat de Sitter solution. The shear scalar is decreasing exponentially, much faster than the power function decrease t^{-2} in the Kasner case. At early times,

$$e^{2\alpha} = \left(\sinh(\sqrt{3\Lambda}t)\right)^{\frac{2}{3}} \approx \left(\sqrt{3\Lambda}t\right)^{\frac{2}{3}}$$
$$\Theta(t) = \tanh\left(\frac{\sqrt{3\Lambda}}{2}t\right) \approx \frac{\sqrt{3\Lambda}}{2}\cdot t$$
$$\sigma^2 \approx \frac{1}{3}\cdot\frac{1}{t^2}. \tag{13.60}$$

and hence, (up to a rescaling) this solution approximates a vacuum Kasner solution at early times.

Dust dominated model

For dust, eq. (13.51) turns into

$$\dot{\alpha}^2 = \frac{K}{3}e^{-3\alpha} + A^2 e^{-6\alpha}. \tag{13.61}$$

Integrating yields

$$e^{3\alpha} = 3At + \frac{3K}{4}t^2. \tag{13.62}$$

Inserting this result into the equation for $\beta_\pm$ leads to the result

$$\dot{\beta}_\pm = \frac{p_\pm}{3At + \frac{3K}{4}t^2}. \tag{13.63}$$

Integration gives

$$\beta_\pm = \frac{p_\pm}{3A}\ln\left[\frac{t}{\frac{3K}{4}t + 1}\right]. \tag{13.64}$$

The line-element can now be written down

$$\begin{aligned} ds^2 &= -dt^2 + \left(3At + \frac{3K}{4}t^2\right)^{\frac{2}{3}} \\ &\times \left[\Theta(t)^{\frac{4}{3}\cos(\phi+\pi/3)}dx^2 + \Theta(t)^{\frac{4}{3}\cos(\phi-\pi/3)}dy^2 + \Theta(t)^{-\frac{4}{3}\cos(\phi)}dz^2\right]. \end{aligned} \tag{13.65}$$

where $\Theta(t)$ is given by

$$\Theta(t) = \frac{t}{\frac{3K}{4}t + 1}. \tag{13.66}$$

The shear is found to be

$$\sigma^2 = \frac{1}{3\left(t + \frac{K}{4A}t^2\right)^2}. \tag{13.67}$$

Hence, the shear decreases faster than the t^{-2} decrease for the Kasner solutions. However, comparing with exponential decrease in the cosmological constants case, the isotropization due to pure dust is by no means as effective as for a cosmological constant.

At late times, we have

$$\left(3At + \frac{3K}{4}t^2\right)^{\frac{2}{3}} \approx \left(\frac{3K}{4}\right)^{\frac{2}{3}}t^{\frac{4}{3}}$$

$$\frac{t}{\frac{3K}{4}t + 1} \approx \text{constant}. \tag{13.68}$$

This is the same as the dust dominated flat FRW universe. At early times, the line-element approaches the Kasner vacuum solutions. Hence, we see that both of these models, even though they start near the initial singularity as anisotropic Kasner solutions, evolve at late times towards the isotropic FRW solutions. The solutions *isotropise* in the future. Especially effective is a cosmological constant which isotropises the universe exponentially, compared to a mere power-law in the dust case.

13.5   Inflation through bulk viscosity

We will in this section investigate a specific type of fluid which has a *bulk viscosity*. The fluid has an effective pressure

$$p_{\text{eff}} = p + \Pi. \tag{13.69}$$

Here, Π is called the *bulk viscous pressure* and is typically on the form $\Pi = -6\xi H$, where H is the Hubble parameter. The positive factor ξ is called the *bulk viscous coefficient*. We will also assume that a cosmological constant is present and that $p = w\rho$.

The equations of motion for this bulk viscous fluid are

$$
\begin{aligned}
-3\left[-\dot{\alpha}^2 + \dot{\beta}_+^2 + \dot{\beta}_-^2\right] &= \rho + \frac{\Lambda}{3}, \\
\ddot{\beta}_\pm + 3\dot{\alpha}\dot{\beta}_\pm &= 0, \\
\dot{\rho} + 3\dot{\alpha}(w+1)\rho &= 18\dot{\alpha}\xi H.
\end{aligned}
\tag{13.70}
$$

The Hubble scalar is given by $H = \dot{\alpha}$. Differentiating the top equation with respect to time, using the latter two to replace $\ddot{\beta}_\pm$ and $\dot{\rho}$, and finally using the uppermost to replace $\dot{\beta}_\pm$, we obtain the expression

$$\dot{H} = 3(H_\Lambda^2 + \xi H - H^2) + \frac{1-w}{2}\rho, \tag{13.71}$$

where $H_\Lambda^2 = \Lambda/3$. Henceforth, we will assume that the bulk viscous coefficient is constant but it should be noted that the above equations are valid for a general ξ. There is one choice of w for which the above equation simplifies, namely $w = 1$. This type of fluid is called a *Zel'dovich fluid* or a *stiff fluid* and we will now assume that we have this type of fluid.

Eq. (13.71) now simplifies to

$$\dot{H} = 3(H_\Lambda^2 + \xi H - H^2). \tag{13.72}$$

Integration leads to

$$H = \frac{\xi}{2} + \hat{H}\frac{e^{6\hat{H}t} - \hat{C}}{e^{6\hat{H}t} + \hat{C}}, \tag{13.73}$$

where $\hat{H}^2 = \xi^2/4 + H_\Lambda^2$, and $\hat{C}$ is an integration constant. We can integrate this equation once more to find α:

$$e^{3\alpha} = e^{\frac{3\xi}{2}t}\left(C_1 e^{3\hat{H}t} + C_2 e^{-3\hat{H}t}\right), \quad \text{with } C_2 = C_1\hat{C}. \tag{13.74}$$

The shear scalar can now be expressed as

$$\sigma^2 = \frac{A e^{\frac{3\xi}{2}t}}{3\left(C_1 e^{3\hat{H}t} + C_2 e^{-3\hat{H}t}\right)}. \tag{13.75}$$

For some values of C_1 and C_2, this model has no initial singularity. If both C_1 and C_2 are positive, then no initial singularity will be present. However, if they have different sign then a singularity will be either in the future or in the past.

At late times, the Hubble scalar approaches

$$\bar{H} = \frac{\xi}{2} + \hat{H}, \tag{13.76}$$

and hence, the late time asymptotics of this solution is a de Sitter solution with $H = \bar{H}$. The bulk viscous pressure makes the universe enter a de Sitter phase at late time. The effective cosmological constant is no longer dictated *only* by the cosmological constant, but has a larger value. This value is non-zero even though $\Lambda = 0$.

Note also that the shear decays exponentially at late times. The effect is the same as for the pure cosmological constant case, but with the effective $\bar{H}$ instead of H_Λ.

For a pure Zel'dovich fluid with vanishing cosmological constant, $\Lambda = 0$, the above solution simplifies to

$$
\begin{aligned}
H &= \frac{\xi e^{3\xi t}}{e^{3\xi t} + \hat{C}}, \\
e^{3\alpha} &= C_1 e^{3\xi t} + C_2, \quad \text{with } C_2 = C_1 \hat{C}.
\end{aligned}
\tag{13.77}
$$

Also in this case, the bulk viscosity drives the universe into inflation at late times. Note also that if C_1 and C_2 are both positive, then there will be no singularity in the past.

This effect is a typical feature of bulk viscous terms. If they are allowed to dominate they drive the universe into a de Sitter-like state. Through these processes, it isotropises the universe indirectly through the massive expansion. The bulk viscous terms do not interact with the shear itself.

13.6   A universe with a dissipative fluid

In this section we will investigate another viscous model which isotropises the universe quite differently. It interacts with the shear and isotropises the universe directly via this interaction. An example of these types of interaction is frictional forces; friction counter-act shear through dissipation. These dissipation processes basically convert the energy in the shear into heat. In the following we will investigate a fluid that has such "frictional forces".

We assume that the anisotropic stress tensor is given by

$$\pi_{\hat{i}\hat{i}} = -2\eta \dot{a}_i, \tag{13.78}$$

where $\eta > 0$ is a constant. We will also assume that the pressure obeys a w-law equation of state

$$p = w\rho. \tag{13.79}$$

The energy-momentum conservation equation is now

$$\dot{\rho} + 3\dot{\alpha}(w+1)\rho = 12\eta \left(\dot{\beta}_+^2 + \dot{\beta}_-^2 \right). \tag{13.80}$$

The left side of this equation is the usual adiabatic expansion for a perfect fluid. The right side is the dissipative term and is manifestly positive. Hence, it expresses the increase of entropy for dissipative processes.

Einstein's field equations are

$$E_{\hat{t}\hat{t}} = -3\left[-\dot{\alpha}^2 + \dot{\beta}_+^2 + \dot{\beta}_-^2\right] = \rho,$$

$$E_\pm = 3\dot{\alpha}\dot{\beta}_\pm + \ddot{\beta}_\pm = -2\eta\dot{\beta}_\pm. \tag{13.81}$$

The $E_\pm$-equations has the first integral

$$\dot{\beta}_\pm e^{3\alpha + 2\eta t} = p_\pm. \tag{13.82}$$

Hence, solving for $\dot{\beta}_\pm$ we find that the anisotropy is exponentially damped:

$$\dot{\beta}_\pm = p_\pm e^{-3\alpha} e^{-2\eta t}. \tag{13.83}$$

Dissipative processes will in general damp the anisotropy quite effectively. The frictional forces in the fluid are reducing the shear exponentially.

Inserting this into the energy-momentum conservation equation leads to

$$\dot{\rho} + 3(w+1)\dot{\alpha}\rho = 12\eta A^2 e^{-6\alpha} e^{-4\eta t}, \tag{13.84}$$

which can be solved in quadrature:

$$\rho = Ke^{-3(w+1)\alpha} + 12\eta A^2 e^{-3(w+1)\alpha} \int e^{-4\eta t} e^{-3\alpha(1-w)} dt. \tag{13.85}$$

The first term is the usual decay of the density due to the expansion of the universe, while the second term is due to the dissipative processes. Unfortunately, for arbitrary w this equation cannot be solved in terms of elementary functions. However, note that for a Zel'dovich fluid ($w = 1$) the dependence on α in the integral disappears. We will for the sake of illustration consider the case where $w = 1$. In that case, the integral can be evaluated to give

$$\rho = e^{-6\alpha}\left(K - 3A^2 e^{-4\eta t}\right). \tag{13.86}$$

Inserting this into the $E_{\hat{t}\hat{t}}$ equation and simplifying, leads to

$$\dot{\alpha}^2 = \frac{K}{3}e^{-6\alpha}. \tag{13.87}$$

This equation can be easily solved, giving

$$e^{3\alpha} = \sqrt{3K}t. \tag{13.88}$$

Here we have set the initial condition $e^{3\alpha} = 0$ at $t = 0$. The energy density is from eq. (13.86)

$$\rho = \frac{1}{3t^2}\left(1 - \frac{3A^2}{K}e^{-4\eta t}\right). \tag{13.89}$$

The density ρ must be positive, thus $K \geq 3A^2$.

The shear is

$$\sigma^2 = \frac{A^2}{K}\frac{1}{t^2}e^{-4\eta t}, \tag{13.90}$$

and hence is exponentially damped, compared to the Kasner case. In this case the dissipative processes are interacting directly with the shear. The frictional forces effectively convert shear into heat.

We believe that these dissipative processes were much more effective than they are today during the early times of our universe. In an era when the universe was much more dense than it is today, the particles would have a much shorter mean-free-path. That means that the particles would collide and interact with each other. These collisions and interactions are strongly non-adiabatic which means that they covert kinetic energy into heat and radiation. Hence, effectively these collisions yielded frictional type of forces. The viscosity seems unavoidable – at least in the early universe – and may have had a significant effect on the evolution of our universe.

In particular, one believes that neutrino viscosity may be one of the most important factors in the isotropisation of our universe. Nevertheless, we know that real fluids are not perfect fluids. Real fluids behave irreversibly and have necessarily viscous terms like the ones we have investigated. However, a complete picture of the effects from viscous fluids has not been given to date. It is important to study these processes because they may be the key to several riddles of our mysterious Universe.

13.7   The Lemaître-Tolman-Bondi universe models

In the previous chapter homogeneous universe models with dark energy were considered. However, we do not know what sort of energy this is or if there is so much of it that it dominates the evolution of the Hubble flow of the universe. One motivation for introducing dark energy into the cosmological models was to explain the present accelerated state of the Hubble flow suggested by the supernova type Ia observations.

However, inhomogeneous universe models may permit an alternative explanation. We observe the supernovae along the backwards light cone. Observing objects farther away is the same as observing them at an earlier time. Maybe the seemingly accelerated expansion is a sort of illusion coming from the inhomogeneity of space? If we are at a position with a larger velocity of the Hubble flow than in the large scale surroundings, then the inhomogeneity causes us to observe a greater value of the Hubble parameter for the supernovae closest to us. Since they are observed at the latest emission times, this would cause us to observe an increase in the Hubble parameter with time even in a universe model with a stationary Hubble flow. If this is how the universe is, we would not need dark energy to explain the supernovae observations. However, it would mean that we are positioned at a very special place in the universe [AAG06].

In order to be able to discuss this and similar problems we need to become familiar with what the general theory of relativity can tell us about inhomogeneous universe models.

The line-element for a spherically symmetric, inhomogeneous universe model may be written

$$ds^2 = -dt^2 + X^2(r,t)dr^2 + R^2(r,t)d\Omega^2, \qquad (13.91)$$

where the coordinates are co-moving with the cosmic fluid, so that the four velocity of the fluid is $u^\mu = \delta_t^\mu$. We shall consider models containing only perfect fluid and LIVE.

The 01-component of the Einstein tensor for the line-element (13.91) is

$$E^0_1 = -2 \left(\frac{\dot{R}'}{R} - \frac{\dot{X}}{X} \frac{R'}{R} \right),$$ (13.92)

where we use the notation $' = \partial/\partial r$ and $\dot{} = \partial/\partial t$. Solving the equation $E^0_1 = 0$ gives

$$X(r,t) = \frac{R'(r,t)}{f(r)},$$ (13.93)

where f is an arbitrary function of the radial coordinate.

We define two Hubble parameters

$$H_\perp \equiv \frac{\dot{R}}{R}, \quad H_r \equiv \frac{\dot{R}'}{R'}.$$ (13.94)

Then the remaining Einstein equations may be written

$$H^2_\perp + 2H_r H_\perp - \frac{\beta}{R^2} - \frac{\beta'}{RR'} = \kappa\rho + \Lambda,$$ (13.95)

$$-6H^2_\perp q_\perp + 2H^2_\perp - 2\frac{\beta}{R^2} - 2H_r H_\perp + \frac{\beta'}{RR'} = -\kappa(\rho + 3p) + 2\Lambda.$$ (13.96)

where $\beta(r) \equiv f^2 - 1$ and

$$q_\perp \equiv -\frac{1}{H^2_\perp}\frac{\ddot{R}}{R}.$$ (13.97)

In order to see more clearly how the Lemaître-Tolman-Bondi (LTB) universe models generalize the homogeneous Friedmann models we introduce a scale factor, $a(r,t)$, and a spatial curvature parameter, $k(r)$, as follows

$$a(r,t) = \frac{R}{r}, \quad k(r) = -\frac{\beta}{r^2}.$$ (13.98)

Then the line-element takes the form

$$ds^2 = -dt^2 + a^2 \left[\left(1 + \frac{ra'}{a}\right)^2 \frac{dr^2}{1 - k(r)r^2} + r^2 d\Omega^2 \right].$$ (13.99)

Although $k(r)$ is related to the spatial curvature, the relationship between the curvature and $k(r)$ is more complicated than in homogeneous universe models. The Ricci curvature scalar of the spatial part of the line element is

$$R_S = 2\left[\frac{k}{a^2} + 2\frac{(r^2 k)'}{(r^2 a^2)'}\right].$$ (13.100)

Adding eqs.(13.95) and (13.96) we obtain for the deceleration parameter,

$$q_\perp = \frac{1}{2}\left(1 + \frac{k(r)}{a^2} + \frac{\kappa p - \Lambda}{H^2_\perp}\right).$$ (13.101)

One may define an effective Hubble parameter

$$H_{\text{eff}} = \frac{1}{3}(H_r + 2H_\perp),$$ (13.102)

and a shear scalar, σ, by

$$\sigma^2 = \frac{1}{3}\left(H_r - H_\perp\right)^2. \tag{13.103}$$

Using eq. (13.101) and the above definitions the Friedmann equation (13.95) may be written

$$3H_{\text{eff}}^2 - \sigma^2 = -\frac{1}{2}R_S + \kappa\rho + \lambda. \tag{13.104}$$

The critical density is the density of a flat, dust dominated universe,

$$\kappa\rho_{\text{cr}} = 3H_{\text{eff}}^2 - \sigma^2. \tag{13.105}$$

Defining a curvature parameter and a density parameter for dust and LIVE by

$$\Omega_K \equiv -\frac{R_S}{2\kappa\rho_{\text{cr}}}, \quad \Omega_m \equiv \frac{\rho}{\rho_{\text{cr}}}, \quad \Omega_\Lambda \equiv \frac{\Lambda}{\kappa\rho_{\text{cr}}}, \tag{13.106}$$

eq. (13.104) may be written

$$\Omega_K + \Omega_m + \Omega_\Lambda = 1. \tag{13.107}$$

Hence one may define a cosmic triangle for the LTB-models in analogy to the triangle of the Friedmann-Lemaitre models.

Dust dominated LTB-universe models

We shall now consider Lemaitre-Tolman-Bondi (LTB)-universe models [Lem33, Tol34, Bon47] with dust only, putting $p = \Lambda = 0$. Adding eqs. (13.95) and (13.96) we then obtain for the deceleration parameter,

$$q_\perp = \frac{1}{2} - \frac{1}{2}\frac{\beta}{\dot{R}^2}. \tag{13.108}$$

The condition for accelerated expansion now takes the form

$$\beta > \dot{R}^2 > 0 \quad \text{or} \quad f^2 > 1 + \dot{R}^2. \tag{13.109}$$

Inserting the expression for the deceleration parameter, equation (13.101) takes the form

$$2R\ddot{R} + \dot{R}^2 = \beta. \tag{13.110}$$

Integration leads to

$$R\dot{R}^2 = \beta R + \alpha(r) \quad \text{or} \quad H_\perp^2 = \frac{\beta}{R^2} + \frac{\alpha}{R^3}. \tag{13.111}$$

Hence, the dynamical effects of α and β are similar to that of curvature and dust, respectively. Therefore $\alpha(r)$ is regarded as a gravitational mass function.
Substituting eq. (13.111) into eq. (13.96) we find

$$\ddot{R} = -\frac{1}{2}\frac{\alpha}{R^2} \quad \text{or} \quad q_\perp = \frac{1}{2}\frac{\alpha}{R\dot{R}^2}. \tag{13.112}$$

Allowing for inhomogeneity with $-\beta R < \alpha(r) < 0$ seems to allow accelerated expansion even for dust dominated universe models. The dynamical effect of $\alpha < 0$ corresponds to that of dust with negative density in a homogeneous universe model. It should be noted, however, that the inequality above forbids accelerated expansion in a 'big bang' model where the scale factor has the initial value $R(r, 0) = 0$ which implies $\alpha(r) \geq 0$. However this initial condition may not be physically realistic. The universe may have started with a finite scale factor, or maybe has collapsed and reached a finite minimum radius. In such models accelerated expansion does not seem to be forbidden.

Integration of eq. (13.111) with $R(r, t_0) = R_0(r)$ gives

$$
\begin{aligned}
\sqrt{\beta R(\alpha + \beta R)} &- \sqrt{\beta R_0(\alpha + \beta R_0)} \\
-\alpha \ln \left(\frac{\sqrt{\beta R} + \sqrt{\alpha + \beta R}}{\sqrt{\beta R_0} + \sqrt{\alpha + \beta R_0}} \right) &= t - t_0.
\end{aligned} \tag{13.113}
$$

Introducing conformal time η by $\sqrt{|\beta|}dt = Rd\eta$ and choosing $t_0 = \eta_0 = 0$ we find

$$
\beta > 0 : \begin{cases} R = \frac{\alpha}{2\beta}(\cosh\eta - 1) + R_0\left[\cosh\eta + \sqrt{\frac{\alpha + \beta R}{\beta R_0}}\sinh\eta\right], \\ \sqrt{\beta}t = \frac{\alpha}{2\beta}(\sinh\eta - \eta) + R_0\left[\sinh\eta + \sqrt{\frac{\alpha + \beta R}{\beta R_0}}(\cosh\eta - 1)\right], \end{cases} \tag{13.114}
$$

$$
\beta = 0 : R = \left(R_0^{\frac{3}{2}} + \frac{3}{2}\sqrt{\alpha}t\right)^{\frac{2}{3}}, \tag{13.115}
$$

$$
\beta < 0 : \begin{cases} R = \frac{\alpha}{2|\beta|}(1 - \cos\eta) + R_0\left[\cos\eta + \sqrt{\frac{\alpha + \beta R}{|\beta|R_0}}\sin\eta\right], \\ \sqrt{|\beta|}t = \frac{\alpha}{2|\beta|}(\eta - \sin\eta) + R_0\left[\sin\eta + \sqrt{\frac{\alpha + \beta R}{|\beta|R_0}}(1 - \cos\eta)\right]. \end{cases} \tag{13.116}
$$

Equations (13.114–13.116) with $R_0 = 0$, represent the form of the LTB-solution of the field equations starting from a Big Bang event at $t = \eta = 0$. The scale factor as a function of time is shown for some typical models in Figure 13.2.

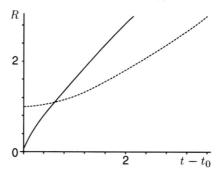

Figure 13.2: Graphs of eq.(13.113) for $\alpha = \beta = 1$ (solid line) and for $\alpha = -1$, $\beta = 1.0001$ (dashed line).

From eqs. (13.114–13.116) we see that the rate of expansion depends on the position through the functions $\alpha(r)$ and $\beta(r)$. Hence, a possible explanation of the supernova type Ia observations which indicate a present state of cosmic acceleration, is that we are positioned at $r = 0$, at a position where the rate of expansion is maximal, and that it decreases with increasing r.

Differentiating eq. (13.111) and comparing the result with eq. (13.95) with $\Lambda = 0$ we obtain the density distribution for this class of models

$$
\kappa\rho = \frac{\alpha'}{R^2 R'} = 3\frac{\alpha'}{V'}. \tag{13.117}
$$

Hence a physically realistic model must have $\alpha' > 0$.

Figure 13.2 shows accelerated cosmic expansion for a dust dominated LTB-universe model with $\alpha(r) < 1$. However this model may not be physically realistic after all. The angular part of the line-element is $R^2 d\Omega^2$, where $d\Omega$ is a solid angle element. It represents the area of a surface extending a certain solid angle. At the origin, $r = 0$, this area must vanish, and thus $R(0, t) = 0$. From eq. (13.101) we then get $\alpha(0) = 0$. Since $\alpha' > 0$, it follows that $\alpha(r) > 0$ for all r. But accelerated expansion is only possible for models with $\alpha < 0$. Hence the dust dominated LTB-universe models have decelerated expansion.

Problems

13.1. *The wonderful properties of the Kasner exponents*
We will in this problem find another useful representation of the Kasner solutions. The exponents in the solution (13.30) have some nice properties as we will see.

(a) For the metric (13.30) call the exponents inside the square brackets x_1, x_2 and x_3. Show that you can represent the Kasner solutions as a cubic equation

$$z^3 + a^3 e^{i\phi} = 0 \tag{13.118}$$

where a is some constant and x_i are the three real parts of the solutions of eq. (13.118).

(b) Write the Kasner solutions as

$$ds^2 = -dt^2 + t^{2p_1} dx^2 + t^{2p_2} dy^2 + t^{2p_3} dz^2. \tag{13.119}$$

Show that $\sum_i p_i = \sum_i p_i^2 = 1$.

13.2. *Dynamical systems approach to a universe with bulk viscous pressure*
We will in this problem consider the type I universe model with a fluid with a bulk viscous pressure, and a cosmological constant. We will assume that the equations of motion are given by eqs. (13.70) and eq. (13.71). These equations imply the following set of equations

$$\dot{\rho} = -3H(w+1)\rho + 18\xi H^2$$
$$\dot{H} = 3(H_\Lambda^2 + \xi H - H^2) + \frac{1-w}{2}\rho \tag{13.120}$$

where $H^2 \geq H_\Lambda^2 + \rho/3 > 0$, and ξ and w are constants. These constants have bounds $\xi \geq 0$ and $-1 \leq w \leq 1$.

(a) Find all the static solutions, $\dot{\rho} = \dot{H} = 0$ to eq. (13.120). What type of solutions do the static solutions correspond to?

(b) Let $\mathbf{X}$ be the column vector with components ρ and H. The system of equations (13.120) can now be written

$$\dot{\mathbf{X}} = \mathbf{F}(\mathbf{X}) \tag{13.121}$$

where $\mathbf{F}$ is a column vector which is a function of $\mathbf{X}$. In a neighbourhood of a fixed point $\mathbf{X}_0$ where $\mathbf{F}(\mathbf{X}_0) = 0$, we can expand the differential equation in a Taylor series to get

$$\dot{\mathbf{X}} \approx \mathbf{F}'(\mathbf{X}_0)\mathbf{X}. \tag{13.122}$$

Whether a point $\mathbf{X}_0$ where $\mathbf{F}(\mathbf{X}_0) = 0$ is stable thus depends on the Jacobian matrix

$$J(\mathbf{X}_0) = \mathbf{F}'(\mathbf{X}_0) \equiv \left(\frac{\partial F^i}{\partial X^j} \right) (\mathbf{X}_0). \tag{13.123}$$

Find the Jacobian $J(\mathbf{X}_0)$ where $\mathbf{X}_0$ are the static points of eq. (13.120). Show that $\det (J(\mathbf{X}_0)) > 0$ and $\mathrm{Tr}\,(J(\mathbf{X}_0)) < 0$. This implies that both eigenvalues of the Jacobian matrix have negative real parts. This implies further that the point $\mathbf{X}_0$ is future stable.

What is been shown in this problem is that the static point found is future stable. Hence, for a large class of possible initial states the universe will end up in the static point. Note this point is stable for any values of w and ξ, not only for the values for which we managed to find an exact solution.

13.3. *Murphy's bulk viscous model*
In this problem we will consider a bulk viscous model of the same type as in section 13.5. We will, however, consider a non-constant bulk viscous coefficient ξ. The eqs. (13.70) and (13.71) shall be used and solved for the choice $\xi = \frac{\alpha_B}{2}\rho$, where α_B is a constant.

(a) Verify eqs. (13.70) and (13.71) in the general case $\xi = \xi(t)$.
  We will in the further assume that $\xi = \alpha_B\rho/2$, and that $\Lambda = 0$.

(b) We will first consider the isotropic case, hence we will assume that $\dot{\beta}_\pm = 0$.
  Show that eqs. (13.70) and (13.71) give

$$\dot{H} = \frac{3}{2}H^2(3\alpha_B H - \gamma), \quad \gamma = w + 1. \tag{13.124}$$

Set $R = e^\alpha$, and show that the general solution to Einstein's field equations is

$$H = \frac{\gamma}{3\alpha_B} \cdot \frac{1}{1 + CR^{\frac{3\gamma}{2}}}, \quad \gamma \neq 0$$

$$CR^{\frac{3\gamma}{2}} + \frac{3\gamma}{2} \ln R = \frac{\gamma^2}{2\alpha_B}(t - t_0), \tag{13.125}$$

where C and t_0 are arbitrary constants.

What is the approximate behaviour for early times, $|\frac{3\gamma}{2} \ln R| \gg CR^{\frac{3\gamma}{2}}$, and for late times, $|\frac{3\gamma}{2} \ln R| \ll CR^{\frac{3\gamma}{2}}$? Compare late time behaviour for $\gamma = 1$ with that of the matter dominated Einstein-de Sitter universe.

(c) We will now generalise the above model to the anisotropic case.
  Show that $A^2 \equiv \sigma^2 R^6$ is a constant of motion, and that this implies

$$A^2 = 3H^2R^6 - \frac{2}{3\alpha_B} \left(\frac{\dot{H}}{H} + 3H \right) R^6. \tag{13.126}$$

If the universe is ever-expanding, we can introduce a new time variable R instead of t. We then can write

$$\frac{d}{dt} = RH\frac{d}{dR}. \tag{13.127}$$

It is also convenient to introduce the function

$$F(R) \equiv H R^3. \tag{13.128}$$

Show that F obeys

$$A^2 = 3F^2 - \frac{2}{3\alpha_B} R^4 F'. \tag{13.129}$$

Integrate this equation and find H, σ^2 and ρ in terms of R. Show that for small R, we have

$$R^3 = \sqrt{3} A(t - t_0). \tag{13.130}$$

Hence, in the anisotropic generalisation there is a singularity for $R = 0$, in contrast to the isotropic model. What is the asymptotic value of ρ as $R \to 0$? Compare this with the isotropic model.

13.4. *Inhomogeneous generalization of the Friedmann-Lemaître model*
An inhomogeneous generalization of the flat ΛCDM-universe model can be found by putting $p = \beta = \beta' = 0$ and adding eqs. (13.95) and (13.96). Show that integration of the resulting equation leads to

$$R(r,t) = \left(\frac{3}{\Lambda}\right)^{\frac{1}{3}} [g(r)]^{\frac{1}{3}} \sinh^{\frac{2}{3}} [(t + h(r))/t_\Lambda], \quad t_\Lambda = \frac{2}{\sqrt{3\Lambda}}. \tag{13.131}$$

Show that the Hubble parameter and the deceleration parameter are, respectively

$$H_\perp = \sqrt{\frac{\Lambda}{3} + \frac{g}{R^3}} = \left(\frac{\Lambda}{3}\right)^{\frac{1}{2}} \coth \frac{t + h}{t_\Lambda}, \quad H_r = \frac{1 - 3\frac{h'}{g'} g H_\perp q_\perp}{1 + 3\frac{h'}{g'} g H_\perp} H_\perp,$$

$$q_\perp = -\frac{1}{H_\perp^2} \frac{\ddot{R}}{R} = \frac{1}{2}\left(1 - 3\tanh^2 \frac{t + h}{t_\Lambda}\right), \tag{13.132}$$

and that the density distribution is

$$\kappa \rho = \frac{\frac{3\rho}{R^3}}{1 + 3\frac{h'}{g'} g H_\perp}. \tag{13.133}$$

Part V

ADVANCED TOPICS

14

Covariant Decomposition, Singularities, and Canonical Cosmology

In this chapter we will perform a 3+1 decomposition of the spacetime. This decomposition is very useful for various applications, in particular, we will use the 3+1 decomposition to derive a Lagrangian and Hamiltonian formalism of general relativity. We will also see how the singularity theorem can be described in this framework.

14.1 Covariant decomposition

Non-relativistic decomposition

We will first consider non-relativistic particles. Consider a velocity field $\mathbf{v}(x^a, t)$ (Latin indices have range 1-3 while Greek indices have range over the full spacetime 0-3), typically this can for example be the flow of material particles in space. We will specialize to Cartesian coordinates; hence, the spatial metric is $h_{ij} = \delta_{ij}$.

A particle is moving along a trajectory $x^a(t)$ with a velocity $v^b(x^a, t)$. The acceleration can be expressed as

$$\mathbf{a} = \frac{D\mathbf{v}}{dt}, \tag{14.1}$$

where $\frac{D\mathbf{v}}{dt}$ is the *total time derivative* of the velocity field,

$$\frac{D\mathbf{v}}{dt} = \frac{\partial \mathbf{v}}{\partial t} + (\mathbf{v} \cdot \nabla)\mathbf{v}. \tag{14.2}$$

Here are $\frac{\partial \mathbf{v}}{\partial t}$ and $(\mathbf{v} \cdot \nabla)\mathbf{v}$ called the *local derivative* and the *convective derivative* respectively. The local derivative describes the change of $\mathbf{v}$ in time at a fixed position. The convective derivative on the other hand describes how the field v depends on position.

By using the chain rule for derivatives we find

$$\frac{Dv^a}{dt} = \frac{\partial v^a}{\partial t} + \frac{\partial v^a}{\partial x^b}\frac{dx^b}{dt} = \frac{\partial v^a}{\partial t} + v^b\frac{\partial v^a}{\partial x^b}. \tag{14.3}$$

This equation shows that the convective derivative can be written on matrix form

$$(\mathbf{v} \cdot \nabla)\,\mathbf{v} = M\mathbf{v}\,, \tag{14.4}$$

where M is the matrix

$$M = (v^i{}_{,j}) = \begin{bmatrix} \frac{\partial v^x}{\partial x} & \frac{\partial v^x}{\partial y} & \frac{\partial v^x}{\partial z} \\ \frac{\partial v^y}{\partial x} & \frac{\partial v^y}{\partial y} & \frac{\partial v^y}{\partial z} \\ \frac{\partial v^z}{\partial x} & \frac{\partial v^z}{\partial y} & \frac{\partial v^z}{\partial z} \end{bmatrix}. \tag{14.5}$$

The entries in this matrix is the gradient of the velocity field.

We can separate this matrix into a symmetric and an anti-symmetric part:

$$v_{i,j} = \theta_{ij} + \omega_{ij}\,, \tag{14.6}$$

where the symmetric part is given by

$$(\theta^i{}_j) = \frac{1}{2}\left(M + M^t\right), \tag{14.7}$$

while the anti-symmetric part is given by

$$(\omega^i{}_j) = \frac{1}{2}\left(M - M^t\right). \tag{14.8}$$

The symmetric part θ_{ij} is called the *expansion tensor* and ω_{ij} the *rotation tensor*. ω_{ij} is sometimes also called the *vorticity tensor*. We can further split the expansion tensor into trace and trace-free parts

$$\theta^i{}_j = \frac{1}{3}\theta\delta^i{}_j + \sigma^i{}_j\,, \tag{14.9}$$

where

$$\begin{aligned} \theta &= \operatorname{Tr} M \\ (\sigma^i{}_j) &= \frac{1}{2}\left(M + M^t\right) - \frac{1}{3}\delta^i{}_j \operatorname{Tr} M. \end{aligned} \tag{14.10}$$

The tensor σ_{ij} is trace-free and is called the *shear tensor*.

The convective derivative can now be written as

$$v_{i,j} = \frac{1}{3}\theta\delta_{ij} + \sigma_{ij} + \omega_{ij}. \tag{14.11}$$

Relativistic decomposition

We will now consider a velocity field $u(x^\mu)$ in spacetime with metric $g_{\mu\nu}$. A particle is moving along a trajectory with four-velocity u. The four-acceleration is given by the covariant derivative along the trajectory

$$\mathbf{a} = \frac{d\mathbf{u}}{d\tau}\,, \tag{14.12}$$

where τ is the proper time of the particle. In coordinate form this is expressed as

$$a_\alpha = u_{\alpha;\mu} u^\mu \equiv \dot{u}_\alpha. \tag{14.13}$$

If this particle is moving freely, then – because free particles move along geodesics – the four-acceleration will vanish.

In chapter 7, problem 7.4, we introduced the projection operator

$$h_{\mu\nu} = g_{\mu\nu} + u_\mu u_\nu, \tag{14.14}$$

which projects tensors onto the plane of simultaneity orthogonal to the four-velocity u^μ. The projection of the tensor $u_{\alpha;\mu}$ is given by

$$(u_{\alpha;\beta})_\perp = u_{\nu;\mu} h^\nu{}_\alpha h^\mu{}_\beta. \tag{14.15}$$

It follows that the relativistic decomposition is

$$\theta = u^\mu{}_{;\mu}, \tag{14.16}$$

$$\sigma_{\alpha\beta} = \frac{1}{2} \left(u_{\mu;\nu} + u_{\nu;\mu} \right) h^\mu{}_\alpha h^\nu{}_\beta - \frac{1}{3} u^\mu{}_{;\mu} h_{\alpha\beta}, \tag{14.17}$$

$$\omega_{\alpha\beta} = \frac{1}{2} \left(u_{\mu;\nu} - u_{\nu;\mu} \right) h^\mu{}_\alpha h^\nu{}_\beta. \tag{14.18}$$

As in the non-relativistic case, θ is called the *expansion scalar*, $\sigma_{\alpha\beta}$ is called the *shear tensor*, and $\omega_{\alpha\beta}$ is called the *vorticity tensor*.

The covariant derivative of the four-velocity can therefore be written as

$$\boxed{u_{\alpha;\beta} = \frac{1}{3} \theta h_{\alpha\beta} + \sigma_{\alpha\beta} + \omega_{\alpha\beta} - \dot{u}_\alpha u_\beta.} \tag{14.19}$$

Due to the four-velocity identity $u^\mu u_\mu = -1$ we have

$$\dot{u}_\mu u^\mu = 0 \quad \text{and} \quad u_{\mu;\beta} u^\mu = 0. \tag{14.20}$$

Thus using the expression $h^\mu{}_\alpha = \delta^\mu{}_\alpha + u^\mu u_\alpha$, the projection of the covariant derivative can be written

$$u_{\nu;\mu} h^\nu{}_\alpha h^\mu{}_\beta = u_{\alpha;\beta} + \dot{u}_\alpha u_\beta. \tag{14.21}$$

Hence, the shear and vorticity tensors can be written as

$$\sigma_{\alpha\beta} = u_{(\alpha;\beta)} - \frac{1}{3} u^\mu{}_{;\mu} h_{\alpha\beta} + \dot{u}_{(\alpha} u_{\beta)},$$

$$\omega_{\alpha\beta} = u_{[\alpha;\beta]} + \dot{u}_{[\alpha} u_{\beta]}. \tag{14.22}$$

Assume that the vector field describes the movement of a physical frame of reference, for example the movement of a collection of particles. We can now (locally) give a covariant characterisation of the following types of reference systems:

Irrotational (non-vortic): $\omega_{\alpha\beta} = 0$.

Stiff: $\sigma_{\alpha\beta} = \theta = 0$.

Static: A system which is stiff and irrotational; i.e. $u_{\mu;\nu}h^{\mu}{}_{\alpha}h^{\nu}{}_{\beta} = 0$.

Inertial: A freely falling static system; i.e. $u_{\alpha;\beta} = 0$.

14.2   Equations of motion

Using Einstein's field equations we will derive the equations of motion using the variables in the relativistic decomposition we described in the previous section. We will assume that the vector field u^{μ} has vanishing four-acceleration $\dot{u}^{\mu} = u^{\mu}{}_{;\nu}u^{\nu} = 0$; hence, they describe the four-velocity of geodesics.

Let us consider an energy-momentum tensor of the form

$$T_{\mu\nu} = \rho u_{\mu}u_{\nu} + ph_{\mu\nu} + \pi_{\mu\nu}. \tag{14.23}$$

The first two terms can be recognized as a usual perfect fluid part. We have also already encountered the last term, which is the *anisotropic stress tensor*. This tensor is symmetric and has the properties

$$\pi^{\mu}{}_{\mu} = 0, \quad u^{\mu}\pi_{\mu\nu} = 0. \tag{14.24}$$

The energy-conservation equation $T^{\nu}{}_{\mu;\nu} = 0$ implies by contraction with u^{μ}

$$u^{\mu}T^{\nu}{}_{\mu;\nu} = 0. \tag{14.25}$$

Using eq. (14.23) this can be written as

$$\begin{aligned}
0 &= u^{\mu}(\rho_{,\nu}u_{\mu}u^{\nu} + \rho u_{\mu;\nu}u^{\nu} + \rho u_{\mu}u^{\nu}{}_{;\nu} \\
&\quad + h^{\nu}{}_{\mu;\nu}p + h^{\nu}{}_{\mu}p_{,\nu} + \pi^{\nu}{}_{\mu;\nu}).
\end{aligned} \tag{14.26}$$

We define an overdot by $\dot{} \equiv u^{\nu}\nabla_{\nu}$. The first term in eq. (14.26) equals $-\dot{\rho}$, the second vanishes because of (14.20), the third equals $-\theta\rho$, and using eq. (14.14) the fourth yields $-\theta p$. The last term can be written (using $u^{\mu}\pi_{\mu\nu} = 0$)

$$\begin{aligned}
u^{\mu}\pi^{\nu}{}_{\mu;\nu} &= -u_{\mu;\nu}\pi^{\mu\nu} = -u_{(\mu;\nu)}\pi^{\mu\nu} \\
&= -(u_{(\mu;\nu)} + \dot{u}_{(\mu}u_{\nu)})\pi^{\mu\nu} = -\sigma_{\mu\nu}\pi^{\mu\nu}
\end{aligned} \tag{14.27}$$

where we have used the symmetry and tracelessness of $\pi_{\mu\nu}$. Hence, the energy-momentum conservation equation can be written

$$\boxed{\dot{\rho} + \theta(\rho + p) + \sigma_{\mu\nu}\pi^{\mu\nu} = 0.} \tag{14.28}$$

In addition to a possible equation of state for the fluid, this equation governs the dynamical evolution of the fluid along the fluid world-lines.

Using eq. (7.50), we can write

$$\begin{aligned}
-u^{\mu}u^{\nu}R_{\alpha\mu\beta\nu} &= u^{\nu}R_{\alpha\mu\nu\beta}u^{\mu} \\
&= u_{\alpha;\beta\nu}u^{\nu} - u_{\alpha;\nu\beta}u^{\nu} \\
&= u_{\alpha;\beta\nu}u^{\nu} + u_{\alpha;\nu}u^{\nu}{}_{;\beta}
\end{aligned} \tag{14.29}$$

where we have also used that $u_{\alpha;\nu}u^{\nu} = 0$. Contracting the above expression (over α and β) leads to

$$-u^{\mu}u^{\nu}R_{\mu\nu} = u^{\beta}{}_{;\beta\nu}u^{\nu} + u_{\alpha;\nu}u^{\nu;\alpha}. \tag{14.30}$$

Using this together with Einstein's field equations and eq. (14.19), we get

$$\boxed{\dot{\theta} + \frac{1}{3}\theta^2 + \sigma_{\mu\nu}\sigma^{\mu\nu} - \omega_{\mu\nu}\omega^{\mu\nu} + \frac{\kappa}{2}(\rho + 3p) - \Lambda = 0.}$$
(14.31)

This equation is called *Raychaudhuri's equation* and tells how the expansion scalar varies along the geodesic curves defined by the vector field u^μ.

Similarly, we can take the symmetric and anti-symmetric part of eq. (14.29) to find a propagation equation for the shear and the rotation respectively. However, the results are not very illuminating at this stage. It is usually more practical to investigate a special case of the above. If we assume that the space-time is foliated into hypersurfaces and that the vector field **u** is the normal vector field to the hypersurfaces, then we must have $\omega_{\mu\nu} = 0 = \dot{u}_\alpha$. The above analysis simplifies in that case and the equations of motion likewise.

In this case, the tensor $u_{\alpha;\beta}$ reduces simply to the extrinsic curvature of the hypersurfaces. Hence,

$$u_{\alpha;\beta} = \theta_{\alpha\beta} = K_{\alpha\beta}.$$
(14.32)

We can now use eq. (7.152), together with Einstein's field equations to obtain

$$\kappa T^{\alpha\beta} u_\alpha u_\beta = \frac{1}{2}\left({}^{(3)}R - K^{\alpha\beta}K_{\alpha\beta} + K^2 \right) - \Lambda.$$
(14.33)

Using the decomposition eq. (14.19) with $\omega_{\alpha\beta} = \dot{u}_\alpha = 0$ and eq. (14.23), we get the generalised Friedmann equation

$$\boxed{\frac{1}{3}\theta^2 = \frac{1}{2}\sigma_{\alpha\beta}\sigma^{\alpha\beta} - \frac{1}{2}{}^{(3)}R + \kappa\rho + \Lambda.}$$
(14.34)

This is the Friedmann equation for spacetimes with shear and a more general geometry of the spatial hypersurfaces. From the above analysis it is clear that the Friedmann equation is essentially the E_{tt}-component of Einstein's field equations. We will see another derivation of the same equation in the next section which gives yet another interpretation of this equation.

Taking the trace-free part of eq. (14.29) we can find the shear propagation equations. Angled brackets mean that the projected and trace-free part should be taken. Thus for a spatial tensor $A_{\alpha\beta}$, we define

$$A_{\langle\alpha\beta\rangle} \equiv A_{\alpha\beta} - \frac{1}{3}h_{\alpha\beta}A^\mu{}_\mu.$$
(14.35)

By projecting Einstein's field equations onto the spatial hypersurfaces and taking the trace-free part we get

$$\begin{aligned}
h^\rho{}_{(\alpha}h^\lambda{}_{\beta)}R_{\rho\lambda} &= h^\rho{}_{(\alpha}h^\lambda{}_{\beta)}\left(\kappa T_{\rho\lambda} + g_{\rho\lambda}(R - \Lambda)\right) \\
&= h^\rho{}_{(\alpha}h^\lambda{}_{\beta)}\kappa T_{\rho\lambda} = \kappa\pi_{\alpha\beta}.
\end{aligned}$$
(14.36)

Projecting and taking the trace-free part of the left side of eq. (14.29), and using eq. (14.36), we get

$$\begin{aligned}
h^\rho{}_{(\alpha}h^\lambda{}_{\beta)}u^\mu u^\nu R_{\rho\mu\lambda\nu} &= h^\rho{}_{(\alpha}h^\lambda{}_{\beta)}(-g^{\mu\nu} + h^{\mu\nu})R_{\rho\mu\lambda\nu} \\
&= -\kappa\pi_{\alpha\beta} + h^\rho{}_{(\alpha}h^\lambda{}_{\beta)}h^{\mu\nu}R_{\rho\mu\lambda\nu}.
\end{aligned}$$
(14.37)

For the second term of this equation we can use eq. (7.83). This, and that the extrinsic curvature is $K_{\alpha\beta} = u_{\alpha;\beta}$, lead to

$$
\begin{aligned}
h^\rho_{\;\langle\alpha}h^\lambda_{\;\beta\rangle}h^{\mu\nu}R_{\rho\mu\lambda\nu} &= {}^{(3)}R_{\langle\alpha\beta\rangle} + KK_{\langle\alpha\beta\rangle} - K_{\mu\langle\alpha}K^\mu_{\;\beta\rangle} \\
&= {}^{(3)}R_{\langle\alpha\beta\rangle} + \theta\sigma_{\alpha\beta} - u_{\mu;\langle\alpha}u^\mu_{\;;\beta\rangle}.
\end{aligned} \tag{14.38}
$$

The trace-free part of the right-hand side of eq. (14.29) is

$$
u_{\langle\alpha;\beta\rangle\mu}u^\mu + u_{\mu;\langle\alpha}u^\mu_{\;;\beta\rangle} = \dot{\sigma}_{\alpha\beta} + u_{\mu;\langle\alpha}u^\mu_{\;;\beta\rangle}. \tag{14.39}
$$

Thus the eq. (14.29) turns into the *shear propagation equations*

$$
\boxed{\dot{\sigma}_{\alpha\beta} + \theta\sigma_{\alpha\beta} + {}^{(3)}R_{\langle\alpha\beta\rangle} = \kappa\pi_{\alpha\beta}.} \tag{14.40}
$$

Note that the "time derivative" $\dot{\sigma}_{\alpha\beta}$ is defined as $\sigma_{\alpha\beta;\mu}u^\mu$. Hence, the expression will in general contain the connection coefficients.

14.3 Singularities

We have presented some cosmological solutions to the Einstein field equations. Some of them begin at a certain cosmic time, which we will for the sake of simplicity set to $t = 0$. This time is usually referred to as the *point of time of the Big Bang*. However, what we have not investigated is what really happens at $t = 0$? Some of the models have clearly no singularity, like the $k = 1$ de Sitter solution. Other models have a singularity, and in chapter 10 we also presented a solution, namely the Schwarzschild solution, that had a singularity. In this section we will be concerned with cosmological singularities.

We will start out by defining a singularity, and in order to do that we need to introduce some technical concepts. If a geodesic has finite affine length[1] then we say that the geodesic is *incomplete*. Hence, if a geodesic is inextendible in at least one direction for a finite affine parameter then the geodesic is incomplete. Singular spacetimes are spacetimes which has at least one incomplete geodesic. There are basically four types of singularities:[2]

1. **Scalar Curvature Singularities:** Singularities where one or more curvature scalar diverges along the geodesic. One example of this singularity is the singularity in the Schwarzschild solution.

2. **Parallelly propagated Curvature Singularities:** Singularities where no scalar blows up, but where one or several components of the Riemann tensor diverges in a parallelly propagated tetrad along the geodesic.

3. **Inextendible non-curvature Singularities:** Singularities where the curvature scalars are everywhere bounded along the geodesic. An example of this is the circular cone. The cone itself is everywhere flat, but the apex of the cone is a singularity that cannot be removed.

4. **Removable Singularities:** Singularities that can be removed by adding for instance a single point. An example of this is a plane with one point removed.

---

[1] Affine length is the length of a geodesic using a unit tangent vector.
[2] There are other kinds of singularities than those mentioned here, for example singularities associated with non-geodesic observers [EK74, LCH07].

Singularities that come under the first category are often easy to spot. Since one or several curvature scalars diverge we can find these by calculating the curvature scalars. The singularities in the second category are also not very difficult to find. The question whether a singularity ends in the third or fourth category is a bit more tricky and troublesome. Usually we only know the intrinsic property of the space and it is often difficult to say whether we can remove the singular point by adding a point or a line etc. However, removable singularities are not very interesting and are rather unphysical; they are constructed from a regular spacetime by artificially removing points. The fourth category consist for example of *coordinate singularities*, while the *physical singularities* are singularities of the first three categories.

Example 14.1 (A coordinate singularity) Examples
Consider the two-dimensional Rindler spacetime (see Fig. 14.1)

$$ds^2 = -x^2 dt^2 + dx^2. \tag{14.41}$$

This metric is similar to the Schwarzschild spacetime in the neighbourhood of the horizon. We have already seen that the horizon is not a true singularity but is merely a result of a certain choice of coordinates.

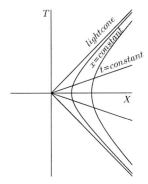

Figure 14.1: The Rindler spacetime is a part of Minkowski spacetime.

The coordinate transformation

$$\begin{aligned} T &= x \sinh t, \\ X &= x \cosh t, \end{aligned} \tag{14.42}$$

turns the Rindler metric into

$$ds^2 = -dT^2 + dX^2. \tag{14.43}$$

This is the Minkowski metric in two dimensions, and $x = 0$ corresponds to the future light-cone of origo. The Minkowski metric has no singularities, and is therefore a regular spacetime. The Rindler spacetime can therefore be embedded isometrically into Minkowski spacetime. Hence, the singularity in the Rindler spacetime is removable.

Example 14.2 (An inextendible non-curvature singularity)
Consider the two-dimensional Milne universe

$$ds^2 = -dt^2 + t^2 dx^2. \tag{14.44}$$

This metric is very similar to the previous metric, just with t and x interchanged. However, as we will see, they have very different physical properties. Consider the variable

x to be an angular variable, thus assume that $0 \leq x < \ell$ for some $\ell > 0$. If we had done the same in the previous example the time variable would have been circular which does not make sense. Here in the Milne universe, on the other hand, it does make sense. This identification of the x variable (identifying $x = 0$ with $x = \ell$) makes it impossible to embed the space isometrically into Minkowski space. Locally it can be embedded into Minkowski space, but not globally.

For $0 < x < \ell$ (not including $x = 0$ and $x = \ell$) we can consider the transformation

$$
\begin{aligned}
T &= t \cosh x \\
X &= t \sinh x.
\end{aligned}
\tag{14.45}
$$

The metric turns into

$$
ds^2 = -dT^2 + dX^2,
\tag{14.46}
$$

which is the flat Minkowski metric. The lines $x = 0$ and $x = \ell$ correspond to the lines $X = 0$ and $X = T \tanh \ell$ in the Minkowski space eq. (14.46), see Fig. 14.2. Saying that x is an angular variable with period ℓ is the same as saying that these two lines should be identified as one. As you go around in the universe and you hit one of the lines and cross it, you are at the same time at the other line and continue from there. In this way you will be always inside the two lines $X = 0$ and $X = T \tanh \ell$. However, these lines intersect at $X = T = 0$. This point will be a singular point in this universe, a conar-like point which cannot be removed. Since the Minkowski space is flat, the whole interior of this space will have vanishing curvature tensor. Hence, this spacetime has a category 3 singularity.

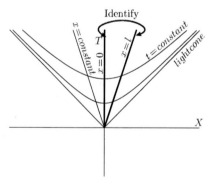

Figure 14.2: The 2D Milne universe with finite spatial sections.

We can now let the identification radius ℓ go to infinity: $\ell \rightarrow \infty$. We will then recover the Milne universe with infinite spatial sections. However, as $\ell \rightarrow \infty$, the initial singularity will remain to be a point-like event. Thus we have to conclude that the Milne universe model has an initial point-like singularity of category 3.

Note that we considered the infinite open universe in section 11.10 from a physical point of view. Nonetheless, we reached the same conclusion. Hence, the above discussion puts mathematical ground to the discussion in section 11.10.

Let us investigate the conditions for a spacetime to have a singularity. This issue is important in the context of cosmology because this can maybe give us the answer to whether we have had a Big Bang or not. We will in this analysis assume that the cosmological constant is part of the energy-momentum tensor as we have seen it can be.

We have already encountered cosmological solutions that had singularities in their past. FRW universes with dust or radiation have a singularity in their past. A solution like the closed de Sitter solution has no singularity as we have seen. So a cosmological constant or a vacuum fluid can avoid such a singularity. Generally, we may ask ourself: What are the necessary criteria for a past singularity?

The weak and strong energy conditions

Two important concepts are linked with this question. We know that most of the matter in our universe has a positive energy-density. An observer with a four-velocity u^μ will measure the energy density $T^{\mu\nu}u_\mu u_\nu$.

The Weak Energy Condition (WEC) *If the energy-momentum tensor obeys the inequality*

$$T^{\mu\nu}u_\mu u_\nu \geq 0, \tag{14.47}$$

for all time-like vectors u^μ then we say that the energy-momentum tensor obeys the weak energy condition.

This condition for the energy-momentum tensor is satisfied by most fluids known, even the vacuum fluid. It is basically saying that all time-like observers will measure a positive energy density.

The Strong Energy Condition (SEC) *If the energy-momentum tensor obeys the inequality*

$$\left(T^{\mu\nu} - \frac{1}{2}Tg^{\mu\nu}\right)u_\mu u_\nu \geq 0, \tag{14.48}$$

for all time-like vectors u^μ then we say that the energy-momentum tensor obeys the strong energy condition.

Note that the SEC is a much stronger restriction on the energy-momentum tensor. For instance if the energy-momentum tensor consists of a single vacuum fluid then the energy-momentum tensor will fail to obey the SEC. The energy-momentum tensor can be diagonalised (with some exceptions) by choosing a frame with the eigenvectors of the energy-momentum tensor. The eigenvalues of the eigenvectors will be ρ and p_i where $i = 1, .., 3$. The eigenvectors p_i are called the *principal pressures*. The WEC is equivalent to

$$\text{WEC} \quad \Leftrightarrow \quad \rho \geq 0 \quad \text{and} \quad \rho + p_i \geq 0 \ (i = 1, 2, 3) \tag{14.49}$$

and the SEC is equivalent to

$$\text{SEC} \quad \Leftrightarrow \quad \rho + \sum_i p_i \geq 0 \quad \text{and} \quad \rho + p_i \geq 0 \ (i = 1, 2, 3). \tag{14.50}$$

If now for instance we have a barotropic perfect fluid

$$p = w\rho \tag{14.51}$$

and all the principal pressures are equal to p, the WEC is equivalent to $w \geq -1$. The SEC on the other hand, put the stronger constraint $w \geq -\frac{1}{3}$. Note from eq.

(11.18) that if SEC is satisfied then gravity is attractive for observers moving along time-like geodesics.

If a spacetime satisfies the Einstein equations then we can replace the energy-momentum tensor with the Ricci tensor. The SEC can therefore be written as

$$R^{\mu\nu} u_\mu u_\nu \geq 0 \tag{14.52}$$

for all time-like u^μ. Hence, the spacetime has a positive curvature for time-like vectors. If we have two neighbouring parallel geodesics, then if the SEC is satisfied, the geodesics will converge and at some point meet.

The singularity theorem

As we have seen, the strong energy condition implies that the space is positively curved for time-like vectors. This turns out to be what we need to have a singularity in the past of a spacetime.

Assume therefore that the matter obeys the SEC, and we will also assume that the geodesics are non-rotating. This implies $\rho + 3p \geq 0$. Hence, from Raychaudhuri's equation (14.31) we get the inequality

$$\dot{\theta} \leq -\frac{1}{3}\theta^2. \tag{14.53}$$

Dividing by θ^2 yields

$$\frac{d}{d\tau}\left(\frac{1}{\theta}\right) \geq \frac{1}{3} \tag{14.54}$$

and hence, integrating

$$\frac{1}{\theta(\tau)} \leq \frac{1}{\theta_0} + \frac{1}{3}\tau. \tag{14.55}$$

Here, θ_0 is the value of θ at $\tau = 0$, and $\tau \leq 0$. Assume further that the geodesic congruences are expanding at $\tau = 0$, i.e. $\theta_0 > 0$ (which would be the case for an expanding universe). Then according to eq. (14.55), the function $\theta^{-1}(\tau)$ must have passed through zero at a finite time τ_s. In particular, τ_s is bounded by the inequality $|\tau_s| \leq 3\theta_0^{-1}$. This means that at the time τ_s, the expansion scalar was infinite $\theta(\tau_s) = \infty$, which indicates that there was a singularity at τ_s. Strictly speaking, this only tells that there is a singularity of the geodesic congruences, but this analysis is one of the key ingredients for proving the singularity theorem stated below. There are also some global aspects that we have to consider, but we refer the reader to Wald [Wal84] or Hawking and Ellis [HE73] for details. Roughly speaking we can say that:

If the matter obeys the SEC and there exist a positive constant $C > 0$ such that $\theta > C$, where H is the Hubble parameter, everywhere in the past of some specific hypersurface, then there exists a past singularity where all past directed geodesics end (see Fig. 14.3).

Note that this is a sufficient criterion, but not necessary. Spacetimes can have singularities even though the SEC is violated.

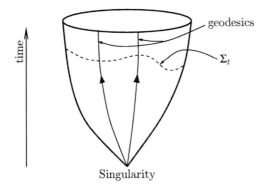

time

geodesics

Σ_t

Singularity

Figure 14.3: An expanding universe containing matter that obeys the SEC, means that the universe has a past singularity.

14.4   Lagrangian formulation of General Relativity

We saw in chapter 8 how Einstein's field equations could be derived using a simple variational principle. In this section we will pursue this idea even further. In classical mechanics the Lagrangian and Hamiltonian formulations are very useful tools in the analysis of the dynamical behaviour of the system. Not only are they important concepts in classical mechanics, but they also proved to be the key to quantum mechanics. The Lagrangians and Hamiltonians link classical mechanics with quantum mechanics in a quite elegant way. We will not dwell upon the possible quantum aspects in this book, but we will introduce the Lagrangian and Hamiltonian formalism for the "classical" gravitational field.

Again we perform a 3+1 split of the spacetime. However, we will here do it in a slightly different way. Consider our spacetime M. We will assume that the spacetime (at least locally) can be foliated with three-dimensional spatial sections. Each of these spatial sections, which will be denoted by Σ_t, is labelled by a time parameter t. It is useful to let the direction of time, denoted by a vector $\mathbf{t}$, be arbitrary; we only demand at this stage that it is non-zero and time-like. We thus have

$$M = \mathbb{R} \times \Sigma_t. \tag{14.56}$$

Let h_{ab} be the metric on the spatial surfaces Σ_t. As the time varies, the metric h_{ab} will also vary describing the dynamical evolution of the spatial surfaces Σ_t.

For each Σ_t, there will be a unit normal vector field $\mathbf{n}$. Since Σ_t is space-like, $\mathbf{n}$ will be time-like. If $\mathbf{t}$ is the time-vector, we can split this into

$$\mathbf{t} = N\mathbf{n} + \mathbf{N}, \tag{14.57}$$

where $\mathbf{N}$ is tangent to Σ_t (and thus orthogonal to $\mathbf{n}$). The function N is called the *lapse* and the vector $\mathbf{N}$ is called the *shift vector*. This is illustrated in Fig. 14.4. We may choose the time vector $\mathbf{t}$ freely; hence, the shift and the lapse can be an arbitrary vector- and scalar function respectively. This is a *gauge freedom* which we have in general relativity, reflecting the general covariance of the theory. This freedom has, as we will see later, interesting consequences for the Lagrangian and Hamiltonian formulation.

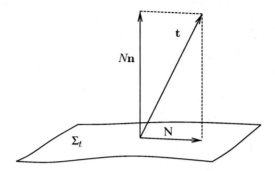

Figure 14.4: The hypersurfaces with the lapse and the shift vector.

The metric components $g_{\mu\nu}$ can now be calculated (raising Latin indices is done using h_{ab}):

$$
\begin{aligned}
g_{tt} &= \mathbf{t}\cdot\mathbf{t} = -N^2 + N^a N_a \\
g_{ta} &= \mathbf{t}\cdot\mathbf{e}_a = N^b h_{ab} = N_a \\
g_{ab} &= \mathbf{e}_a\cdot\mathbf{e}_b = h_{ab}.
\end{aligned}
\tag{14.58}
$$

This shows that instead of using the metric components $g_{\mu\nu}$ as variables, we may equally well use the set of variables (N, N_a, h_{ab}).

The determinant of the metric tensor is

$$
\sqrt{-g} = N\sqrt{h},
\tag{14.59}
$$

where h is the determinant of the spatial metric h_{ab}. We define the time-derivative using the Lie-derivative with respect to the time-vector $\mathbf{t}$. In particular, the time derivative of the metric h_{ab} is

$$
\dot{h}_{ab} \equiv \mathcal{L}_t h_{ab}.
\tag{14.60}
$$

Henceforth, we will use a mixture or Latin and Greek indices. When we want to emphasise that the tensor is purely spatial, we will use Latin indices. In general we will use Greek indices to emphasise the covariant nature of the equations.

We introduce the extrinsic curvature $K_{\mu\nu}$ from section 7.4 which in these variables is given by

$$
K_{\mu\nu} = -\left(\mathbf{e}_\alpha\cdot\nabla_\beta\mathbf{n}\right)h^\alpha{}_\mu h^\beta{}_\nu = \frac{1}{2}\mathcal{L}_\mathbf{n} h_{\mu\nu}.
\tag{14.61}
$$

In the previous sections we used an arbitrary vector field $\mathbf{u}$. In the special case where $\mathbf{u} = \mathbf{n}$ we see that the covariant derivative of the vector $\mathbf{u}$ can be written as

$$
n_{\mu;\nu} = u_{\mu;\nu} = K_{\mu\nu}.
\tag{14.62}
$$

Thus the extrinsic curvature splits in a trace-free part, and a trace part

$$
K_{\mu\nu} = \frac{1}{3}K h_{\mu\nu} + \sigma_{\mu\nu}
\tag{14.63}
$$

where $K = \theta = K^\mu_\mu$. Hence, by comparing with eq. (14.19), the vector field n is found to be non-rotating.

Why this has to be so, is seen quite clearly when we remind ourself of the analysis done in chapter 5. If the normal vector field is rotating, the planes of simultaneity have a discontinuity along some line from the centre. In our case the surfaces Σ_t are assumed to be smooth everywhere and hence, the normal vector field has to be non-rotating.

The covariant derivative on the hypersurfaces Σ_t will be denoted by $\bar{\nabla}_\mu$ and is the projection of the covariant derivative in the four-dimensional space-time. Thus

$$\bar{\nabla}_\mu \mathbf{e}_\nu = h^\alpha_\mu h^\beta_\nu \nabla_\alpha \mathbf{e}_\beta. \tag{14.64}$$

The time derivative of $h_{\mu\nu}$ can now be calculated to give

$$
\begin{aligned}
\dot{h}_{\mu\nu} &= \pounds_t h_{ab} = \pounds_{N\mathbf{n}+\mathbf{N}} h_{ab} \\
&= N\pounds_\mathbf{n} h_{ab} + \pounds_\mathbf{N} h_{ab} \\
&= 2NK_{\mu\nu} + \bar{\nabla}_\mu N_\nu + \bar{\nabla}_\nu N_\mu.
\end{aligned} \tag{14.65}
$$

The Einstein-Hilbert action for pure gravity reads[3]

$$S_G = \frac{1}{2\kappa} \int_M \mathcal{L}_G d^4 x \tag{14.66}$$

where

$$\mathcal{L}_G = (R - 2\Lambda)\sqrt{-g}. \tag{14.67}$$

We will now express the Ricci scalar R in the new variables. The Ricci scalar can be written as

$$R = 2(E_{\mu\nu} n^\mu n^\nu - R_{\mu\nu} n^\mu n^\nu) \tag{14.68}$$

by contracting the Einstein tensor with the vector n^μ. We can write the twice contracted Gauss' equation, eq. (7.152), as

$$E_{\mu\nu} n^\mu n^\nu = \frac{1}{2}\left(\bar{R} - K^{ab}K_{ab} + K^2\right) \tag{14.69}$$

where $\bar{R}$ is the Ricci scalar of the three-spaces Σ_t. As for the term $R_{\mu\nu} n^\mu n^\nu$ in eq. (14.68), we can use the definition of the Riemann tensor:

$$
\begin{aligned}
R_{\mu\nu} n^\mu n^\nu &= R^\alpha_{\ \mu\alpha\nu} n^\mu n^\nu \\
&= -n^\mu \left(\nabla_\mu \nabla_\alpha - \nabla_\alpha \nabla_\mu\right) n^\alpha \\
&= (\nabla_\mu n^\mu)(\nabla_\alpha n^\alpha) - (\nabla_\mu n^\alpha)(\nabla_\alpha n^\mu) \\
&\quad - \nabla_\mu (n^\mu \nabla_\alpha n^\alpha) + \nabla_\mu (n^\alpha \nabla_\alpha n^\mu) \\
&= K^2 - K^{ab}K_{ab} - \nabla_\mu (n^\mu \nabla_\alpha n^\alpha) + \nabla_\mu (n^\alpha \nabla_\alpha n^\mu).
\end{aligned} \tag{14.70}
$$

The last two terms in the above expression are total derivatives and will there-fore only yield boundary terms to the action integral, eq. (14.66). Thus these terms can be omitted from the Lagrangian.

---

[3]It is common in the Lagrangian and Hamiltonian formulation of general relativity to set $2\kappa = 1$. Henceforth we will do the same.

By using eqs. (14.68), (14.69), (14.70) and (14.59), the Lagrangian eq. (14.67) can be written

$$\mathcal{L}_G = N\sqrt{h}\left[\bar{R} + K^{ab}K_{ab} - K^2 - 2\Lambda\right] \qquad (14.71)$$

where the extrinsic curvature is given by

$$K_{ab} = \frac{1}{2N}\left(\dot{h}_{ab} - \bar{\nabla}_a N_b - \bar{\nabla}_b N_a\right). \qquad (14.72)$$

Variation of the Einstein-Hilbert action with respect to the variables (N, N_a, h_{ab}) will now yield Einstein's field equations.

Note that the Lagrangian eq. (14.71) does not contain time derivatives of the variables N or N_a. Hence, variation with respect to these variables immediately yield constants of motion. This fact will be even more apparent in the Hamiltonian formulation which we will introduce now.

14.5   Hamiltonian formulation

In this section we will assume that there are no matter sources in the spacetime. Hence, the total Lagrangian coincide with the Lagrangian for pure gravity $\mathcal{L}_G$.

Assume we have a Lagrangian $L = L(q^A, \dot{q}^A; t)$ where q^A are the generalised coordinates. We define the *canonical momenta*, p_A, by

$$p_A = \frac{\partial L}{\partial \dot{q}^A}. \qquad (14.73)$$

The Hamiltonian is now defined as

$$H = \dot{q}^A p_A - L, \qquad (14.74)$$

where we consider H to be a function of q^A, p_A, and possibly a time-variable t: $H(q^A, p_A; t)$. One can now show that the variational principle implies *Hamilton's equations*[4]:

$$\dot{q}^A = \frac{\partial H}{\partial p_A}, \qquad \dot{p}_A = -\frac{\partial H}{\partial q^A}. \qquad (14.75)$$

We will now apply this to general relativity starting with the Lagrangian eq. (14.71).
Define the canonical momenta

$$\Pi^{ab} \equiv \frac{\partial \mathcal{L}_G}{\partial \dot{h}_{ab}} = \sqrt{h}(K^{ab} - h^{ab}K). \qquad (14.76)$$

The canonical momenta to N and N_a vanish identically:

$$\Pi^N \equiv \frac{\partial \mathcal{L}_G}{\partial \dot{N}} = 0$$

$$\Pi^a \equiv \frac{\partial \mathcal{L}_G}{\partial \dot{N}_a} = 0. \qquad (14.77)$$

---

[4]For the theory of the Hamiltonian formulation in classical mechanics, see, for example, the book by Goldstein [Gol50].

The variables N and N_a have to be interpreted as Lagrange-multipliers, and hence, cannot be considered as real dynamical variables. The only dynamical variables are therefore h_{ab}. As we mentioned in the last section, we can freely choose the functions N and N_a. They correspond to a free choice of time vector, and thus should not be considered as dynamical variables. These variables can be chosen arbitrarily; they reflect a choice of *gauge*.

More specifically, the choice of N says how close two subsequent hypersurfaces Σ_{t_1} and Σ_{t_2} are in time. N represents the free choice of *time rescaling*; it is a generator for time evolution. Similarly, the vector N_a is a generator for *coordinate transformations for the spatial hypersurfaces Σ_t*.

We define the Hamiltonian by

$$H_G = \int_M \mathcal{H}_G d^4x \qquad (14.78)$$

where $\mathcal{H}_G$ is the Hamiltonian density given by

$$\mathcal{H}_G = \dot{h}_{ab}\Pi^{ab} - \mathcal{L}_G. \qquad (14.79)$$

Inserting the expression for the Lagrangian density (14.71) and using eq. (14.76), we get

$$\boxed{\mathcal{H}_G = N\mathfrak{H}_G + N_a\mathfrak{H}_G^a} \qquad (14.80)$$

where

$$\mathfrak{H}_G = \sqrt{h}\left[2\Lambda - \bar{R} + h^{-1}\left(\Pi^{ab}\Pi_{ab} - \frac{1}{2}\Pi^2\right)\right] \qquad (14.81)$$

$$\mathfrak{H}_G^a = -2\sqrt{h}\bar{\nabla}_b\left(h^{-1/2}\Pi^{ba}\right) \qquad (14.82)$$

and $\Pi = \Pi^a_a$. When we vary the Hamiltonian with respect to N and N_a, we get the following interesting result:

$$\boxed{\mathfrak{H}_G = 0, \qquad \mathfrak{H}_G^a = 0.} \qquad (14.83)$$

These equations are called the *Hamiltonian constraint* and the *momentum constraint* respectively. We can recognize $\mathfrak{H} = 0$ and $\mathfrak{H}^a = 0$ as the twice contracted eq. (7.83) and the contracted Codazzi equations for a vacuum spacetime with a cosmological constant.

These two constraints are inevitable in a Hamiltonian formulation and expresses the gauge freedom that we have in the general theory of relativity. It also poses a problem for the ordinary concept of time. Time is quite arbitrary in this formulation, the choice of time is an unphysical gauge freedom. These two constraints therefore manifest a very deep and profound problem: The problem of Time. In a quantum theory of gravity, this is indeed a very serious problem. In ordinary relativistic quantum mechanics, the background spacetime is something fixed. For a quantum theory of gravity, the spacetime is dynamical and the problem of time inevitable pops up and has to be resolved in some way. We will not dwell any further on these deep and difficult questions here; many books have been written on this problem (see for example [Dav74, Dav83, HPMZ94, Pen79, Sav95]).

The rest of the vacuum Einstein field equations can now be derived:

$$\dot{h}_{ab} = \frac{\delta H_G}{\delta \Pi^{ab}}$$

$$= 2h^{-\frac{1}{2}}N\left(\Pi_{ab} - \frac{1}{2}h_{ab}\Pi\right) + 2\bar{\nabla}_{(a}N_{b)} \tag{14.84}$$

$$\dot{\Pi}^{ab} = -\frac{\delta H_G}{\delta h_{ab}}$$

$$= -Nh^{\frac{1}{2}}\left(\bar{R}^{ab} - \frac{1}{2}\bar{R}h^{ab} + h^{ab}\Lambda\right) + \frac{1}{2}Nh^{-\frac{1}{2}}h^{ab}\left(\Pi_{cd}\Pi^{cd} - \frac{1}{2}\Pi^2\right)$$

$$-2Nh^{-\frac{1}{2}}\left(\Pi^{ac}\Pi_c^b - \frac{1}{2}\Pi\Pi^{ab}\right) + h^{\frac{1}{2}}\left(\bar{\nabla}^a\bar{\nabla}^bN - h^{ab}\bar{\nabla}^c\bar{\nabla}_cN\right)$$

$$+h^{\frac{1}{2}}\bar{\nabla}_c\left(h^{-\frac{1}{2}}N^c\Pi^{ab}\right) - 2\Pi^{c(a}\bar{\nabla}_cN^{b)} \tag{14.85}$$

where we have used eq. (14.83) and ignored boundary terms to simplify the equations. The equations (14.83), (14.84) and (14.85) are equivalent to the vacuum Einstein's field equations with a cosmological constant.

14.6   Canonical formulation with matter and energy

If we want to include matter, then we have to include a matter term. The matter action can be written as

$$S_m = \int_M \mathcal{L}_m d^4x. \tag{14.86}$$

From eq. (8.36) the energy momentum tensor is defined via the action by

$$T^{\mu\nu} = -\frac{2}{\sqrt{-g}}\frac{\delta S_m}{\delta g_{\mu\nu}}. \tag{14.87}$$

The total Lagrangian density is now just the sum of the two Lagrangians

$$\mathcal{L}_T = \mathcal{L}_G + \mathcal{L}_m. \tag{14.88}$$

We have already mentioned the electromagnetic case where

$$\mathcal{L}_{EM} = -\frac{1}{4}\sqrt{-g}F^{\mu\nu}F_{\mu\nu}. \tag{14.89}$$

Another important example is the Klein-Gordon Lagrangian

$$\mathcal{L}_{KG} = -\frac{1}{2}\sqrt{-g}\left(\nabla_\mu\phi\nabla^\mu\phi + m^2\phi^2\right). \tag{14.90}$$

All the equations of motion can now be derived similarly as in the vacuum case, except that we in addition get matter degrees of freedom. These matter degrees of freedom are dealt with in the same way as in ordinary Lagrangian and Hamiltonian formulation. For example, if the matter Lagrangian contains only a single matter field ϕ, then the canonically conjugated momentum is

$$\Pi^\phi \equiv \frac{\partial \mathcal{L}_T}{\partial \dot{\phi}}. \tag{14.91}$$

The Hamiltonian density is similarly

$$\mathcal{H}_T = \dot{h}_{ab}\Pi^{ab} + \dot{\phi}\Pi^\phi - \mathcal{L}_T. \tag{14.92}$$

Again, the Hamiltonian density will be a sum

$$\mathcal{H}_T = N\mathfrak{H}_T + N_a\mathfrak{H}_T^a \tag{14.93}$$

where each part is a sum of contributions from pure gravity and the matter

$$\mathfrak{H}_T = \mathfrak{H}_G + \mathfrak{H}_m, \qquad \mathfrak{H}_T^a = \mathfrak{H}_G^a + \mathfrak{H}_m^a. \tag{14.94}$$

One can imagine more complicated theories for which the total Hamiltonian is not purely a direct sum, however, the total Hamiltonian will always be a constraint due to the diffeomorphism invariance of the theory. Hence,

$$\mathfrak{H}_T = 0, \qquad \mathfrak{H}_T^a = 0. \tag{14.95}$$

Example 14.3 (Canonical formulation of the Bianchi type I universe model)   **Example**
Let us consider a simple but nevertheless, illumination example. We studied in chapter 13 the Bianchi type I universe. We will in this example apply the canonical formulation to this model.

In chapter 13 we calculated all the necessary connection coefficients and curvature tensors for this model. The extrinsic curvature is found by using the connection forms. In the calculation we used an orthonormal frame, hence $N = 1$ in that case. We can calculate the invariants and afterwards include a non-trivial N.

The extrinsic curvature is

$$K_{\hat{i}\hat{i}} = \Gamma^{\hat{t}}_{\hat{i}\hat{i}} = \Omega^{\hat{t}}_{\hat{i}}(\mathbf{e}_{\hat{i}}) = \dot{a} + \dot{a}_i \tag{14.96}$$

while the off-diagonal components are zero. Note that from this equation, we can find the volume expansion factor and the shear

$$\begin{aligned} \theta &= K^{\hat{a}}_{\hat{a}} = 3\dot{\alpha} & (14.97)\\ \sigma_{\hat{i}\hat{j}} &= \mathrm{diag}(\dot{\beta}_+ + \sqrt{3}\dot{\beta}_-, \dot{\beta}_+ - \sqrt{3}\dot{\beta}_-, -2\dot{\beta}_+) & (14.98) \end{aligned}$$

where we have used $K_{ab} = \frac{1}{3}\theta h_{ab} + \sigma_{ab}$. Using this we get

$$K_{\hat{a}\hat{b}}K^{\hat{a}\hat{b}} - \left(K^{\hat{a}}_{\hat{a}}\right)^2 = 6(-\dot{\alpha}^2 + \dot{\beta}_+^2 + \dot{\beta}_-^2). \tag{14.99}$$

The three-curvature can be found from eq. (7.152):

$$\bar{R} = 2E_{\hat{i}\hat{i}} + (K^{\hat{a}\hat{b}}K_{\hat{a}\hat{b}} - K^2) = 0. \tag{14.100}$$

This means that the spatial three-hypersurfaces have vanishing Ricci scalar. Actually, one can show that the three-dimensional Riemann tensor vanishes for the Bianchi type I model. The type I model has flat spatial sections; the Bianchi type I generalises the flat FRW model. We find the shear scalar to be

$$\sigma^2 \equiv \frac{1}{2}\sigma_{ab}\sigma^{ab} = 3(\dot{\beta}_+^2 + \dot{\beta}_-^2). \tag{14.101}$$

The type I model reduces to the flat FRW model if and only if $\sigma^2 = 0$.
From eq. (14.71), the Lagrangian for the Bianchi type I model is

$$\mathcal{L}_I = \frac{6e^{3\alpha}}{N}\left(-\dot{\alpha}^2 + \dot{\beta}_+^2 + \dot{\beta}_-^2\right) - 2Ne^{3\alpha}\Lambda. \tag{14.102}$$

We can now easily check that the Euler-Lagrange equations for this Lagrangian reduces to the vacuum Einstein field equations with a cosmological constant.

We go a step further and define the canonical momenta

$$p_\alpha \equiv \frac{\partial \mathcal{L}_I}{\partial \dot{\alpha}} = -\frac{12e^{3\alpha}}{N}\dot{\alpha}$$

$$p_\pm \equiv \frac{\partial \mathcal{L}_I}{\partial \dot{\beta}_\pm} = \frac{12e^{3\alpha}}{N}\dot{\beta}_\pm. \tag{14.103}$$

Using eq. (14.79), the Hamiltonian becomes

$$\mathcal{H}_I = \frac{N}{24}\left[e^{-3\alpha}\left(-p_\alpha^2 + p_+^2 + p_-^2\right) + 12\Lambda e^{3\alpha}\right]. \tag{14.104}$$

Note that since the variables $\beta_\pm$ are cyclic, their conjugated momenta $p_\pm$ are constants of motion. In addition to this, the Hamiltonian must identically vanish

$$\mathcal{H}_I = 0. \tag{14.105}$$

The remaining equations (for α) can be found and integrated without any difficulty. The solutions are of course the same as the solutions in chapter 13.

Note that the Lagrangian for the type I model is the same as for a particle moving in a curved space with metric

$$ds^2 = 12\frac{e^{3\alpha}}{N}(-d\alpha^2 + d\beta_+^2 + d\beta_-^2) \tag{14.106}$$

and with a "time-dependent" potential

$$V(\alpha) = 2N\Lambda e^{3\alpha}. \tag{14.107}$$

The function α acts as a "time"-variable in this space, and the state of the universe can be regarded as a point in this space. The evolution of the universe traces out a world-line in this space. The metric (14.106) is called *DeWitt's supermetric* for the Bianchi type I model. This analogy is often useful because it is often easier to understand the motion of a point particle than the abstract behaviour of the dynamical universe directly.

14.7   The space of three-metrics: Superspace

As we saw in the example in the canonical formulation of the Bianchi type I universe model, we could interpret the evolution of the model as a point particle in a space with a metric given by eq. (14.106).

Such an interpretation can in general be done, and the space in which the point particle moves is called *superspace*. Superspace is the space of all three-dimensional metrics and each point in this space corresponds to a certain spatial metric h_{ab}.

We define *DeWitt's supermetric* as

$$G^{abij} = \frac{1}{4}\sqrt{h}\left(h^{ai}h^{bj} + h^{aj}h^{bi} - 2h^{ab}h^{ij}\right). \tag{14.108}$$

The canonical momenta can now be defined by

$$\Pi^{ab} = -2G^{abij}K_{ij}. \tag{14.109}$$

This definition makes it possible to write the Hamiltonian as

$$\mathfrak{H}_G = \frac{1}{2}G_{abcd}\Pi^{ab}\Pi^{cd} + V(h_{ab}) \tag{14.110}$$

where G_{abcd} is given by

$$G_{abij} = \frac{1}{\sqrt{h}} \left(h_{ai}h_{bj} + h_{aj}h_{bi} - 2h_{ab}h_{ij} \right) \tag{14.111}$$

so that

$$G_{abij} G^{cdij} = \delta^c_{(a} \delta^d_{b)} \tag{14.112}$$

and

$$V(h_{ab}) = \sqrt{h}(2\Lambda - {}^{(3)}R). \tag{14.113}$$

Note that the Hamiltonian has a very simple form. The metric G^{abij} acts as a metric in superspace and $V(h_{ab})$ mimics a potential. The Hamiltonian constraint implies that the total energy is zero, hence

$$\frac{1}{2}G_{abcd}\Pi^{ab}\Pi^{cd} + V(h_{ab}) = 0. \tag{14.114}$$

The universe "point" moves in superspace on zero-level-curves of the Hamiltonian. Including matter fields (for example Klein-Gordon fields), will increase the dimension of superspace; one dimension for each matter degree of freedom.

This analogy between the dynamics of the universe and a point particle dynamics in superspace is very prosperous and useful. The point particle picture is easier to visualise and it is easier to understand the dynamical behaviour of a point particle than the abstract behaviour of the spatial hypersurfaces. In principle, superspace is infinite-dimensional, but in many applications we reduce the system by assuming that the model has a finite number of degrees of freedom. For example, the FRW universe models have only one variable: the scale factor. In this case the vacuum FRW superspace has only one dimension. Other models which has a finite number of degrees of freedom are the homogeneous Bianchi models which we will introduce in the next chapter. We have already investigated the Bianchi type I model, which has 3 degrees of freedom. We call the canonical formulation of such reduced systems by the name *minisuperspace* models.

The Mixmaster Universe

We will here consider one such minisuperspace model. The model we will investigate is the so-called vacuum Bianchi type IX minisuperspace model with $\Lambda = 0$. It was termed the *mixmaster universe* by Misner [Mis69] due to its oscillatory behaviour near the initial singularity.

The metric for this model can be written

$$ds^2 = -N^2 dt^2 + h_{ij}\sigma^i\sigma^j, \tag{14.115}$$

where

$$\begin{aligned}
\sigma^1 &= \cos\psi d\theta + \sin\psi\sin\theta d\phi, \\
\sigma^2 &= -\sin\psi d\theta + \cos\psi\sin\theta d\phi, \\
\sigma^3 &= d\psi + \cos\theta d\phi, \tag{14.116} \\
& 0 \leq \psi < 4\pi, \quad 0 \leq \theta \leq \pi, \quad 0 \leq \phi < 2\pi.
\end{aligned}$$

We can assume that the metric h_{ij} is diagonal and, using the Misner variables, can be written as

$$h_{ij} = e^{-2\Omega}\text{diag}\left(e^{2(\beta_+ + \sqrt{3}\beta_-)}, e^{2(\beta_+ - \sqrt{3}\beta_-)}, e^{-4\beta_+}\right). \tag{14.117}$$

The variables $\beta_\pm$ describe the anisotropy of the spacetime, and, in particular, if $\beta_\pm = 0$, the model reduces to the ordinary closed FRW model. Here, we shall assume that $\beta_\pm \neq 0$ which will, as we will see, result in a very interesting behaviour near the initial singularity as $\Omega \to \infty$.

Using the forms σ^i as basis one-forms, the extrinsic curvature is

$$K_{ij} = \frac{1}{2N}\frac{d}{dt}h_{ij}. \tag{14.118}$$

Hence, we get

$$K^2 - K_{ij}K^{ij} = \frac{6}{N^2}\left(-\dot{\Omega}^2 + \dot{\beta}_+^2 + \dot{\beta}_-^2\right). \tag{14.119}$$

The three-curvature $^{(3)}R$ can be calculated to be

$$
\begin{aligned}
^{(3)}R &= \frac{1}{2}e^{2\Omega}\left[2e^{4\beta_+}\left(1 - \cosh 4\sqrt{3}\beta_-\right) + 4e^{-2\beta_+}\cosh 2\sqrt{3}\beta_- - e^{-8\beta_+}\right] \\
&\equiv -\frac{1}{2}e^{2\Omega}V(\beta_+, \beta_-). \tag{14.120}
\end{aligned}
$$

Finally, we have

$$\sqrt{h} = e^{-3\Omega}. \tag{14.121}$$

Hence, the integrand in the action integral is only dependent on time and thus we can perform the integration over the spatial hypersurfaces. This time-independence reflects the fact that the model we consider is *spatially homogeneous*. By integration, we have

$$\int \sigma^1 \wedge \sigma^2 \wedge \sigma^3 = (4\pi)^2. \tag{14.122}$$

The Lagrangian for the Mixmaster universe is thus

$$\mathcal{L}_{\text{IX}} = \frac{6e^{-3\Omega}}{N}\left(-\dot{\Omega}^2 + \dot{\beta}_+^2 + \dot{\beta}_-^2\right) - \frac{Ne^{-\Omega}}{2}V(\beta_+, \beta_-), \tag{14.123}$$

and the Hamiltonian is

$$\mathcal{H}_{\text{IX}} = \frac{e^{3\Omega}N}{24}\left[-p_\Omega^2 + p_+^2 + p_-^2 + 12e^{-4\Omega}V(\beta_+, \beta_-)\right]. \tag{14.124}$$

Note that this is the Hamiltonian of a particle moving in a curved space with a non-trivial potential. Note that of the potential vanishes, the behaviour is exactly the same as in the Bianchi type I with $\Lambda = 0$ (see Example 14.3). Hence, if $e^{-4\Omega}V \approx 0$, then the behaviour describes Kasner solutions. The function $V(\beta_+, \beta_-)$ has a triangular shape with a minimum at $\beta_\pm = 0$. The function $V(\beta_+, \beta_-)$ is illustrated in Fig. 14.5. The minimum of V is -3; hence,

$$e^{-4\Omega}V(\beta_+, \beta_-) \geq -3e^{-4\Omega}. \tag{14.125}$$

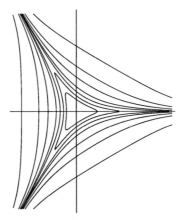

Figure 14.5: The potential for the mixmaster universe. Drawn are equipotential curves for the function $V(\beta_+, \beta_-)$.

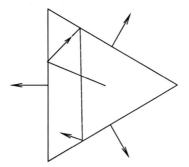

Figure 14.6: The Mixmaster Universe: the Universe point bounces between different Kasner epochs. The Kasner epochs are represented with straight lines with velocity 1. The walls form a triangular-shaped region where the walls recede with velocity $1/2$.

Thus sufficiently close to $\beta_\pm = 0$, the potential will be $e^{-4\Omega}V \approx 0$ as $\Omega \to \infty$. These periods are therefore *Kasner epochs* in the evolution of the universe.

The potential $e^{-4\Omega}V$ has exponentially steep triangularly shaped walls, as can be seen from Fig. 14.5. Consider the special case $\beta_- = 0$. In this case the potential simplifies to

$$e^{-4\Omega}V(\beta_+, 0) = e^{-4\Omega - 8\beta_+} - 4e^{-4\Omega - 2\beta_+}. \qquad (14.126)$$

The case $\beta_+ \to \infty$ represents the narrow channel going out to infinity while $\beta_+ \to -\infty$ represents the wall. From this we can see that the wall recedes with a "minisuperspace velocity" $d\beta_+/d\Omega \approx -1/2$. A "universe particle" travelling with this velocity would experience a constant value of the potential as $\beta_+ \to -\infty$. However, for $e^{-4\Omega}V \approx 0$ the universe will be approximately Kasner-like. The Kasner solutions have $|d\beta_+/d\Omega| = 1$, and hence, if the universe point moves in the negative β_+ direction then it would eventually hit the potential wall and bounce back into a new Kasner epoch. This evolution is schematically illustrated in Fig. 14.6.

The above description is the general behaviour as $\Omega \to \infty$. The universe will go through a succession of Kasner epochs separated by sharp bounces.

Due to the shape of the potential, the universe will generally bounce back and forth inside the triangular shaped area. This oscillatory behaviour of the Bianchi type IX model gave the "Mixmaster universe" its name.

Problems

14.1. *FRW universes with and without singularities*
In this problem we will investigate various FRW models with a perfect fluid. The perfect fluid obeys the barotropic equation of state

$$p = w\rho. \tag{14.127}$$

(a) Write the density ρ as a function of the scale factor a and the parameter w. Assume also that $\rho = \rho_0$ for $a = 1$. Write down the Friedmann equation and solve the equation for $w = -\frac{1}{3}$ for $k = 1$, $k = 0$ and $k = -1$ with the boundary condition $a(1) = 1$.

(b) In one of the above cases, there does not exist a t_0 such that $a(t_0) = 0$. Which case is that? What criterion for a singularity mentioned in section 14.3 does not hold in this case? Does the criteria hold in the other cases?

(c) With the same boundary condition as before, solve the Friedmann equation for $w = -\frac{2}{3}$. In this case the SEC is violated, but are there cases for which there are a singularity? Are there cases for which there are no singularity? Draw a diagram of the typical evolution for the various values of k.

14.2. *A magnetic Bianchi type I model*
In this problem we will consider a Bianchi type I universe (see chapter 13) with a cosmic magnetic field and a vanishing cosmological constant, $\Lambda = 0$. A pure magnetic field has the energy-momentum tensor (see section 8.6)

$$T_{\hat{\mu}\hat{\nu}} = (\rho + p)u_{\hat{\mu}}u_{\hat{\nu}} + pg_{\hat{\mu}\hat{\nu}} + \pi_{\hat{\mu}\hat{\nu}}$$

where

$$\rho = 3p = \frac{1}{2}B^2$$

and $\pi_{\hat{\mu}\hat{\nu}}$ is given by

$$\pi_{ij} = -B_i B_j + \frac{1}{3}B^2\delta_{ij}$$

$$\pi_{0i} = \pi_{i0} = \pi_{00} = 0.$$

Upon a choice of orientation we can assume that the magnetic field is aligned with the z-axis. The anisotropic stress tensor will in that case be diagonal

$$\pi_{\hat{\mu}\hat{\nu}} = B^2\mathrm{diag}\left(0, -\frac{1}{3}, -\frac{1}{3}, \frac{2}{3}\right).$$

We will also consider the case where the shear tensor is diagonal; thus we can write

$$\sigma_{\hat{\mu}\hat{\nu}} = \mathrm{diag}(0, \sigma_+ + \sqrt{3}\sigma_-, \sigma_+ - \sqrt{3}\sigma_-, -2\sigma_+).$$

We will further assume that the time-like vector field u^μ is non-rotating and can be chosen so that $u^\mu\nabla_\mu = \frac{\partial}{\partial t}$. Furthermore, the Bianchi type I model has flat spatial three-geometry: $^{(3)}R_{\mu\nu} = 0$.

(a) Use an orthonormal frame, and show that the equations of motion in section 14.2 reduces to the following set of equations ($\kappa = 1$)

$$
\begin{aligned}
\dot{B} &= -\frac{2}{3}\theta B - 2\sigma_+ B \\
\dot{\theta} &= -\frac{1}{3}\theta^2 - 6(\sigma_+^2 + \sigma_-^2) - \frac{1}{2}B^2 \\
\dot{\sigma}_+ &= -\theta\sigma_+ + \frac{1}{3}B^2 \\
\dot{\sigma}_- &= -\theta\sigma_-
\end{aligned}
\tag{14.128}
$$

with the constraint

$$
\frac{1}{3}\theta^2 = 3(\sigma_+^2 + \sigma_-^2) + \frac{1}{2}B^2.
\tag{14.129}
$$

Which of these correspond to Maxwell's equations for the magnetic field?

(b) It is convenient to introduce a new set of variables by

$$
\Sigma_\pm = \frac{3\sigma_\pm}{\theta}, \quad \mathcal{H} = \frac{B}{\theta},
\tag{14.130}
$$

and a new time variable τ by requiring

$$
\frac{dt}{d\tau} = \frac{3}{\theta}.
\tag{14.131}
$$

Show that in these variables, the equations of motion can be written

$$
\begin{aligned}
\theta' &= -(1+q)\theta \\
\mathcal{H}' &= (q - 1 - 2\Sigma_+)\mathcal{H} \\
\Sigma_+' &= (q - 2)\Sigma_+ + 3\mathcal{H}^2 \\
\Sigma_-' &= (q - 2)\Sigma_-
\end{aligned}
\tag{14.132}
$$

where

$$
\begin{aligned}
q &= 1 + \Sigma_+^2 + \Sigma_-^2 \\
1 &= \Sigma_+^2 + \Sigma_-^2 + \frac{3}{2}\mathcal{H}^2,
\end{aligned}
\tag{14.133}
$$

and prime denotes derivative with respect to τ. Note that one of the variables can in principle be obtained from the constraint equation. Hence, the equation of motion for this variable is redundant.

(c) Show that the solutions corresponding to $\Sigma_+^2 + \Sigma_-^2 = 1$, $\mathcal{H} = 0$, are the Kasner solutions, eq. (13.30).

(d) The set of equations can, as a matter of fact, be solved exactly in full generality (the solutions are called the *Rosen solutions*). However, we will here consider the axisymmetric case where $\Sigma_- = 0$. Show that the equation for Σ_+ in this case can be written

$$
\Sigma_+' = (1 - \Sigma_+^2)(2 - \Sigma_+).
\tag{14.134}
$$

Solve this equation. Find also $\mathcal{H}$ and θ. What are the late-time and early-time asymptotes?

14.3. *FRW universe with a scalar field*

Let us consider a FRW universe with a Klein-Gordon scalar field. The Klein-Gordon field has a Lagrangian given by eq. (14.90). We will assume that the model is isotropic and homogeneous and that the scalar field only depends on the time variable t.

(a) Write down the total Lagrangian (pure gravity + matter fields) for the FRW minisuperspace model with a scalar field.

(b) What is the supermetric for this model? Is it Lorentzian or Riemannian? Find the total Hamiltonian.

(c) Derive the equations of motion from the Hamiltonian equations.

14.4. *The Kantowski-Sachs universe model*

In this problem we shall derive the equations of motion for an anisotropic model called the *Kantowski-Sachs* universe model. The Kantowski-Sachs universe model has the line-element

$$ds^2 = -dt^2 + a(t)^2 dz^2 + b(t)^2 (d\theta^2 + \sin^2\theta d\phi^2). \tag{14.135}$$

Assume also that the universe is empty apart from the presence of a cosmological constant, Λ.

(a) Derive the equations of motion using the generalized Friedmann equation, the shear propagation equations and Raychaudhuri's equation given in section 14.2.

(b) Introduce a non-zero lapse function, $N(t)$, and find the Lagrangian and Hamiltonian for the Kantowski-Sachs model. Find the equations of motion using the Hamiltonian equations. Set $N = 1$ and compare with what you found in (a).

(You can check your answers by comparing your results with Example 15.3 on page 422 in the following chapter.)

15

Spatially Homogeneous Universe Models

In this section we will explore the concept of symmetries even further. We introduced some of the basics in chapter 6, and we will pursue the ideas further here. In doing so, we will generalise the FRW models to the Bianchi models which are in general spatially homogeneous but not necessarily isotropic.

15.1 Lie groups and Lie algebras

First we will introduce some very important concepts used in mathematics and physics. Whenever we talk about continuous symmetries, the words Lie groups and Lie algebra are usually mentioned.

We saw earlier that the Killing vectors generate a special class of diffeomorphisms; Killing vectors generate *isometries*. The isometries of a space form a group. For instance, let us take the sphere, S^2, with the usual round metric. The isometry group of the sphere is all the rotations in three-dimensional space that leaves the sphere invariant. These (orientation-preserving[1]) rotations form the group $SO(3)$. What is so special about this group, is that the group itself, can be considered as a manifold! Since the dimension of the group is three, the group $SO(3)$ can be considered as a three-dimensional manifold. The group $SO(3)$ is an example of a *Lie group*. We define Lie groups as follows.

Definition: Lie Group. *A Lie group, G, is a topological space that has the following properties:*

1. *G is a manifold.*

2. *The group multiplication $m: G \times G \longmapsto G$ is smooth.*

---

[1] We will always assume that we are talking about orientation-preserving isometries, unless stated otherwise.

3. *Inversion* $i : G \longmapsto G$ *is smooth.*

To show that $SO(3)$ has these properties is not difficult. We already know that multiplication and inversion are continuous operations. Each element in $SO(3)$ corresponds to a rotation, and rotations are continuous operations. We can show that $SO(3)$ is actually equal to the manifold $\mathbb{P}^3$ and hence, $SO(3)$ is a manifold.

Let us also define what we mean by *Lie algebra*.

Definition: Lie Algebra.   *A real (or complex) Lie algebra,* $\mathfrak{g}$, *is a (finite dimensional) vector space equipped with a bilinear map* $[-, -] : \mathfrak{g} \times \mathfrak{g} \longmapsto \mathfrak{g}$ *which satisfies the following properties:*

1. $[\mathbf{X}, \mathbf{X}] = 0$ *for all* $\mathbf{X} \in \mathfrak{g}$.

2. *The Jacobi identity:*

$$[\mathbf{X}, [\mathbf{Y}, \mathbf{Z}]] + [\mathbf{Y}, [\mathbf{Z}, \mathbf{X}]] + [\mathbf{Z}, [\mathbf{X}, \mathbf{Y}]] = 0, \tag{15.1}$$

for all $\mathbf{X}, \mathbf{Y}, \mathbf{Z} \in \mathfrak{g}$.

Note that 1. implies that the bilinear map $[-, -]$ is skew-symmetric:

$$[\mathbf{X}, \mathbf{Y}] = -[\mathbf{Y}, \mathbf{X}]. \tag{15.2}$$

An example of a Lie algebra is the space of all $n \times n$ matrices $\mathfrak{gl}(n)$. The bracket $[-, -]$ is in this case simply defined by

$$[A, B] = AB - BA, \tag{15.3}$$

for all matrices A and B. The bracket is in this case the usual commutator multiplication of matrices.

There is actually a deep connection between these two concepts. A Lie algebra is a vector space, while a Lie group is a group manifold. Amazingly we have the following theorem.

Theorem: *Let G be a Lie group. Then the tangent space of G at the identity element,* $T_e G$, *is a Lie algebra; i.e.,*

$$\mathfrak{g} = T_e G.$$

This gives a very interesting connection between Lie algebras and Lie groups. We can, by calculating the tangent space of a Lie group, find a corresponding Lie algebra.

Let us take the example $SO(3)$ again. $SO(3)$ can be considered as the 3×3 matrices obeying $R^t R = 1$ and $\det(R) = 1$ where 1 is the identity element. We consider a curve $R(t)$ in $SO(3)$ going through the identity element. For sake of simplicity, we assume $R(0) = 1$. We denote the tangent vector of this curve at the identity element as A, i.e. $R'(0) = A$. From

$$R^t R = 1,$$

we get by differentiating

$$R'^t R + R^t R' = 0.$$

Hence, at $t = 0$ we get

$$A^t = -A. \tag{15.4}$$

The Lie algebra of $SO(3)$, which is usually written $\mathfrak{so}(3)$, consists of all skew-symmetric matrices.

We have seen how these Lie algebras are the tangent vector space over the unit element of the Lie group. Hence, each element of the Lie algebra can be considered as a vector at the unit element of a manifold. If $\mathbf{X}$ is a vector in the Lie algebra, then we can define the local flow ϕ_t of the vector $\mathbf{X}$ as in section 6.9. The flow will now be a flow on the Lie group G, so each element $\phi_t \in G$. The vector $\mathbf{X}$ is only defined over the unit element, so we have to parallel transport the $\mathbf{X}$ to the point ϕ_t by the group action: $\phi_t \cdot \mathbf{X}$. We usually write the differential equation that defines the flow ϕ_t as

$$(\phi_t)^{-1} \frac{\partial \phi_t}{\partial t} = \mathbf{X},$$
$$\phi_0(e) = e, \tag{15.5}$$

where e is the unit element. This differential equation has, as we saw in section 6.9, the exponential map as the solution. Hence,

$$\phi_t(e) = \exp(t\mathbf{X}). \tag{15.6}$$

We can therefore define the exponential map $\exp : \mathfrak{g} \longmapsto G$ by the action on a Lie algebra element as follows

$$\exp(\mathbf{X}) = \phi_1(e) \in G. \tag{15.7}$$

Thus there is an intimate relation between the Lie algebra, the Lie group and the exponential map. By exponentiation, we can get from the Lie algebra to its Lie group. The inverse function of $\exp$, called $\log$ can also be defined in a neighbourhood of the identity element. For a neighbourhood $U \subset G$ of e, we can define

$$\log : U \longmapsto \mathfrak{g},$$
$$\log \equiv \exp^{-1}\big|_U, \tag{15.8}$$

where $\exp^{-1}$ means the inverse function of $\exp$.

We choose a basis $\{\mathbf{X}_i\}$ for the Lie algebra $\mathfrak{g}$ and define the *structure constants* C_{ij}^k by

$$[\mathbf{X}_i, \mathbf{X}_j] = C_{ij}^k \mathbf{X}_k. \tag{15.9}$$

Note that the structure constants are antisymmetric in the lower indices, $C_{ij}^k = -C_{ji}^k$, and that by a change of basis, we can change the structure constants, without changing the Lie algebra.

Example 15.1 (The Lie Algebra $\mathfrak{so}(3)$)   **Example**
We have shown that the Lie algebra $\mathfrak{so}(3)$ consists of skew-symmetric matrices:

$$\mathfrak{so}(3) = \{A | A^t = -A, A\ 3 \times 3\ \text{matrix.}\}. \tag{15.10}$$

Let us choose the following basis for $\mathfrak{so}(3)$:

$$\mathbf{X}_1 = \begin{bmatrix} 0 & 0 & 0 \\ 0 & 0 & -1 \\ 0 & 1 & 0 \end{bmatrix}, \quad \mathbf{X}_2 = \begin{bmatrix} 0 & 0 & 1 \\ 0 & 0 & 0 \\ -1 & 0 & 0 \end{bmatrix}, \quad \mathbf{X}_3 = \begin{bmatrix} 0 & -1 & 0 \\ 1 & 0 & 0 \\ 0 & 0 & 0 \end{bmatrix}. \tag{15.11}$$

By calculating the commutators, we find

$$[\mathbf{X}_1, \mathbf{X}_2] = \mathbf{X}_3, \quad [\mathbf{X}_2, \mathbf{X}_3] = \mathbf{X}_1, \quad [\mathbf{X}_3, \mathbf{X}_1] = \mathbf{X}_2. \tag{15.12}$$

We note that this can be written

$$[\mathbf{X}_i, \mathbf{X}_j] = \epsilon_{ijk} \mathbf{X}_k. \tag{15.13}$$

The structure constants are therefore

$$C_{ij}^k = \epsilon_{ijk}. \tag{15.14}$$

15.2   Homogeneous spaces

We are now ready to introduce the concept of homogeneous spaces. Roughly speaking, *a homogeneous space is a space where you can get from one point to any other point using an isometry* (see Fig. 15.1).

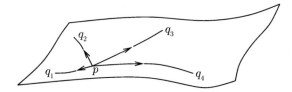

Figure 15.1: In a homogeneous space you can get anywhere on the manifold using an isometry.

Consider a space M with metric g. We define the *isometry group* Isom(M) by

$$\mathrm{Isom}(M) \equiv \{\phi : M \longmapsto M | \phi \text{ is an isometry}\}. \tag{15.15}$$

Recall that ϕ is an isometry if $\phi^*\mathbf{g} = \mathbf{g}$. The isometry group will in general be a Lie group. Since a Killing vector field generates an isometry, a Killing vector field corresponds to an element of the Lie algebra of Isom(M). The Killing vector fields forms a finite dimensional vector space, isomorphic to the Lie algebra of Isom(M).

We can now go on and define the *isotropy subgroup* of a point $p \in M$ by

$$\mathcal{I}_p(M) = \{\phi \in \mathrm{Isom}(M) | \phi(p) = p\}. \tag{15.16}$$

Hence, the isotropy subgroup is the subgroup of the isometry group that leaves the point p fixed. Sometimes the word stabilizer is used for the isotropy subgroup.

The definition of a homogeneous space now goes as follows.

Definition (Homogeneous space):   *If for each pair of points $p, q \in M$ there exists a $\phi \in \mathrm{Isom}(M)$ so that $\phi(p) = q$, then we say that M is a* homogeneous space.

Sometimes we use the word *transitive* for a homogeneous space.

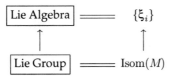

Figure 15.2: The relation between the Isometry group and the set of Killing vectors, and the concepts of Lie groups and Lie algebras.

Let the dimension of $\mathrm{Isom}(M)$ be n and $\mathcal{I}_p(M)$ be m. A necessary condition for the space to be homogeneous is that $n \geq \dim(M)$. We call M *simply transitive* if M is homogeneous and $n = \dim(M)$, and *multiply transitive* if M is homogeneous and $n > \dim(M)$. This implies that for a simply transitive space, $m = 0$, while for a multiply transitive space , $m > 0$. For example, the maximally symmetric spaces are multiply transitive.

Consider the subspace of M given by

$$H_p = \{q \in M | q = \phi(p) \text{ for a } \phi \in \mathrm{Isom}(M)\}, \qquad (15.17)$$

for a point p in M. The subspace H_p is called the *orbit* of p under the isometry group. Hence, all the points we can reach by the action of an isometry on p, is in the orbit of p. If the orbit of p is the whole space M, i.e. $H_p = M$, then the space is transitive (and hence homogeneous).

We have seen how the Lie algebra, Lie groups and symmetries are linked together, see Fig. 15.2. Now we will go a step further and show how we can choose a Lie algebra and then go on and define a space having this symmetry, and that are by construction, simply transitive.

In a simply transitive space there exists a set of Killing vector fields that obey

$$[\boldsymbol{\xi}_i, \boldsymbol{\xi}_j] = D_{ij}^k \boldsymbol{\xi}_k, \qquad (15.18)$$

which can be taken to be the basis vectors. However, it is more convenient to define a basis set as follows. At a point p we choose a basis $\mathbf{e}_i$ and define a left invariant frame by Lie transporting this basis around the space. Hence, we require that

$$\mathcal{L}_{\boldsymbol{\xi}_j} \mathbf{e}_i = [\boldsymbol{\xi}_j, \mathbf{e}_i] = 0. \qquad (15.19)$$

Due to the relation

$$\mathcal{L}_{\boldsymbol{\xi}_j} [\mathbf{e}_i, \mathbf{e}_k] = \left[\mathcal{L}_{\boldsymbol{\xi}_j} \mathbf{e}_i, \mathbf{e}_k \right] + \left[\mathbf{e}_i, \mathcal{L}_{\boldsymbol{\xi}_j} \mathbf{e}_k \right] = 0, \qquad (15.20)$$

the frame $\mathbf{e}_i$ itself spans a Lie algebra. Hence, for some constants C_{ij}^k, we have

$$[\mathbf{e}_i, \mathbf{e}_j] = C_{ij}^k \mathbf{e}_k. \qquad (15.21)$$

Note that for a homogeneous space these structure constants are real constants on each orbit. This is not necessarily true for the "structure constants" we defined for the commutator between basis vectors in an arbitrary basis.

The left invariant frame is an invariant frame under the action of $\boldsymbol{\xi}_i$. The opposite is also true; $\boldsymbol{\xi}_i$ is an invariant basis under the action of $\mathbf{e}_j$ because

$$\mathcal{L}_{\mathbf{e}_j} \boldsymbol{\xi}_i = -\mathcal{L}_{\boldsymbol{\xi}_i} \mathbf{e}_j = 0. \qquad (15.22)$$

Since the isometry group is the only group of this space, we might wonder if these two Lie algebras are two different representations of the Lie algebra of the isometry group. The answer is yes, and this can be seen as follows. Let us for the sake of simplicity, assume that the vector fields $\boldsymbol{\xi}_i$ and $\mathbf{e}_i$ coincide at a point p. Such a choice can always be done, since they are linearly independent by assumption and they both span the tangent space at every point. Thus there will exist an invertible matrix α_j^i such that

$$\mathbf{e}_j = \alpha_j^i \boldsymbol{\xi}_i, \quad \alpha_j^i\big|_p = \delta_j^i \tag{15.23}$$

In general the matrix is dependent on the position. The structure constants are, on the contrary, not dependent on the position. In general we have

$$\begin{aligned}\pounds_{\boldsymbol{\xi}_j}\mathbf{e}_i &= [\boldsymbol{\xi}_j, \mathbf{e}_i] = \alpha_i^k[\boldsymbol{\xi}_j, \boldsymbol{\xi}_k] + \boldsymbol{\xi}_j(\alpha_i^k)\boldsymbol{\xi}_k \\ &= \left(\alpha_i^l D_{jl}^k + \boldsymbol{\xi}_j(\alpha_i^k)\right)\boldsymbol{\xi}_k = 0,\end{aligned} \tag{15.24}$$

using eqs. (15.18) and (15.22). Hence,

$$D_{ij}^k = -\beta_j^l\boldsymbol{\xi}_i(\alpha_l^k), \tag{15.25}$$

where $\beta_j^l = (\alpha^{-1})_j^l$. Similarly,

$$C_{ij}^k = -\alpha_j^l\mathbf{e}_i(\beta_l^k). \tag{15.26}$$

Since the structure constants are not dependent on the position, we can evaluate these at the point p. At p we have $\beta_j^l = \alpha_j^l = \delta_j^l$ and $\boldsymbol{\xi}_i = \mathbf{e}_i$. The derivative of β_l^k can be written in terms of the derivative of α_l^k:

$$\boldsymbol{\xi}_i(\beta_l^k) = -\beta_n^k\beta_l^m\boldsymbol{\xi}_i(\alpha_m^n). \tag{15.27}$$

Thus, at p this is simply

$$\boldsymbol{\xi}_i(\beta_l^k) = -\boldsymbol{\xi}_i(\alpha_l^k). \tag{15.28}$$

Hence, we can write eq. (15.26) at p as

$$C_{ij}^k = -\alpha_j^l\mathbf{e}_i(\beta_l^k) = \boldsymbol{\xi}_i(\alpha_j^k) = -D_{ij}^k. \tag{15.29}$$

Here we see that these structure constants are just different representations of the same Lie algebra. If we choose frames where $\boldsymbol{\xi}_i$ and $\mathbf{e}_i$ coincide at one point, then the structure constants will differ only by a sign.

We say that the frame $\mathbf{e}_j$ defines a *left invariant frame*, while the frame $\boldsymbol{\xi}_i$ defines a *right invariant frame*.

We can therefore construct a homogeneous space as follows. Take the structure constants of a Lie algebra, C_{ij}^k, and define a left invariant frame as

$$[\mathbf{e}_i, \mathbf{e}_j] = C_{ij}^k\mathbf{e}_k. \tag{15.30}$$

If $\boldsymbol{\omega}^k$ is the dual basis to $\mathbf{e}_k$, then according to eq. (6.177)

$$d\boldsymbol{\omega}^k = -\frac{1}{2}C_{ij}^k\boldsymbol{\omega}^i \wedge \boldsymbol{\omega}^j. \tag{15.31}$$

These basis one-forms will also be left invariant:

$$\pounds_{\boldsymbol{\xi}_i}\boldsymbol{\omega}^k = 0, \tag{15.32}$$

as can easily be checked. Using these invariant forms we can equip the space with a left invariant metric given by

$$ds^2 = g_{ij}\omega^i \otimes \omega^j, \tag{15.33}$$

where the metric coefficients g_{ij} are constants. This is a *homogeneous metric* on the space. By construction, the Killing vectors of the metric are ξ_i, and the basis e_j is a left invariant frame.

Example 15.2 (The Poincaré half-plane)   Example
Let us take a two-dimensional example and consider the two-dimensional Lie algebra

$$[\mathbf{X}_1, \mathbf{X}_2] = \mathbf{X}_1, \tag{15.34}$$

where all other commutators are zero. It is arbitrary whether we define the Killing vectors to be the representatives of this Lie algebra or the corresponding left invariant basis vectors. Let us choose the Killing vectors. By inspection we note that

$$[\xi_1, \xi_2] = \xi_1, \tag{15.35}$$

where

$$\begin{aligned}
\xi_1 &= \frac{\partial}{\partial x}, \\
\xi_2 &= x\frac{\partial}{\partial x} + y\frac{\partial}{\partial y}.
\end{aligned} \tag{15.36}$$

We define the left invariant vector fields by

$$[\xi_i, e_j] = 0. \tag{15.37}$$

By solving a set of differential equations we can find the general form of the left invariant fields. One of the solutions is

$$\begin{aligned}
e_1 &= y\frac{\partial}{\partial x}, \\
e_2 &= y\frac{\partial}{\partial y}.
\end{aligned} \tag{15.38}$$

This frame coincides with the frame of Killing vectors at $(x,y) = (0,1)$. Hence,

$$[e_1, e_2] = -e_1, \tag{15.39}$$

which can be shown by direct calculation. The invariant one-forms are the dual to the invariant frame and are given by

$$\begin{aligned}
\omega^1 &= \frac{dx}{y}, \\
\omega^2 &= \frac{dy}{y}.
\end{aligned} \tag{15.40}$$

Thus, an invariant metric can be obtained by

$$ds^2 = \left(\omega^1\right)^2 + \left(\omega^2\right)^2 = \frac{dx^2 + dy^2}{y^2}. \tag{15.41}$$

This is the so-called Poincaré half-plane which we have encountered before in problem 6.3. By construction it has the symmetry group compatible with the Lie algebra given by eq. (15.34).

15.3 The Bianchi models

We have seen how we can construct a homogeneous space, given a Lie algebra In cosmology we are mainly interested in three-dimensional spatial sections. The Bianchi models are cosmological models that have spatially homogeneous sections, invariant under the action of a three-dimensional Lie group.

We assume that the four-dimensional space can be foliated with three-dimensional spatial sections

$$M = \mathbb{R} \times \Sigma_t. \qquad (15.42)$$

The $\mathbb{R}$ is the time variable, and each Σ_t is labelled with a time variable. By construction, each Σ_t is a homogeneous space of dimension three.

For Σ_t homogeneous, we have three different possibilities

1. $\dim \mathrm{Isom}(M) = 6$: Σ_t is a multiply transitive space of maximal symmetry. These are the FRW models.

2. $\dim \mathrm{Isom}(M) = 4$: Σ_t is a multiply transitive space with an isotropy subgroup $\mathcal{I}_p(M) = SO(2)$.

3. $\dim \mathrm{Isom}(M) = 3$: Σ_t is a simply transitive space.

It turns out that except in one case, all of the spaces in category 1 and 2 have a subgroup $H \subset \mathrm{Isom}(M)$ such that H acts simply transitive on Σ_t. The exception is if Σ_t has the covering space[2] $\mathbb{R} \times S^2$. Apart from this single case, we can consider a space in categories 1 and 2 as a special symmetric case of category 3. We will therefore first consider the category 3 case: We will assume that Σ_t is a simply transitive space.

We can therefore wonder: What possibilities do we have for Σ_t under these conditions? The answer to this question depends upon how many different Lie algebras we have in three dimensions. A classification of the three-dimensional Lie algebras is therefore necessary.

The classification of the three-dimensional Lie algebra is called the Bianchi classification, and each Lie algebra is labelled by a number I-IX. By using one of these Lie algebras, we can construct a spatially homogeneous cosmological model. The corresponding cosmological model is called *a Bianchi model*. If, for example, a Bianchi model has the symmetry from the type III algebra, we say that it is a *Bianchi type III model*.

The Bianchi models are listed in terms of their structure constants in Table 15.1. In column 2 and 3 the Bianchi types are written in terms of the so-called Behr decomposition in which the structure constants are decomposed in terms of the trace-free part and trace part

$$C_{ij}^k = \epsilon_{ijl} n^{lk} + a_l \left(\delta^l{}_i \delta^k_j - \delta^l{}_j \delta^k_i \right), \qquad (15.43)$$

where a_i is the "vector" part of the Lie algebra. The trace of C_{ij}^k is

$$C_{ij}^j = 2a_i. \qquad (15.44)$$

We can always choose a basis such that $a_i = a\delta^3_i$. This vector is written in the second column in Table 15.1. We usually call the models with $C_{ij}^j = 0$ for *class A models*. The ones with $C_{ij}^j \neq 0$ are called *class B models*.

There are a couple of things we can note.

---

[2]If M is a covering space of H, then $H = M/\Gamma$ where Γ is a discrete group. For more details of how this quotient is defined, see the later section 15.6.

Bianchi Type	a_i	n	Structure constants
I	0	0	$C^i_{jk} = 0$
II	0	$\mathrm{diag}(1,0,0)$	$C^1_{23} = -C^1_{32} = 1,$ rest of $C^i_{jk} = 0$
III	$\frac{1}{2}\delta^3_i$	$-\frac{1}{2}\mathsf{A}$	$C^1_{13} = -C^1_{31} = 1,$ rest of $C^i_{jk} = 0$
IV	δ^3_i	$\mathrm{diag}(1,0,0)$	$C^1_{13} = -C^1_{31} = 1,$ $C^1_{23} = -C^1_{32} = 1,$ $C^2_{23} = -C^2_{32} = 1,$ rest of $C^i_{jk} = 0$
V	δ^3_i	0	$C^1_{13} = -C^1_{31} = 1,$ $C^2_{23} = -C^2_{32} = 1,$ rest of $C^i_{jk} = 0$
VI$_h$	$\frac{\tilde{h}}{2}\delta^3_i$	$\frac{1}{2}(\tilde{h}-2)\mathsf{A}$	$C^1_{13} = -C^1_{31} = 1,$ $C^2_{23} = -C^2_{32} = (\tilde{h}-1),$ rest of $C^i_{jk} = 0$
VII$_h$	$\frac{\tilde{h}}{2}\delta^3_i$	$\mathrm{diag}(-1,-1,0) + \frac{\tilde{h}}{2}\mathsf{A}$	$C^2_{13} = -C^2_{31} = 1,$ $C^1_{23} = -C^1_{32} = -1,$ $C^2_{23} = -C^2_{32} = \tilde{h},$ rest of $C^i_{jk} = 0$
VIII	0	$\mathrm{diag}(-1,1,1)$	$C^1_{23} = -C^1_{32} = -1,$ $C^2_{31} = -C^2_{13} = 1,$ $C^3_{12} = -C^3_{21} = 1,$ rest of $C^i_{jk} = 0$
IX	0	1	$C^i_{jk} = \varepsilon_{ijk}$

where $\mathsf{A} = \begin{bmatrix} 0 & 1 & 0 \\ 1 & 0 & 0 \\ 0 & 0 & 0 \end{bmatrix}$ and 1 denotes the identity matrix. The parameter $\tilde{h}$ is

related to the group parameter h as follows.

$$\text{VI}_h: \quad h = -\frac{\tilde{h}^2}{(\tilde{h}-2)^2}, \quad \text{VII}_h: \quad h = \frac{\tilde{h}^2}{4-\tilde{h}^2}$$

Table 15.1: The classification scheme of the 3-dimensional Lie algebras.

- Bianchi type I corresponds to flat spatial sections. Thus, it generalizes the flat FRW model.

- Bianchi type IX corresponds to the Lie algebra $\mathfrak{so}(3)$.

- The class A models are: I, II, VI_0, VII_0 and IX.

- We have $\text{VI}_{-1}=\text{III}$.

The Bianchi models are therefore constructed as follows. For the specific Bianchi type, we choose an invariant basis $\{\boldsymbol{\omega}^i\}$ that satisfies

$$\mathbf{d}\boldsymbol{\omega}^k = -\frac{1}{2}C_{ij}^k\boldsymbol{\omega}^i \wedge \boldsymbol{\omega}^j. \tag{15.45}$$

The Bianchi model of the corresponding type can now be written

$$ds^2 = -dt^2 + g_{ij}(t)\boldsymbol{\omega}^i \otimes \boldsymbol{\omega}^j. \tag{15.46}$$

This metric will in general have the symmetries of the corresponding Bianchi group. This *metric approach* is useful when we would like to introduce the Lagrangian and Hamiltonian formulations for the Bianchi models. This has to be done with care though, because it turns out that the canonical formulation only works well for the class A models. This fact is intimately related to the fact that the the class A models are "trace-free".

The Kantowski-Sachs model

The Kantowski-Sachs model is the only spatially homogeneous model that has not a three-dimensional transitive subgroup. It has spatial sections $\mathbb{R} \times S^2$ with a four dimensional symmetry group. Its metric can be written as

$$ds^2 = -dt^2 + a(t)^2dz^2 + b(t)^2(d\theta^2 + \sin^2\theta d\phi^2). \tag{15.47}$$

The functions $a(t)$ and $b(t)$ are functions to be determined by the Einstein field equations.

Example

Example 15.3 (A Kantowski-Sachs universe model)
We will now solve Einstein's field equation for one special case for a Kantowski-Sachs universe. Using the metric (15.47), we can write the vacuum equations as

$$2\frac{\dot{a}\dot{b}}{ab} + \frac{\dot{b}^2}{b^2} + \frac{1}{b^2} = \Lambda$$

$$2\frac{\ddot{b}}{ab} + \frac{\dot{b}^2}{b^2} + \frac{1}{b^2} = \Lambda$$

$$\frac{\ddot{a}}{a} + \frac{\ddot{b}}{b} + \frac{\dot{a}\dot{b}}{ab} = \Lambda. \tag{15.48}$$

Note that there is a special solution where $b(t) = b_0 = \text{constant}$ and

$$\frac{1}{b_0^2} = \Lambda, \qquad \frac{\ddot{a}}{a} = \Lambda. \tag{15.49}$$

This equation can be solved to yield

$$a(t) = e^{\sqrt{\Lambda}t}. \tag{15.50}$$

Thus the metric for this solution is

$$ds^2 = -dt^2 + e^{2\sqrt{\tilde{\Lambda}}t}dz^2 + \frac{1}{\Lambda}(d\theta^2 + \sin^2\theta d\phi^2). \tag{15.51}$$

This metric describes a universe with two spherical dimensions having a fixed size during the cosmic evolution. The third dimension, on the other hand, is expanding exponentially. A closer analysis of this solution shows that this solution is unstable, hence it is unphysical and must be considered as a mathematical artifact.

15.4   The orthonormal frame approach to the non-tilted Bianchi models

A very useful and powerful way to study the dynamical behaviour of the Bianchi models is by using orthonormal frames. This approach was first applied to the Bianchi models in a pioneering work by Ellis and MacCallum [EM69].

We will assume that the energy-momentum tensor has the form

$$T_{\mu\nu} = \rho u_\mu u_\nu + p h_{\mu\nu} + \pi_{\mu\nu} \tag{15.52}$$

where u_μ is the four-velocity of the fluid flow. We will also assume that the four-velocity is orthogonal to the hypersurfaces Σ_t spanned by the action of the isometry group. If this is the case for a model, then we say that the fluid is *non-tilted*. If the fluid four-velocity is not orthogonal to the hypersurfaces Σ_t, then the fluid is *tilted*.

The above assumption implies that the vorticity tensor and the four-acceleration of the fluid are zero:

$$\omega_{\mu\nu} = 0, \quad u_{\mu;\nu}u^\nu = 0. \tag{15.53}$$

This allows us to use the equations of motion derived in chapter 14. We split the expansion tensor into trace and trace-free parts

$$\theta_{\mu\nu} = u_{\mu;\nu} = \frac{1}{3}\theta h_{\mu\nu} + \sigma_{\mu\nu}. \tag{15.54}$$

The commutator functions[3] $c^\alpha_{\mu\nu}$ are given by

$$[\mathbf{e}_\mu, \mathbf{e}_\nu] = c^\alpha_{\mu\nu}\mathbf{e}_\alpha. \tag{15.55}$$

These functions are related to the connection coefficients via eq. (6.137) on page 127. In an orthonormal frame, the rotation forms possess the anti-symmetry $\Omega_{\mu\nu} = -\Omega_{\nu\mu}$ which makes it possible to write the connection coefficients in terms of the structure coefficients

$$\Gamma_{\alpha\mu\nu} = \frac{1}{2}(g_{\alpha\beta}c^\beta_{\nu\mu} + g_{\mu\beta}c^\beta_{\alpha\nu} - g_{\nu\beta}c^\beta_{\mu\alpha}). \tag{15.56}$$

We note that, since the vector u_μ is orthogonal to the hypersurfaces Σ_t, we have $\theta_{\mu\nu} = \Gamma^t{}_{\mu\nu}$ and hence

$$c^t{}_{ta} = c^t{}_{ab} = 0. \tag{15.57}$$

---

[3]We will use the notation where lowercase c's in the structure coefficients mean they are general functions while uppercase mean they are real constants.

For the structure coefficients $c^a{}_{tb}$, we can write

$$c^a{}_{tb} = -(\Gamma^a{}_{tb} - \Gamma^a{}_{bt}) = -\Gamma^t{}_{ab} + \Gamma^a{}_{bt}. \tag{15.58}$$

The first part of the right-hand side is symmetric in a, b and is equal to the expansion tensor: $\Gamma^t{}_{ab} = \theta_{ab}$. The antisymmetry of the rotation forms implies that

$$\Gamma_{abt} = -\Gamma_{bat} \equiv \epsilon_{abc}\Omega^c \tag{15.59}$$

where we have defined a rotation vector Ω^c by

$$\Omega^\alpha = \frac{1}{2}\epsilon^{\alpha\beta\gamma\delta}u_\beta \mathbf{e}_\gamma \cdot \dot{\mathbf{e}}_\delta. \tag{15.60}$$

It is easy to check that this vector is spatial and that eq. (15.59) holds. The structure coefficients $c^a{}_{tb}$ can therefore be written as

$$c^a{}_{tb} = -\theta^a{}_b + \epsilon^a{}_{bc}\Omega^c. \tag{15.61}$$

The vector Ω^c can be interpreted as the local angular velocity, in the rest-frame of an observer with four-velocity u^μ, of a set of Fermi-propagated axes with respect to the spatial frame $\{\mathbf{e}_a\}$.

The remaining structure coefficients are all purely spatial and hence, they must correspond to one of the Bianchi Lie algebras. We write the spatial structure coefficients as

$$c^k{}_{ij} = \epsilon_{ijl}n^{lk} + a_l\left(\delta^l{}_i\delta^k{}_j - \delta^l{}_j\delta^k{}_i\right), \tag{15.62}$$

where n^{lk} is a *symmetric* matrix. Note that these structure coefficients are constants along each orbit of transitivity. Thus n^{lk} and a_i are only functions of time. The spatial frame $\{\mathbf{e}_a\}$ will be a set of left invariant vectors on the hypersurfaces Σ_t. In this orthonormal approach we let the structure coefficients vary as a function of time. The type of Lie algebra therefore has to be classified in terms of invariant properties of the matrix n^{lk} and the vector a_i.

We can find evolution equations for these functions by noting that for all vectors, the Jacobi identity eq. (15.1) holds. In particular, it must hold for the set of vectors $(\mathbf{u}, \mathbf{e}_a, \mathbf{e}_b)$. Thus we must have

$$\begin{aligned}
0 &= [\mathbf{u}, [\mathbf{e}_a, \mathbf{e}_b]] + [\mathbf{e}_a, [\mathbf{e}_b, \mathbf{u}]] + [\mathbf{e}_b, [\mathbf{u}, \mathbf{e}_a]] \\
&= [\mathbf{u}, c^\mu{}_{ab}\mathbf{e}_\mu]] + [\mathbf{e}_a, [\mathbf{e}_b, c^\mu{}_{bt}\mathbf{e}_\mu] + [\mathbf{e}_b, c^\mu{}_{ta}\mathbf{e}_\mu] \\
&= \left(\mathbf{u}(c^\nu{}_{ab}) + c^\nu{}_{t\mu}c^\mu{}_{ab} + c^\nu{}_{a\mu}c^\mu{}_{bt} + c^\nu{}_{b\mu}c^\mu{}_{ta}\right)\mathbf{e}_\nu. \tag{15.63}
\end{aligned}$$

Using eq. (15.57) we get the identity

$$\mathbf{u}(c^k{}_{ab}) + c^k{}_{td}c^d{}_{ab} + c^k{}_{ad}c^d{}_{bt} + c^k{}_{bd}c^d{}_{ta} = 0. \tag{15.64}$$

Applying the Jacobi's identity to the three spatial vectors, and then contracting, we get

$$n^{ij}a_i = 0. \tag{15.65}$$

Using eqs. (15.57), (15.61), (15.62) and (15.65) we can find the evolution equations for the structure constants. Taking the trace of eq. (15.64), we get the propagation equation for a_i

$$\boxed{\mathbf{u}(a_i) + \frac{1}{3}\theta a_i + \sigma_{ij}a^j + \epsilon_{ijk}a^j\Omega^k = 0.} \tag{15.66}$$

Class	Type	a	n_1	n_2	n_3
A	I	0	0	0	0
	II	0	+	0	0
	VI_0	0	+	−	0
	VII_0	0	+	+	0
	VIII	0	+	+	−
	IX	0	+	+	+
B	V	+	0	0	0
	IV	+	+	0	0
	VI_h	+	+	−	0
	VII_h	+	+	+	0

Table 15.2: The Bianchi types in terms of the algebraic properties of the structure coefficients.

and the trace-free part of eq. (15.64) is

$$\mathbf{u}(n_{ab}) + \frac{1}{3}\theta n_{ab} + 2n^k_{(a}\epsilon_{b)kl}\Omega^l - 2n_{k(a}\sigma^k_{b)} = 0. \tag{15.67}$$

In order that the structure coefficients eq. (15.62) shall correspond to a Lie algebra, the vector a_i must, according to eq. (15.65), be in the kernel[4] of the matrix n^{ij}. For the class A model, $a_i = 0$, and this equation is identically satisfied, and for the class B models, a_i must be an eigenvector of the matrix n^{ij} with zero eigenvalue. In any case, since n^{ij} is a symmetric matrix, we can diagonalise it using a specific orientation of the spatial frame. Thus, without any loss of generality we can assume that

$$n_{ij} = \text{diag}(n_1, n_2, n_3), \quad a^i = (0,0,a), \tag{15.68}$$

by a suitable choice of frame. The Jacobi identity then implies $n_3 a = 0$.

The eigenvalues of a matrix are invariant properties of a matrix under conjugation with respect to rotations. The Bianchi models can now be characterised by the relative signs of the eigenvalues n_1, n_2, n_3 and a. In Table 15.2 the classification of the Bianchi types in terms of these eigenvalues is listed. For the types VI_h and VII_h the group parameter is defined by the equation

$$h n_1 n_2 = a^2. \tag{15.69}$$

In this table III=VI_{-1}.

Note that for some of the Bianchi types, two or three eigenvalues are equal to zero. Hence, for these we have unused degrees of freedom to choose the orientation of the spatial frame. For example, the type I case has vanishing structure coefficients. Thus we have an unused $SO(3)$ rotation for the spatial frame. Since the shear is symmetric, we can choose to diagonalise σ_{ab} instead. So *for a Bianchi type I universe model we can without any loss of generality choose the shear to be diagonal.*

Einstein's Field Equations for the non-tilted Bianchi type universes

We can use the results from the previous chapter to find the field equations for the Bianchi type universe models. The Ricci tensor can be found from

---

[4]Consider a matrix M and a vector **v**. The vector **v** is in the *kernel* of M if and only if M**v** = 0.

contracting the Riemann tensor eq. (7.45). Using the four-dimensional Ricci tensor we can show that the tt-equation yields Raychaudhuri's equation, eq. (14.31), and the spatial ab-equations yield the shear propagation equations, eq. (14.40), and the generalised Friedmann equation, eq. (14.34). The off-diagonal ta-equations yield a non-trivial constraint:

$$3a^b \sigma_{ba} - \epsilon_{abc} n^{cd} \sigma^b_{\ d} = 0. \tag{15.70}$$

All the spatial derivatives vanish because the structure coefficients are constant along each surface of transitivity. Hence, the three-dimensional Ricci tensor is given by

$$^{(3)}R_{ab} = \Gamma^d_{ab}\Gamma^c_{dc} - \Gamma^d_{ac}\Gamma^c_{db} - c^d_{cb}\Gamma^c_{ad} = \Gamma^d_{ab}\Gamma^c_{dc} - \Gamma^d_{bc}\Gamma^c_{ad}, \tag{15.71}$$

where we have used eq. (6.137). Using equations (15.56) and (15.62) we get

$$^{(3)}R_{ab} = -2\epsilon^{cd}_{\ (a} n_{b)c} a_d + 2n_{ad}n^d_{\ b} - nn_{ab} - h_{ab}\left(2a^2 + n_{cd}n^{cd} - \frac{1}{2}n^2\right), \tag{15.72}$$

where $n = n^d_{\ d}$. In equations (14.40), the overdot is defined by $\dot{} = u^\mu \nabla_\mu$, thus we have

$$\dot{\sigma}_{ab} = \mathbf{u}(\sigma_{ab}) - \Gamma^\mu_{a\nu}\sigma_{\mu b}u^\nu - \Gamma^\mu_{b\nu}\sigma_{a\mu}u^\nu. \tag{15.73}$$

Using eq. (15.59) we can write this as

$$\dot{\sigma}_{ab} = \mathbf{u}(\sigma_{ab}) - 2\sigma^d_{\ (a}\epsilon_{b)cd}\Omega^c. \tag{15.74}$$

Thus, using equations (14.40), (14.31) and (14.34), Einstein's field equations imply the shear propagation equations

$$\mathbf{u}(\sigma_{ab}) + \theta\sigma_{ab} - 2\sigma^d_{\ (a}\epsilon_{b)cd}\Omega^c + {}^{(3)}R_{ab} - \frac{1}{3}h_{ab}{}^{(3)}R = \kappa\pi_{ab}, \tag{15.75}$$

Raychaudhuri's equation

$$\dot{\theta} + \frac{1}{3}\theta^2 + \sigma_{ab}\sigma^{ab} + \frac{\kappa}{2}(\rho + 3p) - \Lambda = 0, \tag{15.76}$$

and the Friedmann equation

$$\frac{1}{3}\theta^2 = \frac{1}{2}\sigma_{ab}\sigma^{ab} - \frac{1}{2}{}^{(3)}R + \kappa\rho + \Lambda, \tag{15.77}$$

where

$$^{(3)}R = {}^{(3)}R^a_{\ a} = -\left(6a^2 + n_{cd}n^{cd} - \frac{1}{2}n^2\right). \tag{15.78}$$

These are the field equations for the Bianchi type universe models in the orthonormal frame approach. There is an interesting thing worth noting. It may be shown that $^{(3)}R \leq 0$ for all Bianchi types except for type IX. Dividing the Friedmann equation by $\theta^2/3$ leads to

$$1 = \Sigma^2 + \Omega_k + \Omega_\rho + \Omega_\Lambda, \tag{15.79}$$

where

$$\Sigma^2 = \frac{3}{2}\frac{\sigma_{ab}\sigma^{ab}}{\theta^2}, \qquad \Omega_k = -\frac{3}{2}\frac{{}^{(3)}R}{\theta^2},$$

$$\Omega_\rho = \frac{3\kappa\rho}{\theta^2}, \qquad \Omega_\Lambda = \frac{3\Lambda}{\theta^2}. \qquad (15.80)$$

The Friedmann equation (with $\Lambda \geq 0$) implies that for all Bianchi types except for IX, the expansion-normalised shear, Σ, is bounded

$$0 \leq \Sigma^2 \leq 1. \qquad (15.81)$$

Equality in the upper limit in the above equation happens only in the Kasner vacuum solutions; they have maximal possible shear. For all other models, this is a strict inequality.

Example 15.4 (The Bianchi type V universe model)   Example
Let us consider the Bianchi type V universe model. This model is of class B with a vanishing matrix n^{lk}. We will choose an orientation of the spatial frame so that it aligns with the vector a^i. Hence, $a^i = a\delta^i{}_3$ and $n^{lk} = 0$. We will also choose a universal time gauge, $\mathbf{u} = \frac{\partial}{\partial t}$.

Jacobi's identity eq. (15.65), and the n^{lk} propagation equation will now be identically satisfied. The constraint eq. (15.70), leads to the three equations

$$a\sigma_{31} = a\sigma_{32} = a\sigma_{33} = 0. \qquad (15.82)$$

Since $a \neq 0$ (or else we would not have a type V algebra), we get

$$\sigma_{31} = \sigma_{32} = \sigma_{33} = 0. \qquad (15.83)$$

We still have a rotation with respect to the axis defined by $\mathbf{e}_3$ which we can freely choose. We can use this freedom of rotation to set $\sigma_{12} = 0$ as well. Hence, there will be only two non-zero shear components

$$\sigma_{ab} = \text{diag}(\sigma_+, -\sigma_+, 0) \qquad (15.84)$$

since $\sigma^a{}_a = 0$. From the a_i-propagation equations eq. (15.66), we get

$$\Omega^2 = \Omega^1 = 0 \qquad (15.85)$$

from the 1- and 2-equations, and for the 3-equation we get

$$\frac{\partial a}{\partial t} + \frac{1}{3}\theta a = 0. \qquad (15.86)$$

The three-curvature turns simply into

$${}^{(3)}R_{ab} = -2a^2 h_{ab}. \qquad (15.87)$$

This implies that the trace-free part of the three-curvature vanishes:

$${}^{(3)}R_{\langle ab\rangle} = -2a^2 h_{\langle ab\rangle} = 0. \qquad (15.88)$$

The anisotropic stress tensor is now to some extent constrained by the shear propagation equation, eq. (15.75). One possibility is that the anisotropic stress tensor vanish identically: $\pi_{ab} = 0$. Consider this to be the case.

The shear equation now reduces to

$$\sigma_+\Omega^3 = 0 \qquad (15.89)$$

from the off-diagonal equations and

$$\frac{\partial \sigma_+}{\partial t} + \theta \sigma_+ = 0 \tag{15.90}$$

from the diagonal equations. Assuming that $\sigma_+ = 0$ leads to the isotropic negatively curved FRW universe. Hence, assuming anisotropy we have to set $\Omega^3 = 0$.

The remaining equations, Raychaudhuri's equation and the Friedmann equation have also got to be satisfied. The full set of equations has simplified considerably and there remains to integrate these for a particular type of fluid. The case of a vacuum fluid is considered in problem 15.3.

15.5   The 8 model geometries

The connection between the various Bianchi types and the geometry of the space is interesting but highly non-trivial. The classification of three dimensional spaces is still unsettled, but central in the discussion is the model geometries. These geometries were defined by W.P. Thurston, and therefore they are sometimes referred to as the "Thurston geometries" [Thu97]. They are defined as follows.

Definition: Model Geometry (à la Thurston)   *A pair (M, G) with M a connected and simply connected manifold, and G is a group acting transitively on M, is called a* model geometry *if the following conditions are satisfied:*

1. *M can be equipped with a G-invariant Riemannian metric.*

2. *G is maximal; i.e. there does not exist a larger group $H \supset G$ which acts transitively on M and requirement 1 is satisfied.*

3. *There exists a discrete subgroup $\Gamma \subset G$ such that M/Γ is compact; i.e. M allows for a compact quotient.*

The last item, is a technical issue which we will discuss in section 15.6.

Some examples of such model geometries can be found among the maximally symmetric spaces. Since they are maximally symmetric, 1 and 2 is trivially satisfied. 3 is more subtle, but it can be shown that S^n, $\mathbb{E}^n$ and $\mathbb{H}^n$ with their maximally symmetric isometry groups are model geometries for all n.

A question now arises: What are the model geometries in dimension three? In dimension two, the maximally symmetric spaces are the only model geometries. In three dimensions, we will have 8 different model geometries. These are

$$S^3 \qquad \mathbb{E}^3 \qquad \mathbb{H}^3$$

$$\mathbb{E}^1 \times S^2 \qquad \mathbb{E}^1 \times \mathbb{H}^2$$

$$\widetilde{SL(2, \mathbb{R})} \qquad \text{Nil}$$

$$\text{Sol}$$

The first three, are already familiar to us. These are the maximally symmetric spaces that we discussed in section 7.6.

$\mathbb{E}^1 \times S^2$: The product between a sphere and a line. The group is in this case four-dimensional, but as we already mentioned, it does not have a simply transitive subgroup. Hence, it is not one of the Bianchi models. An invariant metric can be written

$$ds^2 = dz^2 + a^2 \left(d\theta^2 + \sin^2\theta d\phi^2\right), \tag{15.91}$$

where a is constant.

$\mathbb{E}^1 \times \mathbb{H}^2$: The product between the hyperbolic plane and a line. The group in this case is also 4 dimensional, but contains a simply transitive subgroup of Bianchi type III. An invariant metric can be written as

$$ds^2 = a^2 \frac{dx^2 + dy^2}{y^2} + dz^2. \tag{15.92}$$

$\widetilde{SL(2, \mathbb{R})}$: The covering space of the matrix Lie group $SL(2, \mathbb{R})$. The group of isometries is of dimension 4 but contains a three-dimensional simply transitive subgroup of Bianchi type VIII or III. An invariant metric is

$$ds^2 = a^2 \frac{dx^2 + dy^2}{y^2} + b^2 \left(2dz + \frac{dx}{y}\right)^2. \tag{15.93}$$

Nil: Nilgeometry, or sometimes also called the Heisenberg group. The group is four-dimensional with an invariant metric

$$ds^2 = dx^2 + dy^2 + a^2 \left[dz + \frac{1}{2}(ydx - xdy)\right]^2. \tag{15.94}$$

Sol: Solvegeometry. The group is 3 dimensional and simply transitive. An invariant metric is

$$ds^2 = e^{2z}dx^2 + e^{-2z}dy^2 + a^2 dz^2. \tag{15.95}$$

Example 15.5 (The Lie algebra of Sol)   Example
We have seen that all the possible three-dimensional Lie algebras are classified in the Bianchi classification. Hence, Sol which has a three dimensional isometry group, must correspond to one of the Bianchi types. The invariant metric is (choosing $a = 1$)

$$ds^2 = e^{2z}dx^2 + e^{-2z}dy^2 + dz^2. \tag{15.96}$$

Let us try the invariant basis

$$\boldsymbol{\omega}^1 = e^z \mathbf{dx}, \quad \boldsymbol{\omega}^2 = e^{-z}\mathbf{dy}, \quad \boldsymbol{\omega}^3 = \mathbf{dz}. \tag{15.97}$$

We calculate their exterior derivatives to find the structure constants, using eq. (6.177). The exterior derivatives are

$$\begin{aligned}
\mathbf{d}\boldsymbol{\omega}^1 &= -\boldsymbol{\omega}^1 \wedge \boldsymbol{\omega}^3 \\
\mathbf{d}\boldsymbol{\omega}^2 &= \boldsymbol{\omega}^2 \wedge \boldsymbol{\omega}^3 \\
\mathbf{d}\boldsymbol{\omega}^3 &= 0.
\end{aligned} \tag{15.98}$$

Model Geometry	$\dim(G)$	Bianchi type
$\mathbb{E}^3$	6	I VII_0
S^3	6	IX
$\mathbb{H}^3$	6	V VII_h
$\mathbb{E}^1 \times S^2$	4	KS
$\mathbb{E}^1 \times \mathbb{H}^2$	4	III
$\widetilde{SL(2,\mathbb{R})}$	4	VIII III
Nil	4	II
Sol	3	VI_0

Table 15.3: The relation between the model geometries and the Bianchi type.

Thus the non-zero structure constants are true constants in this case:

$$C_{13}^1 = -1, \qquad C_{23}^2 = 1. \tag{15.99}$$

Comparing this with the Table 15.1 we see that Sol is a Bianchi type VI_0 geometry.

Each Bianchi model defines a transitive group G_B on some three dimensional simply connected space Σ. Hence, by going to a maximal group G that acts on Σ such that $G_B \subset G$, the pair (Σ, G) will satisfy the first two conditions for a model geometry. It can by construction, only fail to satisfy the third condition; it does not necessarily allow a compact quotient. Note that there can be two different simply transitive groups G_1 and G_2 such that $G_1 \subset G$ and $G_2 \subset G$. This can happen in all the cases where the model geometry has a group of dimension larger than three. For example, the Euclidean space, $\mathbb{E}^3$, is both Bianchi type I invariant and VII_0 invariant. The question of a compact quotient will be addressed in the next section.

Let us finish this section with a table that gives the relation between the Bianchi types and the model geometries (see Table 15.3). Listed are also the dimension of the largest symmetry group possible (the group G). The types IV and VI_h for $h \neq 0, -1$ are not on the list. Thus, this means that there does not exist a compact quotient of these geometries. Interestingly, the Bianchi type III, can correspond to two different model geometries, namely $\mathbb{E}^1 \times \mathbb{H}^2$ and $\widetilde{SL(2,\mathbb{R})}$.

15.6 Constructing compact quotients

In this section we will give a short introduction to how we can construct compact quotients of the model geometries. The method is highly general, so we will not restrict ourselves to the three dimensional case.

The isometry group tells us what points in our spacetime are "equal". Using an isometry you can travel from one to another equivalent point. When we construct quotients of spaces, we use this property of the isometry group. Let us start out by constructing a compact quotient of the Euclidean line to illustrate the idea. The Euclidean line has the metric

$$ds^2 = dx^2, \tag{15.100}$$

with the Killing vector field

$$\xi = \frac{\partial}{\partial x}.$$
(15.101)

The isometries are therefore translations in the x-direction:

$$x \longmapsto x + \ell,$$
(15.102)

for any $\ell \in \mathbb{R}$. This isometry says that any point on the line is equivalent to any other. This we can use to construct a compact space. What we do is to say that every point that is separated by the distance ℓ for some $\ell \neq 0$, is the same point. Thus we identify the points x and $x + \ell$. The variable x now turns into an angular variable, and by introducing the variable $\theta \in [0, 2\pi)$, we can write the metric after the identification as

$$ds^2 = \left(\frac{\ell}{2\pi}\right)^2 d\theta^2.$$
(15.103)

Hence, the quotient is the circle with radius $R = \frac{\ell}{2\pi}$. From the infinite Euclidean line $\mathbb{E}^1$ we have constructed a compact quotient which is a circle.

We will now leap to the general case; we will give a recipe of how we in general can construct such compact quotients. We will thereafter go on and construct some compact spaces using this recipe.

Recipe for constructing Compact Quotients

Consider a space M with a group G acting transitively on M. This could well be the isometry group, but it does not necessary need to be so. However, in most practical problems this will be the case, as it is in this book.

1. Find a discrete subgroup $\Gamma \subset G$ which acts properly discontinuous on M.

2. Construct the quotient M/Γ, given by the identification of points in M under the action of Γ. Hence, define an equivalence relation $\sim$: $p \sim q$ if there exists a $\gamma \in \Gamma$ such that $\gamma(p) = q$. The quotient M/Γ is then the quotient $M/\sim$.

3. If the action of Γ is free, then M/Γ is a smooth manifold.

That the action is free means that the Γ "moves all points". Hence, for all $p \in M$ there does not exist an element $\gamma \in \Gamma$, apart from the identity element, such that $\gamma(p) = p$.

Properly discontinuous mapping means that for any point $p \in M$, there exists a neighbourhood U of p such that $\gamma_1(U) \cap \gamma_2(U) = \emptyset$ for $\gamma_1, \gamma_2 \in \Gamma$, except for $\gamma_1 = \gamma_2$.

For the case of the Euclidean line, $\mathbb{E}$, with $G = \mathbb{R}$, we can choose the discrete subgroup $\Gamma = \mathbb{Z}$. This group identifies any point $x \in \mathbb{Z}$ with the lattice (or grid) $L_1(x) = \{x + 2\pi nR | n \in \mathbb{Z}\}$ for any $R > 0$.

For the higher dimensional Euclidean spaces, we can similarly construct higher dimensional tori. Since $\mathbb{E}^n$ is translation invariant, given a basis $\{\mathbf{e}_i\}$, we can define the lattice

$$L_n = \{\mathbf{v} \in \mathbb{R}^n | \mathbf{v} = v^i \mathbf{e}_i, v^i \in \mathbb{Z}\} \cong \mathbb{Z}^n.$$
(15.104)

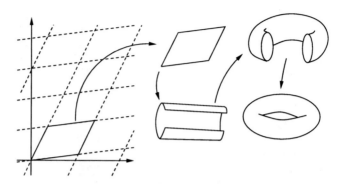

Figure 15.3: How to construct a torus from a lattice in the plane.

This lattice defines an action Γ which has the right properties. Therefore we set $\Gamma \cong L_n$ and identifies points in the Euclidean space to obtain the torus:

$$\mathbb{T}^n = \mathbb{E}^n / L_n \tag{15.105}$$

For the two-dimensional case, the idea is illustrated in Fig. 15.3.
Topologically, the torus $\mathbb{T}^n$ is a product of circles, $\left(S^1\right)^n$, and hence can be parametrised by n angular variables. The torus $\mathbb{T}^n$ is a compact quotient of the Euclidean space. It is, as the Euclidean space, flat; all of the curvature tensors are inherited from the original space. After the identification is obtained, we can deform the torus such that it is no longer flat. The torus $\mathbb{T}^2$ embedded in $\mathbb{E}^3$ is an example of a non-flat torus.

Examples

Example 15.6 (Lens spaces)
The 3-sphere, S^3, is already compact, but we can construct a whole series of topologically different spaces by taking the quotient of S^3.
    We start by considering S^3 embedded in the complex 2-dimensional space, $\mathbb{C}^2$:

$$S^3 = \left\{ (z_1, z_2) \in \mathbb{C}^2 \middle|\ |z_1|^2 + |z_2|^2 = 1 \right\}. \tag{15.106}$$

We note that the mapping $z_j \longmapsto z_j e^{i\lambda}$ for $\lambda \in \mathbb{R}$ leaves the sphere invariant, and hence, is an isometry. We go on and define a subgroup $\Gamma_{p,q}$ generated by the mapping

$$(z_1, z_2) \longmapsto (z_1 \exp(2\pi i/p), z_2 \exp(2\pi i q/p)) \tag{15.107}$$

where p and q are integers with no common divisors. The spaces defined by

$$L(p, q) = S^3 / \Gamma_{p,q}, \tag{15.108}$$

are called *Lens spaces*. These spaces are compact quotients of S^3 and are manifolds. Note that $L(2, 1)$ is the same as projective space, $\mathbb{P}^3$.

Example 15.7 (The Seifert-Weber Dodecahedral space)
We have already claimed that the hyperbolic space, $\mathbb{H}^3$, is a model geometry, thus it must admit a compact quotient. This result might be surprising perhaps, but the hyperbolic space admits a huge number of compact quotients. In fact, the hyperbolic space turns out to be the richest of all the 8 model geometries. Contrary to the other model geometries, all of the possible compact quotients have not been classified.

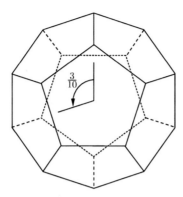

Figure 15.4: The Seifert-Weber dodecahedral space.

We will mention one of the compact hyperbolic spaces here as an example. It is called the *Seifert-Weber Dodecahedral space* and was the first known example of a three-dimensional compact hyperbolic manifold. We use a solid dodecahedron as a "fundamental cube", see Fig. 15.4. Each face of a dodecahedron are pentagons, and they come in pairs; each member of the pair is opposite to the other. Twist one of them by 3/10 and identify the two pentagons in that pair. Do this for all pairs of the dodecahedron. It can be shown that the resulting space is a manifold and is a quotient of $\mathbb{H}^3$. Hence, we have constructed a compact hyperbolic manifold.

In principle, there is nothing in Einstein's field equations that forbid us to make such identifications. On the contrary, Einstein's field equations are *local*; they tell us only about the local geometry. Regarding the global properties of spacetime, very much is left unsaid. Hence, if we find by local measurements that the local geometry is flat, say, then, even though we would assume that the geometry is flat everywhere, could not say anything precise about the global structure of the universe. We do not know if the universe is infinite, like an infinite sheet, or compact like a flat torus. To find out about the global topology of the universe, we have to make different measurements which can reveal to us the global structure of the universe we live in.

Problems

15.1. *A Bianchi type II universe model*
In this problem we will study the Bianchi type II universe model. The Bianchi type II Lie algebra is defined by the single non-trivial commutator

$$[\mathbf{X}, \mathbf{Y}] = \mathbf{Z}. \tag{15.109}$$

Using the orthonormal frame approach we will derive the equations of motion for this model, and find a particular solution.

(a) Let e_μ be an orthonormal frame. The vectors $\mathbf{X}$, $\mathbf{Y}$ and $\mathbf{Z}$ are linearly independent, so there exists coefficients $\lambda^I{}_j$ such that $e_i = \lambda^I{}_i \mathbf{X}_I$ where $\mathbf{X}_I = (\mathbf{X}, \mathbf{Y}, \mathbf{Z})$ and i are spatial indices. For simplicity, choose $e_3 = \lambda^3{}_3 \mathbf{Z}$. Show that

$$[e_1, e_2] = k e_3, \tag{15.110}$$

where k is a constant on each hypersurface. Hence, since the vectors e_i form a spatial basis, there will be a preferred direction in the spatial hypersurfaces. Therefore we choose the orientation of the frame such that e_3 points in this direction. Explain that the most general form of the structure constants C_{ij}^k under the above assumptions is

$$C_{ij}^k = \epsilon_{ijl} n^{lk}, \quad n^{lk} = \text{diag}(0, 0, n), \quad n \neq 0. \tag{15.111}$$

(b) We will assume that the matter content in this universe is that of non-tilted dust and $\Lambda = 0$. Hence,

$$T_{\mu\nu} = \rho u_\mu u_\nu \tag{15.112}$$

where $\mathbf{u} = \mathbf{e}_0$. Show that the $(0, i)$ equations imply that

$$\sigma_{13} = \sigma_{23} = 0. \tag{15.113}$$

This implies that the only non-zero off-diagonal shear component is σ_{12}. However, we still have an unused rotation of the vectors $\mathbf{e}_1$ and $\mathbf{e}_2$ (we have only fixed the direction of $\mathbf{e}_3$) with respect to the axis defined by $\mathbf{e}_3$. Thus we can use this freedom to diagonalise σ_{ab} completely. Hence, we can without loss of generality assume that the shear is diagonal.

Show, using the equations for n_{ab}, that we must have

$$\Omega^a = 0. \tag{15.114}$$

(c) Set $\sigma_{ab} = \text{diag}(\sigma_+ + \sqrt{3}\sigma_-, \sigma_+ - \sqrt{3}\sigma_-, -2\sigma_+)$ (so that σ_{ab} is trace-free). Choose the universal time gauge $\mathbf{u} = \frac{\partial}{\partial t}$. Set down the equations of motion for $n, \sigma_\pm, \rho$. Write also down Raychaudhuri's equation and the generalised Friedmann equation in these variables.

(d) We will now find a particular solution to these equations. We will search for a solution where the variables have the time-dependence

$$\theta \propto t^{-1}$$
$$\sigma_\pm \propto t^{-1}$$
$$n \propto t^{-1}$$
$$\rho \propto t^{-2}. \tag{15.115}$$

To avoid that $n = 0$ (which would not yield a Bianchi type II spacetime) we must assume that $\sigma_- = 0$. Find a solution of this form. The obtained solutions is called the Collins-Stewart solution for dust. Show that the solution corresponds to the metric

$$ds^2 = -dt^2 + t^{3/2}(dx^2 + dy^2) + t \left(dz + \frac{1}{2} x dy \right)^2. \tag{15.116}$$

15.2. A homogeneous plane wave

We will consider a solution of Einstein's field equations given by

$$ds^2 = -dt^2 + t^2 dx^2 + t^{2r} e^{2rx} \left[e^\beta (\omega^2)^2 + e^{-\beta} (\omega^3)^2 \right], \tag{15.117}$$

where

$$\omega^2 = \cos[b(x + \ln t)] \mathbf{dy} + \sin[b(x + \ln t)] \mathbf{dz},$$
$$\omega^3 = -\sin[b(x + \ln t)] \mathbf{dy} + \cos[b(x + \ln t)] \mathbf{dz} \tag{15.118}$$

and β, r and b are constants satisfying

$$b^2 \sinh^2 \beta = r(1 - r). \tag{15.119}$$

(a) Show that this metric has a null Killing vector given by

$$\xi = \frac{\partial}{\partial v} \tag{15.120}$$

where $v = te^{-x}$. (Hint: introduce the coordinates $v = te^{-x}$, $u = te^x$)

(b) Introduce an orthonormal basis η^μ, where

$$\eta^0 = dt, \qquad \eta^1 = tdx,$$
$$\eta^2 = t^r e^{rx} e^{\beta/2} \omega^2, \qquad \eta^3 = t^r e^{rx} e^{-\beta/2} \omega^3. \tag{15.121}$$

Show that the structure constants obey the relation

$$C^A_{0B} = C^A_{1B}, \qquad A, B = 2, 3. \tag{15.122}$$

Find the rest of the structure constants. What is $a_i = -\frac{1}{2} C^j_{ij}$?

(c) Show that the matrix n^{ab} is

$$n^{ab} = \frac{1}{t} \operatorname{diag}(0, -be^\beta, be^{-\beta}). \tag{15.123}$$

Show further that the volume expansion tensor is

$$\theta^a{}_b = \frac{1}{t} \begin{bmatrix} 1 & 0 & 0 \\ 0 & r & -b\cosh\beta \\ 0 & -b\cosh\beta & r \end{bmatrix} \tag{15.124}$$

and

$$\Omega^a = b \sinh\beta \, \delta^a{}_1. \tag{15.125}$$

(d) Is this spacetime spatially homogeneous? If so, to which Bianchi type does it belong?

The metric above, in fact, satisfies the vacuum Einstein field equations ($\Lambda = 0$) and describes a gravitational *plane wave*. Since this spacetime also, in addition to the 3 spatially Killing vectors spanning the Bianchi type, has a null Killing vector, it is homogeneous in spacetime (not only in the spatially directions).

15.3. *Vacuum dominated Bianchi type V universe model*
Use the results of example 15.4 and solve Einstein's field equations for a type V universe model with a cosmological constant. Also, write down the metric of the resulting solutions.

15.4. *The exceptional case, $VI^*_{-1/9}$*
In this problem we will consider a special case of the Bianchi models which has to be treated separately. This is called the *exceptional case*.

(a) For all Bianchi models except type I, the constraint (15.70), is a non-trivial constraint. Consider the class B models where $a^b = a\delta^b{}_3$. Assume also that a choice of frame is chosen so that n^{ab} is diagonal. Show that this constraint leads to

$$\begin{aligned} 3a\sigma_{33} + (n^1 - n^2)\sigma_{21} &= 0, \\ 3a\sigma_{31} + n^2\sigma_{32} &= 0, \\ 3a\sigma_{32} - n^1\sigma_{31} &= 0. \end{aligned} \tag{15.126}$$

(b) This means that in general, three components of the shear have to be constrained. However, show that in the special case of

$$9a^2 + n^1 n^2 = 0, \tag{15.127}$$

one of the above constraints vanishes identically (they are not linearly independent). Hence, we can have an additional shear degree of freedom in this case. According to eq. (15.69), this happens in type $\text{VI}_{-1/9}$. Models for which this extra shear degree of freedom is included are denoted with a star; i.e. $\text{VI}^*_{-1/9}$.

15.5. *Symmetries of hyperbolic space*
We will in this problem consider the hyperbolic space, $\mathbb{H}^3$, given in Poincaré coordinates:

$$ds^2 = \frac{1}{z^2}(dx^2 + dy^2 + dz^2). \tag{15.128}$$

(a) Show that the following vector fields are Killing vector fields.

$$\boldsymbol{\xi}_1 = \tfrac{\partial}{\partial x}, \quad \boldsymbol{\xi}_2 = \tfrac{\partial}{\partial y}, \quad \boldsymbol{\xi}_3 = y\tfrac{\partial}{\partial x} - x\tfrac{\partial}{\partial y},$$
$$\boldsymbol{\xi}_4 = x\tfrac{\partial}{\partial x} + y\tfrac{\partial}{\partial y} + z\tfrac{\partial}{\partial z}. \tag{15.129}$$

Indicate on a figure the flow of each of these vector fields.

(b) Verify that the Killing vectors $\boldsymbol{\xi}_1$, $\boldsymbol{\xi}_2$, and $\boldsymbol{\xi}_4$ are non-vanishing everywhere (except possibly at the boundary), while $\boldsymbol{\xi}_3$ vanishes along a line. Also, verify that the set

$$\left\{ \boldsymbol{\xi}_1, \; \boldsymbol{\xi}_2, \; \tilde{h}\boldsymbol{\xi}_3 + \boldsymbol{\xi}_4 \right\}$$

where $\tilde{h}$ is any real number, forms a basis.

(c) Show that for $\tilde{h} = 0$, and $\tilde{h} \neq 0$, this set of Killing vectors corresponds to the Bianchi type V, and VII_h Lie algebras, respectively.

(d) Find the corresponding left-invariant frame $\{e_1, e_2, e_3\}$ which coincides with the frame of Killing vectors given in (b) at $(x, y, z) = (0, 0, 1)$.
Indicate on a figure the flow of each of the left-invariant basis vectors.

(e) Find the corresponding left-invariant one-forms.

15.6. *The matrix group $SU(2)$ is the sphere S^3*
We will in this problem consider the group of 2×2 matrices with complex entries given by

$$SU(2) = \left\{ A \in GL(2, \mathbb{C}) | A^\dagger A = 1, \quad \det A = 1 \right\},$$

and the three-sphere embedded in $\mathbb{R}^4$,

$$S^3 = \left\{ (X, Y, U, V) \in \mathbb{R}^4 | X^2 + Y^2 + U^2 + V^2 = 1 \right\}.$$

Here, dagger $\dagger$ means the adjoint matrix; i.e. transpose and complex conjugate.

Given a 2×2 matrix A with complex entries. What are the conditions on the entries of the matrix in order for the matrix to be in $SU(2)$?

We consider the matrix

$$A = \begin{bmatrix} X + iY & U + iV \\ -U + iV & X - iY \end{bmatrix}.$$

Show that this matrix is in $SU(2)$ if and only if (X, Y, U, V) are coordinates on the three-sphere S^3.

This implies, firstly, that the group $SU(2)$ is a manifold since S^3 is one, and, secondly, the sphere S^3 admits a group structure. This is directly related to the fact that the sphere admits a simply transitive group; the sphere is acting simply transitively on itself. Verify that the Bianchi type IX algebra, corresponding to the sphere S^3, is the same as the Lie algebra of $SU(2)$.

15.7. *Left-invariant one-forms of Bianchi type VIII and IX*

(a) Consider the set of one-forms:

$$\omega^1 = a\left(dx - \frac{dz}{y}\right),$$

$$\omega^2 = \frac{b}{y}\left(\cos x\, dy + \sin x\, dz\right),$$

$$\omega^3 = \frac{c}{y}\left(-\sin x\, dy + \cos x\, dz\right). \qquad (15.130)$$

Verify that these one-forms can be considered to be a set of left-invariant one-forms for the Bianchi type VIII model. Assume that these one-forms are orthonormal. In general, there will be only 3 Killing vectors for the Bianchi type VIII metric. Under what conditions will the type VIII metric simplify to the Thurston geometry eq. (15.93), and hence, acquire an additional rotational Killing vector?

(b) Consider the set of one-forms given by eq. (14.116) on page 407. Verify that these one-forms are indeed left-invariant one-forms on the Bianchi type IX Lie group.

16

Israel's Formalism:
The Metric Junction Method

A question that often arises in gravitational theory is what happens to the geometry of space when there is a jump discontinuity in the energy-momentum tensor along a surface. For example, what is the connection between the curvature properties for the interior Schwarzschild solution and the exterior Schwarzschild solution? Here, along the boundary of some surface, the energy density experiences a jump discontinuity. Another analogous scenario is for example a shock wave propagating outwards from an exploding star. In models of such shock waves the density can be infinite.

To investigate these problems, W. Israel [Isr66] developed a mathematical framework which is called *Israel's formalism*.

16.1 The relativistic theory of surface layers

Consider a spacetime which is separated into two different regions. This can for instance be the interior and the exterior region of a star, or it can be a domain wall dividing the spacetime in two. Assume therefore that spacetime $\mathcal{M}$ is split in two,

$$\mathcal{M} = \mathcal{M}^+ \cup \mathcal{M}^-, \tag{16.1}$$

with a common boundary Σ:

$$\partial \mathcal{M}^+ \cap \partial \mathcal{M}^- = \Sigma. \tag{16.2}$$

This is illustrated in Fig. 16.1.

Assume also that this surface is a hypersurface of dimension 3. In the interior of each of the two different regions $\mathcal{M}^\pm$, Einstein's equations are assumed to be satisfied. Thus

$$E_{\mu\nu}^\pm = \kappa T_{\mu\nu}^\pm, \tag{16.3}$$

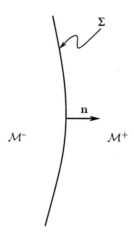

Figure 16.1: The hypersurface Σ divides the spacetime into two regions $\mathcal{M}^+$ and $\mathcal{M}^-$.

where $+$ and $-$ means the tensor evaluated in $\mathcal{M}^+$ and $\mathcal{M}^-$, respectively. The line-elements of the two regions are given by

$$ds^2 = g^{\pm}_{\mu\nu} dx^{\mu}_{\pm} dx^{\nu}_{\pm}, \tag{16.4}$$

and the induced line-element on Σ is

$$d\sigma^2 = h_{ij} dx^i dx^j. \tag{16.5}$$

Define the unit normal vector $\mathbf{n}$ to Σ to be the vector pointing from $\mathcal{M}^-$ to $\mathcal{M}^+$. The surface Σ can be both space-like and time-like which is given by the norm of $\mathbf{n}$:

$$\mathbf{n} \cdot \mathbf{n} = g_{\mu\nu} n^{\mu} n^{\nu} \equiv \epsilon = \begin{cases} 1, & \text{if } \Sigma \text{ is time-like} \\ -1, & \text{if } \Sigma \text{ is space-like.} \end{cases} \tag{16.6}$$

The case $\epsilon = 0$ will not be treated here[1].

The geometry of each of the regions are reflected in how the hypersurface Σ is embedded in the different regions $\mathcal{M}^{\pm}$. Using the extrinsic curvature of Σ induced by the two different regions we can compare the geometries in which Σ is embedded.

Define therefore $K^{\pm}_{\mu\nu}$ as the n-component of the covariant derivative in region $\mathcal{M}^{\pm}$ on a vector e_{μ} in Σ. Hence,

$$K^{\pm}_{\mu\nu} = \mathbf{n} \cdot \nabla^{\pm}_{\mu} e_{\nu} = \epsilon n_{\alpha} \left. \Gamma^{\alpha}_{\mu\nu} \right|^{\pm}. \tag{16.7}$$

The question is now how these two extrinsic curvature tensors relate. We require the induced metric, $h_{\mu\nu}$, on Σ from $\mathcal{M}^{\pm}$ to agree. However, the embeddings, and thus the extrinsic curvature tensors, do not need to agree. The induced metric is given by the projection

$$h^{\pm}_{\mu\nu} = g^{\pm}_{\mu\nu} - \epsilon n^{\pm}_{\mu} n^{\pm}_{\nu}, \tag{16.8}$$

---

[1] For null shells, see, for example, Poisson's book [Poi04].

and hence, there must be a coordinate transformation on Σ which relates $h^+_{\mu\nu}$ with $h^-_{\mu\nu}$. Thus we can set $h^\pm_{\mu\nu} = h_{\mu\nu}$.

From eq. (7.83), and the Codazzi equation (7.84) we have

$$E_{\mu\nu}n^\mu n^\nu|^\pm = -\frac{1}{2}\epsilon^{(3)}R + \frac{1}{2}\left(K^2 - K_{\alpha\beta}K^{\alpha\beta}\right)^\pm \tag{16.9}$$

$$E_{\mu\nu}h^\mu_\alpha n^\nu|^\pm = -\left({}^{(3)}\nabla_\mu K^\mu_\alpha - {}^{(3)}\nabla_\alpha K\right)^\pm \tag{16.10}$$

$$E_{\mu\nu}h^\mu_\alpha h^\nu_\beta|^\pm = {}^{(3)}E_{\alpha\beta} + \epsilon n^\mu \nabla_\mu \left(K_{\alpha\beta} - h_{\alpha\beta}K\right)^\pm - 3\epsilon K_{\alpha\beta}K|^\pm$$

$$+ 2\epsilon K^\mu_\alpha K_{\mu\beta}|^\pm + \frac{1}{2}\epsilon h_{\alpha\beta}(K^2 + K^{\mu\nu}K_{\mu\nu})^\pm. \tag{16.11}$$

16.2 Einstein's field equations

We will now relate the curvature tensors to the energy-momentum tensor according to Einstein's field equations. The energy-momentum tensor is allowed to be discontinuous at Σ, but is continuous elsewhere. The metric tensor is required to be continuous across the whole spacetime.

The Einstein tensor contains second derivatives of the metric tensor, but we will allow the derivative $g_{\mu\nu;\alpha}n^\alpha$ to be discontinuous. The second derivative can therefore be a delta-function since $\theta'(x) = \delta(x)$ where $\theta(x)$ is the step-function

$$\theta(x) = \begin{cases} 0, & x < 0 \\ 1, & x > 1. \end{cases} \tag{16.12}$$

Let y be an orthogonal coordinate so that $\frac{\partial}{\partial y} = \mathbf{n}$ and $y = 0$ at the surface Σ. The most general energy-momentum tensor across the boundary is therefore

$$T_{\alpha\beta} = S_{\alpha\beta}\delta(y) + T^+_{\alpha\beta}\theta(y) + T^-_{\alpha\beta}\theta(-y). \tag{16.13}$$

The energy-momentum tensor of the surface $S_{\alpha\beta}$ can be defined as the integral over the thickness of the surface Σ as the thickness goes to zero

$$S_{\alpha\beta} = \lim_{\tau\to 0} \int_{-\tau/2}^{\tau/2} T_{\alpha\beta}dy. \tag{16.14}$$

For this to be well defined the tensor $S_{\alpha\beta}$ has to live on the hypersurface so that

$$h^\alpha_\mu h^\beta_\nu S_{\alpha\beta} = S_{\mu\nu}. \tag{16.15}$$

This way of defining the tensor $S_{\alpha\beta}$ is called the *thin shell approximation*.

Introduce a set of coordinates so that x^i are coordinates on the hypersurface and y is the coordinate in the orthogonal direction (as before). Using the thin shell approximation we can find an expression for S_{ij}.

From equation (16.11), we have

$$\lim_{\tau\to 0} \int_{-\tau/2}^{\tau/2} E_{ij}dy = \lim_{\tau\to 0} \int_{-\tau/2}^{\tau/2} \left[\epsilon n^\mu \nabla_\mu \left(K_{ij} - h_{ij}K\right) + U_{ij}\right]dy, \tag{16.16}$$

where U_{ij} is containing quadratic terms in K_{ab} and the three-curvature. Thus this term is assumed to be bounded. The remainder of the integrand is a total derivative so we get

$$\lim_{\tau \to 0} \int_{-\tau/2}^{\tau/2} E_{ij} dy = \epsilon \left([K_{ij}] - h_{ij}[K] \right), \tag{16.17}$$

where we have defined the bracket operation as

$$[T] \equiv T^+ - T^-, \tag{16.18}$$

for a general tensor T. Using Einstein's field equations we get

$$\boxed{[K_{ij}] - h_{ij}[K] = \epsilon \kappa S_{ij}.} \tag{16.19}$$

This equation is called the *Lanczos equation*. The remaining components of $S_{\mu\nu}$ vanish

$$S_{nn} = S_{ni} = 0. \tag{16.20}$$

For a given tensor $T^{\pm}$ we define

$$\{T\} = \frac{1}{2}(T^+ + T^-). \tag{16.21}$$

Note the following identities

$$[TS] = [T]\{S\} + \{T\}[S], \tag{16.22}$$

$$\{TS\} = \{T\}\{S\} + \frac{1}{4}[T][S]. \tag{16.23}$$

These will be useful later on.

By contracting the Lanczos equation (16.19) and substituting into the same equation, we get

$$[K_{ij}] = \kappa \epsilon \left(S_{ij} - \frac{1}{2} h_{ij} S \right). \tag{16.24}$$

This equation connects the difference of embeddings of the surface Σ through the energy-momentum tensor of the surface. It is one of the equations of motion of the surface. Henceforth, we will, for simplicity, consider only time-like hypersurfaces; thus we assume $\epsilon = 1$.

The remaining equations of motion is obtained by replacing the right-hand side of eqs. (16.9) and (16.10), using Einstein's equations, with the energy-momentum tensor (16.13). Applying the [ ] operation and using the Lanczos equation (16.19), yield the equations

$$\boxed{{}^{(3)}\nabla_j S^j_{\ i} + [T_{in}] = 0,} \tag{16.25}$$

and

$$\boxed{S_{ij}\{K^{ij}\} + [T_{nn}] = 0.} \tag{16.26}$$

Note that we can go to a more general coordinate system (not necessary orthogonal) by letting all Latin indices go to their projected versions. Hence, let

S_{ij} go to $S_{\alpha\beta}h^\alpha_\mu h^\beta_\nu$, and $T_{nn} = T_{\alpha\beta}n^\alpha n^\beta$. We readily see that these are tensor equations and independent upon the choice of coordinates.

Applying { } to the eqs. (16.9) and (16.10), using the Lanczos equation (16.19), we get two constraints

$$^{(3)}R - \{K\}^2 + \{K_{ij}\}\{K^{ij}\} = -\frac{\kappa^2}{4}\left(S_{ij}S^{ij} - \frac{1}{2}S^2\right) - 2\kappa\{T_{nn}\}, \quad (16.27)$$

$$\{^{(3)}\nabla_j K^j_i\} - \{^{(3)}\nabla_i K\} = -\kappa\{T_{in}\}. \quad (16.28)$$

16.3   Surface layers and boundary surfaces

There are two important concepts related to this formalism. If, say, we have an exploding star sending out a thin shell of matter, the energy-momentum tensor will have a sharp peak at the location of the shell. In the approximation where the shell is infinitesimally thin we can describe the energy-momentum tensor by a delta-function at the shell. This is called a *surface layer*. More precisely, a surface layer is a thin layer of matter where the energy-momentum tensor has a non-zero $S_{\alpha\beta}$.

A *boundary surface* is a surface where $S_{\alpha\beta} = 0$. For example, the surface of a star is a boundary surface. Here the energy-momentum tensor has only a discontinuity at the surface and is everywhere bounded.

Let us elaborate a bit more about these cases. Consider a surface layer on which there is a flow of particles. The particles have four-velocity $\mathbf{u} = u^\alpha \mathbf{e}_\alpha$ and are confined to the surface. Thus the velocity is orthogonal to $\mathbf{n}$:

$$u^\alpha n_\alpha|^\pm = 0. \quad (16.29)$$

The geodesic equation reads

$$u^\alpha \nabla_\alpha u^\beta|^\pm = \left(u^{\alpha(3)}\nabla_\alpha u^\beta + \Gamma^n_{\alpha\gamma}u^\alpha u^\gamma n^\beta\right)^\pm. \quad (16.30)$$

By contracting this equation with n_β and using $K_{ij} = n_\alpha \Gamma^\alpha_{ij}$ the orthogonal component of the four-acceleration is

$$n_\alpha a^\alpha|^\pm = K^\pm_{ij} u^i u^j. \quad (16.31)$$

This shows that non-zero difference in the embeddings implies non-zero four-acceleration of the particles of the surface layer. The tangential acceleration is

$$a^j = u^{i(3)}\nabla_i u^j. \quad (16.32)$$

We can split the energy-momentum tensor S_{ij} into

$$S_{ij} = \sigma u_i u_j + t_{ij}, \quad t_{ij}u^i u^j = 0. \quad (16.33)$$

Here, σ is called the mass-energy density of the layer and t_{ij} is called the stress tensor of the layer. Using the Lanczos equation (16.19), and eq. (16.31), we can write

$$[a^\alpha n_\alpha] = \frac{\kappa}{2}\sigma. \quad (16.34)$$

Hence, the orthogonal component of the four-acceleration is determined by the mass-energy density of the surface layer.

Inspecting eq. (16.25) we see that it is similar to the energy-conservation equation for particles on the surface layer. An observer comoving with the particles on the layer with velocity $u = u^i e_i$ observes a momentum-flux given by the contraction of eq. (16.25) with u

$$u^{i\,(3)}\nabla_j S^j_{\ i} = -u^i[T_{in}].$$

$$(16.35)$$

Hence, the bulk energy-momentum tensor may exert a force on the particle in the surface layer.

Consider now eq. (16.26). This equation can also be written as an energy-conservation equation. First note that

$$n_\alpha \nabla_\beta S^{\alpha\beta}\big|^{\pm} = S_{ij} K^{ij}\big|^{\pm}.$$

$$(16.36)$$

This leads to

$$S_{ij}\{K^{ij}\} = \{n_\alpha \nabla_\beta S^{\alpha\beta}\},$$

$$(16.37)$$

and hence eq. (16.26) can be written as

$$\{n_\alpha \nabla_\beta S^{\alpha\beta}\} + [T_{\alpha\beta} n^\alpha n^\beta] = 0.$$

$$(16.38)$$

The term $[T_{\alpha\beta} n^\alpha n^\beta]$ is the difference in pressures exerted normal to the surface Σ. If no such pressure exist (for example for a surface layer in vacuum), the energy-momentum tensor of the surface will obey $\{n_\alpha \nabla_\beta S^{\alpha\beta}\} = 0$. Also, using eq. (16.36) and the Lanczos equation (16.19), we get

$$[n_\alpha \nabla_\beta S^{\alpha\beta}] = \kappa(S_{ij} S^{ij} - S^2).$$

$$(16.39)$$

The equation of continuity of the shell, eq. (16.35), may be written in the case of vanishing energy-momentum tensor outside the shell as

$$u^{i\,(3)}\nabla_j S^j_{\ i} = 0.$$

$$(16.40)$$

Inserting the expression (16.33) into this equation, and noting that $u^{i\,(3)}\nabla_i = d/d\tau$, gives

$$\dot{\sigma} = -\sigma^{(3)}\nabla_i u^i + u^{i\,(3)}\nabla_j t^j_{\ i}.$$

$$(16.41)$$

For boundary surfaces these equations will simplify. As mentioned, the boundary surfaces are characterized by $S_{ij} = 0$ which – by the Lanczos equation – implies

$$[K_{ij}] = 0.$$

$$(16.42)$$

This shows that the embeddings of the surface Σ have to be the same for the two regions. Furthermore, eqs. (16.9) and (16.10) imply that for boundary surfaces we have

$$[T_{\alpha\beta} n^\alpha n^\beta] = [E_{\alpha\beta} n^\alpha n^\beta] = 0,$$

$$(16.43)$$

$$[T_{\alpha\beta} h^\alpha_{\ \mu} n^\beta] = [E_{\alpha\beta} h^\alpha_{\ \mu} n^\beta] = 0.$$

$$(16.44)$$

16.4 Spherical shell of dust in vacuum

Consider a surface energy-momentum tensor of the form

$$S^{ij} = \sigma u^i u^j, \quad u^i u_i = -1, \qquad (16.45)$$

which describes a shell of dust. We will assume that the energy-momentum tensor in $\mathcal{M}^\pm$ is that of vacuum; i.e., $T^\pm_{\mu\nu} = 0$. The comoving velocity u^i of the dust is tangent to the surface Σ and – according to eq. (16.25)– we have

$$^{(3)}\nabla_j(\sigma u^j u^i) = u^{i\,(3)}\nabla_j(\sigma u^j) + \sigma u^{j\,(3)}\nabla_j u^i = 0. \qquad (16.46)$$

Contracting this equation with u_i, and using that $u_i u^{j\,(3)}\nabla_j u^i = u_i a^i = 0$, yields

$$^{(3)}\nabla_j(\sigma u^i) = 0, \qquad (16.47)$$

which shows that the particle number is conserved, and

$$u^{j\,(3)}\nabla_j u^i = 0, \qquad (16.48)$$

which shows that the dust particles are freely falling and their world-lines are geodesics in Σ.

Eq. (16.26) implies

$$S_{ij}\{K^{ij}\} = \sigma u_i u_j\{K^{ij}\} = 0, \qquad (16.49)$$

since we assumed $T^\pm_{\mu\nu} = 0$.

Consider a spherically symmetric spacetime with a shell of dust embedded in vacuum. Outside the shell there is Schwarzschild metric

$$
\begin{aligned}
(ds^2)^+ &= g^+_{\mu\nu} dx^\mu_+ dx^\nu_+ \\
&= -\left(1 - \frac{2M}{r}\right) dt^2 + \frac{dr^2}{1 - \frac{2M}{r}} + r^2(d\theta^2 + \sin^2\theta d\phi^2), \quad (16.50)
\end{aligned}
$$

while inside the metric is that of flat spacetime

$$
\begin{aligned}
(ds^2)^- &= g^-_{\mu\nu} dx^\mu_- dx^\nu_- \\
&= -dT^2 + dr^2 + r^2(d\theta^2 + \sin^2\theta d\phi^2). \quad (16.51)
\end{aligned}
$$

Note that the interior and exterior coordinates do not join smoothly at Σ. This does not really matter since the equations for the junction conditions are coordinate independent tensor equations.

The line-element for the 3-spacetime of the shell is

$$ds^2 = -d\tau^2 + R^2(\tau)(d\theta^2 + \sin^2\theta d\phi^2), \qquad (16.52)$$

where τ is the proper time of the shell. Eq. (16.25) now gives

$$u^{i\,(3)}\nabla_j(\sigma u_i u^j) = 0, \qquad (16.53)$$

which leads to

$$\dot\sigma = -\sigma^{(3)}\nabla_j u^j = -\sigma \frac{1}{\sqrt{|h|}}\left(\sqrt{|h|}u^j\right)_{,j}, \qquad (16.54)$$

where the dot denotes differentiation with respect to the proper time of the shell. With

$$h = -R^4(\tau)\sin^2\theta, \tag{16.55}$$

and $\mathbf{u} = u^\tau \mathbf{e}_\tau = \mathbf{e}_\tau$, this leads to

$$\dot{\sigma} = -\sigma\frac{1}{R^2}\left(R^2\right)^\cdot, \; \tau = -2\sigma\frac{\dot{R}}{R}. \tag{16.56}$$

Integration gives

$$\sigma R^2 = \text{constant}, \tag{16.57}$$

and hence, the rest mass,

$$\mu = 4\pi\sigma R^2, \tag{16.58}$$

of the shell is constant.

The four-velocity of the particles measured from outside the shell is

$$u^\alpha_+ = \frac{dx^\alpha}{d\tau} = (\dot{t}, \dot{R}, 0, 0). \tag{16.59}$$

The vector n_α can be seen by inspection to be

$$n^+_\alpha = (-\dot{R}, \dot{t}, 0, 0). \tag{16.60}$$

The expression for $\dot{t}$ can be found from the four-velocity identities

$$u^\alpha u_\alpha|^+ = -n_\alpha n^\alpha|^+ = \dot{t}^2 g^+_{tt} + \dot{R}^2 g^+_{rr} = -1. \tag{16.61}$$

Thus

$$\dot{t} = \frac{\sqrt{1 - \frac{2M}{R} + \dot{R}^2}}{1 - \frac{2M}{R}}. \tag{16.62}$$

Taking the covariant derivative, $u^\beta\nabla_\beta$, of the identity $u_\alpha u^\alpha = -1$, we obtain

$$u_\alpha a^\alpha|^+ = u_\alpha u^\beta\nabla_\beta u^\alpha|^+ = u_t u^\beta\nabla_\beta u^t|^+ + u_r u^\beta\nabla_\beta u^r|^+, \tag{16.63}$$

which can be used to substitute $u^\beta\nabla_\beta u^t$ into

$$\begin{aligned} n_\alpha a^\alpha|^+ &= u_\alpha u^\beta\nabla_\beta u^\alpha|^+ = n_t u^\beta\nabla_\beta u^t|^+ + n_r u^\beta\nabla_\beta u^r|^+ \\ &= \left(n_r - n_t\frac{u_r}{u_t}\right)u^\beta\nabla_\beta u^r\bigg|^+. \end{aligned} \tag{16.64}$$

Writing the covariant derivative using the connection coefficients

$$u^\beta\nabla_\beta u^r|^+ = u^r{}_{,\alpha}u^\alpha|^+ + \Gamma^r{}_{\alpha\beta}u^\alpha u^\beta|^+, \tag{16.65}$$

and using the expression for the connection coefficients, eq. (6.111) and the metric (16.50), we get

$$\begin{aligned} \Gamma^r{}_{\alpha\beta}u^\alpha u^\beta|^+ &= \frac{1}{2}g^{rr}\left(2g_{r\alpha,\beta} - g_{\alpha\beta,r}\right)u^\alpha u^\beta|^+ \\ &= \frac{1}{2}g^{rr}\left(2g_{rr,r}u^r u^r - g_{tt,r}u^t u^t\right)|^+ \\ &= \frac{M}{R^2}. \end{aligned} \tag{16.66}$$

Thus,

$$u^\beta \nabla_\beta u^r \big|^+ = \ddot{R} + \frac{M}{R^2}. \tag{16.67}$$

Using eqs. (16.59), (16.60) and (16.62) together with the metric (16.50), yield

$$\left(n_r - n_t \frac{u_r}{u_t}\right)\Bigg|^+ = \left(n_r - n_t \frac{g_{rr} u^r}{g_{tt} u^t}\right)\Bigg|^+$$

$$= \frac{1}{\sqrt{1 - \frac{2M}{R} + \dot{R}^2}}. \tag{16.68}$$

Hence, eq. (16.64) turns into

$$n_\alpha a^\alpha \big|^+ = \frac{\ddot{R} + \frac{M}{R^2}}{\sqrt{1 - \frac{2M}{R} + \dot{R}^2}}. \tag{16.69}$$

To get the expression for the inner region, we can set $M = 0$ and obtain

$$n_\alpha a^\alpha \big|^- = \frac{\ddot{R}}{\sqrt{1 + \dot{R}^2}}. \tag{16.70}$$

The equation of motion can now be found by using eqs. (16.31) and (16.49):

$$n_\alpha a^\alpha \big|^+ + n_\alpha a^\alpha \big|^- = 0, \tag{16.71}$$

which – using eqs. (16.69) and (16.70) – leads to

$$\frac{\ddot{R}}{\sqrt{1 + \dot{R}^2}} + \frac{\ddot{R} + \frac{M}{R^2}}{\sqrt{1 - \frac{2M}{R} + \dot{R}^2}} = 0. \tag{16.72}$$

This is the equation of motion for the expanding shell. Multiplying by $\dot{R}$, the expression turns into a total derivative

$$\frac{d}{d\tau}\left[\sqrt{1 + \dot{R}^2} + \sqrt{1 - \frac{2M}{R} + \dot{R}^2}\right] = 0. \tag{16.73}$$

The equation has a first integral

$$\sqrt{1 + \dot{R}^2} + \sqrt{1 - \frac{2M}{R} + \dot{R}^2} = 2a, \tag{16.74}$$

which can be rearranged to yield

$$\sqrt{1 + \dot{R}^2} = a + \frac{M}{2aR}. \tag{16.75}$$

Here, a is a constant of integration. The physical interpretation of a is as follows. Note first that if $\dot{R} = 0$ as $R \to \infty$, then $a = 1$. From eq. (16.34) we get, using eq. (16.75),

$$4\pi R^2 \sigma = \frac{M}{a}. \tag{16.76}$$

The left-hand side is the rest mass of the particles in the shell. The gravitational mass M of the Schwarzschild solution gives the *total mass* of the shell. Hence, both the rest mass and the kinetic energy contributes to M. The difference

$$\frac{M}{a} - M = \frac{M(1-a)}{a}, \qquad (16.77)$$

therefore gives the "binding energy" of the shell. A shell which reaches zero velocity at infinity, has zero binding energy.

Example

Example 16.1 (A source for the Kerr field)

Here we will consider the Kerr metric which in Boyer-Lindquist coordinates is given in (10.165) on page 239. The following source for the Kerr field was first found by Israel himself [Isr70].

In the following we need both the covariant and contravariant components of the Kerr metric. These are (order of the diagonal, $(g_{tt}, g_{rr}, g_{\theta\theta}, g_{\phi\phi})$)

$$(g_{\mu\nu}) = \begin{bmatrix} -\left(1 - \frac{2M}{\Sigma}\right) & 0 & 0 & -\frac{2Mar\sin^2\theta}{\Sigma} \\ 0 & \frac{\Sigma}{\Delta} & 0 & 0 \\ 0 & 0 & \Sigma & 0 \\ -\frac{2Mar\sin^2\theta}{\Sigma} & 0 & 0 & \left(r^2 + a^2 + \frac{2Ma^2r\sin^2\theta}{\Sigma}\right)\sin^2\theta \end{bmatrix}, \quad (16.78)$$

$$(g^{\mu\nu}) = \begin{bmatrix} -\left(r^2 + a^2 + \frac{2Ma^2r\sin^2\theta}{\Sigma}\right)\frac{1}{\Delta} & 0 & 0 & -\frac{2Mar}{\Sigma\Delta} \\ 0 & \frac{\Delta}{\Sigma} & 0 & 0 \\ 0 & 0 & \frac{1}{\Sigma} & 0 \\ -\frac{2Mar}{\Sigma\Delta} & 0 & 0 & \frac{\Delta - a^2\sin^2\theta}{\Sigma\Delta\sin^2\theta} \end{bmatrix}, \quad (16.79)$$

where

$$\Sigma = r^2 + a^2\cos^2\theta, \quad \Delta = r^2 + a^2 - 2Mr.$$

Consider the unit vector given by $\mathbf{n} = n^r \mathbf{e}_r$, $g_{rr}(n^r)^2 = 1$. Using the metric (16.78), gives

$$\mathbf{n} = \sqrt{\frac{\Delta}{\Sigma}}\mathbf{e}_r. \qquad (16.80)$$

We will now find the exterior curvature of a surface given by $r = $ constant. Eq. (7.75) gives the following non-zero components

$$K_{\theta\theta} = n^r \Gamma_{r\theta\theta} = -\frac{1}{2}n^r\frac{\partial g_{\theta\theta}}{\partial r} = -rn^r = -r\sqrt{\frac{\Delta}{\Sigma}},$$

$$K_{\phi\phi} = -\frac{1}{2}\sqrt{\frac{\Delta}{\Sigma}}\frac{\partial g_{\phi\phi}}{\partial r} = -\left[r + \left(1 - 2\frac{r^2}{\Sigma}\right)\frac{Ma^2}{\Sigma}\sin^2\theta\right]\sqrt{\frac{\Delta}{\Sigma}}\sin^2\theta,$$

$$K_{tt} = -\frac{1}{2}\sqrt{\frac{\Delta}{\Sigma}}\frac{\partial g_{tt}}{\partial r} = \left(1 - 2\frac{r^2}{\Sigma}\right)M\sqrt{\frac{\Delta}{\Sigma^3}},$$

$$K_{t\phi} = -\frac{1}{2}\sqrt{\frac{\Delta}{\Sigma}}\frac{\partial g_{t\phi}}{\partial r} = \left(1 - 2\frac{r^2}{\Sigma}\right)Ma\sqrt{\frac{\Delta}{\Sigma^3}}. \qquad (16.81)$$

We consider the surface $r = 0$. The metric for $r = 0$ is diagonal

$$g_{\mu\nu} = \text{diag}(-1, \cos^2\theta, a^2\cos^2\theta, a^2\sin^2\theta),$$

$$g^{\mu\nu} = \frac{1}{a^2\cos^2\theta}\text{diag}(-a^2\cos^2\theta, a^2, 1, \cot^2\theta), \qquad (16.82)$$

so the extrinsic curvature on the surface simplifies to

$$K_{\theta\theta} = 0, \qquad K_{\phi\phi} = -M\frac{\sin^4\theta}{\cos^3\theta},$$

$$K_{tt} = -\frac{M}{a^2\cos^3\theta}, \qquad K_{t\phi} = \frac{M\sin^2\theta}{a\cos^3\theta}. \qquad (16.83)$$

We also need the mixed components, which – using the metric (16.82) – can be found to be

$$K^\theta_{\ \theta} = 0, \quad K^\phi_{\ \phi} = -M \frac{\sin^2\theta}{a^2 \cos^3\theta}, \quad K^t_{\ t} = aK^\phi_{\ t} = \frac{M}{a^2\cos^3\theta}. \tag{16.84}$$

Thus,

$$K = K^t_{\ t} + K^\theta_{\ \theta} + K^\phi_{\ \phi} = \frac{M}{a^2\cos\theta}.$$

We will assume that two identical spacetimes is glued along the surface $r = 0$. The different extrinsic curvatures $K^{i\pm}_{\ j}$ will then differ by only a sign: $K^{i+}_{\ j} = -K^{i-}_{\ j}$. The energy-momentum tensor of the surface layer can then be found using the Lanczos equation (16.19). The expressions (16.84) yield (using $G = 1 = c$)

$$
\begin{aligned}
S^t_{\ t} &= \frac{1}{4\pi}\frac{\sin^2\theta}{a^2\cos^3\theta}M, & S^\theta_{\ \theta} &= \frac{1}{4\pi}\frac{M}{a^2\cos\theta}, \\
S^\phi_{\ \phi} &= -\frac{1}{4\pi}\frac{M}{a^2\cos^3\theta}, & S^\phi_{\ t} &= \frac{1}{4\pi}\frac{M}{a^3\cos^3\theta}.
\end{aligned}
\tag{16.85}
$$

These expressions are encompassed in the single equation

$$S^i_{\ j} = \sigma(u^i u_j + k^i k_j), \quad \sigma = -\frac{1}{4\pi}\frac{M}{a^2\cos\theta}, \tag{16.86}$$

and

$$
\begin{aligned}
u^i &= (u^t, u^\theta, u^\phi) = \left(\tan\theta, 0, \frac{1}{a\sin\theta\cos\theta}\right), \\
k^i &= (k^t, k^\theta, k^\phi) = \left(0, \frac{1}{a\cos\theta}, 0\right).
\end{aligned}
\tag{16.87}
$$

This is Israel's source for the Kerr spacetime. The surface layer consist of matter with negative energy-density, and a stress $t^i_{\ j} = \sigma k^i k_j$ where the only non-zero component is $t^\theta_{\ \theta}$.

The coordinate velocity of particles comoving with the surface, v^i, can be found from

$$u^i = \frac{dx^i}{d\tau} = \frac{dt}{d\tau}\frac{dx^i}{dt} = u^t v^i, \tag{16.88}$$

which gives

$$v^\phi = \frac{1}{a\sin^2\theta}. \tag{16.89}$$

The coordinate velocity of light moving in the ϕ-direction, can be found by inserting $ds = dr = 0$ in the metric of the surface layer. This gives

$$c^\phi = \frac{1}{a\sin\theta}, \tag{16.90}$$

and hence,

$$v^\phi = \frac{c^\phi}{\sin\theta}, \tag{16.91}$$

which shows that the particles are moving at tachyonic speeds.

Further details and an extension to the Kerr-Newman spacetime (rotating black hole with an electric charge) can be found in [Lop84, Grø85].

Problems

16.1. *Energy equation for a shell of dust*
Use the first integral eq. (16.75) for a shell of dust and write the equation on
the form

$$\frac{1}{2}\dot{R}^2 + V(R) = E,$$ (16.92)

where E is a constant such that $E = 0$ for $a = 1$. What is $V(R)$?

What is the condition on a for recollapse of the shell? What is the condition
for ever-expansion?

16.2. *Charged shell of dust*

(a) Show that the equation of motion of a thin, charged shell of dust is

$$\ddot{R} + \frac{\mu^2 - Q^2}{2\mu}\sqrt{1 + \dot{R}^2}\frac{1}{R^2} = 0,$$ (16.93)

where μ is the rest mass of the shell and Q its charge.

(b) Show that the energy equation of the shell may be written

$$M = \mu\sqrt{1 + \dot{R}^2} - \frac{\mu^2 - Q^2}{2R},$$

where M is the total mass of the shell, which appears in the Reissner-
Nordström metric, eq. (10.120). Give a physical interpretation of the terms.

16.3. *A spherical domain wall*
In this problem we will consider a spherical domain wall (or a shell) and assume that spherical coordinates are used. The energy-momentum tensor of
the shell is

$$t_{ij} = -\sigma(h_{ij} + u_i u_j).$$ (16.94)

Hence the only non-vanishing components are

$$t^\theta_{\ \theta} = t^\phi_{\ \phi} = -\sigma.$$

(a) Show that the equation of continuity (16.44), as applied to a spherical
domain wall reduces to $\dot{\sigma} = 0$. What happens to the rest mass of the
domain wall during expansion? Try to find a physical reason for the
result you found.

(b) Show that the energy equation of the domain wall may be written

$$M = 4\pi\sigma R^2 \left[\sqrt{1 + \dot{R}^2} - 2\pi\sigma R\right].$$

(c) Calculate, in terms of σ, the radius R_S of a static domain wall with radius
equal to its Schwarzschild radius. Can the domain wall have greater
radius than R_S?

16.4. *Dynamics of spherical domain walls*
We shall consider a spherical domain wall in the Schwarzschild-de Sitter spacetime with line element

$$ds^2 = -f(r)dt^2 + \frac{dr^2}{f(r)} + r^2\left(d\theta^2 + \sin^2\theta d\phi^2\right),$$

$$f(r) = 1 - \frac{2M}{r} - \frac{\Lambda}{3}r^2.$$ (16.95)

The values of Λ and M can be different inside and outside the domain wall. The values of f inside and outside are denoted, respectively, by f^- and f^+. The motion of the domain wall is given by $r = R(\tau)$ where τ is the proper time of the wall. The line element of the domain wall is

$$ds_d^2 = -d\tau^2 + R(\tau)^2 \left(d\theta^2 + \sin^2\theta d\phi^2\right). \qquad (16.96)$$

Hence, observers on the wall will perceive $R(\tau)$ as an expansion factor.

(a) Show that the four velocity of a fixed particle on the wall is

$$\mathbf{u} = \frac{\sqrt{f + \dot{R}^2}}{f}\mathbf{e}_t + \dot{R}\mathbf{e}_r.$$

(b) Show that the unit normal vector is

$$\mathbf{n} = \frac{\dot{R}}{f}\mathbf{e}_t - \sqrt{f + \dot{R}^2}\,\mathbf{e}_r.$$

(c) Show that the $\theta\theta$-component of the extrinsic curvature is

$$K_{\theta\theta} = -\sqrt{f + \dot{R}^2}\,R.$$

(d) Use the Israel junction condition, eq. (16.33), with the energy momentum tensor of a domain wall, $S_{ij} = -\sigma h_{ij}$, to show that the equation of motion of the domain wall is

$$\sqrt{f^+ + \dot{R}^2} + \sqrt{f^- + \dot{R}^2} = 4\pi G\sigma R. \qquad (16.97)$$

17

Brane-worlds

In 1999, Lisa Randall and Raman Sundrum presented a five-dimensional model for our universe [RS99b, RS99a]. They imagined our four-dimensional world as a *brane-world* or a surface layer in a five-dimensional bulk. This bulk may be infinite in size, but due to the special properties of the bulk the gravitational fields are effectively localised to the brane. The other standard model fields are confined to the brane; only gravity is allowed to propagate in the fifth dimension.

Here we will shortly review the idea behind the brane-world models. The interest for brane-worlds has been enormous the following years after Randall and Sundrum's papers appeared. This focus on brane-worlds has also renewed the interest for the metric junction method, which we introduced in the previous chapter, and this application is a prime example of the diversity and the generality of the metric junction method. The brane-worlds are models with an extra dimension, and hence, we cannot use all the former equations directly without special consideration of the dimensionality. However, the Lanczos equation (16.19) is valid without any further adjustments.

17.1 Field equations on the brane

(Shiromizu *et al.* [SMS00] and Maartens [Maa00])
In the brane-world scenario our four-dimensional world is described as a four-dimensional surface – the brane – in a five-dimensional spacetime – the bulk. In order to deduce the field equations on the brane we start with eq. (7.83) written on the form

$$^{(4)}R^{\alpha}{}_{\mu\beta\nu} = {}^{(5)}R^{\lambda}{}_{\delta\rho\sigma}h^{\alpha}{}_{\lambda}h^{\delta}{}_{\mu}h^{\rho}{}_{\beta}h^{\sigma}{}_{\nu} + K^{\alpha}{}_{\beta}K_{\mu\nu} - K^{\alpha}{}_{\nu}K_{\mu\beta}, \tag{17.1}$$

where $h_{\alpha\beta} = g_{\alpha\beta} - n_{\alpha}n_{\beta}$ is the metric on the brane. Contracting α with β we find

$$^{(4)}R_{\mu\nu} = {}^{(5)}R_{\rho\sigma}h^{\rho}{}_{\mu}h^{\sigma}{}_{\nu} - {}^{(5)}R^{\alpha}{}_{\beta\gamma\delta}n_{\alpha}h^{\beta}{}_{\mu}n^{\gamma}h^{\delta}{}_{\nu} + KK_{\mu\nu} - K^{\alpha}{}_{\mu}K_{\alpha\nu}, \tag{17.2}$$

where $K = K^\alpha_\alpha$. Calculating the Einstein tensor this gives

$$
\begin{aligned}
{}^{(4)}E_{\mu\nu} &= {}^{(5)}E_{\rho\sigma}h^\rho_\mu h^\sigma_\nu - {}^{(5)}R_{\alpha\beta}n^\alpha n^\beta h_{\mu\nu} + KK_{\mu\nu} - K^\alpha_\mu K_{\alpha\nu} \\
&\quad - \frac{1}{2}h_{\mu\nu}\left(K^2 - K^{\alpha\beta}K_{\alpha\beta}\right) - {}^{(5)}R^\alpha_{\beta\gamma\delta}n_\alpha h^\beta_\mu n^\gamma h^\delta_\nu.
\end{aligned} \tag{17.3}
$$

Decomposing the Riemann tensor into the Ricci tensor, the Ricci scalar and the Weyl curvature tensor, $C_{\mu\alpha\nu\beta}$, according to

$$
\begin{aligned}
{}^{(5)}R_{\mu\alpha\nu\beta} &= \frac{2}{3}\left(g_{\mu[\nu}{}^{(5)}R_{\beta]\alpha} - g_{\alpha[\nu}{}^{(5)}R_{\beta]\mu}\right) - \frac{1}{6}g_{\mu[\nu}g_{\beta]\alpha}{}^{(5)}R \\
&\quad + {}^{(5)}C_{\mu\alpha\nu\beta},
\end{aligned} \tag{17.4}
$$

and replacing the Ricci tensor with the Einstein tensor, eq. (17.3) may be rewritten as

$$
\begin{aligned}
{}^{(4)}E_{\mu\nu} &= \frac{2}{3}\left[{}^{(5)}E_{\rho\sigma}h^\rho_\mu h^\sigma_\nu - h_{\mu\nu}\left({}^{(5)}E_{\alpha\beta}n^\alpha n^\beta - \frac{1}{4}{}^{(5)}E\right)\right] \\
&\quad + KK_{\mu\nu} - K^\alpha_\mu K_{\alpha\nu} - \frac{1}{2}h_{\mu\nu}\left(K^2 - K^{\alpha\beta}K_{\alpha\beta}\right) - \mathcal{E}_{\mu\nu}.
\end{aligned} \tag{17.5}
$$

Here,

$$
\boxed{\mathcal{E}_{\mu\nu} \equiv {}^{(5)}C_{\alpha\beta\gamma\delta}n^\alpha h^\beta_\mu n^\gamma h^\delta_\nu,} \tag{17.6}
$$

is the so-called[1] "electric part" of the Weyl tensor. Eq. (17.5) is a geometrical identity without physical contents.

We now apply Einstein's five-dimensional field equations[2]

$$
{}^{(5)}E_{\mu\nu} = \kappa_5\, {}^{(5)}T_{\mu\nu}. \tag{17.7}
$$

The energy-momentum tensor has contributions from the brane, $T_{b\mu\nu} = \delta(y)S_{\mu\nu}$, where $y = 0$ is the position of the brane, and from the bulk, $T_{B\mu\nu}$, i.e. ${}^{(5)}T_{\mu\nu} = T_{b\mu\nu} + T_{B\mu\nu}$. Using eq. (17.7) the five-dimensional Einstein tensor can be replaced by the energy-momentum tensor in eq. (17.5). Next, the extrinsic curvature tensor can be replaced by the stress-energy tensor, $S_{\mu\nu}$, of the brane by means of Israel's junction conditions

$$
\boxed{[K_{ij}] = \kappa_5\left(S_{ij} - \frac{1}{3}Sh_{ij}\right).} \tag{17.8}
$$

Note that the factor $1/2$ in eq. (16.19) has been replaced by $1/3$ and we have assumed $\epsilon = 1$. This is due to the four spacetime dimensions of the brane, which implies that $h^i_i = 4$. Assuming mirror symmetry, or $\mathbb{Z}_2$-symmetry, across the brane, we can replace the jump in the extrinsic curvature by twice the value of the extrinsic curvature at the location of the brane. Hence (dropping the sup-script $+$)

$$
K_{ij} = \frac{\kappa_5}{2}\left(S_{ij} - \frac{1}{3}Sh_{ij}\right). \tag{17.9}
$$

---

[1] Usually the electric part of the Weyl tensor is defined for n^α time-like; however, it is common in the brane literature to call $\mathcal{E}_{\mu\nu}$ the electric part although n^α is space-like.

[2] In the literature on brane cosmology it has become usual to denote Einstein's gravitational constant by κ_5^2 and not by κ_5 as we do in this book.

We shall assume that the bulk is empty except for LIVE represented by a cosmological constant, Λ_B. The stress-energy tensor of the brane is written as

$$S_{ij} = -\lambda h_{ij} + \tilde{T}_{ij}, \qquad (17.10)$$

where λ and $\tilde{T}_{ij}$ are the vacuum energy density and energy-momentum tensor, respectively, on the brane. From a five-dimensional point of view λ is interpreted as the tension of the brane. Using eqs. (17.7), (17.9) and (17.10), eq. (17.5) can be written as

$$\boxed{{}^{(4)}E_{ij} + \Lambda h_{ij} = \kappa_4 \tilde{T}_{ij} + \kappa_5^2 \tau_{ij} - \mathcal{E}_{ij},} \qquad (17.11)$$

where

$$\Lambda = \frac{1}{2}\left(\Lambda_B + \frac{\kappa_5^2 \lambda^2}{6}\right) = \frac{1}{2}\left(\Lambda_B + \kappa_4 \lambda\right), \qquad (17.12)$$

is the ordinary cosmological constant measured by brane inhabitants, and[3]

$$8\pi G_N = \kappa_4 = \frac{\kappa_5^2}{6}\lambda. \qquad (17.13)$$

Furthermore,

$$\tau_{ij} = -\frac{1}{4}\tilde{T}_{ia}\tilde{T}^a_{\;j} + \frac{1}{12}\tilde{T}\tilde{T}_{ij} - \frac{1}{24}h_{ij}\left(3\tilde{T}_{ab}\tilde{T}^{ab} - \tilde{T}^2\right), \qquad (17.14)$$

and $\mathcal{E}_{ij}$ is the electric part of the Weyl tensor defined in eq. (17.6).

Equation (17.11) is the brane generalization of the four-dimensional Einstein equations. Note that for the Newtonian gravitational constant to be non-zero and positive there must exist a positive vacuum energy (or brane tension) on the brane.

If the matter on the brane is a perfect fluid,

$$\tilde{T}_{ij} = \rho u_i u_j + p \tilde{h}_{ij}, \qquad (17.15)$$

where $\tilde{h}_{ij} = h_{ij} + u_i u_j$ is the spatial metric tensor on the brane. The effective energy-momentum tensor coming from the Israel matching conditions associated with the external curvature of the brane, is

$$\tau_{ij} = \frac{1}{12}\rho^2 u_i u_j + \frac{1}{12}\rho(\rho + 2p)\tilde{h}_{ij}. \qquad (17.16)$$

If the fluid obeys the equation of state $p = w\rho$ this tensor takes the form

$$\tau_{ij} = \frac{1}{12}\rho^2\left[u_i u_j + (1 + 2w)\tilde{h}_{ij}\right]. \qquad (17.17)$$

The term $\mathcal{E}_{ij}$ in eq. (17.11) represents the effect on the brane of the free gravitational field in the bulk. This term vanishes if the bulk spacetime is

---

[3]Eq. (17.13) can also be expressed as a relation between the four- and five-dimensional Planck masses. The four-dimensional Planck mass is given by $m_{\text{Pl}} = \sqrt{\hbar c/G}$. Using units so that $\hbar = c = 1$, Newton's gravitational constant may be expressed by $G = m_{\text{Pl}}^{-2}$ or $\kappa_4 = 8\pi m_{\text{Pl}}^{-2}$. In a similar way the five-dimensional gravitational constant and Planck mass are related by $G_5 = m_5^{-3}$. Hence, $\kappa_5 = 8\pi m_5^{-3}$. Inserting these expressions into eq. (17.13) gives $m_{\text{Pl}}^2 = 3m_5^6/4\pi\lambda$.

purely anti-de Sitter. Also, if there are several branes they interact gravitationally via the Weyl curvature that they generate. The effective energy-density
on the brane, arising from the free gravitational field in the bulk, is defined as

$$\mathcal{U} = -\frac{\kappa_4 \lambda}{6} \mathcal{E}_{ij} u^i u^j. \tag{17.18}$$

Furthermore, the tensor $\mathcal{E}_{ij}$ can be covariantly decomposed as

$$\mathcal{E}_{ij} = -\frac{6}{\kappa_4 \lambda} \left[\mathcal{U} \left(u_i u_j + \frac{1}{3} \tilde{h}_{ij} \right) + \mathcal{P}_{ij} + 2\mathcal{Q}_{(i} u_{j)} \right]. \tag{17.19}$$

Here, $\mathcal{P}_{ij}$ is a trace-less and symmetric tensor called the non-local anisotropic
stress tensor, and $\mathcal{Q}_i$ is the non-local energy flux. This tensor is very similar to
the energy-momentum tensor of a radiation fluid. This correspondence goes
even further. From Bianchi's second identity, eq. (7.58), we have

$$\nabla^i \mathcal{E}_{ij} = \kappa_5^2 \nabla^i \tau_{ij}. \tag{17.20}$$

In the case of an isotropic brane with no energy flux on the brane, i.e. a brane
that may be described by the Robertson-Walker line-element, the electric part
of the Weyl tensor may be written as

$$\mathcal{E}_{ij} = -\frac{6}{\kappa_4 \lambda} \mathcal{U} \left(u_i u_j + \frac{1}{3} \tilde{h}_{ij} \right). \tag{17.21}$$

For a perfect fluid, the right hand side of eq. (17.20) vanishes due to the
energy-momentum conservation of the fluid. Hence, in this case the non-local
energy-density obeys the radiation-like energy-conservation equation

$$\dot{\mathcal{U}} + 4H\mathcal{U} = 0, \tag{17.22}$$

where H is the Hubble parameter on the brane. However, unlike radiation,
the non-local energy-density may be negative. Also, it is worth noting that the
limit $\lambda \to \infty$ while keeping κ_4 fixed makes $\kappa_5 \to 0$ and $\mathcal{E}_{ij} \to 0$. In this limit
the non-local density $\mathcal{U}$ decouples the brane and we recover the conventional
Friedmann equations of four-dimensional cosmology.

17.2   Five-dimensional brane cosmology

Let us now consider some universe models resulting from a brane picture of
the world which is assumed to be five-dimensional (see also [Lan03, MPLP01]).
The line-element of the five-dimensional spacetime may then be written

$$ds^2 = -n^2(t,y)dt^2 + a^2(t,y) \left[\frac{dr^2}{1 - kr^2} + r^2(d\theta^2 + \sin^2\theta d\phi^2) \right] + b^2(t,y)dy^2. \tag{17.23}$$

The brane has zero thickness and is localized at $y = 0$. The functions $a(t,y)$,
$b(t,y)$ and $n(t,y)$ are continuous at the brane, but their derivatives are discontinuous. The metric in the brane is

$$ds^2_{\text{Brane}} = -n^2(t,0)dt^2 + a^2(t,0) \left[\frac{dr^2}{1 - kr^2} + r^2(d\theta^2 + \sin^2\theta d\phi^2) \right]. \tag{17.24}$$

If t is the proper time on the brane then $n(t, 0) = 1$.

Einstein's equations of the five-dimensional world are

$$^{(5)} R_{\mu\nu} - \frac{1}{2} {}^{(5)} R g_{\mu\nu} = \kappa_5 \left(T_{b\mu\nu} + T_{B\mu\nu} \right), \tag{17.25}$$

where $^{(5)} R_{\mu\nu}$ is the five-dimensional Ricci tensor and $^{(5)} R \equiv {}^{(5)} R^{\mu}_{\mu}$ its trace, $\kappa_5 = 8\pi G_5$ is the gravitational constant of five-dimensional spacetime Furthermore, $T_{b\mu\nu}$ and $T_{B\mu\nu}$ are the energy-momentum tensors of the brane and bulk, respectively.

We shall consider isotropic perfect bulk and brane fluids. Then the energy-momentum tensors of the brane and bulk are

$$T_{b}{}^{\mu}{}_{\nu} = S^{\mu}{}_{\nu} \delta(y) = \mathrm{diag}(-\rho_b, p_b, p_b, p_b, 0) \delta(y) \tag{17.26}$$

where ρ_b is the brane energy density and p_b the brane pressure, and

$$T_{B}{}^{\mu}{}_{\nu} = \mathrm{diag}(-\rho_B, p_B, p_B, p_B, p_B), \tag{17.27}$$

respectively.

Einstein's field equations in the bulk are (using an orthonormal frame)

$$E_{\hat{t}\hat{t}} = \frac{3}{n^2} \left(\frac{\dot{a}^2}{a^2} + \frac{\dot{a}\dot{b}}{ab} \right) - \frac{3}{b^2} \left(\frac{a''}{a} + \frac{a'^2}{a^2} - \frac{a'b'}{ab} \right) + \frac{3k}{a^2} = \kappa_5 \rho_B, \tag{17.28}$$

$$E_{\hat{i}\hat{i}} = \frac{1}{b^2} \left[2\frac{a''}{a} + \frac{n''}{n} + \frac{a'^2}{a^2} + 2\frac{a'n'}{an} - \frac{b'}{b} \left(2\frac{a'}{a} + \frac{n'}{n} \right) \right]$$
$$+ \frac{1}{n^2} \left[2\frac{\dot{a}\dot{n}}{an} - 2\frac{\ddot{a}}{a} - \frac{\dot{a}^2}{a^2} + \frac{\dot{b}}{b} \left(\frac{\dot{n}}{n} - 2\frac{\dot{a}}{a} \right) - \frac{\ddot{b}}{b} \right] - \frac{k}{a^2} = \kappa_5 p_B, \tag{17.29}$$

$$E_{\hat{t}\hat{y}} = 3 \left(\frac{n'\dot{a}}{na} - \frac{\dot{a}'}{a} + \frac{a'\dot{b}}{ab} \right) = 0, \tag{17.30}$$

$$E_{\hat{y}\hat{y}} = \frac{3}{b^2} \left(\frac{a'^2}{a^2} + \frac{a'n'}{an} \right) - \frac{3}{n^2} \left(\frac{\ddot{a}}{a} + \frac{\dot{a}^2}{a^2} - \frac{\dot{a}\dot{n}}{an} \right) - \frac{3k}{a^2} = \kappa_5 p_B, \tag{17.31}$$

where a dot denotes derivative with respect to t and a prime with respect to y. Eq. (17.30) is due to the assumption that there is no energy flux in the bulk.

The Bianchi identity implies the energy-momentum conservation law for the bulk fluid

$$T_{B}{}^{\mu}{}_{\nu;\mu} = 0, \tag{17.32}$$

which gives the equations

$$\dot{\rho}_B + \left(3\frac{\dot{a}}{a} + \frac{\dot{b}}{b} \right) (\rho_B + p_B) = 0, \tag{17.33}$$

$$p'_B + \frac{n'}{n} (\rho_B + p_B) - 3\frac{a'}{a} (\rho_B - p_B) = 0. \tag{17.34}$$

In the case of a time-like brane, $\epsilon = 1$. From eq. (16.7), and using that the unit normal vector to the brane is $\mathbf{n} = \mathbf{e}_y$, we find the non-vanishing components of the extrinsic curvature tensor of the brane

$$K_{tt} = -\frac{1}{2}n^y\frac{\partial g_{tt}}{\partial y} = \frac{n_0}{b_0}n_0',$$

$$K_{ii} = -\frac{1}{2}n^y\frac{\partial g_{ii}}{\partial y} = -\frac{a_0}{b_0}a_0', \qquad (17.35)$$

where the index 0 means that the quantity shall be evaluated at the brane. If the brane is identified with our world, then $a_0 \equiv a(t,0)$ is the expansion factor of the Friedmann-Robertson-Walker models. Note that the non-vanishing of the extrinsic curvature of the brane means that the five-dimensional metric depends necessarily on the coordinate of the fifth dimension, in contrast to the usual assumption in the Kaluza-Klein approach which we will review in the next chapter.

Substituting eqs. (17.26) and (17.34) into eq. (17.9) gives the relations

$$\frac{n_0'}{n_0} = \frac{\kappa_5}{6}b_0\left(2\rho_b + 3p_b\right), \qquad \frac{a_0'}{a_0} = -\frac{\kappa_5}{6}b_0\rho_b. \qquad (17.36)$$

Inserting these expressions into eq. (17.31) and letting t be the proper time on the brane, so that $n_0 = 1$, we get

$$\frac{\ddot{a}_0}{a_0} + \frac{\dot{a}_0^2}{a_0^2} = -\frac{\kappa_5^2}{36}\rho_b\left(\rho_b + 3p_b\right) - \frac{\kappa_5}{3}p_B - \frac{k}{a_0^2}. \qquad (17.37)$$

We shall now solve Einstein's vacuum equations with a cosmological constant $\Lambda_B = \kappa_5\rho_B$ in the bulk outside the brane[4]. In the main text we shall assume that the scale factor of the fifth dimension is constant and normalized to 1. (Some models with variable $b(t,y)$ will be considered in problem 17.4.) With $b(y,t) = 1$ the combination $E_{\hat{t}\hat{t}} + 2E_{\hat{y}\hat{y}} - 3E_{\hat{i}\hat{i}}$ yields

$$3\frac{a''}{a} + \frac{n''}{n} = \frac{\kappa_5}{3}\left(p_B - \rho_B\right), \qquad (17.38)$$

and the equation $E_{\hat{t}\hat{y}} = 0$ leads to

$$\frac{n'}{n} = \frac{\dot{a}'}{\dot{a}}. \qquad (17.39)$$

Integration gives

$$\dot{a} = f(t)n, \qquad (17.40)$$

where $f(t)$ is an arbitrary function of t. Note that $f(t) = \dot{a}_0$ since $n_0 = 1$. Furthermore, eq. (17.28) gives

$$(aa')' - f^2 - k + \frac{\Lambda_B}{3}a^2 = 0. \qquad (17.41)$$

Multiplying by aa' and integrating, one obtains

$$(aa')^2 - f^2a^2 - ka^2 + \frac{\Lambda_B}{6}a^4 = U. \qquad (17.42)$$

---

[4]Some authors identify Λ_B with ρ_B. If this is done, one should multiply Λ_B by κ_5 or alternatively by $8\pi m_5^{-3}$ in the equations below.

Using eq. (17.31) one finds that U cannot depend on the time; i.e. U is a constant. Evaluating the terms at the position of the brane, $y = 0$, where $f = \dot{a}_0$, and inserting the second of eq. (17.36), we arrive at

$$H_0^2 = \left(\frac{\dot{a}_0}{a_0}\right)^2 = \frac{\kappa_5^2}{36}\rho_b^2 + \frac{\Lambda_B}{6} - \frac{k}{a_0^2} + \frac{U}{a_0^4}. \tag{17.43}$$

This equation relates the Hubble parameter to the energy density. However, it is different from the usual Friedmann equation. In particular H^2 depends quadratically upon the density and not linearly as usual. As long as the five-dimensional Planck scale m_5 is larger than 10TeV the effect of the ρ^2 term will be negligible from the time of neutrino decoupling (at 1MeV, i.e. about 1s after the big bang) onwards. The last term in eq. (17.43) reminds of a radiation term, but there is no contribution from radiation in the energy-momentum tensor. If non-vanishing it would constitute a sort of dark radiation. Later, in section 17.4, it is explicitly shown that this is exactly the radiation-like term that arises from the tensor $\mathcal{E}_{ij}$ defined in eq. (17.21).

Problem with perfect fluid brane world in an empty bulk

Eqs. (17.36) and (17.39) lead to the energy conservation equation on the brane

$$\dot{\rho}_b + 3H_0\left(\rho_b + p_b\right) = 0. \tag{17.44}$$

Integration of this equation for a perfect fluid with equation of state $p_b = w\rho_b$ gives

$$\rho_b = \rho_0 a_0^{-3(1+w)}, \tag{17.45}$$

with $\rho_b(t_0) = \rho_0$ and the normalization $a_0(t_0) = 1$. In the simplest case where $\Lambda_B = k = U = 0$, eq. (17.43) can be integrated to yield the result

$$\begin{aligned} a_0 &\propto t^{\frac{1}{3(1+w)}}, \quad w \neq -1, \\ a_0 &\propto \exp\left(\frac{\kappa_5}{6}\rho_b t\right), \quad w = -1. \end{aligned} \tag{17.46}$$

This is the expansion factor in the brane, i.e. in our four-dimensional world. In the cases with radiation ($w = 1/3$) and dust ($w = 0$) the evolution of the expansion factor is $a_0 \propto t^{1/4}$ and $a_0 \propto t^{1/3}$, respectively, instead of the usual $a_0 \propto t^{1/2}$ and $a_0 \propto t^{2/3}$. The new cosmological equation thus leads typically to slower evolution. This behaviour is problematic. When it is inserted into the theory of cosmic nucleosynthesis the predictions of the abundances of the lightest elements are different from the observed ones. Hence, the five-dimensional brane universe models with perfect fluid in a single brane embedded in an empty bulk with vanishing cosmological constant come in conflict with observations.

17.3   Solutions in the bulk

Due to the presence of the brane the spacetime of the bulk is curved. We shall now find solutions with vanishing bulk matter describing the geometry of the bulk in an empty bulk and in a bulk with a cosmological constant. Eq. (17.41) may be written

$$(a^2)'' + \frac{2\Lambda_B}{3} a^2 = 2(f^2 + k).$$ (17.47)

In the case of an empty universe with $\Lambda_B = 0$ which is mirror symmetric about $y = 0$, integration of this equation with respect to y and use of eq. (17.40) gives

$$
\begin{aligned}
a^2(t, y) &= (f^2 + k)y^2 + A(t)|y| + a_0^2, \\
n(t, y) &= \frac{a_0}{a} + \frac{\dot{A}}{2af}|y| + \frac{\dot{f}}{a}y^2,
\end{aligned}
$$ (17.48)

where A is an arbitrary function of t and $f(t) = \dot{a}_0$. Determining the function A by applying eq. (17.36) yields

$$
\begin{aligned}
a^2(t, y) &= a_0^2 \left(1 - \frac{\kappa_5}{3}\rho_b|y|\right) + (f^2 + k)y^2, \\
n(t, y) &= \frac{a_0}{a}\left[1 + \frac{\kappa_5}{3}(2\rho_b + 3p_b)|y|\right] + \frac{\dot{f}}{a}y^2
\end{aligned}
$$ (17.49)

where ρ_b and p_b obey the adiabatic energy conservation equation

$$\frac{d}{dt}\left(a_0^3\rho_b\right) + p_b\frac{d}{dt}\left(a_0^3\right) = 0.$$ (17.50)

We now consider the case that there is a negative Lorentz invariant vacuum energy in the bulk, corresponding to a cosmological constant $\Lambda_B < 0$. Then $\kappa_5 p_B = -\kappa_5\rho_B = -\Lambda_B$. Defining a parameter μ by

$$\mu^2 = -\frac{2\Lambda_B}{3},$$ (17.51)

assuming mirror symmetry about $y = 0$, and integrating eq. (17.47) with respect to y gives

$$a^2(t, y) = A(t)\cosh(\mu y) + B(t)\sinh(\mu|y|) + \frac{3(f^2 + k)}{\Lambda_B}.$$ (17.52)

Utilizing eq. (17.36) together with eq. (17.40) and the normalization $n_0 = 1$ leads to

$$
\begin{aligned}
a^2(t, y) &= \left[a_0^2 - \frac{3(f^2 + k)}{\Lambda_B}\right]\cosh(\mu y) - \frac{\kappa_5 a_0^2 \rho_b}{3\mu}\sinh(\mu|y|) + \frac{3(f^2 + k)}{\Lambda_B}, \\
n(t, y) &= \frac{a_0}{a}\left(1 - \frac{3\dot{f}}{\Lambda_B}\right)\cosh(\mu y) + \frac{\kappa_5}{6\mu a f}(2\rho_b + 3p_b)\sinh(\mu|y|) + \frac{3\dot{f}}{\Lambda_B}.
\end{aligned}
$$ (17.53)

If the bulk cosmological constant is positive the hyperbolic functions in the above equations should be replaced by trigonometric ones. The functions a_0 and $f = \dot{a}_0$ can be found by integrating eq. (17.43).

17.4 Towards a realistic brane cosmology

We shall now consider a brane with total energy density $\rho_b = \lambda + \rho$ where λ is the tension of the brane which is assumed to be constant in time, and ρ is the energy density of ordinary cosmic matter. Then eq. (17.43) takes the form

$$H_0^2 = \frac{\kappa_5^2}{36}\rho^2 + \frac{\kappa_5^2}{18}\rho\lambda + \frac{\kappa_5^2}{36}\lambda^2 + \frac{\Lambda_B}{6} - \frac{k}{a_0^2} + \frac{U}{a_0^4}. \tag{17.54}$$

Inserting the four-dimensional cosmological constant Λ defined in eq. (17.12) and the four-dimensional gravitational constant defined in eq.(17.13), eq. (17.54) takes the form of a four-dimensional generalized Friedmann equation,

$$\boxed{H_0^2 = \frac{\Lambda}{3} + \frac{\kappa_4}{3}\rho\left(1 + \frac{\rho}{2\lambda}\right) - \frac{k}{a_0^2} + \frac{U}{a_0^4}.} \tag{17.55}$$

We now assume that the cosmic matter on the brane obeys the equation of state $p = w\rho$. From eq. (17.44) we have $\rho = \rho_0 a_0^{-q}$, $q = 3(1 + w)$. Hence, eq. (17.55) takes the form

$$H_0^2 = \frac{\Lambda}{3} + \frac{\kappa_4}{3}\frac{\rho_0}{a_0^q} + \frac{\kappa_4}{6\lambda}\frac{\rho_0^2}{a_0^{2q}} - \frac{k}{a_0^2} + \frac{U}{a_0^4}. \tag{17.56}$$

A critical brane has $\Lambda = 0$, i.e.,

$$\Lambda_B = -\frac{\kappa_5^2}{6}\lambda^2. \tag{17.57}$$

The Friedmann equation of a critical brane with $U = 0$ reduces to

$$H_0^2 = \frac{\kappa_4}{3}\rho\left(1 + \frac{\rho}{2\lambda}\right) - \frac{k}{a_0^2}. \tag{17.58}$$

Hence, we have recovered the usual Friedmann equation, but with a high energy correction which becomes significant only when the energy density of the matter approaches the tension of the brane.

Subtracting eq. (17.43) with $k = U = 0$ from eq. (17.37) gives

$$\frac{\ddot{a}_0}{a_0} = -\frac{\kappa_5^2}{36}\rho_b\left(2\rho_b + 3p_b\right) - \frac{\kappa_5}{6}\left(\rho_B + 2p_B\right). \tag{17.59}$$

Inserting $\rho_b = \lambda + \rho$, $p_b = -\lambda + p$, $\kappa_5\rho_B = -\kappa_5 p_B = \Lambda_B$, and using eq.(17.13), leads for a critical brane to

$$\frac{\ddot{a}_0}{a_0} = -\frac{\kappa_4}{6}\left[\rho + 3p + (2\rho + 3p)\frac{\rho}{\lambda}\right]. \tag{17.60}$$

Thus the condition for accelerated expansion on the brane is

$$\ddot{a}_0 > 0, \quad \text{if} \quad p < -\left(\frac{\lambda + 2\rho}{\lambda + \rho}\right)\frac{\rho}{3}. \tag{17.61}$$

In the low energy limit, $\rho \ll \lambda$, there is accelerated expansion if $p < -\rho/3$, while in the high energy limit, $\rho \gg \lambda$, there is accelerated expansion if $p < -2\rho/3$.

We shall now consider cosmological solutions where the cosmic matter is a perfect fluid with equation of state $p = w\rho$. Equation (17.44) then has the solution (17.45). In this case eq. (17.58) (i.e. $k = U = \Lambda = 0$) may be written

$$\dot{x}^2 = q^2(\beta x + \xi), \quad \beta = \frac{\kappa_4}{3}\rho_0, \quad \xi = \frac{\kappa_4}{6\lambda}\rho_0^2, \quad q = 3(1+w), \qquad (17.62)$$

where we have introduced a new variable $x = a_0^q$. The solution with $a_0(0) = 0$ is

$$a_0^q = \frac{q^2}{4}\beta t^2 + q\sqrt{\xi}t. \qquad (17.63)$$

This expression shows that there is a transition, at a typical time of the order $t_\lambda \approx 1/\sqrt{\kappa_4\lambda}$, between a high energy regime characterized by the behaviour $a_0 \propto t^{1/q}$ and a low energy regime characterized by the standard evolution $a_0 \propto t^{2/q}$.

For a non-critical brane the Friedmann equation takes the form

$$\dot{x}^2 = q^2\left(\frac{\Lambda}{3}x^2 + \beta x + \xi\right). \qquad (17.64)$$

Integration with $a_0(0) = 0$ gives

$$a_0^q = \sqrt{\frac{3\xi}{\Lambda}}\sinh\left(q\sqrt{\frac{\Lambda}{3}}t\right) + \frac{3\beta}{2\Lambda}\left[\cosh\left(q\sqrt{\frac{\Lambda}{3}}t\right) - 1\right], \qquad \Lambda > 0,$$

$$a_0^q = \sqrt{\frac{3\xi}{|\Lambda|}}\sin\left(q\sqrt{\frac{|\Lambda|}{3}}t\right) + \frac{3\beta}{2\Lambda}\left[\cos\left(q\sqrt{\frac{|\Lambda|}{3}}t\right) - 1\right], \qquad \Lambda < 0.$$

$$(17.65)$$

The evolution of the expansion factors is shown in Fig. 17.1. Note that in the

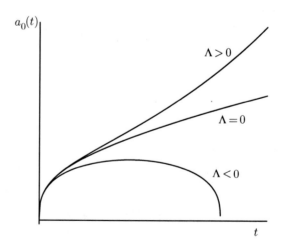

Figure 17.1: Evolution of the expansion factor for $\Lambda > 0$, $\Lambda = 0$ and $\Lambda < 0$.

case $\Lambda > 0$, which admits a positive cosmological constant, the universe will enter an era with accelerated expansion.

As we have seen (problem 10.3 with $M = 0$, section 12.2, example 14.2 and Appendix C) the Minkowski spacetime and the de Sitter spacetime can be represented both by static metrics and as expanding universes. Minkowski spacetime described with reference to an expanding reference frame is the Milne universe, and the de Sitter spacetime is the exponentially accelerated universe of the inflationary era. Similarly, the bulk of the brane world considered above is in fact a Schwarzschild–anti-de Sitter spacetime ($\Lambda_B < 0$) and can also be represented by a static metric, namely

$$ds^2 = -f(R)dT^2 + \frac{dR^2}{f(R)} + R^2 \gamma_{ij} dx^i dx^j, \quad f(R) = 1 - \frac{U}{R^2} - \frac{\Lambda_B}{6} R^2, \quad (17.66)$$

where γ_{ij} is the 3-dimensional spatial metric

$$\gamma_{ij} dx^i dx^j = R^{-2} \tilde{h}_{ij} dx^i dx^j = \frac{dr^2}{1 - r^2} + r^2 \left(d\theta^2 + \sin^2 \theta d\phi^2 \right). \qquad (17.67)$$

For simplicity, we have assumed that the FRW model is closed, i.e. $k = 1$. The expression (17.66) shows that the constant U is the five-dimensional analogue of the Schwarzschild mass. The R^{-2} dependence instead of the usual R^{-1} is due to the fourth spatial dimension.

The metric (17.66) corresponds to a description of the brane-world from a bulk point of view, while the metric (17.23) represents the description from the brane point of view. While the brane is at rest in the coordinate system of eq. (17.23), it moves in the static reference frame. The trajectory of the brane can be defined in parametric form $T = T(\tau)$, $R = R(\tau)$ where τ is the proper time of the brane. The five-velocity identity $g_{ab}u^a u^b = -1$ then takes the form

$$-f\dot{T}^2 + \frac{\dot{R}^2}{f} = -1, \qquad (17.68)$$

where the dot denotes differentiation with respect to τ. This yields

$$\dot{T} = \frac{\sqrt{f + \dot{R}^2}}{f}. \qquad (17.69)$$

The unit normal vector of the brane is defined by

$$n_a u^a = 0, \quad n_a n^a = 1. \qquad (17.70)$$

Up to a sign ambiguity this leads to

$$\mathbf{n} = -\frac{\dot{R}}{f} \mathbf{e}_T - \sqrt{f + \dot{R}^2} \mathbf{e}_R. \qquad (17.71)$$

The four-dimensional metric on the brane is

$$ds^2 = -d\tau^2 + R(\tau)^2 \left[\frac{dr^2}{1 - r^2} + r^2 (d\theta^2 + \sin^2 \theta d\phi^2) \right]. \qquad (17.72)$$

This expression shows that the expansion factor of the brane, denoted by a_0 previously, can be identified with the radial coordinate of the brane, $R(\tau)$. The $\theta\theta$-component of eq. (16.7) gives

$$K_{\theta\theta} = -\frac{1}{2} n^R \frac{\partial g_{\theta\theta}}{\partial R} = -R\sqrt{f + \dot{R}^2}. \qquad (17.73)$$

Using the junction conditions (17.8) with the energy-momentum tensor (17.26), we get

$$\frac{\sqrt{f + \dot{R}^2}}{R} = \frac{\kappa_5}{6}\rho_b. \tag{17.74}$$

Taking the square of this equation and substituting for $f(R)$ from eq. (17.66) yields

$$\frac{\dot{R}^2}{R^2} = \frac{\kappa_5^2}{36}\rho_b^2 + \frac{\Lambda_B}{6} - \frac{1}{R^2} + \frac{U}{R^4}, \tag{17.75}$$

which is exactly the Friedmann equation (17.43) with $k = 1$.

This embedding of the brane in a Schwarzschild–anti-de Sitter bulk can be used to show explicitly the correspondence between the dark energy term $\mathcal{U}$, defined by eq. (17.18), and the radiation-like term U/R^4 in the generalized Friedmann equation. Using the definition of the Weyl tensor, eq. (17.4), one can find the independent non-zero components of the Weyl tensor for the Schwarzschild–anti-de Sitter spacetime (17.66) to be (for $i, j \neq T, R$)

$$C_{TRTR} = -3\frac{U}{R^4}, \quad C_{TiTj} = \frac{U}{R^4}f\tilde{h}_{ij}, \quad C_{RiRj} = -\frac{U}{R^4}\frac{1}{f}\tilde{h}_{ij},$$

$$C_{r\theta r\theta} = \frac{U}{R^4}\tilde{h}_{rr}\tilde{h}_{\theta\theta}, \quad C_{r\phi r\phi} = \frac{U}{R^4}\tilde{h}_{rr}\tilde{h}_{\phi\phi}, \quad C_{\theta\phi\theta\phi} = \frac{U}{R^4}\tilde{h}_{\phi\phi}\tilde{h}_{\theta\theta}. \tag{17.76}$$

From the definition of $\mathcal{E}_{ij}$, eq. (17.6), we can find the components of the electric part of the Weyl tensor. For $i, j \neq \tau$, we have

$$\begin{aligned}
\mathcal{E}_{ij} &= C_{TiTj}(n^T)^2 + C_{RiRj}(n^R)^2 = -\frac{U}{R^4}\tilde{h}_{ij}, \\
\mathcal{E}_{\tau\tau} &= C_{TRTR}\left[(n^T)^2(u^R)^2 + (n^R)^2(u^T)^2 - 2n^T u^R n^R u^T\right] \\
&= C_{TRTR}\left[n^T u^R - n^R u^T\right]^2 = -3\frac{U}{R^4}.
\end{aligned} \tag{17.77}$$

Here we have also used that the Weyl tensor possesses the same symmetries as the Riemann tensor. The components can also be written as

$$\mathcal{E}_{ij} = -\frac{3U}{R^4}\left(u_i u_j + \frac{1}{3}\tilde{h}_{ij}\right). \tag{17.78}$$

Comparing this result with eq. (17.21), we see that we have to identify

$$\frac{U}{R^4} = \frac{2}{\kappa_4\lambda}\mathcal{U}. \tag{17.79}$$

Since U is a integration constant and that $H = \frac{\dot{R}}{R}$, $\mathcal{U}$ obeys eq. (17.22) as it should do.

17.5   Inflation in the brane

(see also [Kal99])

We shall briefly consider inflationary universe models within the framework of brane cosmology. The simplest model is that of a brane with constant vacuum energy density, $\rho_b = \rho_\lambda$, and with $k = U = 0$ in an empty bulk. The solution for this case is given in eqs. (17.46) and (17.49) giving the line element

$$ds^2 = (1 - H|y|)^2 \left(-dt^2 + e^{2Ht}[dr^2 + r^2(d\theta^2 + \sin^2\theta d\phi^2)]\right) + dy^2, \quad (17.80)$$

where $H = \frac{\kappa_5\lambda}{6}$. On the brane, where $y = 0$ this reduces to the ordinary de Sitter metric. This line-element describes an inflating brane in a five dimensional bulk with a Rindler-like horizon at $y_H = \pm H^{-1}$. The bulk is not singular at the horizon, only at the surface $y = 0$ due to the presence of the brane.

From eq. (17.43) with $\rho = k = 0$ follows that the Hubble parameter for an inflating brane in a bulk with a negative cosmological constant is

$$H_0^2 = \frac{\kappa_5^2}{36}\lambda^2 + \frac{\Lambda_B}{6} = \frac{\kappa_5^2}{36}\lambda^2 - \mu^2. \quad (17.81)$$

For sufficiently small vacuum energy on the brane the Hubble parameter is imaginary. Then one can analytically continue the solution by a coordinate transformation $t = -ix'$, $x = it'$, $y = y'$, $z = z'$. Writing $H_0 = i\mathcal{H}_0$ and considering a brane with negative spatial curvature the line-element takes the form (see Appendix C)

$$ds^2 = a(y)^2 \left[-dt^2 + |\mathcal{H}_0|^2 \cos^2|\mathcal{H}_0 t| \left(\frac{dr^2}{1+r^2} + r^2(d\theta^2 + \sin^2\theta d\phi^2)\right)\right] + dy^2, \quad (17.82)$$

where we have omitted the prime on the coordinates.

If there is a negative cosmological constant in the bulk, the geometry of the bulk is given by eq. (17.53). Furthermore, if $-\kappa_5^2\lambda^2/6 \leq \Lambda_B \leq 0$ the Hubble-parameter is real and the expansion factor in the brane $a(t,0)$, is still an exponential function of time as in eq. (17.80). The line-element is then

$$ds^2 = \left(\cosh\mu y - \frac{\kappa_5\lambda}{6\mu}\sinh\mu|y|\right)^2$$
$$\times \left(-dt^2 + e^{2Ht}\left[dr^2 + r^2(d\theta^2 + \sin^2\theta d\phi^2)\right]\right) + dy^2, \quad (17.83)$$

where μ is given in eq. (17.51). This describes an inflating brane in an anti-de Sitter bulk.

A notable property of this solution is the existence and location of a bulk event horizon. Its position is found by putting $g_{tt} = 0$ which gives

$$y_H = \pm\ell \, \text{artanh}\left(\frac{6\mu}{\kappa_5\lambda}\right), \quad (17.84)$$

where

$$\ell = \sqrt{-\frac{6}{\Lambda_B}}, \quad (17.85)$$

is the anti-de Sitter curvature radius. In the limit $\Lambda_B \to 0$ this reproduces $y_H = \pm 6/(\kappa_5\lambda)$ found in the solution (17.80) for an inflating 3-brane in a flat bulk. On the other hand, when $\Lambda_B \to -\kappa_5\lambda/6$, then $y_H \to \infty$. Hence, as the brane expansion decreases, either by increasing the bulk cosmological constant or by decreasing the density of the vacuum energy in the brane, the Rindler horizon moves farther away from the brane.

If $\Lambda_B > 0$ the inflation of the 3-brane is even more vigorous than in the case $\Lambda_B < 0$. In this case the hyperbolic functions of eq. (17.83) are replaced by the corresponding trigonometric functions, and the position of the event horizon is

$$y_H = \bar{\ell}\arctan\left(\frac{6\bar{\mu}}{\kappa_5\lambda}\right), \quad \bar{\mu} = \sqrt{\frac{\Lambda_B}{6}}. \tag{17.86}$$

In this case the location of the Rindler horizon as a function of $\sqrt{\Lambda_B}/6$ oscillates and can be arbitrarily close and far from the brane.

If there is a negative vacuum energy in the brane, $\lambda < 0$, the solution is still given by eq. (17.83). As can be seen from eq. (17.81) the brane is still inflating although the negative vacuum energy in the brane is gravitationally attractive. However, in this case there is no Rindler horizon in the bulk. In contrast to the case $\lambda > 0$ the brane is now gravitationally attractive for particles in the bulk rather than repulsive. Hence, any free particle in the bulk will fall onto the brane in a finite time and contribute with positive energy to the brane. This indicates that the solution with negative brane energy is unstable.

Following Maartens et al. [MWBH00] we shall now deduce the brane generalization of the slow-roll parameters η and ε of eq. (12.46). From eq. (17.58) with $k = 0$ and $\rho = V$ where $V(\phi)$ is the potential of a scalar field ϕ we have

$$H_0^2 = \frac{\kappa_4}{3}V\left(1 + \frac{V}{2\lambda}\right), \tag{17.87}$$

which generalizes eq. (12.43). Inserting this expression into eq. (12.44) using eq. (12.42) and writing the resulting equation in the form (12.45), we obtain

$$\eta = \frac{1}{\kappa_4}\frac{V''}{V}\frac{2\lambda}{2\lambda + V}, \quad \varepsilon = \frac{2}{\kappa_4}\left(\frac{V'}{V}\right)^2\frac{\lambda(\lambda + V)}{(2\lambda + V)^2}. \tag{17.88}$$

The slow-roll approximation requires $|\eta|, \varepsilon \ll 1$. At low energies, $V \ll \lambda$, the slow-roll parameters reduce to the expressions in eq. (12.46). However at high energies, $V \gg \lambda$, the new factors become $\approx \lambda/V \ll 1$. Hence, the brane effects makes it easier for the scalar field to roll slowly for a given potential.

Eq. (12.53) for the number of e-folds during inflation is now replaced by

$$N = -\kappa_4 \int_{\phi_i}^{\phi_f} \frac{V}{V'}\left(1 + \frac{V}{2\lambda}\right)d\phi. \tag{17.89}$$

The effect of the modified Friedmann equation at high energies is to increase the rate of expansion by the term $V/(2\lambda)$. Hence, there is more inflation between any two values of ϕ in brane cosmology than in standard cosmology for a given potential. Thus we can obtain a given number of e-folds for a smaller initial value, ϕ_i, of the inflaton field.

Let us consider a simple model of an inflationary universe, driven by a scalar field with potential $V = (1/2)m^2\phi^2$. Then eq. (17.89), together with eq. (17.58) leads to

$$N = \frac{4\pi}{m_{\mathrm{Pl}}^2}\left(\phi_i^2 - \phi_f^2\right) + \frac{\pi^2 m^2}{3m_5^6}\left(\phi_i^4 - \phi_f^4\right). \tag{17.90}$$

The new "brane-term", compared to the four-dimensional equation (12.54), means that in the brane universe models we get more inflation for a given initial value ϕ_i of the scalar field.

17.6 Dynamics of two branes

Some of the most important applications of the theory of brane cosmology have been made to brane universe models with two branes. We shall therefore extend the theory of the previous sections to such models. The dynamics of a brane-world with two branes have been developed by Binetruy *et al.* [BDL01].

We shall use a coordinate system where the metric of the bulk is given in eq. (17.23). One brane representing our four-dimensional world, is called the visible brane. It is at rest at $y = 0$. The other, called the hidden brane, has a time dependent position $y = R(t)$. The function $R = R(t)$ is often called the *radion*. The time coordinate t is chosen to be the proper time of the visible brane. The induced metric on the visible brane is then

$$ds_{\text{vis}}^2 = -dt^2 + a_0(t)^2 \left[\frac{dr^2}{1 - kr^2} + r^2 \left(d\theta^2 + \sin^2 \theta d\phi^2 \right) \right]. \tag{17.91}$$

The induced metric on the hidden brane depends upon its velocity like the proper time of a moving clock as given in eq. (10.57), and has the form

$$
\begin{aligned}
ds_{\text{hid}}^2 \;=\; & -\left[n(t, R(t))^2 - \dot{R}^2 \right] dt^2 \\
& + a(t, R(t))^2 \left[\frac{dr^2}{1 - kr^2} + r^2 \left(d\theta^2 + \sin^2 \theta d\phi^2 \right) \right]
\end{aligned}
\tag{17.92}
$$

where a dot denotes differentiation with respect to the proper time of the visible brane. In terms of the proper time τ of the hidden brane this can be written

$$ds_{\text{hid}}^2 = -d\tau^2 + a_2(\tau)^2 \left[\frac{dr^2}{1 - kr^2} + r^2 \left(d\theta^2 + \sin^2 \theta d\phi^2 \right) \right], \tag{17.93}$$

where $a_2 = a(t, R(t))$ is the expansion factor of the hidden brane. The proper time of the hidden brane is related to the proper time of the visible brane by

$$d\tau = n(t, R(t)) \sqrt{1 - \frac{\dot{R}^2}{n^2}} \, dt = n_2 \gamma^{-1} dt, \tag{17.94}$$

where

$$\gamma \equiv \frac{1}{\sqrt{1 - \frac{\dot{R}^2}{n^2}}}. \tag{17.95}$$

Due to the local character of gravity according to the general theory of relativity eq. (17.43) of the Hubble parameter in the visible brane is still valid without any changes. It will be useful to define an expansion rate, $\mathcal{H}_2$, for the hidden brane by

$$\mathcal{H}_2 \equiv \frac{\dot{a}_2}{a_2} = \left(\frac{\dot{a}}{a} + \frac{a'}{a} \dot{R} \right)_{y=R}. \tag{17.96}$$

Note that $\mathcal{H}_2$ does not coincide with the standard definition of the Hubble parameter for an observer in the hidden brane because it is defined with respect to the proper time of the visible brane and not of the hidden brane. The Hubble parameter of the hidden brane is

$$H_2 = \frac{1}{a_2} \frac{da_2}{d\tau} = \frac{\gamma}{n_2} \mathcal{H}_2. \tag{17.97}$$

The four-velocity of a comoving observer in the hidden brane is

$$u^\mu = \left(\frac{dt}{d\tau}, 0, 0, 0, \frac{dy}{d\tau} \right) = \frac{\gamma}{n_2} \left(1, 0, 0, 0, \dot{R} \right).$$

(17.98)

We shall now use the junction conditions to relate the motion of the hidden brane to its matter content. The unit normal vector to the hidden brane is

$$\mathbf{n} = \gamma \left(\frac{\dot{R}}{n^2} \mathbf{e}_t + \mathbf{e}_y \right).$$

(17.99)

From eq. (16.24) we now find the following non-vanishing components of the extrinsic curvature tensor

$$K^0_{\ 0} = \frac{\gamma^5}{n^2} \left(\ddot{R} + nn' - 2\frac{n'}{n}\dot{R}^2 - \frac{\dot{n}}{n}\dot{R} \right),$$

$$K^i_{\ j} = \gamma \left(\frac{a'}{a} + \frac{\dot{a}}{a}\frac{\dot{R}}{n^2} \right) \delta^i_{\ j},$$

$$K^5_{\ 0} = \dot{R}K^0_{\ 0}, \qquad K^0_{\ 5} = -\frac{\dot{R}}{n^2}K^0_{\ 0}, \qquad K^5_{\ 5} = -\frac{\dot{R}^2}{n^4}K^0_{\ 0}, \qquad (17.100)$$

where all quantities are evaluated on the brane.

The energy-momentum tensor of the visible and hidden branes are, respectively

$$T^\mu_{\ \nu\text{vis}} = S^\mu_{\ \nu\text{vis}}\delta(y) = \text{diag}(-\rho_{\text{vis}}, p_{\text{vis}}, p_{\text{vis}}, p_{\text{vis}}, 0)\delta(y),$$
$$T^\mu_{\ \nu\text{hid}} = S^\mu_{\ \nu\text{hid}}\delta(y - R(t)) = \text{diag}(-\rho_{\text{hid}}, p_{\text{hid}}, p_{\text{hid}}, p_{\text{hid}}, 0)\delta(y - R(t)).$$

(17.101)

Defining

$$\hat{S}_{\mu\nu} \equiv S_{\mu\nu} - \frac{1}{3}Sh_{\mu\nu},$$

(17.102)

we find

$$\hat{S}^0_{\ 0} = -\frac{1}{3}\gamma^2 \left(2\rho_{\text{hid}} + 3p_{\text{hid}} \right),$$

$$\hat{S}^i_{\ j} = \frac{1}{3}\rho_{\text{hid}}\delta^i_{\ j},$$

$$\hat{S}^5_{\ 0} = \dot{R}\hat{S}^0_{\ 0}, \qquad \hat{S}^0_{\ 5} = -\frac{\dot{R}}{n^2}\hat{S}^0_{\ 0}, \qquad \hat{S}^5_{\ 5} = -\frac{\dot{R}^2}{n^4}\hat{S}^0_{\ 0}. \qquad (17.103)$$

Inserting these expressions into the Israel junction conditions, eq. (17.8), leads to only two equations

$$\frac{\ddot{R}}{n^2} + \frac{n'}{n}\left(1 - 2\frac{\dot{R}^2}{n^2} \right) - \frac{\dot{n}}{n}\frac{\dot{R}}{n^2} = -\frac{1}{6}\kappa_5 \left(2\rho_{\text{hid}} + 3p_{\text{hid}} \right) \left(1 - \frac{\dot{R}^2}{n^2} \right)^{\frac{3}{2}},$$

$$\frac{a'}{a} + \frac{\dot{a}}{a}\frac{\dot{R}}{n^2} = \frac{1}{6}\kappa_5\rho_{\text{hid}} \left(1 - \frac{\dot{R}^2}{n^2} \right)^{\frac{1}{2}}, \qquad (17.104)$$

where the metric functions are to be evaluated on the brane. These equations generalize eq. (17.36). By differentiation one can show that the left hand side of the upper eq. (17.104) is just the four-acceleration in the y-direction of a co-moving particle in the hidden brane having four-velocity (17.98). The equation shows that the matter of the brane causes this motion to deviate from geodesic motion. Also one can show that the upper equation in (17.104) follows by differentiating the lower and using eq. (17.39) together with the energy conservation equation

$$\dot{\rho}_{hid} + 3H_2(\rho_{hid} + p_{hid}) = 0. \tag{17.105}$$

Solving eq. (17.104) with respect to $\dot{R}$ one obtains

$$\dot{R} = \frac{n\left(-\frac{a'\dot{a}}{a^2 n} \pm \frac{\kappa_5}{6}\rho_{hid}\sqrt{\frac{\dot{a}^2}{a^2 n^2} - \frac{a'^2}{a^2} + \frac{\kappa_5^2 \rho_{hid}^2}{36}}\right)}{\frac{\dot{a}^2}{a^2 n^2} + \frac{\kappa_5^2 \rho_{hid}^2}{36}}. \tag{17.106}$$

By means of eq. (17.94) the equations of motion of the hidden brane relative to the visible brane can also we rewritten in terms of the proper time τ of the hidden brane instead of the proper time t of the visible brane. Then eqs. (17.104) take the form

$$\frac{d^2 R}{d\tau^2} + \frac{n'}{n}\left(1 + \left(\frac{dR}{d\tau}\right)^2\right) = -\frac{\kappa_5}{6}(2\rho_{hid} + 3p_{hid})\sqrt{1 + \left(\frac{dR}{d\tau}\right)^2},$$

$$\sqrt{1 + \left(\frac{dR}{d\tau}\right)^2}\frac{a'}{a} + \frac{\dot{a}}{an}\left(\frac{dR}{d\tau}\right) = \frac{\kappa_5}{6}\rho_{hid}. \tag{17.107}$$

If the branes are at rest relative to each other, i.e. with $\dot{R} = 0$ in a bulk without matter, but with a non-vanishing cosmological constant, then eqs. (17.104) reduce to eq. (17.36), and the geometry of the bulk is given by eq. (17.53). Using these equations one can express the energy density and pressure of the hidden brane in terms of the energy density an pressure of the visible brane and the position $y = R_2$ of the hidden brane

$$p_{hid} = \frac{6\mu}{\kappa_5}\left(\frac{\sinh\mu R - \frac{\kappa_5 \rho_{vis}}{6\mu}\cosh\mu R}{\cosh\mu R - \frac{\kappa_5 \rho_{vis}}{6\mu}\sinh\mu R}\right). \tag{17.108}$$

In the limit $\mu R \approx 0$, for example if the branes are very close to each other, or if the positions of the branes are identified with another one obtains $\rho_{hid} \approx -\rho_{vis}$.

17.7   The hierarchy problem and the weakness of gravity

In our universe there seems to exist two fundamental energy scales: The electroweak scale, $m_{EW} \sim 10^3 \text{GeV}$, and the Planck scale, $m_{Pl} \sim 10^{19} \text{GeV}$. The hierarchy problem is in essence: Why is there such a vast difference between the two scales? A related question is: Why is gravity so weak? At the Planck energy scale one expects gravity to be as strong as the gauge interactions.

One way of answering these questions has been by so-called Kaluza-Klein compactification[5], where one or more additional compact dimensions are introduced. Gravity is postulated to be fundamentally strong. Expanding the

---

[5]This will be the subject of the next chapter.

metric as a Fourier series one get an infinite number of field modes in four dimensions. Modes with $n \neq 0$ correspond to massive modes with mass n/R, where R is the radius of an extra dimension. The zero mode corresponds to massless gravitons. As we take R to be smaller and smaller the mass of the first massive mode becomes very large. This means that if the compact dimension has sufficiently small extension, only the zero mode has been probed by gravitational experiments up to the present time. Hence effectively gravity is weak at the observed scales. In order that effects of the fundamental strength of gravity shall not be observed, the extension of the compact dimension must be less than about 10^{-18}m.

The questions above can also be answered without demanding that the extra dimensions have so extremely small extension. Assume that the electroweak scale, characterized by the mass m_{EW}, is the only fundamental short distance scale in nature. Furthermore, suppose that there are n extra compact dimensions of radius R. In the brane-world scenarios it is also assumed that the electromagnetic, weak and strong forces, as well as the matter in the universe, is trapped in ordinary four-dimensional space, i.e. on our 3-brane. Only gravity is able to spread out in the extra dimensions.

The Planck scale $m_{Pl(4+n)}$ of this $(4 + n)$-dimensional theory is taken to be the electroweak scale m_{EW}. The gravitational potential at a distance r from a point mass m in ordinary four dimensional spacetime is

$$V(r) = G\frac{m}{r}. \tag{17.109}$$

Using units so that $\hbar = c = 1$, Newton's gravitational constant is given by $G = m_{Pl}^{-2}$. Hence, the ordinary Newtonian gravitational potential takes the form

$$V(r) = \frac{1}{m_{Pl}^2}\frac{m}{r}. \tag{17.110}$$

Suppose now that the particle is in a space with n extra compact dimensions with radius R. The gravity is spreading in all these dimensions, and the gravitational potential measured at a distance $r \ll R$ from the particle is

$$V(r) \approx \frac{1}{m_{Pl(4+n)}^{n+2}}\frac{m}{r^{n+1}}. \tag{17.111}$$

On the other hand, if one measures the potential at a distance $r \gg R$ from the particle one does not recognize that part of gravity which spreads in the extra dimensions. Then one measures an effective $(1/r)$-potential. Eq. (17.111) is, however, still valid, but with r^n – due to the extra dimensions – replaced by R^n. Hence one measures a potential

$$V(r) \approx \frac{1}{m_{Pl(4+n)}^{2+n}R^n}\frac{m}{r}. \tag{17.112}$$

Comparing with eq. (17.110) our effective four-dimensional Planck mass is given by [AHDD98]

$$m_{Pl}^2 \approx m_{Pl(4+n)}^{2+n}R^n. \tag{17.113}$$

According to this picture the gravitational force is so weak because it is diluted by the extra dimensions. Viewed from the higher-dimensional bulk there might be only one fundamental scale.

Putting $m_{Pl(4+n)} \approx m_{EW}$ and demanding that R be chosen to reproduce the observed m_{Pl} yields

$$R \approx 10^{\frac{30}{n}-17} \mathrm{cm} \times \left(\frac{1\,\mathrm{TeV}}{m_{EW}} \right)^{1+\frac{2}{n}}. \tag{17.114}$$

For $n = 1$ the typical radius of the compact dimensions is $R = 10^{13}\mathrm{cm}$ implying deviations from Newtonian gravity over solar system distances, so this case is empirically excluded. However, for $n = 2$ one gets $R = 10^{-2}\mathrm{cm}$. Measurements of deviations from Newton's law at such scales are feasible in experiments to be performed in the near future.

The Kaluza-Klein requirement on the extension of the compact dimensions, mentioned above, appears in a different way in this scenario. From high-energy accelerator experiments we know that the strong, weak and electromagnetic forces cannot be modified at distances larger than about $10^{-18}\mathrm{m}$. If the 3-brane representing our world has a finite thickness R in the higher dimensional bulk, one should be able to measure deviations of the usual force laws at distances less than R. If these forces are trapped in a brane, the thickness of the brane must therefore be less than $10^{-18}\mathrm{m}$.

However, there is a problem. While this scenario eliminates the hierarchy between the electroweak scale m_{EW} and the Planck scale m_{Pl}, it introduces a new hierarchy, namely the one between the compactification scale and the electroweak scale. This motivated L. Randall and R. Sundrum to explore alternative solutions to the hierarchy problem and to search for another reason for the weakness of gravity.

17.8   The Randall-Sundrum models

Two five-dimensional static universe models have been constructed by L. Randall and R. Sundrum [RS99b, RS99a] (see also [Pad02, Räs02]) to explain the hierarchy problem and the weakness of gravity.

In the first model there are two parallel branes, the visible brane is at $y = 0$ and the hidden at $y = y_h$. The bulk coordinate is taken to be periodic with period equal to $2y_h$. Also, the surface (x^i, y) is identified with the surface $(x^i, -y)$. This is usually referred to as the $\mathbb{Z}_2$-symmetry in the literature. Furthermore it is assumed that the branes are domain walls with equal and opposite tension interpreted as vacuum energy by brane inhabitants. Hence

$$p_{vis} = -\rho_{vis}, \quad \text{and } p_{hid} = -\rho_{hid} \quad \text{with } \rho_{hid} = -\rho_{vis} = -\lambda, \tag{17.115}$$

where $\lambda < 0$ is the tension of the visible brane. The branes are separated by an anti-de Sitter bulk with a cosmological constant $\Lambda_B < 0$, and are supposed to be critical. Hence, from (17.12) with $\Lambda = 0$ follows

$$\Lambda_B = -\frac{\kappa_5^2}{6}\lambda^2. \tag{17.116}$$

Thus, the cosmological constant in the bulk and the tension of the bulk are negative, and there is a fine-tuning between these which secures the vanishing of the four-dimensional cosmological constant observed by habitants of the visible brane.

It is assumed that there exists a solution that respects four-dimensional Poincare invariance in ordinary spacetime. A five-dimensional metric satisfying this ansatz takes the form

$$ds^2 = a(y)^2 \eta_{\alpha\beta} dx^\alpha dx^\beta + dy^2, \tag{17.117}$$

where $\eta_{\alpha\beta}$ is the Minkowski metric on the brane and $0 \le y \le y_h$ is the coordinate of a compact extra dimension with a finite size set by y_h. In the present case eq. (17.36) takes the form

$$\frac{a_i'}{a_i} = \pm\sqrt{-\frac{\Lambda_B}{6}} = \pm\frac{1}{\ell}, \tag{17.118}$$

with $i = 1$ and $i = 2$ for the visible and the hidden brane, respectively. Choosing the negative sign and imposing $\mathbb{Z}_2$-symmetry about $y = 0$, the solution is

$$a = e^{-|y|/\ell}, \tag{17.119}$$

This function is called the *warp factor*. Hence the line-element of the bulk between the branes is

$$ds^2 = e^{-2|y|/\ell} \eta_{\alpha\beta} dx^\alpha dx^\beta + dy^2, \tag{17.120}$$

which represents a slice of anti-de Sitter space, and the branes are Minkowski branes. The warp factor of the RS-I model is shown in Fig. 17.2.

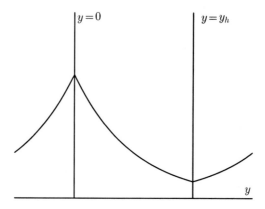

Figure 17.2: The warp factor in the RS-I model.

The most important quality of the RS-I model is that it provides an ingenious approach to the hierarchy problem. In the RS-I scenario the fundamental Planck scale is equal to the fundamental electroweak scale. However, the scales separate when we consider the effective interactions on the brane. By a field renormalization invoking a Higgs field calculation Randall and Sundrum arrive at the following result: Any mass m_0 in five-dimensional spacetime on the brane representing our world corresponds to a physical mass

$$m = e^{-y_h/\ell} m_0. \tag{17.121}$$

If $e^{y_h/\ell}$ is of order 10^{15}, which only requires $y_h/\ell \approx 50$, this mechanism produces weak scale, i.e,. TeV, physical masses from masses around the Planck scale, 10^{19} GeV, in the five-dimensional spacetime. Saying this means that the Planck scale is considered fundamental and the TeV scale as derived. However one could equally well have regarded the TeV scale as fundamental and the Planck scale as derived since the ratio of the two is the only physical dimensionless quantity. From this point of view, which is the one of an observer in the brane representing our four-dimensional world, the Planck scale, i.e. the weakness of gravity, arises because of the small overlap of the graviton wave function in the fifth dimension with our brane.

From a phenomenological point of view this result is particularly exciting. If the fundamental scale of gravity is as low as a few TeV then we would expect quantum gravity effects to start showing up in forthcoming collider experiments.

It should be pointed out that for the RS-I model to work we need a radius stabilization mechanism. As can be seen $y_h/\ell \approx 50$ seems somewhat arbitrary and therefore we are in need for something that stabilizes the distance between the branes. In the original RS-I model such a mechanism was absent. However, later Goldberger and Wise [GW99] suggested a model where the presence of bulk fields stabilize the extra dimensions. Such a mechanism can therefore provide a model which solves the hierarchy problem without fine-tuning the parameters.

J. Garriga and T. Tanaka [GT00] have considered the gravitational field of a point mass, m, surrounded by spherically symmetric static space in the Randall-Sundrum brane, in the weak field approximation. They found the Newtonian gravitational potential

$$V(r) = -\frac{Gm}{r}\left(1 + \frac{2\ell^2}{3r^2}\right). \tag{17.122}$$

Thus, deviations from Newton's gravitational law should be apparent at distances of the order of the characteristic scale of the cosmological constant of the bulk. Hence this distance cannot be greater then about a tenth of a millimetre.

In these brane models, although gravity is allowed to propagate in the bulk, the standard model fields are confined to the brane. Hence, electromagnetism, and the weak and strong forces are fields living on the brane only. Interactions involving these fields will not directly feel the extra dimension and will therefore remain almost entirely unmodified. Only gravity is modified in these scenarios.

It should be noted, however, that the RS-I model is unstable. As noted after eq. (17.86), matter in the bulk outside a brane with negative energy density will fall towards the brane and make its energy positive. Randall and Sundrum have constructed a second brane universe model that does not suffer from such an instability. However, the second model does not provide a resolution of the hierarchy problem, although it gives an explanation for the weakness of gravity in our world.

In the second RS-model there is only one brane in an anti-de Sitter bulk of infinite extension. The brane has positive vacuum energy density which is again fine tuned against the bulk cosmological constant to ensure Poincaré invariance on the brane. The warp factor is similar to that of the RS-I model, but there is now global symmetry about the position of the brane. The warp factor of this model is shown in Fig. 17.3.

Standard Kaluza-Klein compactification ensures that gravity looks four-dimensional by stating that the extra dimensions should be small. In the RS-II

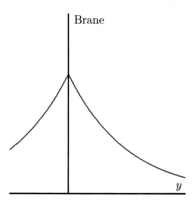

Figure 17.3: The warp factor in the RS-II model.

model the extra dimension is infinite, and gravity is allowed to propagate into the extra dimension so we would expect it to look five-dimensional even to an observer on the brane. However, the exponential warp factor causes the gravitational interaction to be damped in the direction away from the brane. This has the effect that gravity looks four-dimensional to a brane-world observer.

The ideas of the RS-I and RS-II models can be combined such that both the hierarchy problem is solved and the weakness of gravity is explained. In the combined model there are two branes with positive vacuum energy density, the Planck brane and the TeV brane. The hierarchy problem is solved in the same way as in the RS-I model provided we live on the TeV brane, and in a similar way to RS-II gravity looks four-dimensional, at least up to a few TeV, on both branes.

Problems

17.1. *Domain wall brane universe models*
We shall here consider the brane cosmological solutions for brane domain walls with an equation of state $p_b = -\rho_b$.

(a) Show that the energy density of a domain wall brane is constant.

(b) A *critical brane* is defined as a brane satisfying

$$\frac{\kappa_5^2}{6}\rho_b + \Lambda_B = 0.$$

Show that the expansion factor in a critical domain wall brane with $a(0) = 0$ is

$$a = \sqrt{2U_1 t - kt^2},$$

where U_1 is a constant of integration.

(c) Show that the solutions for non-critical domain wall branes with $\alpha^2 \equiv \frac{\kappa_5}{6}\rho_b + \Lambda_B$ are

$$a^2 = \sqrt{\frac{\beta}{\alpha}} \sinh\left(2\sqrt{\alpha}t\right) + \frac{k}{2\alpha}, \quad \beta > 0,$$

$$a^2 = e^{2\sqrt{\alpha}t} + \frac{k}{2\alpha}, \quad \beta = 0,$$

$$a^2 = \pm\sqrt{\frac{-\beta}{\alpha}} \cosh\left(2\sqrt{\alpha}t\right) + \frac{k}{2\alpha}, \quad \beta < 0, \qquad (17.123)$$

where $\beta = U - \frac{k^2}{4\alpha}$.

Plot the expansion factors as functions of time for the following cases:
(i) $k = 1, \beta < 0$; (ii) $k = 1, \beta > 0$; (iii) $k = -1, \beta < 0$; (iv) $k = -1, \beta > 0$.

17.2. A brane without $\mathbb{Z}_2$-symmetry

In this problem it is not assumed – in contrast to what is usually assumed for brane worlds – that there is a $\mathbb{Z}_2$-symmetry across the brane. Assume therefore that the metric on either side of the brane is given by the Schwarzschild–anti-de Sitter metric

$$ds_\pm^2 = -f^\pm(R)dT^2 + \frac{dR^2}{f^\pm(R)} + R^2\gamma_{ij}dx^i dx^j, \quad f^\pm(R) = 1 - \frac{U^\pm}{R^2} - \frac{\Lambda_B^\pm}{6}R^2,$$

where γ_{ij} is the metric on the three-sphere.

(a) Assume that the metric on the brane is that of a FRW model

$$ds^2 = -d\tau^2 + R(\tau)^2\gamma_{ij}dx^i dx^j, \qquad (17.124)$$

where τ is the proper time of the brane. Write the junction conditions in term of the functions f and show that

$$\frac{\sqrt{f^+ + \dot{R}^2}}{R} + \frac{\sqrt{f^- + \dot{R}^2}}{R} = \frac{\kappa_5}{3}\rho_b. \qquad (17.125)$$

(b) Show that the Friedmann equation on the brane is

$$\frac{\dot{R}^2}{R^2} = \frac{\kappa_5^2}{36}\rho_b^2 + \frac{\Lambda_B^+ + \Lambda_B^-}{12} + \frac{(\Lambda_B^+ - \Lambda_B^-)^2}{16\kappa_5^2\rho_b^2} - \frac{1}{R^2} + \frac{9(U^+ - U^-)^2}{4\kappa_5^2\rho_b^2}\frac{1}{R^8}$$

$$+ \left(U^+ + U^- + \frac{3(U^+ - U^-)(\Lambda_B^+ - \Lambda_B^-)}{2\kappa_5^2\rho_b^2}\right)\frac{1}{2R^4}.$$

Write also down the Friedmann equation in the case $\rho_b = \lambda + \rho$ and $U^\pm = 0$. What is the cosmological constant on the brane?

This model requires a severe fine-tuning of the values of Λ_B on both sides of the brane in order to be consistent with observations.

17.3. Warp factors and expansion factors for bulk and brane domain walls with factorizable metric functions (I. Brevik et al. [BGOY02])

Assume that the metric functions $n(t, y)$ and $a(t, y)$ of the line element (17.23) obey the conditions $a(t, y) = a_0(t)n(y)$, $n(t, y) = n(y)$, and that the bulk is filled with a perfect fluid with equation of state $p_B = w\rho_B$.

(a) Use eq. (17.38) to show that the only type of perfect fluid allowing a time dependent density in the brane is the so-called stiff fluid with $p_B = w\rho_B$. In following problems we shall assume that both the bulk and the brane are empty except for a cosmological constant Λ_B in the bulk and a tension λ on the brane.

(b) Show that in this case eq. (17.28) leads to the equations

$$\frac{\dot{a}_0^2}{a_0^2} + \frac{k}{a_0^2} = \frac{1}{2}(n^2)'' + \frac{\Lambda_B}{3}n^2 = D, \qquad (17.126)$$

where D is a constant. Show from eq.(17.43) with $U = 0$ that $D = \Lambda/3$, where Λ is the four-dimensional cosmological constant given in eq. (17.12).

Show also that eq. (17.38) now reduces to

$$n'' + \frac{\Lambda_B}{6}n = 0. \qquad (17.127)$$

Hence, the assumption that the metric function a is separable requires that the function n has to obey two differential equations.

(c) Assuming mirror symmetry about $y = 0$, and normalising the warp factor n so that $n(0) = 1$ at the brane, show that the equations have the following solutions:

$$\Lambda = 0 : \begin{cases} k = 0, & a_0 = 0, \; n = e^{-|y|/\ell}, \\ k = -1, & a_0 = t, \; n = e^{-|y|/\ell}, \end{cases} \qquad (17.128)$$

where ℓ is given in eq. (17.118).

$$\Lambda > 0 : \begin{cases} \Lambda_B < 0, & n = \ell\sqrt{\frac{\Lambda}{3}}\sinh\left(\frac{y_h - |y|}{\ell}\right), \\ \Lambda_B > 0, & n = \ell\sqrt{\frac{\Lambda}{3}}\sin\left(\frac{y_h - |y|}{\ell}\right), \end{cases} \qquad (17.129)$$

where y_h is a constant defining the position of the horizon in the bulk. The expansion function is the same for the latter two cases, but depends upon the spatial curvature on the brane. The solutions for $a_0(t)$ are the same as for the de Sitter solutions (12.10) with different spatial geometry. For $\Lambda < 0$, there is only one solution, for $k = -1$

$$n = \ell\sqrt{-\frac{\Lambda}{3}}\sinh\left(\frac{y_h - |y|}{\ell}\right), \qquad a_0 = \sqrt{-\frac{3}{\Lambda}}\sin\left(\sqrt{-\frac{\Lambda}{3}}t\right). \qquad (17.130)$$

17.4. Solutions with variable scale factor in the fifth dimension
Assume that the bulk is filled with vacuum energy with density λ and a perfect fluid with density ρ and pressure p obeying an equation of state $p = w\rho$, where w is constant. We shall consider models with $n = 1$ in the bulk.

(a) Show that in this case $a' = bh(y)$ where $h(y)$ is an arbitrary function.

(b) Show that when the bulk is empty except for vacuum energy given by a cosmological constant, Λ_B, the function a obeys the differential equation

$$\left(a^2\right)^{\cdot\cdot} - \frac{2\Lambda_B}{3}a^2 = 2(b^2h^2 - k). \qquad (17.131)$$

(c) Find $a^2(t,y)$ in terms of arbitrary functions of y appearing in the integration, for the cases $\Lambda_B = 0$ and $\Lambda_B < 0$.

(d) Use eq. (17.36) to show that the "gravitational constant" is given by

$$8\pi G = -\frac{\kappa_5}{3}\left(\frac{2+3w}{1+w}\right)\frac{h(0)}{a_0(t)}. \tag{17.132}$$

Note that a positive G requires $h(0) < 0$. Does this equation allow a constant "gravitational constant"?

18

Kaluza-Klein Theory

Already in 1914 – before Einstein had fulfilled the construction of the general theory of relativity – Gunnar Nordström[1] had published a five-dimensional scalar-tensor theory of gravitation in an effort to unify gravitation and electromagnetism. Since it was based upon his own theory of gravitation which was soon surpassed by Einstein's theory, this work was neglected for several decades.

However, in 1919, Theodor Kaluza constructed a similar unified theory of gravity and electromagnetism based on the linearized version of the general theory of relativity. The full theory was worked out by Oscar Klein in 1926. Later Einstein became interested in this theory and developed it further together with Peter Bergmann.

During the last thirty years more general versions of multidimensional theories have been constructed in order to find a scheme for unifying the four fundamental forces. There are now a large class of such theories, and the introduction of several spatial dimensions is part of the superstring theories and M-theory that many physicists now hold as promising in the effort towards working out a quantum theory of gravitation.

In the present chapter we shall consider the version of the theory presented by Oscar Klein and show how it provides a geometrical unified theory of gravity and electromagnetism.

18.1 A fifth extra dimension

The idea is quite simple. Let us assume that there is – in addition to the four spacetime dimensions – one compact extra spatial dimension. This extra dimension has to be small, or else we would have been able to see it. We will investigate what a such dimension means to the physics of the observable four-dimensional spacetime, following [Weh01, WR04].

---

[1]English translations of this and the other works mentioned in this section are found in the book *Modern Kaluza-Klein Theories* [ACF87].

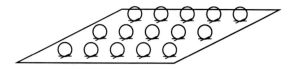

Figure 18.1: In the Kaluza-Klein theory we assume that every point in spacetime has a small extra dimension.

Assume that our world is a five-dimensional manifold with metric

$$ds^2 = G_{ab}dx^a dx^b, \qquad (18.1)$$

where Latin indices have the range 0-4, Greek have range 0-3. Assume also that there is one spatial Killing vector $\boldsymbol{\xi}$. This makes it possible to compactify the space in that direction, and make it as small as needed. We can therefore interpret this – if the extra dimension is small enough – as if every point in our four-dimensional world has an extra dimension attached to it. In Fig. 18.1 we have illustrated this idea.

The physical implications of this "small internal dimension" can be seen if we project the fifth dimension onto the orthogonal complement of the Killing vector $\boldsymbol{\xi}$ as follows. We choose a set of basis vectors such that $\mathbf{e}_4$ coincides with this Killing vector. The remaining vectors are chosen to be vectors that are Lie-transported around the manifold. Hence, we choose $\mathbf{e}_\mu$ to be an invariant basis

$$[\mathbf{e}_\mu, \mathbf{e}_4] = 0. \qquad (18.2)$$

This implies that

$$\mathbf{e}_4(G_{ab}) = 0; \qquad (18.3)$$

the metric is independent of the fifth dimension.

The vectors $\mathbf{e}_\mu$ will not in general be orthogonal to $\mathbf{e}_4$. Thus in general

$$\mathbf{e}_\mu \cdot \mathbf{e}_4 = G_{\mu 4}. \qquad (18.4)$$

We decompose our vectors $\mathbf{e}_\mu$ into a parallel and orthogonal part

$$\mathbf{e}_\mu = \mathbf{e}_{\mu\perp} + \mathbf{e}_{\mu||}, \quad \mathbf{e}_{\mu\perp} \cdot \mathbf{e}_4 = 0. \qquad (18.5)$$

(see Fig. 18.2)

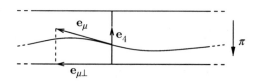

Figure 18.2: The projection of the extra dimension onto the orthogonal complement. Here, π is the projection map.

Proceeding along similar lines as in section 4.7, we write the line element as

$$ds^2 = g_{\mu\nu}dx^\mu dx^\nu + G_{44}\left(dy + \frac{G_{4\mu}}{G_{44}}dx^\mu\right)^2, \qquad (18.6)$$

where $g_{\mu\nu}$ is the projection of G_{ab} onto the orthogonal complement of e_4. The projection tensor can in this case be written as

$$g_{ab} = G_{ab} - \xi_a\xi_b,\tag{18.7}$$

where $\boldsymbol{\xi} = e_4 = \xi^a e_a$. Define $\mathbf{A}$ to be the one-form with components

$$A_\mu = \frac{G_{4\mu}}{G_{44}},\tag{18.8}$$

and ϕ to be the scalar

$$\phi = \sqrt{G_{44}}.\tag{18.9}$$

The scalar ϕ defines the size of the extra dimension while the vector A_μ defines the "tilt" of the extra dimension.

18.2   The Kaluza-Klein action

A physical interpretation of the theory can be given by finding the action of this theory. We start by assuming that the five-dimensional action has the same form as four-dimensional Einstein gravity. Assume therefore that the Kaluza-Klein action is

$$S_{KK} = \frac{1}{2\kappa_5} \int {}^{(5)}R\sqrt{-G}d^5x.\tag{18.10}$$

Here, κ_5 is the five-dimensional gravitational constant. We will relate the five-dimensional Ricci scalar to the four-dimensional one. This can be done using a similar method as in the derivation of eq. (7.83), but we cannot use this result directly. The reason for this can be seen as follows. Using an orthonormal frame, we choose

$$\boldsymbol{\omega}^{\hat{4}} = \phi\left(\mathbf{dy} + A_{\hat{\mu}}\boldsymbol{\omega}^{\hat{\mu}}\right).\tag{18.11}$$

Taking the exterior derivative, we get

$$\begin{aligned}\mathbf{d}\boldsymbol{\omega}^{\hat{4}} &= \frac{\phi_{;\hat{\nu}}}{\phi}\boldsymbol{\omega}^{\hat{\nu}} \wedge \boldsymbol{\omega}^{\hat{4}} + \phi A_{\hat{\mu};\hat{\nu}}\boldsymbol{\omega}^{\hat{\nu}} \wedge \boldsymbol{\omega}^{\hat{\mu}} \\ &= (\ln\phi)_{;\hat{\nu}}\boldsymbol{\omega}^{\hat{\nu}} \wedge \boldsymbol{\omega}^{\hat{4}} + \phi A_{\hat{\mu};\hat{\nu}}\boldsymbol{\omega}^{\hat{\nu}} \wedge \boldsymbol{\omega}^{\hat{\mu}}.\end{aligned}\tag{18.12}$$

We define the antisymmetric tensor

$$F_{\mu\nu} = A_{\mu;\nu} - A_{\nu;\mu},\tag{18.13}$$

so that

$$\mathbf{d}\boldsymbol{\omega}^{\hat{4}} = (\ln\phi)_{;\hat{\nu}}\boldsymbol{\omega}^{\hat{\nu}} \wedge \boldsymbol{\omega}^{\hat{4}} - \frac{1}{2}\phi F_{\hat{\nu}\hat{\mu}}\boldsymbol{\omega}^{\hat{\mu}} \wedge \boldsymbol{\omega}^{\hat{\nu}}.\tag{18.14}$$

If the space at a point orthogonal to the vector e_4 spans a hypersurface in the five dimensional world, then the exterior derivative $\mathbf{d}\boldsymbol{\omega}^{\hat{4}}$ must vanish. Thus ϕ is constant, and $F_{\alpha\beta} = 0$. This makes the situation trivial, as can be shown. However, we are interested in the case when e_4 is not orthogonal to the hypersurface; i.e., when $A_\mu \neq 0$. We will see that this assumption yields interesting physics in the four-dimensional world.

Using Cartan's first structural equation, eq. (6.182), we can find the five-dimensional rotation forms. From eq. (18.14) we get

$$^{(5)}\Omega^{\hat{4}}_{\ \hat{\mu}} = \frac{1}{2}\phi F_{\hat{\mu}\hat{\alpha}}\omega^{\hat{\alpha}} + (\ln\phi)_{;\hat{\mu}}\omega^{\hat{4}}. \tag{18.15}$$

The five-dimensional version of Cartan's first structural equation yields

$$\mathbf{d}\omega^{\hat{\nu}} = -{}^{(5)}\Omega^{\hat{\nu}}_{\ \hat{\alpha}}\wedge\omega^{\hat{\alpha}} - {}^{(5)}\Omega^{\hat{\nu}}_{\ \hat{4}}\wedge\omega^{\hat{4}} \tag{18.16}$$

while the four-dimensional version states

$$\mathbf{d}\omega^{\hat{\nu}} = -{}^{(4)}\Omega^{\hat{\nu}}_{\ \hat{\alpha}}\wedge\omega^{\hat{\alpha}}. \tag{18.17}$$

This implies that

$$^{(5)}\Omega^{\hat{\nu}}_{\ \hat{\alpha}} = {}^{(4)}\Omega^{\hat{\nu}}_{\ \hat{\alpha}} - \frac{1}{2}\phi F^{\hat{\nu}}_{\ \hat{\alpha}}\omega^{\hat{4}}. \tag{18.18}$$

From this we can read off the connection coeffients

$$\Gamma^{\hat{4}}_{\ \hat{\alpha}\hat{\beta}} = -\Gamma^{\hat{\alpha}}_{\ \hat{4}\hat{\beta}} = \frac{1}{2}\phi F_{\hat{\alpha}\hat{\beta}}. \tag{18.19}$$

Note that generally we have

$$\left({}^{(5)}\Omega^{\hat{\nu}}_{\ \hat{\alpha}}\right)_{\perp} = {}^{(4)}\Omega^{\hat{\nu}}_{\ \hat{\alpha}}. \tag{18.20}$$

However, taking the exterior derivative, and then projecting, yields

$$\begin{aligned}
\left(\mathbf{d}^{(5)}\Omega^{\hat{\nu}}_{\ \hat{\alpha}}\right)_{\perp} &= \left(\mathbf{d}^{(4)}\Omega^{\hat{\nu}}_{\ \hat{\alpha}} - \frac{1}{2}\mathbf{d}(\phi F^{\hat{\nu}}_{\ \hat{\alpha}}\omega^{\hat{4}})\right)_{\perp} \\
&= \mathbf{d}^{(4)}\Omega^{\hat{\nu}}_{\ \hat{\alpha}} - \frac{1}{4}\phi^2 F^{\hat{\nu}}_{\ \hat{\alpha}}F_{\hat{\beta}\hat{\gamma}}\omega^{\hat{\beta}}\wedge\omega^{\hat{\gamma}}. \tag{18.21}
\end{aligned}$$

Following the procedure in section 7.4 we will calculate the projected Riemann tensor. On one hand, we have

$$\left(\mathbf{d}^2\mathbf{e}_{\hat{\nu}}\right)_{\perp} = \left(\frac{1}{2}{}^{(5)}R^{\hat{d}}_{\ \hat{\nu}\hat{a}\hat{b}}\mathbf{e}_{\hat{d}}\otimes\omega^{\hat{a}}\wedge\omega^{\hat{b}}\right)_{\perp} = \frac{1}{2}{}^{(5)}R^{\hat{\mu}}_{\ \hat{\nu}\hat{\alpha}\hat{\beta}}\mathbf{e}_{\hat{\mu}}\otimes\omega^{\hat{\alpha}}\wedge\omega^{\hat{\beta}}. \tag{18.22}$$

On the other hand, using the Riemann tensor in the four dimensional spacetime, we have

$$\begin{aligned}
\left(\mathbf{d}^2\mathbf{e}_{\hat{\nu}}\right)_{\perp} &= \left(\mathbf{d}\left[\mathbf{e}_{\hat{a}}\otimes{}^{(5)}\Omega^{\hat{a}}_{\ \hat{\nu}}\right]\right)_{\perp} \\
&= \left(\mathbf{d}\mathbf{e}_{\hat{a}}\otimes{}^{(5)}\Omega^{\hat{a}}_{\ \hat{\nu}}\right)_{\perp} + \left(\mathbf{e}_{\hat{a}}\otimes\mathbf{d}^{(5)}\Omega^{\hat{a}}_{\ \hat{\nu}}\right)_{\perp} \\
&= \mathbf{e}_{\hat{\mu}}\otimes\left(\mathbf{d}^{(5)}\Omega^{\hat{\mu}}_{\ \hat{\nu}} + {}^{(5)}\Omega^{\hat{\mu}}_{\ \hat{a}}\wedge{}^{(5)}\Omega^{\hat{a}}_{\ \hat{\nu}}\right)_{\perp}. \tag{18.23}
\end{aligned}$$

We can now use Cartan's second structural equation, eq. (7.47), by decomposing the wedge product

$$\begin{aligned}
\left(\mathbf{d}^2\mathbf{e}_{\hat{\nu}}\right)_{\perp} &= \mathbf{e}_{\hat{\mu}}\otimes\left(\mathbf{d}^{(5)}\Omega^{\hat{\mu}}_{\ \hat{\nu}} + {}^{(5)}\Omega^{\hat{\mu}}_{\ \hat{\lambda}}\wedge{}^{(5)}\Omega^{\hat{\lambda}}_{\ \hat{\nu}} + {}^{(5)}\Omega^{\hat{\mu}}_{\ \hat{n}}\wedge{}^{(5)}\Omega^{\hat{n}}_{\ \hat{\nu}}\right)_{\perp} \\
&= \mathbf{e}_{\hat{\mu}}\otimes\left({}^{(4)}R^{\hat{\mu}}_{\ \hat{\nu}} - \frac{1}{4}\phi^2 F^{\hat{\mu}}_{\ \hat{\nu}}F_{\hat{\alpha}\hat{\beta}}\omega^{\hat{\alpha}}\wedge\omega^{\hat{\beta}} + \left[{}^{(5)}\Omega^{\hat{\mu}}_{\ \hat{4}}\wedge{}^{(5)}\Omega^{\hat{4}}_{\ \hat{\nu}}\right]_{\perp}\right) \\
&= \frac{1}{2}\left({}^{(4)}R^{\hat{\mu}}_{\ \hat{\nu}\hat{\alpha}\hat{\beta}} - \frac{1}{4}\phi^2\left[2F^{\hat{\mu}}_{\ \hat{\nu}}F_{\hat{\alpha}\hat{\beta}} + F^{\hat{\mu}}_{\ \hat{\alpha}}F_{\hat{\nu}\hat{\beta}}\right]\right)\mathbf{e}_{\hat{\mu}}\otimes\omega^{\hat{\alpha}}\wedge\omega^{\hat{\beta}} \tag{18.24}
\end{aligned}$$

where we have used eq. (18.15). From equations (18.22) and (18.24) it follows that

$$
{}^{(5)}R^{\hat{\mu}}{}_{\hat{\nu}\hat{\alpha}\hat{\beta}} = {}^{(4)}R^{\hat{\mu}}{}_{\hat{\nu}\hat{\alpha}\hat{\beta}} - \frac{1}{4}\phi^2\left(2F^{\hat{\mu}}{}_{\hat{\nu}}F_{\hat{\alpha}\hat{\beta}} - F^{\hat{\mu}}{}_{\hat{\alpha}}F_{\hat{\beta}\hat{\nu}} + F^{\hat{\mu}}{}_{\hat{\beta}}F_{\hat{\alpha}\hat{\nu}}\right).
\tag{18.25}
$$

This is eq. (7.83) in a different form than the hypersurface case. The difference is exactly due to the different properties of the connection coefficients.

By contracting this equation twice we get an expression relating the Ricci scalars in four and five dimensions. The right side of eq. (18.25) contracts to

$$
{}^{(4)}R^{\hat{\alpha}\hat{\beta}}{}_{\hat{\alpha}\hat{\beta}} - \frac{1}{4}\phi^2\left(2F^{\hat{\alpha}\hat{\beta}}F_{\hat{\alpha}\hat{\beta}} - F^{\hat{\alpha}}{}_{\hat{\alpha}}F^{\hat{\beta}}{}_{\hat{\beta}} + F^{\hat{\alpha}\hat{\beta}}F_{\hat{\alpha}\hat{\beta}}\right)
$$
$$
= {}^{(4)}R - \frac{3}{4}\phi^2 F^{\hat{\alpha}\hat{\beta}}F_{\hat{\alpha}\hat{\beta}}.
\tag{18.26}
$$

We first express the left hand side of eq. (18.25) in terms of the the projection tensor g_{ab}. Upon contraction over the 5 dimensional space we get

$$
\begin{aligned}
{}^{(5)}R^{e}{}_{fij}g^{a}{}_{e}g^{fb}g^{i}{}_{a}g^{j}{}_{b} &= {}^{(5)}R^{e}{}_{fij}g^{e}{}_{i}g^{fj} \\
&= {}^{(5)}R^{e}{}_{fij}(G^{e}{}_{i} - \xi^{e}\xi_{i})(G^{fj} - \xi^{f}\xi^{j}) \\
&= {}^{(5)}R - 2R_{ab}\xi^{a}\xi^{b},
\end{aligned}
\tag{18.27}
$$

where we have used the antisymmetry of the Riemann tensor. It remains to find the contraction $R_{ab}\xi^{a}\xi^{b}$. Using the same trick as in eq. (14.70) with $\xi_{a} = n_{a}$ as the normal vector, we can find an expression for $R_{ab}\xi^{a}\xi^{b}$. The covariant derivative of ξ_{b} is

$$
\nabla_{a}\xi_{b} = -\xi_{c}\Gamma^{c}{}_{ba} = -\Gamma^{4}{}_{ba}.
\tag{18.28}
$$

This yields

$$
\begin{aligned}
\nabla^{a}\xi_{a} &= 0 \\
\nabla^{b}(\xi^{a}\nabla_{a}\xi_{b}) &= -(\ln\phi)^{;\mu}{}_{;\mu} \\
(\nabla^{a}\xi^{b})(\nabla_{b}\xi_{a}) &= -\frac{1}{4}\phi^2 F^{\alpha\beta}F_{\alpha\beta} - (\ln\phi)_{;\mu}(\ln\phi)^{;\mu}.
\end{aligned}
\tag{18.29}
$$

Hence, from the second to last line in eq. (14.70), we get

$$
\begin{aligned}
R_{ab}\xi^{a}\xi^{b} &= \frac{1}{4}\phi^2 F^{\alpha\beta}F_{\alpha\beta} + (\ln\phi)_{;\mu}(\ln\phi)^{;\mu} - (\ln\phi)^{;\mu}{}_{;\mu} \\
&= \frac{1}{4}\phi^2 F^{\alpha\beta}F_{\alpha\beta} - \frac{1}{\phi}\Box\phi.
\end{aligned}
\tag{18.30}
$$

Thus from equations (18.26), (18.27) and (18.30), we get

$$
{}^{(5)}R = {}^{(4)}R - \frac{1}{4}\phi^2 F^{\alpha\beta}F_{\alpha\beta} - \frac{2}{\phi}\Box\phi.
\tag{18.31}
$$

Amazingly, we have obtained a Lagrangian which looks very much like four-dimensional gravity plus electromagnetism. We also have a scalar field which couples to the electromagnetic field. Five-dimensional Einstein gravity from a four-dimensional point of view, looks like four-dimensional Einstein gravity plus electromagnetism and a scalar field. This is the "miracle" of

Kaluza-Klein theory; it connects the theory of electromagnetism and gravity in a fascinating way.

The determinant of the five-dimensional metric can be written as

$$\sqrt{-G} = \phi\sqrt{-g},\tag{18.32}$$

thus the five-dimensional Lagrangian takes the form

$$\mathcal{L}_J = \phi\sqrt{-g}\left({}^{(4)}R - \frac{1}{4}\phi^2 F^{\alpha\beta}F_{\alpha\beta} - \frac{2}{\phi}\Box\phi\right).\tag{18.33}$$

This is the Kaluza-Klein Lagrangian in the *Jordan frame*. In this frame the Lagrangian has an overall scaling in the ϕ-field.

We first consider the simplest case where the scalar field is constant and equal to unity.

Thus assume for a while that the scalar field is

$$\phi = 1.$$

The Lagrangian simplifies and the action can be written

$$S_{KK} = \frac{1}{2\kappa_5}\int\sqrt{-g}\left({}^{(4)}R - \frac{1}{4}F^{\alpha\beta}F_{\alpha\beta}\right)d^5x.\tag{18.34}$$

We want to relate this to the four-dimensional action. The fifth dimension is spanned by a Killing vector, hence the integrand is independent of the fifth coordinate, $x^4 \equiv y$. This makes it possible to integrate the action over the coordinate y. If the length of the fifth dimension (or the compactification length) is ℓ, then we have

$$
\begin{aligned}
S_{KK} &= \frac{1}{2\kappa_5}\int\sqrt{-g}\left({}^{(4)}R - \frac{1}{4}F^{\alpha\beta}F_{\alpha\beta}\right)d^4x\,dy\\
&= \frac{\ell}{2\kappa_5}\int\sqrt{-g}\left({}^{(4)}R - \frac{1}{4}F^{\alpha\beta}F_{\alpha\beta}\right)d^4x.
\end{aligned}\tag{18.35}
$$

For this to correspond to the four-dimensional action, we have to identify the four-dimensional gravitational constant with

$$\kappa_4 = \frac{\kappa_5}{\ell}.\tag{18.36}$$

If we further rescale the field A_μ by

$$A_\mu \longmapsto \sqrt{2\kappa_4}A_\mu,\tag{18.37}$$

then the action can be written in the usual four-dimensional form

$$\boxed{S_{KK} = \int\sqrt{-g}\left(\frac{1}{2\kappa_4}{}^{(4)}R - \frac{1}{4}F^{\alpha\beta}F_{\alpha\beta}\right)d^4x.}\tag{18.38}$$

This action describes Einstein gravity in four dimensions coupled to an electromagnetic field. Hence, Einstein's theory of relativity and electromagnetism have been unified.

18.3   Implications of a fifth extra dimension

In the same way as gravitation is reduced to a geometric property of the spacetime in Einstein's theory of general relativity, gravitation and electromagnetism are reduced to geometric properties of a five-dimensional spacetime in the Kaluza-Klein theory.

Let us consider geodesic curves in the five-dimensional spacetime $^{(5)}\mathcal{M}$.[2] They are given by the equation

$$\frac{d^2 u^a}{ds^2} + \Gamma^a{}_{ij} u^i u^j = 0 \tag{18.39}$$

where s is the proper time in $^{(5)}\mathcal{M}$. Since x^4 is a cyclic coordinate, p_4, defined by

$$p_4 = \frac{\partial L}{\partial \dot{x}^4} = m_0 G_{4a} u^a = m_0 (A_\mu u^\mu + u^4) = m_0 u_4, \tag{18.40}$$

is a constant of motion. Here, m_0 is an invariant mass for the particle, and p_4 is the component of the momentum of the particle in the e_4-direction. Solving eq. (18.40) with respect to u^4 yields

$$u^4 = u_4 - A_\mu u^\mu. \tag{18.41}$$

The μ-component of the geodesic equation (18.39) is

$$\frac{d^2 u^\mu}{ds^2} + \Gamma^\mu{}_{\alpha\beta} u^\alpha u^\beta + 2\Gamma^\mu{}_{4\beta} u^4 u^\beta + \Gamma^\mu{}_{44} u^4 u^4 = 0, \tag{18.42}$$

where we have used a coordinate basis. From eq. (18.15) we see that $\Gamma^\mu{}_{44} = 0$. Also, since $F_{\alpha\beta}$ transforms as a four-dimensional tensor, we have

$$\frac{d^2 u^\mu}{ds^2} + \Gamma^\mu{}_{\alpha\beta} u^\alpha u^\beta = \frac{p_4}{m_0} F^\mu{}_\beta u^\beta. \tag{18.43}$$

If we take into account the rescaling (18.37) and compare this equation with the movement of a charged particle in an electromagnetic field, we see that the charge of the particle is

$$q = 8\pi \sqrt{\varepsilon_0 G} \frac{p_4}{c}. \tag{18.44}$$

A neutral particle has $p_4 = 0$. In the Kaluza-Klein theory a charge in the four-dimensional spacetime corresponds to a covariant momentum component in the fifth dimension. The charge of a particle is conserved since p_4 is a constant of motion.

The parameter s is the invariant interval in $^{(5)}\mathcal{M}$. We introduce the proper time of the particle as a parameter. The line-element has the form

$$-\epsilon ds^2 = -d\tau^2 + (dy + A_\mu dx^\mu)^2 \tag{18.45}$$

where $d\tau$ is the proper time in the four-dimensional spacetime (we will only consider time-like curves in four-space), and $\epsilon = 1, 0$ and -1 for time-like, null and space-like curves in $^{(5)}\mathcal{M}$, respectively. This implies

$$-\epsilon = -\left(\frac{d\tau}{ds}\right)^2 + (u^4 + A_\mu u^\mu)^2 = -\left(\frac{d\tau}{ds}\right)^2 + (u_4)^2. \tag{18.46}$$

---

[2]In this section we will set $\phi = 1$.

Thus eq. (18.43) can be written

$$\frac{d^2 u^\mu}{d\tau^2} + \Gamma^\mu{}_{\alpha\beta} \frac{dx^\alpha}{d\tau} \frac{dx^\beta}{d\tau} = \frac{q}{m_0 \sqrt{(u_4)^2 + \epsilon}} F^\mu{}_\beta u^\beta. \tag{18.47}$$

Hence, it follows that the particles physical (as measured in the four-dimensional spacetime) rest mass is

$$\bar{m}_0 = m_0 \sqrt{(u_4)^2 + \epsilon} = \sqrt{m_q^2 + \epsilon m_0^2} \tag{18.48}$$

where we have used eq. (18.44) and

$$m_q = \frac{|q|}{8\pi\sqrt{\varepsilon_0 G}}. \tag{18.49}$$

The smallest possible m_q is for $q = e$ (e is the elementary charge). This gives $m_q = 10^{-9}$kg. For null or time-like ($\epsilon = 0, 1$) geodetic curves the particle mass is equal or larger than this. For space-like curves in $^{(5)}\mathcal{M}$ – which perfectly well can be time-like in the four-dimensional spacetime – $\bar{m}_0$ can be arbitrary small. Thus the trajectories of charged particles with mass less than 10^{-9}kg are space-like in $^{(5)}\mathcal{M}$. Using eqs. (18.44) and (18.49) we can write eq. (18.48) as

$$\bar{m}_0 = m_q \sqrt{1 + \epsilon \frac{c^2}{(u_4)^2}}. \tag{18.50}$$

This shows that for a particle with large charge-to-mass ratio, for example an electron ($q/\bar{m}_0 = -2.9 \cdot 10^{20}$), the world-line is tachyonic ($\epsilon = -1$) and $u_4 \approx c$.

The five-dimensional world is neutral and without any electromagnetic fields. One may wonder, then, what is the five-dimensional field which corresponds to the Coulomb field of a charge from the four-dimensional point of view? The nature of the five-dimensional field may be identified by noting that what we perceive as charge is the motion of a neutral particle around a closed fifth dimension. Such motion generates an inertial dragging field. A detailed calculation [Grø86] shows that the Coulomb field is indeed the projection of the inertial dragging field into our four-dimensional world. Hence, if gravity was correctly decribed by a theory like that of Newton involving no inertial dragging field, there would not exist any electromagnetic fields. From the five-dimensional point of view electromagnetism is a general relativistic gravitational effect which vanishes in the Newtonian limit.

The 5-dimensional wave-equation and the Klein-Gordon equation

Consider the five-dimensional wave-equation

$$\Box_5 \psi = 0 \tag{18.51}$$

where ψ represents a wave-function, and $\Box_5$ is the five-dimensional d'Alembert operator

$$\Box_5 \psi = \frac{1}{\sqrt{-G}} \frac{\partial}{\partial x^a} \left(\sqrt{-G} G^{ab} \frac{\partial \psi}{\partial x^b} \right). \tag{18.52}$$

Since the fifth dimension is closed and periodic, ψ must be a periodic function in x^4. Hence it can be expanded in Fourier modes

$$\psi(x^a) = \sum_n \psi_n(x^\mu)e^{iny/\ell}, \qquad x^4 = y. \qquad (18.53)$$

The inverse metric G^{ab} is

$$G^{\mu\nu} = g^{\mu\nu}, \quad G^{4\mu} = A^\mu, \quad G^{44} = 1 + A_\mu A^\mu, \qquad (18.54)$$

so – using $\sqrt{-G} = \sqrt{-g}$ – we get

$$\sqrt{-g}\Box_5\psi = \frac{\partial}{\partial x^\mu}\left(\sqrt{-g}g^{\mu\nu}\frac{\partial\psi}{\partial x^\nu}\right) + \frac{\partial}{\partial y}\left(\sqrt{-g}G^{4\mu}\frac{\partial\psi}{\partial x^\mu}\right)$$
$$+ \frac{\partial}{\partial x^\mu}\left(\sqrt{-g}G^{\mu 4}\frac{\partial\psi}{\partial y}\right) + \frac{\partial}{\partial y}\left(\sqrt{-g}G^{44}\frac{\partial\psi}{\partial y}\right). \qquad (18.55)$$

Substituting eqs. (18.53) and (18.54) we can write eq. (18.51) as

$$\Box_4\psi_n + \frac{in}{\ell}\left[\frac{\partial}{\partial x^\mu}(\sqrt{-g}A^\mu\psi_n) + \sqrt{-g}A^\mu\frac{\partial}{\partial x^\mu}\psi_n\right]$$
$$- \frac{n^2}{\ell^2}\sqrt{-g}(1 + A^\mu A_\mu)\psi_n = 0. \qquad (18.56)$$

Let us assume that the metric in the observable spacetime is the Minkowski metric. Also, introduce a charge[3], q_n, and mass, m_n, by

$$q_n = n\frac{\hbar\sqrt{16\pi G}}{c\ell}, \qquad m_n = n\frac{\hbar c}{\ell}. \qquad (18.57)$$

Eq. (18.56) can now – after the rescaling $A_\mu \mapsto \sqrt{16\pi G}A_\mu$ – be written

$$g^{\mu\nu}\left(\frac{\partial}{\partial x^\mu} - i\frac{q_n}{\hbar}A_\mu\right)\left(\frac{\partial}{\partial x^\nu} - i\frac{q_n}{\hbar}A_\nu\right)\psi_n - \frac{m_n^2}{\hbar^2}\psi_n = 0. \qquad (18.58)$$

This is the Klein-Gordon equation for particles with charge q_n in the presence of an electromagnetic field.

The expectation value of the momentum in the fifth dimension, p_4, for the eigenfunction $\psi(x^a) = \psi_n(x^\mu)e^{iny/\ell}$, is given by

$$p_4 = \frac{1}{\ell}\int dy \int d^4x\psi^*\left(-i\hbar\frac{\partial}{\partial y}\right)\psi = n\frac{\hbar}{\ell}. \qquad (18.59)$$

Hence, the momentum in the fifth dimension is quantised. Eq. (18.44) can – using eq. (18.59)– be written as

$$q = n \cdot 8\pi\frac{\sqrt{\varepsilon_0 G}}{c}\frac{\hbar}{\ell}. \qquad (18.60)$$

This shows that the charge of the particle is quantised. In the Kaluza-Klein theory the quantisation of the charge is a result of the quantisation of the momentum.

---

[3]This charge must be interpreted as the charge as expressed in CGS units. In SI units one chooses Ampere (A) to be a fundamental unit and the Coulomb (C) to be a derived unit: $1C = 1As$. However, in Kaluza-Klein theory there is no reason why the unit of Ampere should appear since this refers to moving charges and no charges have yet been defined! In SI units the charge is given below, eq. (18.60). To get from SI to CGS units swap ε_0 with $1/(4\pi)$.

Substituting $q = ne$ we get

$$\ell = 8\pi\sqrt{\varepsilon_0 G}\frac{\hbar}{ce} = 4\sqrt{\frac{\pi}{\alpha}}\ell_{\text{Pl}} \tag{18.61}$$

where $\alpha = e^2/(4\pi\varepsilon_0\hbar c)$ is the fine structure constant, and $\ell_{\text{Pl}} = (\hbar G/c^3)^{1/2}$ is the Planck length.

This means that $\ell = 10^{-33}$cm. That the fifth dimension is so incredible tiny explains why we do not have any physical experience of it. The quantisation of charge in Kaluza-Klein theory is reformulated in terms of quantisation of the radius of the compact dimension. Hence, if this can be explained in a quantum theory of gravity, we can also explain the quantisation of charge.

18.4   Conformal transformations

The scalar field ϕ is called a *dilaton* field. If we want to include a non-constant ϕ in Kaluza-Klein theory, then we have to rescale the Lagrangian such that we can identify $\phi^{(4)}R$ with a Ricci scalar $^{(4)}\tilde{R}$. In order to do this we will introduce the notion of *conformal transformations*.

Definition: Conformal transformations   *Let N and M be two manifolds with metrics $\tilde{g}$ and g respectively. A smooth function $f : N \longmapsto M$ is said to be a conformal transformation if*

$$\Omega^{-2}\tilde{\mathbf{g}} = f^*\mathbf{g} \tag{18.62}$$

where Ω is some non-zero function.

If such a map exists for two manifolds N and M, then we say that they are *conformally equivalent*.

A conformal transformation *rescales* the metric, thus conformal transformations relates manifolds where the metric is the same up to a rescaling. In particular, we see that isometries are conformal transformations with $\Omega = 1$.

Let $\mathbf{v}$ and $\mathbf{u}$ be two vectors. Isometries will preserve both the lengths and the angles of these vectors. For conformal transformations we have

$$\tilde{\mathbf{g}}(\mathbf{v}, \mathbf{v}) = \Omega^2\mathbf{g}(\mathbf{v}, \mathbf{v}) \tag{18.63}$$

while

$$\begin{aligned}\tilde{\angle}(\mathbf{v}, \mathbf{u}) &\equiv \frac{\tilde{\mathbf{g}}(\mathbf{v}, \mathbf{u})}{\sqrt{\tilde{\mathbf{g}}(\mathbf{v}, \mathbf{v})\tilde{\mathbf{g}}(\mathbf{u}, \mathbf{u})}} \\ &= \frac{\mathbf{g}(\mathbf{v}, \mathbf{u})}{\sqrt{\mathbf{g}(\mathbf{v}, \mathbf{v})\mathbf{g}(\mathbf{u}, \mathbf{u})}} = \angle(\mathbf{v}, \mathbf{u}).\end{aligned} \tag{18.64}$$

Thus, *angles are preserved under conformal transformations*.

A manifold M is said to be *conformally flat* if, for every point $p \in M$, there exists an open neighbourhood $U \subset M$ such that U is conformally equivalent to a flat manifold. Note that conformal flatness is only defined locally, which is of course not as restrictive as the global requirement.

An important result is that *any 2-dimensional Lorentzian or Riemannian manifold is conformally flat*. For manifolds of higher dimensions, we have to investigate the properties of the curvature tensors under conformal transformations more carefully.

Example 18.1 (Hyperbolic space is conformally flat)   *Example*
Let us provide an example of two conformally related manifolds.
     We will consider hyperbolic space $\mathbb{H}^3$. The metric for $\mathbb{H}^3$ can be written as

$$ds^2 = \frac{dr^2}{1 + r^2} + r^2(d\phi^2 + \sin^2 \phi d\theta^2). \tag{18.65}$$

We will show that this metric is conformally flat. To show this, we try to find a coordinate transformation $R(r)$ such that the metric takes the form

$$ds^2 = \Omega^2 \left[dR^2 + R^2(d\phi^2 + \sin^2 \phi d\theta^2) \right]. \tag{18.66}$$

The metric inside the square brackets is flat, thus if a function $R(r)$ exists, then the metric is conformally flat. From the above we clearly require

$$\begin{aligned} r &= \Omega R \\ \frac{dr}{\sqrt{1 + r^2}} &= \Omega dR. \end{aligned} \tag{18.67}$$

These equations can be solved to yield

$$\begin{aligned} R &= \frac{r}{\sqrt{1 + r^2} + 1} \\ \Omega &= \frac{2}{1 - R^2}. \end{aligned} \tag{18.68}$$

Note that R is bounded by $0 \le R < 1$; hence hyperbolic space is conformally equivalent to the open Euclidean disk

$$\mathbb{D}^3 = \left\{ (R, \phi, \theta) \in \mathbb{E}^3 \mid 0 \le R < 1 \right\}. \tag{18.69}$$

Thus the hyperbolic space is conformally flat.
     Remember that we have already encountered hyperbolic space in a different form, namely the Poincaré half-space model

$$ds^2 = \frac{1}{z^2} \left(dx^2 + dy^2 + dz^2 \right). \tag{18.70}$$

From this metric we can easily see that hyperbolic space is conformally flat. The scale factor in this case is simply $\Omega = z$. The disk model of the hyperbolic space, which we described above, is called *the Poincaré disk*. Thus since two successive conformal transformations are also a conformal transformation, then we have to conclude that *the upper half Euclidean plane is conformally equivalent to the Euclidean disk.*

_____

     We call a vector field $\boldsymbol{\xi}_C$ generating a conformal transformation a *conformal Killing vector field*. We can define a conformal Killing vector field by the requirement

$$\boxed{ \mathcal{L}_{\boldsymbol{\xi}_C} \mathbf{g} = 2\kappa \mathbf{g}, } \tag{18.71}$$

where κ is in general a function. If κ happens to be *constant* then we call $\boldsymbol{\xi}_C$ a *homothety*. The homotheties generate a subclass of the conformal transformations, usually called the similarity group. The similarity group are special conformal transformations where the function Ω in eq. (18.62) is a constant.
     In component form, the metric will transform under a conformal transformation as

$$g_{\mu\nu} \longmapsto \Omega^2 g_{\mu\nu}. \tag{18.72}$$

We can use this to investigate the transformation properties for the curvature tensors under such conformal transformations. Through a rather lengthy calculation, we can compare the Riemann tensors of $\tilde{g}_{\mu\nu}$ and $g_{\mu\nu}$. The result is as follows

$$
\begin{aligned}
\tilde{R}^{\delta}{}_{\alpha\beta\gamma} = \ & R^{\delta}{}_{\alpha\beta\gamma} + 2\delta^{\delta}{}_{[\alpha}\nabla_{\beta]}\nabla_{\gamma}\ln\Omega - 2g^{\delta\lambda}g_{\gamma[\alpha}\nabla_{\beta]}\nabla_{\lambda}\ln\Omega \\
& + 2\left(\nabla_{[\alpha}\ln\Omega\right)\delta^{\delta}{}_{\beta]}\nabla_{\gamma}\ln\Omega - 2\left(\nabla_{[\alpha}\ln\Omega\right)g_{\beta]\gamma}g^{\delta\lambda}\nabla_{\lambda}\ln\Omega \\
& - 2g_{\gamma[\alpha}\delta^{\delta}{}_{\beta]}g^{\lambda\kappa}\left(\nabla_{\lambda}\ln\Omega\right)\nabla_{\kappa}\ln\Omega.
\end{aligned}
\tag{18.73}
$$

Contracting once, we obtain the Ricci tensor

$$
\begin{aligned}
\tilde{R}_{\alpha\gamma} = \ & R_{\alpha\gamma} - (n-2)\nabla_{\alpha}\nabla_{\gamma}\ln\Omega - g_{\alpha\gamma}g^{\delta\lambda}\nabla_{\delta}\nabla_{\lambda}\ln\Omega \\
& + (n-2)\left(\nabla_{\alpha}\ln\Omega\right)\nabla_{\gamma}\ln\Omega \\
& - (n-2)g_{\alpha\gamma}g^{\delta\lambda}\left(\nabla_{\delta}\ln\Omega\right)\nabla_{\lambda}\ln\Omega,
\end{aligned}
\tag{18.74}
$$

where n is the dimension of the manifold. Contracting with $\tilde{g}^{\alpha\gamma} = \Omega^{-2}g^{\alpha\gamma}$ we obtain the Ricci scalar

$$
\begin{aligned}
\tilde{R} = \ & \Omega^{-2}\left[R - 2(n-1)g^{\alpha\beta}\nabla_{\alpha}\nabla_{\beta}\ln\Omega\right. \\
& \left. - (n-2)(n-1)g^{\alpha\beta}\left(\nabla_{\alpha}\ln\Omega\right)\nabla_{\beta}\ln\Omega\right].
\end{aligned}
\tag{18.75}
$$

It is useful to define the *Weyl tensor* $C^{\alpha}{}_{\beta\gamma\delta}$ for dimensions $n \geq 3$ as

$$
\begin{aligned}
C_{\alpha\beta\gamma\delta} = \ & R_{\alpha\beta\gamma\delta} - \frac{2}{n-2}\left(g_{\alpha[\gamma}R_{\delta]\beta} - g_{\beta[\gamma}R_{\delta]\alpha}\right) \\
& + \frac{2}{(n-1)(n-2)}Rg_{\alpha[\gamma}g_{\delta]\beta}.
\end{aligned}
\tag{18.76}
$$

This tensor has many interesting properties. First of all, it has the same symmetries as the Riemann tensor concerning permutations of the indices. Secondly, the trace over any pair of indices vanishes:

$$
C^{\alpha}{}_{\beta\alpha\delta} = 0.
\tag{18.77}
$$

This tensor is completely trace-free. Thirdly, it transforms very nicely under conformal transformations

$$
\tilde{C}^{\alpha}{}_{\beta\gamma\delta} = C^{\alpha}{}_{\beta\gamma\delta}.
\tag{18.78}
$$

Whether or not a space is conformally flat relies on the Weyl tensor through the following theorem:[4]

Theorem: Conformal flatness *A manifold of dimension $n \geq 4$ is conformally flat if and only if its Weyl tensor vanish.*

For these reasons the Weyl tensor is also called *the conformal curvature tensor*.

Since the Weyl tensor is trace-less, it will not contribute to the Ricci tensor. The Weyl tensor is basically the part of the Riemann tensor which is not determined by Einstein's field equations. For example, in a vacuum spacetime, the Ricci part of the Riemann tensor will be zero due to the field equations. The remaining non-zero components of the Riemann tensor must therefore correspond to non-zero components of the Weyl tensor. This is the case in, for example, the Schwarzschild spacetime where the Ricci tensor vanishes.

---

[4]For the case of dimension 3, see problem 18.5.

Example 18.2 (Homotheties for the Euclidean plane)
Let us consider the Euclidean plane with metric

$$ds^2 = dx^2 + dy^2 = \delta_{\mu\nu} dx^\mu dx^\nu. \tag{18.79}$$

We will try to find all possible homotheties for the Euclidean plane. To find these we solve the equation

$$\pounds_\xi g = 2\kappa g. \tag{18.80}$$

Since in Cartesian coordinates all the connection coefficients vanish, the conformal Killing equation reduces to

$$\xi_{\mu,\nu} + \xi_{\nu,\mu} = 2\kappa\delta_{\mu\nu}. \tag{18.81}$$

The diagonal equations are

$$\xi_{1,1} = \kappa, \qquad \xi_{2,2} = \kappa, \tag{18.82}$$

which have in general the solutions

$$\xi_1 = \kappa x + f_1(y), \qquad \xi_2 = \kappa y + f_2(x). \tag{18.83}$$

Inserting this into the off-diagonal equations we get

$$f_1'(y) = -f_2'(x). \tag{18.84}$$

This shows that $f_1(y)$ and $f_2(x)$ can at most be linear in their respective variables. For $\kappa = 0$ we get the usual Killing vector fields

$$\boldsymbol{\xi}_1 = \frac{\partial}{\partial x}, \quad \boldsymbol{\xi}_2 = \frac{\partial}{\partial y}, \quad \boldsymbol{\xi}_3 = y\frac{\partial}{\partial x} - x\frac{\partial}{\partial y}. \tag{18.85}$$

We note that there is only one linearly independent vector field for $\kappa \neq 0$. We choose $\kappa = 1$, and the homothety can be written

$$\boldsymbol{\xi}_4 = x\frac{\partial}{\partial x} + y\frac{\partial}{\partial y}. \tag{18.86}$$

This is a radial vector field, each vector pointing away from the origin. As we move along the vector field we "expand" the space radially.

18.5   Conformal transformation of the Kaluza-Klein action

In the Jordan frame, the Kaluza-Klein Lagrangian has an overall scaling factor given by the scalar field ϕ. In this section we will introduce a different frame, the so-called *Einstein frame*. This frame is related to the Jordan frame via a conformal transformation.

Let us therefore perform a conformal transformation of the Kaluza-Klein action to illuminate the effect this extra dimension has upon our four-dimensional world.

It turns out that we get the same result whether we transform the four-dimensional spacetime or the five-dimensional spacetime. Let us choose a transformation of the four-dimensional metric. Henceforth we will skip the label [(4)] on the tensors because we will only consider four-dimensional objects.

The Ricci scalar transforms as

$$
\begin{aligned}
\tilde{R} &= \Omega^{-2}\left[R - 6g^{\alpha\beta}\nabla_\alpha\nabla_\beta \ln\Omega - 6g^{\alpha\beta}\left(\nabla_\alpha \ln\Omega\right)\nabla_\beta \ln\Omega\right] \\
&= \Omega^{-2}R - 6\Omega^{-3}\square\Omega.
\end{aligned}
\tag{18.87}
$$

The determinant of the metric $\sqrt{-g}$ will transform as

$$
\sqrt{-\tilde{g}} = \sqrt{-\Omega^8 g} = \Omega^4\sqrt{-g},
\tag{18.88}
$$

thus the pure gravity term in the action will transform as

$$
\sqrt{-\tilde{g}}\phi\tilde{R} = \sqrt{-g}\phi\left[\Omega^2 R - 6\Omega\square\Omega\right].
\tag{18.89}
$$

Hence, by choosing

$$
\Omega = \phi^{-\frac{1}{2}},
\tag{18.90}
$$

we can get rid of the ϕ in front of the Ricci tensor. The action will then turn into the sought after form.

Using $\Omega = \phi^{-\frac{1}{2}}$ we get

$$
F_{\mu\nu}F_{\alpha\beta}\tilde{g}^{\mu\alpha}\tilde{g}^{\nu\beta} = \phi^2 F_{\mu\nu}F_{\alpha\beta}g^{\mu\alpha}g^{\nu\beta} = \phi^2 F_{\alpha\beta}F^{\alpha\beta},
\tag{18.91}
$$

and

$$
\begin{aligned}
\tilde{\square}\phi &= \frac{1}{\sqrt{-\tilde{g}}}\left(\sqrt{-\tilde{g}}\tilde{g}^{\mu\nu}\phi_{,\mu}\right)_{,\nu} \\
&= \frac{\phi^2}{\sqrt{-g}}\left(\phi^{-1}\sqrt{-g}g^{\mu\nu}\phi_{,\mu}\right)_{,\nu} \\
&= -\phi_{,\nu}\phi^{,\nu} + \phi\square\phi.
\end{aligned}
\tag{18.92}
$$

Also, using

$$
\square\Omega = \frac{3}{4}\phi^{-\frac{5}{2}}\phi_{,\nu}\phi^{,\nu} - \frac{1}{2}\phi^{-\frac{3}{2}}\square\phi,
\tag{18.93}
$$

we get the Lagrangian into the form

$$
\begin{aligned}
\mathcal{L}_{KK} &= \sqrt{-\tilde{g}}\phi\left[\tilde{R} - \frac{1}{4}\phi^2 F_{\mu\nu}F_{\alpha\beta}\tilde{g}^{\mu\alpha}\tilde{g}^{\nu\beta} - \frac{2}{\phi}\tilde{\square}\phi\right] \\
&= \sqrt{-g}\left[R - \frac{1}{4}\phi^3 F_{\alpha\beta}F^{\alpha\beta} - \frac{5}{2}\frac{\phi_{,\nu}\phi^{,\nu}}{\phi^2} + \frac{\square\phi}{\phi}\right].
\end{aligned}
\tag{18.94}
$$

Since the action is the integral of this Lagrangian, we can perform a partial integration of the $\square\phi$-term. The total derivative yields only boundary terms and can therefore be disposed of. For convenience, it is useful to define φ by

$$
\phi^3 = e^{-\sqrt{6\kappa_4}\varphi},
\tag{18.95}
$$

and to rescale A_μ as $A_\mu \mapsto \sqrt{2\kappa_4}A_\mu$. The Kaluza-Klein action in the *Einstein frame* can then be written as

$$
\boxed{S_{KK} = \int \sqrt{-g}\left(\frac{1}{2\kappa_4}R - \frac{1}{4}e^{-\sqrt{6\kappa_4}\varphi}F_{\alpha\beta}F^{\alpha\beta} - \frac{1}{2}\varphi_{,\nu}\varphi^{,\nu}\right)d^4x.}
\tag{18.96}
$$

Hence, in addition to pure Einstein gravity, we have an electromagnetic field and a scalar field which is coupled to the kinetic term of the electromagnetic field. As the scalar field varies, the field strength of the electromagnetic field varies.

Upon variation of the above action, we find the following equations of motion:

$$\Box\varphi \;=\; \sqrt{\frac{3\kappa_4}{8}}\,e^{-\sqrt{6\kappa_4}\varphi}F_{\alpha\beta}F^{\alpha\beta}, \qquad (18.97)$$

$$\nabla^\alpha\left(e^{-\sqrt{6\kappa_4}\varphi}F_{\alpha\beta}\right) \;=\; 0, \qquad (18.98)$$

$$E_{\mu\nu} \;=\; \kappa_4\left[e^{-\sqrt{6\kappa_4}\varphi}T^{EM}_{\mu\nu} + T^{\varphi}_{\mu\nu}\right], \qquad (18.99)$$

where

$$T^{EM}_{\mu\nu} \;=\; F^\alpha{}_\mu F_{\alpha\nu} - \frac{1}{4}F_{\alpha\beta}F^{\alpha\beta}g_{\mu\nu},$$

$$T^{\varphi}_{\mu\nu} \;=\; (\nabla_\mu\varphi)(\nabla_\nu\varphi) - \frac{1}{2}(\nabla^\alpha\varphi)(\nabla_\alpha\varphi)g_{\mu\nu} \qquad (18.100)$$

are the usual energy-momentum tensors for the electromagnetic field and scalar field, respectively.

Jordan frame or Einstein frame?

We have so far introduced two choices of frames which seem to give different actions and thus different physical interpretations. However, the two frames, the Jordan and Einstein frames, are mathematically equivalent; they are related via a conformal transformation.

Suppose we have defined the speed of light to be unity and the unit of time to be the inverse of some atomic transition frequency. We can now measure distances in space by sending light signals and determine the travel time using the proper time as defined by this atomic transition. This will correspond to a measurement done in the Jordan frame. In the Jordan frame the matter Lagrangian is independent of the dilation, ϕ; i.e. it will have the form $\mathcal{L}_m[\hat{g}_{\mu\nu}, \psi_m]$ where ψ_m is some matter field. An atomic clock will measure a proper time with respect to the metric $\hat{g}_{\mu\nu}$, and hence, any measurement where the time is defined in terms of the atomic clock will determine the spacetime geometry in the Jordan frame.

Suppose, on the other hand, that one desides to define the unit of time in terms of a purely general relativistic object, like a black hole (the speed of light is still unity). A unit of time can, for example, be the inverse of the fundamental quasinormal frequency of a certain "standard" non-spinning black hole. The proper time of the black hole then defines a unit of time which measures the geometry of spacetime in terms of a metric for which the gravitational action takes the standard Einstein-Hilbert form; hence, the result is the Einstein frame metric.

By choosing a frame, i.e. conventions and units of time etc., only one of the actions can be physically correct even though they might be mathematically equivalent. However, if we consider any arbitrary conventions for adjustable, not fixed, then two mathematically equivalent theories can also be physically equivalent[5].

---

[5]These issues have been discussed more elaborately by Flanagan [Fla04].

18.6 Kaluza-Klein cosmology

Here, we will give some applications of Kaluza-Klein theory to cosmology. The extra dimension can alter the evolution of the observable universe, as we will see. Cosmology may also be the arena which can explain why the fifth dimension is so incredibly smaller than the 3 spatial dimensions we observe.

5-dimensional Kasner universe

(see also Chodos and Detweiler, and Hervik [CD80, Her01])
For five-dimensional spacetime the Kasner solutions have the form

$$ds^2 = -dt^2 + \sum_{i=1}^{4} t^{2p_i}(dx^i)^2 \tag{18.101}$$

where

$$\sum_{i=1}^{4} p_i = \sum_{i=1}^{4} p_i^2 = 1. \tag{18.102}$$

We can give a geometrical meaning of the exponents in this case as well, similarly to the four-dimensional Kasner solutions, see Fig. 18.3. Consider the

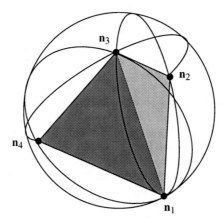

Figure 18.3: A tetrahedron inscribed in a sphere.

sphere centered at $(1/4, 0, 0)$ with radius $3/4$. Inscribe a regular tetrahedron inside this sphere with its four vertices on the sphere. If the vertices are called n_i, $i = 1, ..., 4$, then the 1-component of the point n_i will give the exponent p_i. All the different orientations of the tetrahedron correspond to the different solutions.

Note that there are two different configurations of the tetrahedron which give an isotropic flat observable universe. One is the case where $p_1 = 1$, $p_2 = p_3 = p_4 = 0$ and has the metric

$$ds^2 = -dt^2 + t^2(dx^1)^2 + (dx^2)^2 + (dx^3)^2 + (dx^4)^2. \tag{18.103}$$

This metric has one expanding direction and the rest are stationary. This is certainly not an accurate description of the universe which we live in. The other solution is more interesting. It has $p_1 = p_2 = p_3 = 1/2$ and $p_4 = -1/2$. This metric has three expanding directions and a contracting one:

$$ds^2 = -dt^2 + t\left[(dx^1)^2 + (dx^2)^2 + (dx^3)^2\right] + t^{-1}(dx^4)^2. \qquad (18.104)$$

Hence, this universe model has the right behaviour. If the fifth direction is closed with $0 \leq x^4 \leq \ell$, the size of the fifth dimension will be

$$L = \frac{\ell}{\sqrt{t}}. \qquad (18.105)$$

The five-dimensional plane symmetric de Sitter universe

(see also Appelquist and Chodos [AC83])
The line-element has the form

$$ds^2 = -dt^2 + f(t)(dx^2 + dy^2 + dz^2) + g(t)(dx^4)^2 \qquad (18.106)$$

with the solution

$$f(t) = A \sinh \omega t, \quad g(t) = B \frac{\cosh^2 \omega t}{\sinh \omega t}. \qquad (18.107)$$

Here, $\omega^2 = 2\Lambda/3$.
  In the limit $t \to 0$, we have

$$f(t) \approx A\omega t, \quad g(t) \approx B(\omega t)^{-1} \qquad (18.108)$$

and thus the solution corresponds to the Kasner solution (18.104). For $\omega t \gg 1$ we have

$$f(t), g(t) \propto e^{\omega t}. \qquad (18.109)$$

This is the isotropic five-dimensional de Sitter universe.

Plane-wave solutions in five dimensions and exact Einstein-Maxwell solutions in four dimensions

We will now provide an example where we generate exact solutions of the four-dimensional Einstein-Maxwell equations from exact solutions for the five-dimensional vacuum field equations [Her03].
  Let $\beta_+, \omega, \beta, Q_1, Q_2$ be parameters and let s be given by

$$s(1 - s) = 2\beta_+^2 + \frac{2}{3}\omega^2 \sinh^2 2\beta + \frac{1}{6}(Q_1^2 + Q_2^2). \qquad (18.110)$$

Also, define the two one-forms

$$\begin{aligned} \boldsymbol{\omega}^2 &= \cos[\omega(w + t)]\mathbf{dy} - \sin[\omega(w + t)]\mathbf{dz}, \\ \boldsymbol{\omega}^3 &= \sin[\omega(w + t)]\mathbf{dy} + \cos[\omega(w + t)]\mathbf{dz}. \end{aligned} \qquad (18.111)$$

Then there exist homogeneous plane-wave solutions given by

$$\begin{aligned} ds_5^2 = \quad & e^{2t}(-dt^2 + dw^2) + e^{2s(w+t)} \\ \times \quad & \left[e^{-4\beta_+(w+t)}\left\{dx + e^{3\beta_+(w+t)}\left(q_1 e^{-\beta}\boldsymbol{\omega}^3 - q_2 e^{\beta}\boldsymbol{\omega}^2\right)\right\}^2 \right. \\ + \quad & \left. e^{2\beta_+(w+t)}\left\{e^{-2\beta}\left(\boldsymbol{\omega}^2\right)^2 + e^{2\beta}\left(\boldsymbol{\omega}^3\right)^2\right\} \right], \qquad (18.112) \end{aligned}$$

where

$$q_1 = \frac{Q_1\omega + 3\beta_+ Q_2 e^{2\beta}}{\omega^2 + 9\beta_+^2},$$

$$q_2 = \frac{Q_2\omega - 3\beta_+ Q_1 e^{-2\beta}}{\omega^2 + 9\beta_+^2}. \tag{18.113}$$

These are solutions to the five-dimensional vacuum field equations and describe a travelling gravitational wave in five dimensions.

Note that $\boldsymbol{\xi} = \frac{\partial}{\partial x}$ is a Killing vector, and hence, we can compactify the space in this direction. Also note that for a specific choice of parameters $s = 2\beta_+$, and hence, for this choice the dilaton will be constant. In this case, the above solutions can be reduced to exact solutions to Einstein's field equations with an electromagnetic field. Let us therefore choose $s = 2\beta_+$, and perform the Kaluza-Klein reduction with respect to the Killing vector $\boldsymbol{\xi} = \frac{\partial}{\partial x}$.

The metric is already written in the right form so we can just read off the electromagnetic vector potential

$$\mathbf{A} = e^{3\beta_+(w+t)} \left(q_1 e^{-\beta} \boldsymbol{\omega}^3 - q_2 e^{\beta} \boldsymbol{\omega}^2 \right). \tag{18.114}$$

The electromagnetic field tensor is thus

$$\mathbf{F} = \mathbf{dA} = e^{-t}(\boldsymbol{\eta}^0 + \boldsymbol{\eta}^1) \wedge (Q_1\boldsymbol{\eta}^2 + Q_2\boldsymbol{\eta}^3), \tag{18.115}$$

where we have introduced an orthonormal frame $\boldsymbol{\eta}^\mu$ so that

$$ds_4^2 = \eta_{\mu\nu}\boldsymbol{\eta}^\mu\boldsymbol{\eta}^\nu, \tag{18.116}$$

where $\eta_{\mu\nu}$ is the four-dimensional Minkowski metric.

The four-dimensional spacetime has the following solution given in terms of the orthonormal frame:

$$\begin{aligned}
\boldsymbol{\eta}^0 &= e^t\mathbf{dt} \\
\boldsymbol{\eta}^1 &= e^t\mathbf{dw} \\
\boldsymbol{\eta}^2 &= e^{s(w+t)}e^{-\beta}\left\{\cos[\omega(w+t)]\mathbf{dy} - \sin[\omega(w+t)]\mathbf{dz}\right\} \\
\boldsymbol{\eta}^3 &= e^{s(w+t)}e^{\beta}\left\{\sin[\omega(w+t)]\mathbf{dy} + \cos[\omega(w+t)]\mathbf{dz}\right\} \\
s(1-s) &= \omega^2\sinh^2 2\beta + \frac{1}{4}(Q_1^2 + Q_2^2).
\end{aligned} \tag{18.117}$$

Here, we have redefined the free parameter s so that its similarity with metric (15.117) on page 434 is more evident. That $\mathbf{F}$ satisfies the source-free Maxwell equations

$$\mathbf{dF} = 0, \qquad \mathbf{d}^\dagger\mathbf{F} = 0, \tag{18.118}$$

and that Einstein's field equations are satisfied (for the specific choice of constants $16\pi G = e = c = 1$) can be readily verified.

The four-dimensional solutions generalise the metric (15.117) to the non-vacuum case, and are homogeneous plane-wave solutions of Bianchi type VII_h. The source is an electromagnetic field which is of a very particular type. The electric and magnetic fields are (in the orthonormal frame)

$$E_i = e^{-t}(0, -Q_1, -Q_2), \qquad B_i = e^{-t}(0, -Q_2, Q_1). \tag{18.119}$$

Thus this is a *null* field where all invariants composed of the two-form field vanish

$$F_{\mu\nu}F^{\mu\nu} = F_{\mu\nu}(\star F)^{\mu\nu} = 0. \qquad (18.120)$$

The energy density of the field is

$$\rho_{EM} = \frac{1}{2}(E^2 + B^2) = e^{-2t}(Q_1^2 + Q_2^2). \qquad (18.121)$$

Problems

18.1. *A five-dimensional vacuum universe*
In this problem we will consider a five-dimensional universe with four-dimensional spatial sections given by the metric

$$d\sigma^2 = \frac{4}{u^2}\left[du^2 + \left(dv + \frac{1}{2}(xdy - ydx)\right)^2 + u\left(dx^2 + dy^2\right)\right]. \qquad (18.122)$$

This is the metric of the two-dimensional complex hyperbolic plane, $\mathbb{H}_{\mathbb{C}}^2$ (2 complex dimensions, but 4 real dimensions). The Ricci tensor for this space is proportional to the four-metric, h_{ij}:

$$R_{ij} = -\frac{3}{8}h_{ij}.$$

(a) Consider the five-dimensional universe model where

$$ds_5^2 = -dt^2 + a^2(t)d\sigma^2. \qquad (18.123)$$

Find the Friedmann equation for the vacuum, using the twice contracted Gauss equation (7.152). Show that there is a solution where the metric takes the form

$$ds_5^2 = -dt^2 + \frac{t^2}{2u^2}\left[du^2 + u\left(dx^2 + dy^2\right)\right.$$
$$\left. + \left(dv + \frac{1}{2}(xdy - ydx)\right)^2\right]. \qquad (18.124)$$

(b) Since $\xi = \frac{\partial}{\partial v}$ is a Killing vector, we can perform a Kaluza-Klein reduction by compactifying the space in this direction. Do this for this model, and write down the expressions for the dilaton field ϕ and the two-form field $\mathbf{F} = \mathbf{dA}$ in the Jordan frame.

(c) In the Jordan frame, show that the underlying four-dimensional space is an open FRW model. (Hint: Perform the coordinate transformation $Z = \sqrt{u}, X = 2x$ and $Y = 2y$, and show that the spatial three-space is the hyperbolic space in Poincaré half-space form, see problem 7.5 on page 173.)

18.2. *Another five-dimensional vacuum spacetime*
Consider the metric

$$ds^2 = 2du\left[dv + \frac{v^2}{a^2}du + 2v(dx + \sin ydz)\right]$$
$$+ a^2(dx + \sin ydz)^2 + \frac{a^2}{2}(dy^2 + \cos^2 ydz^2), \qquad (18.125)$$

where a is a constant. This metric is a solution to the five-dimensional Einstein equations for vacuum; i.e., $R_{\mu\nu} = 0$.

(a) Verify that $\boldsymbol{\xi} = \frac{\partial}{\partial x}$ is a Killing vector. Consider also the vector $\mathbf{k} = \frac{\partial}{\partial v}$. Show that

$$k^{\mu}k_{\mu} = k^{\mu}{}_{;\mu} = k^{\mu;\nu}k_{(\mu;\nu)} = k^{\mu;\nu}k_{[\mu;\nu]} = 0,$$

which means that $\mathbf{k}$ is null, expansion-free, shear-free, and twist-free.

(b) Using $\boldsymbol{\xi}$ as the five-dimensional Kaluza-Klein Killing vector, write the metric on standard form. Find the two-form $\mathbf{F}$ for this metric. Verify that $\mathbf{F}$ satisfies $d\mathbf{F} = 0$ and $d^{\dagger}\mathbf{F} = 0$. What is the four-dimensional reduced metric?

(c) Consider again the vector $\mathbf{k}$ for the four-dimensional reduced metric. Is $\mathbf{k}$ also null, expansion-free, shear-free, and twist-free for the reduced metric?

18.3. *A five-dimensional cosmological constant*
Show that a five-dimensional cosmological constant (i.e., in the Jordan frame) implies a potential for the scalar field φ in the Einstein frame. Show that this potential has the form

$$V(\varphi) = Ae^{\lambda\varphi}, \tag{18.126}$$

where A and λ are constants.

18.4. *Homotheties and Self-similarity*
In this problem we will consider the plane-wave solution given in eq. (18.117). We will show that this spacetime is a so-called *self-similar spacetime*.

(a) Show that the basis one-forms defined in eq. (18.117) have the property

$$\pounds_{\mathbf{X}}\eta^{\mu} = \eta^{\mu} \tag{18.127}$$

where $\mathbf{X}$ is the vector-field

$$\mathbf{X} = \frac{\partial}{\partial t} - \frac{\partial}{\partial w} + y\frac{\partial}{\partial y} + z\frac{\partial}{\partial z}. \tag{18.128}$$

Show further that this implies that $\mathbf{X}$ is a homothety.

Homotheties (including the isometries) form what is called the *similarity group* of a spacetime. If the similarity group acts transitive on the spacetime, then we call the spacetime *self-similar*. In particular, this means that the plane-wave spacetime (18.117) is self-similar.

(b) How do the matter fields transform under such homotheties? More specifically, find $\pounds_{\mathbf{X}}\mathbf{A}$, $\pounds_{\mathbf{X}}\mathbf{F}$, and $\pounds_{\mathbf{X}}\rho$, where $\mathbf{A}$, $\mathbf{F}$ and ρ are the electromagnetic one-form potential, field strength, and energy-density, respectively.

18.5. *Conformal flatness for three-manifolds*
All two-dimensional manifolds are conformally flat, and for $n \geq 4$ an n-manifold is conformally flat if and only if the Weyl tensor vanishes. For three-dimensional manifolds we have the following (see eg. [GHL90]).
A three-dimensional Riemannian space is conformally flat if and only if the covariant derivative of the tensor

$$S_{\mu\nu} \equiv R_{\mu\nu} - \frac{1}{4}Rg_{\mu\nu} \tag{18.129}$$

is a symmetric 3-tensor.

(a) Show that for three-dimensional spaces the number of independent components of the Riemann and the Ricci tensors are both 6. This shows that all the components of the Riemann tensor survive the contraction when one forms the Ricci tensor. Thus the Weyl tensor have to be identically zero for three-dimensional spaces.

(b) Show that the maximally symmetric Riemannian spaces S^3, $\mathbb{E}^3$ and $\mathbb{H}^3$, are conformally flat.

(c) Show that the Thurston geometry, Sol, with metric given in eq. (15.95), is not conformally flat.

Part VI

APPENDICES

Constants of Nature

Fundamental constants

Speed of light	$c = 2.9979 \cdot 10^8 \mathrm{m/s}$
Newton's gravitational constant	$G = 6.673 \cdot 10^{-11} \mathrm{Nm^2/kg^2}$
Elementary charge	$e = 1.602 \cdot 10^{-19} \mathrm{C}$
Electron volt	$1 \mathrm{eV} = 1.602 \cdot 10^{-19} \mathrm{J}$
Planck's constant	$\hbar = 1.055 \cdot 10^{-34} \mathrm{J\,s}$
Magnetic constant	$\mu_0 = 4\pi \cdot 10^{-7} \mathrm{N/A^2}$
Permittivity in vacuum	$\varepsilon_0 = \frac{1}{\mu_0 c^2} = 8.854 \cdot 10^{-12} \mathrm{C^2/Nm^2}$
Boltzmann's constant	$k_B = 1.381 \cdot 10^{-23} \mathrm{J/K}$
Stefan-Boltzmann's constant	$\sigma = 5.67 \cdot 10^{-8} \mathrm{J/sm^2 K^4}$
Fine-structure constant	$\alpha = \frac{e^2}{4\pi\varepsilon_0 \hbar c} \approx \frac{1}{137}$
Mass of electron	$m_e = 9.109 \cdot 10^{-31} \mathrm{kg}$
Mass of proton	$m_p = 1.673 \cdot 10^{-27} \mathrm{kg}$

The Solar System

Mass of the Earth	$M_{\mathrm{Earth}} = 6.0 \cdot 10^{24} \mathrm{kg}$
Mass of the Sun	$M_{\mathrm{Sun}} = 2.0 \cdot 10^{30} \mathrm{kg}$
Distance Earth-Sun	$a = 1 AU$
	$a = 1.5 \cdot 10^{11} \mathrm{m}$
Radius of the Earth	$R_{\mathrm{Earth}} = 6.4 \cdot 10^6 \mathrm{m}$
Radius of the Sun	$R_{\mathrm{Sun}} = 7.0 \cdot 10^8 \mathrm{m}$
Acceleration of gravity at Earth's surface	$g = 9.8 \mathrm{m/s^2}$
Lunar mass	$M_{\mathrm{Moon}} = 7.4 \cdot 10^{22} \mathrm{kg}$
Distance Mercury-Sun	$a_{\mathrm{Mercury}} = 5.8 \cdot 10^{10} \mathrm{m}$
Orbital period of Mercury	$T_{\mathrm{Mercury}} = 88 \text{ days}$
Eccentricity of Mercurian orbit	$e = 0.17$
Perihelion precession of Mercurian orbit	$\Delta\phi = 43'' \text{ per century}$

Astrophysical/Cosmological parameters

Only approximate values are given. Some values must be handled with care.

Hubble constant	$H_0 = 72\text{km s}^{-1}\text{Mpc}^{-1}$
CMB temperature	$T = 2.725\text{K}$
CMB fluctuations	$\delta T/T \approx 10^{-5}$
Age of the universe	$t_0 = 13,7 \cdot 10^9$ years
Curvature	$\Omega_k = -0.02 \pm 0.02$
Vacuum energy	$\Omega_\Lambda = 0.7$
Ordinary matter	$\Omega_m = 0.04$
Dark matter	$\Omega_{DM} = 0.28$

Planckian units

Planck length	$\ell_{\text{Pl}} = \sqrt{\frac{\hbar G}{c^3}}$	$= 1.62 \cdot 10^{-35}\text{m}$
Planck time	$t_{\text{Pl}} = \sqrt{\frac{\hbar G}{c^5}}$	$= 5.39 \cdot 10^{-44}\text{s}$
Planck mass	$m_{\text{Pl}} = \sqrt{\frac{\hbar c}{G}}$	$= 2.18 \cdot 10^{-8}\text{kg}$
Planck energy	$E_{\text{Pl}} = \sqrt{\frac{\hbar c^5}{G}}$	$= 1.22 \cdot 10^{19}\text{GeV}$

B

Penrose Diagrams

In this appendix we will review the concept of *Penrose diagrams*. They provide a useful geometric picture of the global and causal structure of the spacetime.

B.1 Conformal transformations and causal structure

Penrose diagrams are concerned with mapping an infinite spacetime onto a finite manifold with a boundary using a conformal transformation. Recall that conformal transformations rescale the metric

$$f^*\mathbf{g} = \Omega^{-2}\mathbf{g}.$$

Such transformations preserve the causal structure; hence they preserve the sign of the norm $\mathbf{g}(\mathbf{v}, \mathbf{v})$ for any given vector $\mathbf{v}$. This means that space-like vectors are mapped to space-like vectors, light-like to light-like vectors, and time-like to time-like vectors. Using conformal transformations we can under some circumstances pull the infinities of spacetime back onto a finite and bounded region. For example, the function $\arctan x$ maps the whole real line $\mathbb{R}$ onto the finite interval $[-\pi/2, \pi/2]$.

Let us consider Minkowski spacetime and see how we can map the infinite Minkowski space onto a diamond-shaped finite region using a conformal transformation. In polar coordinates Minkowski space takes the form

$$ds^2 = -dt^2 + dr^2 + r^2 \left(d\theta^2 + \sin^2\theta d\phi^2 \right). \tag{B.1}$$

Introducing null coordinates by

$$u = \frac{1}{2}(t - r), \quad v = \frac{1}{2}(t + r), \tag{B.2}$$

gives

$$ds^2 = -4dudv + (v - u)^2 \left(d\theta^2 + \sin^2\theta d\phi^2 \right). \tag{B.3}$$

Introducing the coordinates

$$U = \arctan u, \quad V = \arctan v, \tag{B.4}$$

the metric is brought onto the form

$$ds^2 = \frac{1}{\cos^2 U \cos^2 V} \left[-4 dU dV + \sin^2(V - U) \left(d\theta^2 + \sin^2 \theta d\phi^2 \right) \right]. \tag{B.5}$$

Note that the range of U and V are finite. Both coordinates lie in the interval $[-\pi/2, \pi/2]$, and hence, Minkowski space is mapped onto the finite region $[-\pi/2, \pi/2] \times [-\pi/2, \pi/2]$. Making the coordinate transformation

$$R = V - U, \quad T = V + U, \tag{B.6}$$

the metric can be expressed as

$$ds^2 = \frac{1}{\cos^2 U \cos^2 V} \left[-dT^2 + dR^2 + \sin^2 R \left(d\theta^2 + \sin^2 \theta d\phi^2 \right) \right]. \tag{B.7}$$

The conformal factor $1/\cos^2 U \cos^2 V$ can be disposed of using a conformal transformation $\Omega^{-2} = 1/\cos^2 U \cos^2 V$. The Penrose diagram of Minkowski space is a diagram of the spacetime given by the regular metric inside the square brackets; i.e., the conformally related metric

$$d\tilde{s}^2 = -dT^2 + dR^2 + \sin^2 R \left(d\theta^2 + \sin^2 \theta d\phi^2 \right). \tag{B.8}$$

Usually the two spherical dimensions are suppressed to make the diagram two-dimensional. The resulting diagram is depicted in Fig. B.1.

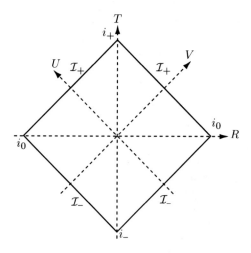

Figure B.1: Penrose diagram of Minkowski space.

Here, $i_\pm$, i_0 and $\mathcal{I}_\pm$ constitute the boundary of the diagram and have the following interpretations.

i_+ (i_-): Time-like future (past) infinity. All maximally extended time-like geodesics end (begin) here.

$\mathcal{I}_+$ ($\mathcal{I}_-$): Light-like future (past) infinity. All maximally extended light-like geodesics end (begin) here.

i_0: Space-like infinity. All maximally extended space-like geodesics end/begin here.

B.2   Schwarzschild spacetime

We can also find the Penrose diagram for the Schwarzschild spacetime. The Schwarzschild spacetime in Kruskal-Szekeres-coordinates is (see eq. (10.113))

$$ds^2 = -\frac{32M^3}{r}e^{-\frac{r}{2M}}\,du\,dv + r^2(d\theta^2 + \sin^2\theta d\phi^2). \tag{B.9}$$

In this case a slightly more complicated function than $\arctan x$ is needed. Here, the function

$$F(x) \equiv \arctan\left[\frac{x}{\sqrt{1+x^2}}\ln(1+x^2)\right], \tag{B.10}$$

will do the trick. We perform the coordinate transformation

$$U = F(u), \quad V = F(v)$$

which maps the analytically extended Schwarzschild solution onto a finite region. The Penrose diagram is depicted in Fig. B.2.

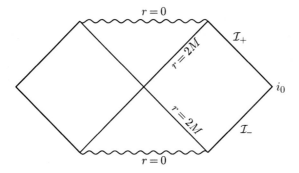

Figure B.2: Penrose diagram of Schwarzschild spacetime.

The lines $U = 0$ and $V = 0$ ($r = 2M$) correspond to the event horizon. The wavy horizontal lines are the future and past singularities in the Schwarzschild spacetime.

B.3   de Sitter spacetime

Consider the de Sitter space with positive spatial curvature in global coordinates

$$ds^2 = -dt^2 + \cosh^2 t\left[\frac{dr^2}{1-r^2} + r^2(d\theta^2 + \sin^2\theta d\phi^2)\right]. \tag{B.11}$$

We introduce the conformal time, η, by

$$\eta = \arctan(e^t) - \frac{\pi}{4}. \tag{B.12}$$

The metric then turns into

$$ds^2 = \cosh^2 t\left[-d\eta^2 + \frac{dr^2}{1-r^2} + r^2(d\theta^2 + \sin^2\theta d\phi^2)\right]. \tag{B.13}$$

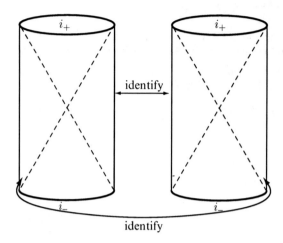

identify

identify

Figure B.3: Penrose diagram of de Sitter spacetime.

We suppress one of the coordinates and represent global de Sitter space as two beer cans with the interior included. This is illustrated in Fig. B.3. The surfaces of the beer cans are identified as indicated on the figure. In this case past and future space-like infinities, $i_{\pm}$, constitute the whole boundary; the spatial sections for the closed de Sitter model are finite and without boundary.

The different sections of de Sitter spacetime are different sections of these beer cans. Some of these are illustrated in Fig. B.4. The flat de Sitter model is the inside of the future light-cone of a point of past time-like infinity. The hyperbolic de Sitter model, on the other hand, is the inside of the future light-cone of the central point of one of the cans. Static de Sitter space is also entirely inside one can. It is the inside of a diamond-shaped region where the light-like boundary is the de Sitter horizon.

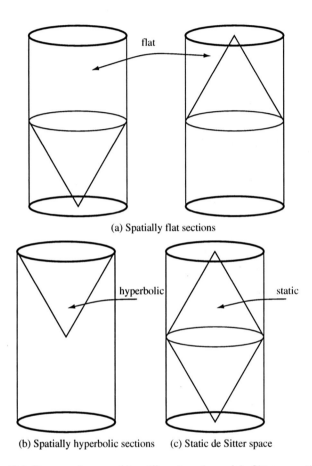

(a) Spatially flat sections

(b) Spatially hyperbolic sections   (c) Static de Sitter space

Figure B.4: Penrose diagram of the different sections of de Sitter spacetime.

C

Anti-de Sitter Spacetime

In the recent years the interest for anti-de Sitter spacetimes – maximally symmetric spacetimes with a negative cosmological constant – has exploded. The main interest for these spaces has come from string theory and M-Theory, but also cosmological models with extra dimensions use properties of the anti-de Sitter spacetimes. We will in this appendix review the construction of these spacetimes and investigate some of their properties.

C.1 The anti-de Sitter hyperboloid

The n-dimensional anti-de Sitter space, denoted AdS_n, can be considered as the hyperboloid

$$-V^2 - U^2 + X_1^2 + X_2^2 + ... + X_{n-1}^2 = -R^2, \tag{C.1}$$

embedded in the flat $(n + 1)$-dimensional ambient space with metric

$$ds^2 = -dV^2 - dU^2 + dX_1^2 + dX_2^2 + ... + dX_{n-1}^2. \tag{C.2}$$

AdS_n is maximally symmetric and is a solution to Einstein's field equations with a negative cosmological constant

$$E_{\mu\nu} = -\Lambda g_{\mu\nu}, \qquad \Lambda < 0, \tag{C.3}$$

where $R^2 = -\frac{(n-1)(n-2)}{2\Lambda}$. That this space is maximally symmetric can be seen if we note that the symmetry group is the matrix group G which leaves the metric (C.2) invariant; i.e., all matrices M such that

$$\mathsf{M}^t \eta \mathsf{M} = \eta, \tag{C.4}$$

where $\eta = \text{diag}(-1, -1, 1, ..., 1)$. This symmetry group G is usually called $O(2, n - 1)$ and has dimension $\frac{1}{2}n(n + 1)$.

Note that the ambient space has signature $(- - +...+)$ which implies that the hyperboloid, eq. (C.1), has closed time-like curves. The time-variable can be periodic, which, in any physical context, would be absurd. To prevent this, one usually considers the *universal cover* of AdS_n – which will be denoted $\widetilde{AdS_n}$ – by unwinding the time-direction. Hence, instead of the hyperboloid model, one gets a infinite sheet model of AdS_n. How this "unwinding" can be achieved will be more clear later on.

C.2   Foliations of AdS_n

Consider the global parameterization of the AdS_n-hyperboloid given by

$$
\begin{aligned}
V &= \sqrt{r^2 + R^2}\sin(t/R) \\
U &= \sqrt{r^2 + R^2}\cos(t/R) \\
X_1^2 + \ldots + X_{n-1}^2 &= r^2,
\end{aligned}
\tag{C.5}
$$

and let $d\Omega_{n-2}^2$ be the metric on the unit $(n-2)$-sphere. The metric on AdS_n can be written in static coordinates

$$
ds^2 = -\left(1 + \frac{r^2}{R^2}\right)dt^2 + \frac{dr^2}{1 + \frac{r^2}{R^2}} + r^2 d\Omega_{n-2}^2.
\tag{C.6}
$$

For this metric the time-variable t is an angular variable with periodicity $2\pi R$. By going to the universal cover $\widetilde{AdS_n}$, the time variable becomes infinite in range, $-\infty < t < \infty$, which will avoid the troublesome closed time-like curves. Generally it is assumed that this metric is the metric on the universal cover $\widetilde{AdS_n}$.

In the four-dimensional case, this metric reduces to the static de Sitter model with a cosmological constant $\Lambda = -\frac{3}{R^2}$.

There are other interesting foliations of AdS_n as well. First, there are two other static versions of the AdS_n, none of them covers the AdS_n hyperboloid globally. Static AdS_n in Poincaré coordinates,

$$
ds^2 = \frac{R^2}{z^2}\left(-dt^2 + dz^2 + dx_1^2 + \ldots + dx_{n-2}^2\right),
\tag{C.7}
$$

is achieved as follows. We introduce the dummy variable, α, by

$$
\alpha = -t^2 + x_1^2 + \ldots + x_{n-2}^2.
\tag{C.8}
$$

The metric (C.7) can now be obtained by

$$
\begin{aligned}
X_{n-1} - U &= \frac{R}{z} \\
X_{n-1} + U &= R\left(-z + \frac{\alpha}{z}\right) \\
V &= \frac{Rt}{z} \\
X_i &= \frac{Rx_i}{z} \quad (i = 1, \ldots, n-2).
\end{aligned}
\tag{C.9}
$$

Static AdS_n on "black hole" form is obtained as follows. Let $\hat{V}$, $\hat{X}_i$, $i = 1, \ldots, n-2$ be coordinates on the hyperbolic space $\mathbb{H}^{n-2}$ (or rather of the ambient space, see section 7.6 for details) such that

$$
-\hat{V}^2 + \hat{X}_1^2 + \ldots + \hat{X}_{n-2}^2 = -1,
\tag{C.10}
$$

and let dH_{n-2}^2 be the metric on $\mathbb{H}^{n-2}$; i.e.,

$$dH_{n-2}^2 = -d\hat{V}^2 + d\hat{X}_1^2 + \dots + d\hat{X}_{n-2}^2. \tag{C.11}$$

The parameterization

$$
\begin{aligned}
U &= \sqrt{r^2 - R^2}\cosh(t/R) \\
X_{n-1} &= \sqrt{r^2 - R^2}\sinh(t/R) \\
V &= r\hat{V} \\
X_i &= r\hat{X}_i \quad (i = 1, \dots, n-2),
\end{aligned}
\tag{C.12}
$$

yields

$$ds^2 = -\left(\frac{r^2}{R^2} - 1\right)dt^2 + \frac{dr^2}{\frac{r^2}{R^2} - 1} + r^2 dH_{n-2}^2. \tag{C.13}$$

This is a static version of AdS_n and has a horizon at $r = R$. All of these static versions of AdS_n are useful for different purposes.

Second, the cosmological form of AdS_n,

$$ds^2 = -dt^2 + R^2 \sin^2(t/R)dH_{n-1}^2, \tag{C.14}$$

can be obtained by the parameterization

$$
\begin{aligned}
U &= R\cos\left(\frac{t}{R}\right)\hat{V} \\
V &= R\sin\left(\frac{t}{R}\right)\hat{V} \\
X_i &= R\hat{X}_i \quad (i = 1, \dots, n-1),
\end{aligned}
\tag{C.15}
$$

where $\hat{V}$, $\hat{X}_i$, $i = 1, \dots, n-1$ are coordinates on the hyperbolic space $\mathbb{H}^{n-1}$.

These are not, of course, all of the different forms of AdS_n, but they illustrate the diversity of the different possibilities of this space. This space has a rich structure which is one of the reasons why it is so interesting.

C.3 Geodesics in AdS_n

In order to explore some of the properties of AdS_n we will study geodesics in this space. More specifically, time-like and light-like geodesics will be found which will provide us with a physical interpretation of this space. It is convenient to use the globally defined coordinates on $\widehat{\mathrm{AdS}}_n$.

Consider radially moving particles and assume $R = 1$ for simplicity. The effective Lagrangian is then

$$L = \frac{1}{2}\left[-(1+r^2)\dot{t}^2 + \frac{\dot{r}^2}{1+r^2}\right]. \tag{C.16}$$

Since t is a cyclic coordinate, p_t will be a constant of motion:

$$p_t \equiv \frac{\partial L}{\partial \dot{t}} = -(1+r^2)\dot{t}. \tag{C.17}$$

Together with the identity $u^\mu u_\mu = -\varepsilon$, where $\varepsilon = 1, 0, -1$ for time-like, light-like and space-like geodesics, respectively, this yields the differential equation

$$\dot{r}^2 + \varepsilon r^2 = p_t^2 - \varepsilon. \qquad (C.18)$$

This equation can easily be integrated

$$\varepsilon = 1: \quad r = \pm\sqrt{p_t^2 - 1}\sin(\tau - \tau_0) \qquad (C.19)$$
$$\varepsilon = 0: \quad r = \pm|p_t|(\tau - \tau_0) \qquad (C.20)$$
$$\varepsilon = -1: \quad r = \pm\sqrt{p_t^2 + 1}\sinh(\tau - \tau_0). \qquad (C.21)$$

The last of these tells us that spatial infinity is at infinite spatial distance; i.e. $\tau \to \infty$ as $r \to \infty$. Massive particles have $\varepsilon = 1$ and will recollapse after a time $\Delta\tau = \pi$. However, photons will reach spatial infinity $r = \infty$. In terms of the coordinate time t, we have

$$p_t = -(1 + r^2)\dot{t}, \qquad (C.22)$$

and hence for light,

$$dt = -\frac{p_t d\tau}{1 + p_t^2(\tau - \tau_0)^2}. \qquad (C.23)$$

Thus, upon integration

$$t = t_0 + \arctan\left[|p_t|(\tau - \tau_0)\right]. \qquad (C.24)$$

Since $t - t_0 < \pi/2$, *photons reach spatial infinity within finite coordinate time.* This is an astonishing result. As we follow a photon, it will reach infinity within a finite coordinate time, and at this time, it will cease to exist! This is one of the major problems with AdS_n, it leaks photons – and thus energy – out through spatial infinity. The space does not have constant energy. In a quantum theory this problem has to be resolved before any sensible field theory can be imposed.

C.4   The BTZ black hole

Consider the anti-de Sitter space on "black hole" form (C.13). We have already noted that the spacetime has a horizon at $r = R$. Thus one might get tempted to think there is a singularity at $r = 0$ as well. However, this cannot be true since the metric (C.13) is only a certain choice of coordinates of the AdS-hyperboloid. This hyperboloid is everywhere regular, and hence, the space (C.13) cannot have a singularity at $r = 0$.

Notwithstanding, it is possible to construct a black hole solution which has a singularity using metric (C.13). This black hole solution is called the BTZ black hole.

For simplicity, assume that $n = 3$, but the construction can be performed in any dimension $n \geq 3$. The metric for AdS_3 can be written

$$ds^2 = -\left(\frac{r^2}{R^2} - 1\right)dt^2 + \frac{dr^2}{\frac{r^2}{R^2} - 1} + r^2 d\phi^2. \qquad (C.25)$$

Note that the variable ϕ has infinite range, and hence, despite its immediate appearance it is not an angular variable. However, the metric has a Killing vector $\boldsymbol{\xi} = \frac{\partial}{\partial \phi}$ which we can use to periodically identify ϕ. After the identification, ϕ will be an angular variable with a certain period and the metric looks more like a black hole metric.

Given $M > 0$, we identify

$$\phi \sim \phi + 2\pi\sqrt{M}. \tag{C.26}$$

Also, it is convenient to rescale ϕ so that it has period 2π. Therefore, define $\hat{\phi} = \phi/\sqrt{M}$, $\hat{r} = \sqrt{M}r$ and $\hat{t} = t/\sqrt{M}$ so that we obtain

$$ds^2_{\text{BTZ}} = -\left(\frac{\hat{r}^2}{R^2} - M\right)d\hat{t}^2 + \frac{d\hat{r}^2}{\frac{\hat{r}^2}{R^2} - M} + \hat{r}^2 d\hat{\phi}^2. \tag{C.27}$$

This is the metric for the BTZ black hole and was first found by Bañados, Teitelboim and Zanelli [BTZ92].

This metric is locally isometric to the anti-de Sitter space with a horizon at $\hat{r} = \sqrt{M}R$. It is a constant curvature Lorentzian space and thus the space cannot have a *curvature singularity* anywhere. Notwithstanding, it *does* have a singularity at $\hat{r} = 0$. How this comes about can be seen as follows.

We identified points in the space given by the metric (C.25), under $\phi \sim \phi + 2\pi\sqrt{M}$. If we go back to the parameterization (C.12), we note that this group action is not free; all points given by $r = 0$ are fixed points under the above identification. This means that we violate requirement 3 on page 431. Thus the resulting manifold does not need to be a smooth manifold. As a matter of fact, the points given by $r = 0$ is a singularity of the same type as the compactified Milne universe in Example 14.2. The BTZ black hole possesses an *inextendible non-curvature singularity*.

Similarly as for the Schwarzschild black hole, we can associate a temperature

$$T = \frac{\sqrt{M}}{2\pi R}, \tag{C.28}$$

and an entropy

$$S = \frac{1}{4}A = \frac{\pi\sqrt{M}R}{2}, \tag{C.29}$$

to the black hole horizon. Furthermore, it is possible to construct a rotating BTZ black hole [BHTZ93, Car95], but we will not consider this case here.

C.5   AdS$_3$ as the group $SL(2, \mathbb{R})$

Interestingly, AdS$_3$ admits a group structure. In a String theory context, this makes this space particularly interesting. We will not dwell upon the stringy aspects of this space here, but we will emphasize on the consequences this group structure has for the geometry.

To establish the isomorphism between AdS$_3$ and the group $SL(2, \mathbb{R})$ we write the matrices in $SL(2, \mathbb{R})$ as

$$\mathsf{A} = \frac{1}{R}\begin{bmatrix} U + X_1 & V + X_2 \\ -V + X_2 & U - X_1 \end{bmatrix}. \tag{C.30}$$

The matrix A is in $SL(2, \mathbb{R})$ if and only if

$$\det(\mathsf{A}) = 1 \quad \Leftrightarrow \quad -U^2 - V^2 + X_1^2 + X_2^2 = -R^2. \tag{C.31}$$

Thus the matrix A is in $SL(2, \mathbb{R})$ if and only if the coordinates (U, V, X_1, X_2) are coordinates on AdS_3. Hence, the isomorphism is established.

In a sloppy notation the metric can be written

$$ds^2 = -R^2 \det(\mathbf{dA}). \tag{C.32}$$

Isometries are therefore mappings that map $SL(2, \mathbb{R})$ onto itself, and leave the determinant fixed. Any $\mathsf{L} \in SL(2, \mathbb{R})$ will do the trick, due to the fact that $SL(2, \mathbb{R})$ is a Lie group and that

$$\det(\mathbf{dA} \cdot \mathsf{L}) = \det(\mathsf{L} \cdot \mathbf{dA}) = \det(\mathbf{dA}) \cdot \det(\mathsf{L}) = \det(\mathbf{dA}). \tag{C.33}$$

Isometries are therefore given by left and right multiplication of the matrices. The isometry group is $SL(2, \mathbb{R}) \times SL(2, \mathbb{R})/\mathbb{Z}_2$: the two copies of $SL(2, \mathbb{R})$ act by left and right multiplication

$$\mathsf{A} \quad \longmapsto \quad \mathsf{L} \cdot \mathsf{A} \cdot \mathsf{R},$$
$$\mathsf{L}, \mathsf{R} \in SL(2, \mathbb{R}) \quad \text{with} \quad (\mathsf{L}, \mathsf{R}) \sim (-\mathsf{L}, -\mathsf{R}). \tag{C.34}$$

Hence, the group structure of AdS_3 immediately provides us with the isometries. Note that we have already considered $SL(2, \mathbb{R})$ with a Riemannian metric in section 15.5. This space does not have the same isometries because the metric in that case cannot be expressed in terms of a group-invariant polynomial. Hence, eq. (C.32) will fail and in general left and right multiplication will not leave the metric invariant.

How to Read This Book

In Fig. D.1 we have indicated different possible course outlines for this book. Depending on the length of the course and what the emphasis would be, different routes may be chosen. The flowchart also indicates different dependencies of the various chapters.

Depending on the background a course in general relativity can start either with chapter 1 or chapter 3. If the students are assumed to have no knowledge of special relativity, it is advised that the course should start with chapter 1. However, if the students have prior knowledge of the special theory of relativity the introductory chapters may be omitted and one can go directly to chapter 3.

A minimal course in general relativity is recommended to consist of the first chapters including chapter 10. Such a course would not include any cosmology since the cosmology chapters start with chapter 11. Regarding chapter 9, this chapter can be omitted since no other chapter depends directly on this chapter (however, note that some sections of chapter 9 might be useful for a student). In a course where cosmology is included we recommend, at least, chapters 11 and 12. By including also chapter 13 the students will get a taste of non-FRW cosmologies and thereby get introduced to some non-standard cosmologies and their properties.

For a more comprehensive course, also some of the more advanced chapters can be included. These chapters include various modern applications of general relativity and can be useful for graduate students, as well as researchers in general relativity.

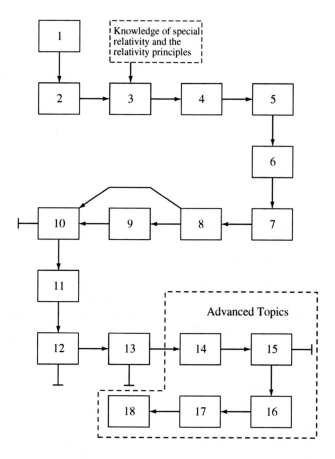

Figure D.1: A rough flowchart of dependencies and possible course outlines for the chapters contained in this book. Here, $\perp$ means a natural stop.

Suggested Further Reading

In this appendix we will provide some references which can act as a springboard into the literature. They come in addition to the ones already cited in the text. There exist a wealth of articles and books related to the field, so this list is by no means complete. Unavoidably it is also biased but we have tried to include references that we think may be relevant.

General reference works

There are a wealth of books written on general relativity since its birth. Some recommended books are Misner, Thorne and Wheeler's treatise [MTW73], Wald's book [Wal84], Stephani's introduction to the field of general relativity [Ste77], and the newer book by Ludvigsen [Lud99]. Stewart's book [Ste91] is recommended for the more experienced reader, and treats other advanced topics than we do in this book, for example, spinors, and asymptopia. Regarding exact solutions of Einstein's Field equations, the books by Stephani *et al.* [SKM⁺03] and Krasiński [Kra97] are unavoidable references.

Chapter 1

In classical mechanics there are many books worth mentioning. In particular the book by Goldstein [Gol50] is worth reading. The jewel in classical mechanics, the canonical transformations and the Hamilton-Jacobi equation, is something a theorist cannot afford to avoid learning. For an application of the theory, and highly relevant for gravitational physics, see for example Roy's book on orbital motion [Roy88].

Chapter 2

There is also quite a large amount of literature concerning the special theory of relativity. Of special historical interest is perhaps the book by Einstein himself [Ein16]. Moreover, [Bar99] is a useful and straightforward introduction to the special theory of relativity.

Chapter 3

The book by Göckeler and Schückler [GS87] deals with vectors and forms more generally. It can also be considered as complementary to the field of differential geometry in general.

Chapter 4

A reference to the world of differential geometry is Spivak's first volume in his comprehensive introduction [Spi75a]. In this book manifolds and differential structure are more rigorously introduced and investigated.

A more advanced and mathematical book in differential geometry is for example [MT97]. It's highly technical but treats also more topological aspects of differential geometry.

Chapter 5

A thorough investigation of non-inertial reference frames are some research articles by Eriksen and Grøn [EG90, EG00a, EG00b, EG00c, EG02, EG04]. For the rotating reference, see also [Grø04].

Chapter 6

For a more mathematical treatment, two volumes of Spivak's comprehensive work is recommended [Spi75a, Spi75b]. Also a quite technical book by Gallot, Hulin and Lafontaine has some nice applications and examples [GHL90].

From a physical point of view, Frankel has written a nice book on the geometry of physics [Fra97]. Also, the book by Nakahara which covers many areas in mathematical physics is highly recommendable [Nak90].

Chapter 7

The second volume of Spivak's work [Spi75b] contains both an introduction to the concept of curvature (very much like the introduction in this book), and some interesting historical sections. Benedetti and Petroni's book on hyperbolic geometry [BP92], is an excellent book to learn more about the many interesting aspects of hyperbolic space. For the theory of hypersurfaces, a whole chapter in Poisson's book [Poi04] is devoted to various mathematical and physical aspects of this topic.

Chapter 8

Carroll's book [Car04] contains a similar approach to the Einstein field equations to the one taken here. Some classical books on Einstein's general theory of relativity are [Foc66, ABS75, DFC90, Møl72, Rin77, Har03]. We should also mention a biography of the man Hilbert [Rei96]. Also some of Minkowski's life is vividly portrayed in this book, due to the fact that they were close friends.

Chapter 9

Gravitational waves and the weak field limit are treated in the book by Schutz [Sch85]. This book can also be considered as a general reference as it is supposed to be an introduction to the field. An article by Ruggiero and Tartagia [RT02] gives an nice introduction to gravitomagnetic effects. It gives also a nice review of the different experimental tests of general relativity which have been performed to date.

Chapter 10

The original articles by Bekenstein [Bek73] and Hawking [Haw75] are classics in terms of black hole radiation and entropy. Also the related article by Gibbons and Hawking [GH77] are worth reading. The famous book by Chandrasekhar on black holes [Cha83] is also a very nice book on the physics of black holes. For a more recent elucidation of the physics of black holes, see Poisson's book [Poi04].

De Felice and Clarke's book [DFC90] treats the Ernst equation and investigates the Kerr solution more throughly. It also discusses the Reissner-Nordström black hole more thoroughly. See also [Hoe93] in this regard.

Chapter 11

As an interesting and easy-to-read book on various aspects of cosmology, is [Col98]. The book by Peacock [Pea98] is very useful as a general reference to the realm of modern cosmology, and the book by Islam is also worth looking into [Isl92]. An interesting, and rather philosophical book is the classic work by Barrow and Tipler [BT86]. This work has been highly debated but is probably unsurpassed when it comes to its depth and richness.

Chapter 12

A nice review of the various sections of de Sitter space is given by Eriksen and Grøn [EG95]. Apart from Guth's paper [Gut81], some of the original papers on inflation are worth noting [AS82, Lin82, Lin83]. See also the two books by Kolb and Turner [KT90] and Linde [Lin90] for a more detailed account of how inflation solves some of the cosmological problems. Furthermore, inflation and early universe cosmology, including CMB anisotropies, are nicely dealt with in the more recent book by Liddle and Lyth [LL00]. Other introductionary articles on these topics are [GB99, Pal00, Pee93].

Chapter 13

Some of the earliest investigations of the Bianchi type I model are from the sixties [Tho67, Jac68, Jac69, Sau69]. These address the mechanism of isotropization of our universe, both in terms isotropic fluids and magnetic fields. Later, these issues have been discussed by other authors [LNSZ76, HP78, LeB97]. For a review of viscous universe models, see [Grø90].

Chapter 14

The complete covariant decomposition of spacetime, including all degrees of freedom, is done in for example [Maa97]. Some applications of these equation of motion are, for example [Bar97, BM98]. An alternative covariant decomposition is the so-called Newman-Penrose-formalism which is treated in for example Stewart's book [Ste91].

The canonical version of General relativity was first formulated by Arnowitt, Deser and Misner [ADM62]. Later, DeWitt used this formulation to formulate "Quantum Cosmology" [DeW67]. After this, numerous papers and books using the canonical formulation [Rya72, RS75] and its quantum version [Mis69, Lou88, CHPW91, Hal91, Haw84, Haw94, Sim01] have appeared. One paper worth pointing out is Hartle and Hawking's paper were they formulate the "No Boundary"-proposal [HH83]. Also, Vilenkin's alternative proposal is worth noting [Vil82, Vil83, Vil94].

Chapter 15

The book by Kobayashi [Kob72], deals with homogeneous spaces in general and states the theorems and their proofs we use in this chapter. Of historical interest is also Bianchi's original article [Bia98]. Kantowski and Sachs treated the remaining multiply transitive case in [KS66]. More recent treatises, which are extremely valuable for anybody interested in the dynamical behaviour of Bianchi models, are the report [BS86] and the book *Dynamical Systems in Cosmology* [WE97]. For the tilted Bianchi models, see [CH05, HvdHC05, HvdHLC06, HL06] and references therein. A book [Thu97] and an article [Thu82] by Thurston give a nice mathematical introduction to the model geometries and their importance in topology and geometry. Kodama's two articles bring these ideas into the field of cosmology [Kod98, Kod02]. Other papers that discuss more physical aspects of these ideas are [LRL95, LSW99, Lev02].

Chapter 16

Apart from the original paper by Israel [Isr66], there are some other research papers reviewing the metric junction method [Kuc68, BI91]. Moreover, Possion's book [Poi04] considers the metric junction method, including the null case.

Chapter 17

There are an enormous number of pages written on brane-worlds the years since its launch. For a couple of reviews on brane-worlds, see [Maa04, Col05]. Moreover, in addition to those already mentioned in the chapter, we would like to emphasize some works on anisotropic branes. A Kasner brane was found by Frolov [Fro01]; anisotropic branes with isotropic fluids have been studied [CS01, Top01, Col02a, Col02b], as well as with magnetic fields [SVF01, BH02]; and exact plane-wave branes in a bulk containing gravitational waves was found in [CH04].

Chapter 18

Apart from a few books [ACF87], the articles on Kaluza-Klein theory is scattered around in the literature. The generalization of the Bianchi models to 4+1 dimensions is done in [Her02].

Bibliography

[AAG06]   H. Alnes, M. Amarzguioui, and Ø. Grøn. An inhomogeneous alterna-
             tive to dark energy. *Phys. Rev.*, D73:083519, 2006.

[ABS75]   R. Adler, M. Bazin, and M. Schiffer. *Introduction to General Relativity*.
             McGraw-Hill, 1975.

[AC83]   T. Appelquist and A. Chodos. The quantum dynamics of Kaluza-Klein
             theories. *Phys. Rev.*, D28:772, 1983.

[ACF87]   T. Appelquist, A. Chodos, and P.G.O. Freund, editors. *Modern Kaluza-
             Klein Theories*. Addison-Wesley, 1987.

[ADM62]   Arnowitt, R., Deser, S., and Misner, C.W. The dynamics of general
             relativity. In L. Witten, editor, *GRAVITATION: an introduction to current
             research*. John Wiley & Sons, 1962.

[AHDD98]   N. Arkani-Hamed, S. Dimopoulos, and G.R. Dvali. The hierarchy prob-
             lem and new dimensions at a millimeter. *Phys. Lett.*, B429:263–272, 1998.

[Ama03]   M. Amarzguioui. Theory of cosmological perturbations. Cand. Scient.
             Thesis, University of Oslo, 2003.

[APMS01]   C. Armendariz-Picon, V. Mukhanov, and P.J. Steinhardt. Essentials of
             k-essence. *Phys.Rev.*, D63:103510, 2001.

[AS82]   A. Albrecht and P.J. Steinhardt. Cosmology for grand unified theories
             with radiately induced symmetry breaking. *Phys. Rev. Lett.*, 48:1220,
             1982.

[ASSS03]   U. Alam, V. Sahni, T.D. Saini, and A.A. Starobinsky. Exploring the ex-
             panding universe and dark energy using the statefinder diagnosic. *Mon.
             Not. Roy. Astron. Soc.*, 344:1057, 2003.

[Bar97]   J.D. Barrow. Cosmological limits on slightly skew stresses. *Phys. Rev.*,
             D55:7451–7460, 1997.

[Bar99]   G. Barton. *Introduction to the relativity principle*. Wiley, 1999.

[BC66]   D.R. Brill and J.M. Cohen. Rotating masses an their effect on inertial
             frames. *Phys. Rev.*, 143:1011, 1966.

[BDL01]   P. Binétruy, C. Deffayet, and D. Langlois. The radion in brane cosmology.
             Nucl. Phys., B615:219–236, 2001.

[BDS62]   O.M.P. Bilaniuk, V.K. Deshpande, and E.C.G. Sudarshan. 'Meta' Rela-
             tivity. *Am. J. Phys.*, 30:718–723, 1962.

[Bek73]   J.D. Bekenstein. Black holes and entropy. *Phys. Rev.*, D7:2333–2346, 1973.

[Bek74]   J.D. Bekenstein. Generalized second law of Thermodynamics in black-
             hole physics. *Phys. Rev.*, D9:3292–3300, 1974.

[Bet. al.96] C.L. Bennett *et al.* Four-year COBE DMR cosmic microwave back-
             ground observations: Maps and basic results. *Astrophys. J.*, 464:L1, 1996.

[BGOY02]   I. Brevik, K. Ghoroku, S.D. Odintsov, and M. Yahiro. Localization of
             gravity on brane embedded in AdS_5 and dS_5. *Phys. Rev.*, D66:064016,
             2002.

[BH02]   J.D. Barrow and S. Hervik. Magnetic brane-worlds. *Class. Quantum Grav.*, 19:155–172, 2002.

[BHTZ93]   M. Bañados, M. Henneaux, C. Teitelboim, and J. Zanelli. Geometry of the (2+1) black hole. *Phys. Rev.*, D48:1506, 1993.

[BI91]   C. Barrabès and W. Israel. Thin shells in general relativity and cosmology: The lightlike limit. *Phys. Rev.*, D43:1129, 1991.

[Bia98]   L. Bianchi. Sugli spazii a tre dimensioni che ammettono un gruppo continuo di movimenti. *Mem. Mat. Fis. Soc. It. Sc.*, Serie Terza 11:267, 1898. Engl. transl. *Gen. Rel. Grav.*, 33:2171, 2001.

[Bir23]   G.D. Birkhoff. *Relativity and Modern Physics*. Harvard University Press, 1923. p.253-256.

[BLT03]   R. Bertotti, L. Less, and P. Tortora. A test of general relativity using radio links with the Cassini spacecraft. *Nature*, 425:375–376, 2003.

[BM98]   J.B. Barrow and R. Maartens. Anisotropic stresses in inhomogeneous universes. *Phys. Rev.*, D59:043502, 1998.

[Bon47]   H. Bondi. Spherically symmetric models in general relativity. *Mon. Not. Roy. Astron. Soc.*, 107:410–426, 1947.

[Bou05]   R. Bousso. Precision cosmology and the landscape. *hep-th/0610211*, 2005. Presented at Amazing Light: Visions for Discovery: An International Symposium in Honor of the 90th Birthday Years of Charles H. Townes, Berkeley, California, 6-8 Oct 2005.

[BP92]   R. Benedetti and C. Petronio. *Lectures on Hyperbolic Geometry*. Springer-Verlag, 1992.

[BPY05]   W. Buchmüller, R. D. Peccei, and T. Yanagida. Leptogenesis as the origin of matter. *Ann. Rev. Nucl. Part. Sci.*, 55:311–355, 2005.

[BS86]   J.D. Barrow and D.H. Sonoda. Asymptotic stability of Bianchi type universes. *Phys. Rept.*, 139:1–49, 1986.

[BT86]   J.D. Barrow and F.J. Tipler. *The Anthropic Cosmological Principle*. Oxford University Press, 1986.

[BTZ92]   M. Bañados, C. Teitelboim, and J. Zanelli. The black hole in three-dimensional spacetime. *Phys. Rev. Lett.*, 69:1849, 1992.

[Car95]   S. Carlip. The (2+1)-dimensional black hole. *Class. Quantum Grav.*, 12:2853–2880, 1995.

[Car04]   S.M. Carroll. *Spacetime and geometry: An introduction to general relativity*. Addison-Wesley, 2004.

[CCV97]   I. Ciufolini, F. Chieppa, and F. Vespe. Test of Lense-Thirring orbital shift due to spin. *Class. Quant. Grav.*, 14:2701, 1997.

[CD80]   A. Chodos and S. Detweiler. Where has the fifth dimension gone? *Phys. Rev.*, D21:2167–2170, 1980.

[CH04]   A.A. Coley and S. Hervik. Brane waves. *Class. Quant. Grav.*, 21:5759–5766, 2004.

[CH05]   A. Coley and S. Hervik. A dynamical systems approach to the tilted Bianchi models of solvable type. *Class. Quant. Grav.*, 22:579–606, 2005.

[Cha83]   S. Chandrasekhar. *The Mathemetical Theory of Black Holes*. Clarendon Press, Oxford, 1983.

[CHPW91]   S. Coleman, J.B. Hartle, T. Piran, and S. Weinberg, editors. *Quantum Cosmology and Baby Universes: Proceedings of the 1989 Jerusalem Winter School*. World Scientific, 1991.

[CIK65]   D.C. Champeney, C.R. Isaak, and A.M. Khan. A time dilatation experiment based on the Mössbauer effect. *Proc. Phys. Soc.*, 85:583, 1965.

[Ciu02]   I. Ciufolini. Test of general relativity: 1995-2002 measurement of frame-dragging. *gr-qc/0209109*, 2002.

[CKW03]   R.R. Caldwell, M. Kamionkowski, and N.N. Weinberg. Phantom energy and cosmic doomsday. *Phys. Rev. Lett.*, 91:071301, 2003.

[Col98]   P. Coles, editor. *The New Cosmology*. Routledge, 1998.

[Col02a]   A.A. Coley. Dynamics of brane-world cosmological models. *Phys. Rev.*, D66:023512, 2002.

[Col02b]   A.A. Coley. No chaos in brane-world cosmology. *Class. Quantum Grav.*, 19:L45–L56, 2002.

[Col05]   A.A. Coley. The dynamics of brane-world cosmological models. *astro-ph/0504226*, 2005.

[CPC$^+$98]   I. Ciufolini, E. Pavlis, F. Chieppa, E. Fernandes Viera, and J. Perez-Mercader. Detection of Lense-Thirring effect due to Earth's spin. *Science*, 279:2100, 1998.

[CS01]   A. Campos and C.F. Sopuerta. Evolution of cosmological models in the brane world scenario. *Phys. Rev.*, D63:104012, 2001.

[Dav74]   P.C.W. Davies. *The Physics of Time Asymmety*. Surrey University Press, 1974.

[Dav83]   P.C.W. Davies. Inflation and time asymmetry in the universe. *Nature*, 301:398–400, 1983.

[DeW67]   B.S. DeWitt. Quantum theory of gravity. I. the canonical theory. *Phys. Rev.*, 160:1113–1148, 1967.

[DFC90]   F. De Felice and J.S. Clarke. *Relativity on curved manifolds*. Cambridge University Press, 1990.

[EG90]   E. Eriksen and Ø. Grøn. Relativistic dynamics in uniformly accelerated reference frames with application to the clock paradox. *Eur. J. Phys.*, 11:39, 1990.

[EG95]   E. Eriksen and Ø. Grøn. The de Sitter universe models. *Int. J. Mod. Phys.*, 4:115–159, 1995.

[EG00a]   E. Eriksen and Ø. Grøn. Electrodynamics of hyperbolically accelerated charges I: The electromagnetic field of a charged particle with hyperbolic motion. *Annals of Physics*, 286:320–342, 2000.

[EG00b]   E. Eriksen and Ø. Grøn. Electrodynamics of hyperbolically accelerated charges II: Does a charged particle with hyperbolic motion radiate? *Annals of Physics*, 286:343–372, 2000.

[EG00c]   E. Eriksen and Ø. Grøn. Electrodynamics of hyperbolically accelerated charges III: Energy-momentum of the field of a hyperbolically moving charge. *Annals of Physics*, 286:373–399, 2000.

[EG02]   E. Eriksen and Ø. Grøn. Electrodynamics of hyperbolically accelerated charges IV: Energy-momentum conservation of radiating charged particles. *Annals of Physics*, 297:243–294, 2002.

[EG04]   E. Eriksen and Ø. Grøn. Electrodynamics of hyperbolically accelerated charges V: The field of a change in the Rindler space and the Milne space. *Annals of Physics*, 313:147–196, 2004.

[Ein16]   A. Einstein. *Relativity*. Routledge, 1916. English translation.

[EK74]   G.F.R. Ellis and A.R. King. Was the big bang a whimper? *Commun. Math. Phys.*, 38:119, 1974.

[EM69]   G.F.R. Ellis and M.A.H. MacCallum. A class of homogeneous cosmological models. *Comm. Math. Phys.*, 12:108, 1969.

[Fla04]   É.É. Flanagan. The conformal frame freedom in theories of gravitation. *Class. Quant. Grav.*, 21:3817, 2004.

[FN94]   J. Foster and J.D. Nightingale. *A short course in General Relativity.* Springer Verlag, 1994.

[Foc66]   V. Fock. *The Theory of Space Time and Gravitation.* Pergamon Press, 1966.

[Fra97]   T. Frankel. *The Geometry of Physics: An Introduction.* Cambridge University Press, 1997.

[Fro01]   A.V. Frolov. Kasner-AdS spacetime and anisotropic brane-world cosmology. *Phys. Lett.*, B514:213–216, 2001.

[FS63]   D.H Frisch and J.H. Smith. Measurement of relativistic time-dilation using mu-mesons. *Am. J. Phys.*, 31:342, 1963.

[GB99]   J. Garcia-Bellido. Astrophysics and cosmology. Lectures at 1999 European School of High Energy Physics, Casta-Papiernicka, Slovak Republic, 22 August - 4 September 1999, hep-ph/0004188, 1999.

[GE07]   Ø. Grøn and Ø. Elgarøy. Is space expanding in the Friedmann universe models? *Am. J. Phys.*, to appear, 2007.

[GH77]   G.W. Gibbons and S.W. Hawking. Cosmological event horizons, thermodynamics, and particle creation. *Phys. Rev.*, D15:2738–2751, 1977.

[GHL90]   Gallot, S., Hulin, D., and Lafontaine, J. *Riemannian Geometry.* Springer Verlag, 2. edition, 1990.

[Gol50]   H. Goldstein. *Classical Mechanics.* Addison-Wesley, 1950.

[GR92]   Ø. Grøn and S. Refsdal. Gravitational lenses and the age of the universe. *Eur. J. Phys.*, 13:178–183, 1992.

[Grø85]   Ø. Grøn. New derivation of Lopez's source of the Kerr-Newman field. *Phys. Rev.*, D32:1588, 1985.

[Grø86]   Ø. Grøn. Classical Kaluza-Klein description of the Hydrogen atom. *Il. Nuovo Cim.*, 91B:57–66, 1986.

[Grø90]   Ø. Grøn. Viscous inflationary universe models. *Astrophys. and Space Science*, 173:191–225, 1990.

[Grø04]   Ø. Grøn. Space geometry in rotating reference frames: A historical approach. In G. Rizzi and M.L. Ruggiero, editors, *Relativity in Rotating Frames.* Kluwer Academic Publishers, 2004.

[GS87]   M. Göckeler and T. Schückler. *Differential geometry, gauge theories, and gravity.* Cambridge University Press, 1987.

[GT00]   J. Garriga and T. Tanaka. Gravity in the Randall-Sundrum brane world. *Phys. Rev. Lett.*, 84:2778–2781, 2000.

[Gul22]   A. Gullstrand. Allgemeine Lösung des statischen Einköpferproblems in der Einsteinschen Gravitationstheorie. *Arkiv Mat. Astron. Fys.*, 16(8): 1–115, 1922.

[Gut81]   A. Guth. The inflationary universe: A possible solution to the horizon and flatness problems. *Phys. Rev.*, D23:347–356, 1981.

[GV99]   Ø. Grøn and K. Vøyenli. On the foundation of the Principle of Relativity. *Found. Phys.*, 29:1695–1733, 1999.

[GW99]   W.D. Goldberger and M.B. Wise. Modulus stabilization with bulk fields. *Phys. Rev. Lett.*, 83:4922–4925, 1999.

[Hal91]   J. Halliwell. Introductury lectures on quantum cosmology. In Coleman et al. [CHPW91].

[Ham]   A.J.S. Hamilton. http://ucsub.colorado.edu/~flournoy/Introduction.html.

[Har70]   E.R. Harrison. Fluctuations at the threshold of classical cosmology. *Phys. Rev.*, D1:2726, 1970.

[Har81]   E.R. Harrison. *Cosmology: The Science of the Universe*. Cambridge University Press, 1981.

[Har03]   J.B. Hartle. *Gravity: An introduction to Einstein's General Relativity*. Addison-Wesley, 2003.

[Haw75]   S.W. Hawking. Particle creation by black holes. *Commun. math. Phys.*, 43:199–220, 1975.

[Haw84]   S.W. Hawking. The quantum state of the universe. *Nuc. Phys.*, B239:257–276, 1984.

[Haw94]   S.W. Hawking. The no boundary condition and the arrow of time. In J.J. Halliwell, J. Pèrez-Mercader, and W.H. Zurek, editors, *Physical Origins of Time Asymmetry*. Cambridge University Press, 1994.

[HE73]   S. W. Hawking and G. F. R. Ellis. *The large scale structure of space-time*. Cambridge University Press, 1973.

[Her01]   S. Hervik. Discrete symmetries in translation invariant cosmological models. *Gen. Rel. Grav.*, 33:2027, 2001.

[Her02]   S. Hervik. Multidimensional cosmology: spatially homogeneous models of dimension 4+1. *Class. Quant. Grav.*, 19:5409–5427, 2002.

[Her03]   S. Hervik. Vacuum plane waves in 4+1 D and exact solutions to Einstein's equations in 3+1 D. *Class. Quant. Grav.*, 20:4315–4327, 2003.

[HH83]   J.B. Hartle and S.W. Hawking. Wave function of the universe. *Phys. Rev.*, D28:2960–2975, 1983.

[HL04]   A.J.S. Hamilton and J.P. Lisle. The river model of black holes. *gr-qc/0411060*, 2004.

[HL06]   S. Hervik and W.C. Lim. The late-time behaviour of vortic Bianchi type VIII universes. *Class. Quant. Grav.*, 23:3017–3035, 2006.

[Hoe93]   C. Hoenselaers. Axisymmetric stationary solutions of Einstein's equations. In F.J. Chinea and González-Romero, editors, *Rotating Objects and Relativistic Physics*, LNP423. Springer, 1993.

[HP78]   B.L. Hu and L. Parker. Anisotropy damping through quantum effects in the early universe. *Phys. Rev.*, D17:933945, 1978.

[HPMZ94]   J.J Halliwell, J. Pérez-Mercader, and W.H. Zurek, editors. *Physical Origins of Time Asymmetry*. Cambridge University Press, 1994.

[HvdHC05]   S. Hervik, R. van den Hoogen, and A. Coley. Future asymptotic behaviour of tilted Bianchi models of type IV and VII$_h$. *Class. Quant. Grav.*, 22:607–634, 2005.

[HvdHLC06]   S. Hervik, R.J. van den Hoogen, W.C. Lim, and A.A. Coley. The futures of Bianchi type VII$_0$ cosmologies with vorticity. *Class. Quant. Grav.*, 23:845–866, 2006.

[Isl92]   J.N. Islam. *An introduction to mathematical cosmology*. Cambridge University Press, 1992.

[Isr66]   W. Israel. Singular hypersurfaces and thin shells in general relativity. *Il Nuovo Cimento*, 44 B:1, 1966.

[Isr70]   W. Israel. Source of the Kerr metric. *Phys. Rev.*, D2:641, 1970.

[Jac68]   K.C. Jacobs. Spatially homogeneous and Euclidean cosmological models with shear. *Astrophy. J.*, 153:661–678, 1968.

[Jac69]   K.C. Jacobs. Cosmologies of Bianchi type I with a uniform magnetic field. *Astroph. J.*, 155:379–391, 1969.

[Jeb21]   J.P. Jebsen. Uber die allgemeinen kugelsymmetrisch e Lösungen der Einsteinchen Gravitationsgleichungen im Vakuum. *Arkiv för matematik, astronomi och fysik*, 15(18), 1921.

[Kal99]   N. Kaloper. Bent domain walls as brane-worlds. *Phys. Rev.*, D60:123506, 1999.

[Kas21]   E. Kasner. Geometrical theorems on Einstein's cosmological equations. *Am. J. Math.*, pages 217–221, 1921.

[Ker63]   R. Kerr. Gravitational field of a spinning mass as an example of algebraically special metrics. *Phys. Rev. Lett.*, 11:237–238, 1963.

[KMP01]   A. Kamenshik, U. Moshella, and V. Pasquier. An alternative to quintessence. *Phys. Lett.*, B511:265, 2001.

[Kob72]   S. Kobayashi. *Transformation Groups in Differential Geometry*. Springer Verlag, 1972.

[Kod98]   H. Kodama. Canonical structure of locally homogeneous systems on compact closed 3-manifolds of type $\mathbb{E}^3$, Nil and Sol. *Prog. Theor. Phys.*, 99:173, 1998.

[Kod02]   H. Kodama. Phase space of compact Bianchi models. *Prog. Theor. Phys.*, 107:305–362, 2002.

[Kra97]   A. Krasiński. *Inhomogeneous Cosmological models*. Cambridge University Press, 1997.

[Kre73]   M.N. Kreisler. Are there faster-than-light particles? *American Scientist*, 61:201–208, 1973.

[KS66]   R. Kantowski and R.K. Sachs. Some spatially homogeneous anisotropic relativistic cosmological models. *J. Math. Phys.*, 7:443, 1966.

[KT90]   E.W. Kolb and M.S. Turner. *The Early Universe*. Addison-Wesley, Redwood City, California, 1990.

[Kuc68]   K. Kuchař. Charged shells in general relativity. *Czech. J. Phys.*, B18:435, 1968.

[Lan03]   D. Langlois. Brane cosmology: an introduction. *Prog. Theor. Phys. Suppl.*, 148:181–212, 2003.

[LC17]   T. Levi-Civita. Nozione di parallelismo in una varietà qualunque e conseguente specificazione geometrica della curvatura Riemanniana. *Rendiconti di Palermo*, 42:173–205, 1917.

[LCH07]   W.C. Lim, A.A. Coley, and S. Hervik. Kinematic and Weyl singularities. *Class. Quantum Grav.*, 24:595–604, 2007.

[LeB97]   V.G. LeBlanc. Asymptotic states of magnetic Bianchi I cosmologies. *Class. Quantum Grav.*, 14:2281–2301, 1997.

[Lem33]   G. Lemaître. The Universe in expansion. *Annals Soc. Sci. Brux. Ser. I Sci. Math. Astron. Phys.*, A53:51–85, 1933.

[Lev02]   J. Levin. Topology and the cosmic microwave background. *Phys. Rept.*, 365:251–333, 2002.

[Lin82]   A.D. Linde. A new inflationary universe scenario: A possible solution of the horizon, flatness, homogeneity, isotropy and primordial monopole problems. *Phys. Lett.*, B108:389–393, 1982.

[Lin83]   A.D. Linde. Chaotic inflation. *Phys. Lett.*, B129:177, 1983.

[Lin90]   A.D. Linde. *Particle Physics and Inflationary Cosmology*. Harwood, Chur, Switzerland, 1990.

[LL00]   A.R. Liddle and D.H. Lyth. *Cosmological Inflation and Large-Scale Structure*. Cambridge University Press, 2000.

[LNSZ76]   V.N. Lukash, I.D. Novikov, A.A. Starobinsky, and Ya.B. Zeldovich. Quantum effects and evolution of cosmological models. *Il Nuovo Cimento*, 35:293–307, 1976.

[Lop84]   C.A. Lopez. Extended model of the electron in general relativity. *Phys. Rev.*, D30:313, 1984.

[Lor67]   L. Lorenz. On the identity of the vibrations of light with electrical currents. *Philos. Mag.*, 34:287, 1867.

[Lou88]   J. Louko. Semiclassical path measure and factor ordering in quantum cosology. *Annals of Physics*, 181:318–373, 1988.

[LRL95]   M. Lachièze-Rey and J.-P. Luminet. Cosmic topology. *Phys. Rep.*, 254:135–214, 1995.

[LSW99]   J.-P. Luminet, G. Starkman, and J. Weeks. Is space finite? *Scientific American*, pages 68–75, 1999.

[LT18]   J. Lense and H. Thirring. Uber den Einfluss der Eigenrotation der Zentralkorper auf die Bewegung der Planeten und Monde nach der Einsteinschen Gravitationstheorie. *Phys. Z.*, 19:156, 1918.

[Lud99]   M. Ludvigsen. *General Relativity: A geometric approach*. Cambridge University Press, 1999.

[Maa97]   R. Maartens. Linearization instability of gravitational waves. *Phys. Rev.*, D55, 1997.

[Maa00]   R. Maartens. Cosmological dynamics on the brane. *Phys. Rev.*, D62:084023, 2000.

[Maa04]   R. Maartens. Brane-world gravity. *Living Rev. Rel.*, 7:7, 2004.

[Met. al.94]   J.C. Mather *et al.* Measurement of the Cosmic Microwave Background spectrum by the COBE FIRAS instrument. *Astrophys. J.*, 420:439, 1994.

[MFS89]   Y. Miller, B. Fort, and G. Soucail, editors. *Gravitational Lensing*, volume 360 of *Lecture Notes in Physics*. Springer, Berlin, 1989.

[MHL89]   J.M. Moran, J.N. Hewitt, and K.Y. Lo, editors. *Gravitational Lenses*, volume 330 of *Lecture Notes in Physics*. Springer, Berlin, 1989.

[Mis69]   C.W. Misner. Quantum cosmology. I. *Phys. Rev.*, 186:1319–1327, 1969.

[MM87]   A.A. Michelson and E.W. Morley. On the relative motion of the Earth and the Luminiferous Aether. *Philos. Mag. S.5*, 24(151):449–463, 1887.

[Møl72]   C. Møller. *The Theory of Relativity*. Oxford Clarendon Press, 1972.

[MPLP01]   R.N. Mohapatra, A. Pérez-Lorenzana, and C.A. de S. Pires. Cosmology of brane-bulk models of five dimensions. *Int. J. Mod. Phys.*, A16:1431, 2001.

[MT97]   I. Madsen and J. Tornehave. *From Calculus to Cohomology*. Cambridge University Press, 1997.

[MTW73]   C.W. Misner, K.S. Thorne, and J.A. Wheeler. *Gravitation*. San Francisco: Freeman, 1973.

[MWBH00]   R. Maartens, D. Wands, B. Basset, and I. Heard. Chaotic inflation on the brane. *Phys. Rev.*, D62:041301, 2000.

[Nak90]   M. Nakahara. *Geometry, Topology and Physics*. Adam Hilger, 1990.

[NCC+65]   E.T. Newman, E. Couch, K. Chinnapared, A. Exton, A. Parkash, and R. Torrence. Metric of a rotating charged mass. *J. Math. Phys.*, 6:918–919, 1965.

[Pad02]   A. Padilla. *Braneworld Cosmology and Holography*. PhD thesis, University of Durham, 2002. Also available at *hep-th/0210217*.

[Pai21]   P. Painlevé. La mécanique classique et la théorie de la relativité. *C.R. Acad. Sci. (Paris)*, 173:677–680, 1921.

[Pal00]   P.B. Pal. Determination of cosmological parameters: an introduction for non-specialists. *Pranama*, 54:79–91, 2000.

[Pea98]   J.A. Peacock. *Cosmological Physics*. Cambridge University Press, 1998.

[Pee93]   P.J.E. Peebles. *Principles of Physical Cosmology*. Princeton University Press, 1993.

[Pen69]   R. Penrose. Gravitational collapse: The role of general relativity. *Nuovo Cimento*, 1:252–276, 1969. special number.

[Pen79]   R. Penrose. Singularities and time-asymmetry. In *General Relativity: An Einstein Centenary Survey*, 1979.

[Pet. al.99]   S. Perlmutter *et al.* Measurements of Omega and Lambda from 42 high-redshift supernovae. *Astrophys. J.*, 517(565-586), 1999.

[Poi04]   E. Poisson. *A Relativist's Toolkit*. Cambridge University Press, 2004.

[PRj60]   P.V. Pound and G.A. Rebka jr. Apparent weight of photons. *Phys. Rev. Lett.*, 4:337, 1960.

[Räs02]   S. Räsänen. A primer on the ekpyrotic scenario. *astro-ph/0208282*, 2002.

[Rec78]   E. Recami, editor. *Tachyons, Monopoles, and related topics*. North-Holland Publ. Comp., 1978.

[Ref64a]   S. Refsdal. The gravitational lens effect. *Mon. Not. R. Astron. Soc.*, 128:295, 1964.

[Ref64b]   S. Refsdal. On the possibility of determining Hubble's parameter and the masses of galaxies from the gravitational lens effect. *Mon. Not. R. Astron. Soc.*, 128:307, 1964.

[Rei96]   C. Reid. *Hilbert*. New York: Copernicus, 1996.

[Ret. al.98]   A.G. Reiss *et. al.* Observational evidence from supernovae for an accelerating universe and a cosmological constant. *Astron. J.*, 116:1009–1038, 1998.

[Ret. al.01]   A.G. Reiss *et. al.* The farthest known supernova: The support of an acceleration universe an the glimpse of the epoch of desceleration. *Astrophys. J*, 560:49, 2001.

[Rin77]   W. Rindler. *Essential Relativity*. Springer, 1977.

[Rin00]   W. Rindler. Finite foliations of open FRW universes and the point-like big bang. *Phys. Lett.*, A276:52, 2000.

[Rip01]   P.D. Rippis. Thin shells in a Universe with an embedded Schwarzschild mass. Cand. Scient. Thesis, University of Oslo, 2001.

[Ros64]   W.G.V. Rosser. *An Introduction to the Theory of Relativity*. Butterworths, London, 1964.

[Roy88]   A.E. Roy. *Orbital Motion*. Institute of Physics Publishing, 1988.

[RS75]   M. Ryan and L. Shepley. *Homogenous Relativistic Cosmologies*. Princeton University Press, 1975.

[RS99a]   L. Randall and R. Sundrum. An alternative to compactification. *Phys. Rev. Lett.*, 83:46900, 1999.

[RS99b]   L. Randall and R. Sundrum. A large mass hierarchy from small extra dimension. *Phys. Rev. Lett.*, 83:3370, 1999.

[RT02]   M.L. Ruggiero and A. Tartaglia. Gravitomagnetic effects. *Nuovo Cimento*, 117B:743–768, 2002.

[Rya72]   M. Ryan. *Hamiltonian Cosmology, Lecture Notes in Physics 13*. Springer Verlag, 1972.

[Ryd03]   B. Ryden. *Introduction to Cosmology*. Addison-Wesley, 2003.

[Sag13]   G. Sagnac. The luminiferous ether demonstrated by the effect of the relative motion of the ether in an interferometer in uniform motion. *C.R. Hebd. Seances Acad. Sci.*, 157:708–710, 1913.

[SAI$^+$71]   I.I. Shapiro, M.E. Ash, R.P. Ingalls, W.B. Smith, D.B. Campbell, R.F. Dyce, R.B. Jurgens, and G.H. Pettengill. Fourth test of general relativity – new radar result. *Phys. Rev. Lett.*, 26:1132–1135, 1971.

[Sau69]   P.T. Saunders. Observations in some simple cosmological models with shear. *Mon. Not. R. Astr. Soc.*, 142:213–227, 1969.

[Sav95]   S.F. Savitt, editor. *Time's Arrows Today*. Cambridge University Press, 1995.

[Sch85]   B.F. Schutz. *A first course in general relativity*. Cambridge University Press, 1985.

[Sch04]   C. Schmid. The cosmological origin of inertia: Mach's principle. *gr-qc/0409026*, 2004.

[Set. al.98] B. Schmidt *et al*. The high-z supernova search: Measuring cosmic deceleration and global curvature of the Universe using type Ia supernovae. *Astrophys. J.*, 507:46–63, 1998.

[Sil02]   A. Silbergleit. On cosmological evolution with the Lambda-term and any linear equation of state. *astro-ph/0208465*, 2002.

[Sim01]   C. Simeone. *Deparametrization and Path Integral Quantazation of Cosmological Models*. World Scientific, 2001.

[SKM$^+$03]   H. Stephani, D. Kramer, M. MacCallum, C. Hoenselaers, and E. Herlt. *Exact Solutions to Einstein's Field Equaltions, Second Ed.* Cambridge University Press, 2003.

[SMS00]   T. Shiromizu, K. Maeda, and M. Sasaki. The Einstein equation on the 3-brane world. *Phys. Rev.*, D62:024012, 2000.

[Spi75a]   M. Spivak. *A Comprehensive Introduction to Differential Geometry*, volume I. Publish or Perish, 1975.

[Spi75b]   M. Spivak. *A Comprehensive Introduction to Differential Geometry*, volume II. Publish or Perish, 1975.

[Ste77]   H. Stephani. *General Relativity*. Cambridge University Press, 1977.

[Ste91]   J. Stewart. *Advanced General Relativity*. Cambridge University Press, 1991.

[SVF01]   M.G. Santos, F. Vernizzi, and P.G. Ferreira. Isotropy and stability of the brane. *Phys. Rev.*, D64:063506, 2001.

[Tho67]   K.S. Thorne. Primordial element formation, primordial magentic fields, and the isotropy of the universe. *Astroph. J.*, 148:51–68, 1967.

[Thu82]   W.P. Thurston. Three dimensional manifolds, Kleinian groups and hyperbolic geometry. *Bull. Amer. Math. Soc.*, 6:357–381, 1982.

[Thu97]   W.P. Thurston. *Three-Dimensional Geometry*, volume 1. Princeton Uni. Press, 1997.

[Tol34]   R.C. Tolman. Effect of inhomogeneity on cosmological models. *Proc. Nat. Acad. Sci.*, 20:169–176, 1934.

[Top01]   A.V. Toporensky. The shear dynamics in Bianchi I cosmological model on the brane. *Class. Quantum Grav.*, 18:2311, 2001.

[TR02]   M.S. Turner and A.G. Reiss. Do SNe Ia provide direct evidence for past deceleration of the Universe? *Astrophys.J.*, 569:18, 2002.

[TW89]   J.H. Taylor and J.M. Weisberg. Further experimental tests of relativistic gravity using binary pulsar PSR B1913+16. *Aph. J.*, 345:434, 1989.

[UKE03]   J.P. Uzan, U. Kirchner, and G.F.R. Ellis. WMAP data and the curvature of space. *Mon. Not. Roy. Astron. Soc.*, 344:L65, 2003.

[Väl99]   J. Väliviita. *An Analytic Apprach to Cosmic Microwave Background Radiation Anisotropies*. PhD thesis, University of Helsinki, 1999.

[Vil82]   A. Vilenkin. Creation of universes from nothing. *Phys. Lett.*, 117B:25–28, 1982.

[Vil83]   A. Vilenkin. Birth of inflationary universes. *Phys. Rev.*, D27:2848–2855, 1983.

[Vil94]   A. Vilenkin. Approaches to quantum cosmology. *Phys. Rev.*, D50:2581–2594, 1994.

[vW81]   C. von Westenholz. *Differential Forms in Mathematical Physics*. North Holland Publishing Company, Rev. Ed. 1981.

[Wal84]   R.M. Wald. *General Relativity*. The University of Chicago Press, 1984.

[WE97]   J. Wainwright and G.F.R. Ellis, editors. *Dynamical Systems in Cosmology*. Cambridge University Press, 1997.

[Weh01]   I.K. Wehus. Ekstra dimensjoner og Kosmologi. Cand. Scient. Thesis, University of Oslo, 2001. In Norwegian.

[Wei72]   S. Weinberg. *Gravitation and Cosmology*. John Wiley & Sons, New York, 1972.

[WR04]   I.K. Wehus and F. Ravndal. Dynamics of the scalar field in 5-dimensional Kaluza-Klein theory. *Int. J. Mod. Phys.*, A19:4671–4686, 2004.

[Zel70]   Ya. B. Zel'dovich. Gravitational instability: An approximate theory for large density perturbations. *Astron. Astrophys.*, 5:84, 1970.

[ZP01]   W. Zimdahl and D. Pavón. Interacting quintessence. *Phys. Lett.*, B521:133–138, 2001.

[ZWS99]   I. Zlatev, L. Wang, and P.J. Steinhardt. Quintessence, cosmic coincidence, and the cosmological constant. *Phys. Rev. Lett.*, 82:896, 1999.

Index

Printed in the United States
106494LV00001B/40/A

9 780387 691992